ATOMIC PHYSICS 4

1969—Atomic Physics 1

Proceedings of the First International Conference on
Atomic Physics, June 3-7, 1968, in New York City
V. W. Hughes, Conference Chairman
B. Bederson, V. W. Cohen, and F. M. J. Pichanik, Editors

1971—Atomic Physics 2

Proceedings of the Second International Conference on
Atomic Physics, July 21-24, 1970, Oxford, England
G. K. Woodgate, Conference Chairman
P. G. H. Sandars, Editor

1973—Atomic Physics 3

Proceedings of the Third International Conference on
Atomic Physics, August 7-11, 1972, Boulder, Colorado
S. J. Smith and G. K. Walters, Conference Chairmen and Editors

1975—Atomic Physics 4

Proceedings of the Fourth International Conference on
Atomic Physics, July 22-26, 1974, Heidelberg, Germany
G. zu Putlitz, Conference Chairman, E. W. Weber and A. Winnacker, Editors

A Continuation Order Plan is available for this series. A continuation order will bring delivery of each new volume immediately upon publication. Volumes are billed only upon actual shipment. For further information please contact the publisher.

ATOMIC PHYSICS 4

Editors
G. zu PUTLITZ
Conference Chairman

E. W. WEBER

and

A. WINNACKER

Physics Institute, University of Heidelberg

PLENUM PRESS • NEW YORK AND LONDON

FICAP

Library of Congress Catalog Card Number 72-176581
ISBN 0-306-37194-4

Proceedings of the Fourth International Conference on
Atomic Physics, July 22-26, 1974, Heidelberg, Germany

© 1975 Plenum Press, New York
A Division of Plenum Publishing Corporation
227 West 17th Street, New York, N.Y. 10011

United Kingdom edition published by Plenum Press, London
A Division of Plenum Publishing Company, Ltd.
Davis House (4th Floor), 8 Scrubs Lane, Harlesden, London NW10 6SE, England

Printed in the United States of America

E. U. CONDON (1902-1974)

Organizing Committee

J. Brossel	Université de Paris, Paris, France
P. Burke	Queen's University, Belfast, Northern Ireland
H. B. G. Casimir	Philips Research Laboratories, Eindhoven, Holland
V. W. Cohen	Brookhaven National Laboratory, Upton, L.I., N.Y., USA
H. Ehrhardt	Universität Trier-Kaiserslautern, Kaiserslautern, Germany
H. Foley	Columbia University, New York, USA
G. Herzberg	National Research Council, Ottawa, Canada
V. W. Hughes	Yale University, New Haven, USA
P. Jacquinot	Lab. Aimé Cotton, Orsay, France
N. Kaliteevsky	Leningrad State University, Leningrad, USSR
A. Kastler	Université de Paris, Paris, France
H. Kleinpoppen	University of Stirling, Stirling, England
I. Lindgren	Chalmers University of Technology, Gothenburg, Sweden
E. E. Nikitin	Institute of Chemical Physics, Academy of Sciences, Moscow, USSR
W. Paul	Universität Bonn, Bonn, Germany
J. C. Pebay-Peyroula	Université Scientifique et Medical de Grenoble, Grenoble, France
I. I. Rabi	Columbia University, New York, USA
N. F. Ramsey	Harvard University, Cambridge, Mass., USA
A. L. Schawlow	Stanford University, Stanford, USA
H. A. Shugart	University of California, Berkeley, USA
S. J. Smith	Joint Institute for Laboratory Astrophysics, Boulder, USA
K. Takayanagi	University of Tokyo, Tokyo, Japan
G. K. Walters	Rice University, Houston, USA
G. K. Woodgate	Oxford University, Oxford, England
F. Zavattini,	CERN, Genève, Switzerland
G. zu Putlitz	Universität Heidelberg, Heidelberg, Germany, Chairman

Local Committee

H. Ackermann	G. zu Putlitz
K. Dorenburg	H. zu Putlitz
P. Heitjans	E. W. Weber
J. Kowalski	H. G. Weber
R. Müller	A. Winnacker
H. Nienstädt	J. Wobbe
H. Orth	

University of Heidelberg

Program Committee

P. Burke	Queen's University, Belfast, Northern Ireland
J. Hüfner	Universität Heidelberg, Heidelberg, Germany
V. W. Hughes	Yale University, New Haven, USA
E.-W. Otten	Universität Mainz, Mainz, Germany
H. Van Regemorter	Dep. Astrophysique Fondamentale, Observatoire de Paris, Meudon, France
N. F. Ramsey	Harvard University, Cambridge, Mass., USA
J. P. Toennies	Max-Planck-Institut f. Strömungsforschung, Göttingen, Germany
G. zu Putlitz	Universität Heidelberg, Heidelberg, Germany, Chairman

Sponsoring Agencies

Deutscher Akademischer Austauschdienst (DAAD)
Deutsche Forschungsgemeinschaft (DFG)
European Research Office
International Union of Pure and Applied Physics (IUPAP)
Land Baden-Württemberg
Universität Heidelberg

Individual Sponsors

BASF, Ludwigshafen
BBC, Mannheim
Boehringer, Mannheim
Borer, Solothurn
Bruker-Physik, Karlsruhe
Deutsche Intertechnique GmbH, Mainz
B. Halle, Nachf., Berlin
Heidelberger Druckmaschinen, Heidelberg
IBM Deutschland
Kupferberg Sektkellerei, Mainz
Ortec, München
PWA Kunststoff GmbH, Mannheim
Schlossquell AG, Heidelberg
Schomandl KG, München
Siemens AG, Erlangen
Springer Verlag, Berlin
Carl Zeiss, Oberkochen

Preface

ATOMIC PHYSICS 4 extends the series of books containing the invited papers presented at each "International Conference on Atomic Physics." FICAP, the fourth conference of this type since its foundation in 1968, was held at the University of Heidelberg. The goal of these conferences, to cover the field of atomic physics with all its different branches, to review the present status of research, to revive the fundamental basis of atomic physics and to emphasize future developments of this field as well as its applications was met by more than thirty invited speakers, leaders in the field of atomic physics. Their talks were supplemented by more than two hundred contributed papers contained in the FICAP Book of Abstracts.

This volume begins with papers given in honour and memory of E. U. Condon, to whom this conference was dedicated. It continues with articles on fundamental interactions in atoms and quantum electrodynamics, on the fast progressing field of high energy heavy ion collisions and quasi-molecules, on electronic and atomic collisions and the structure of electronic and μ-mesic atoms. The volume closes with contributions concerning the application of lasers in atomic physics, a new field of vastly increasing importance to fundamental experiments as well as applications. We feel that this book contains a very stimulating account of the present main streams of research in atomic physics and its possible future directions.

Gisbert zu Putlitz
Physikalisches Institut
Universität Heidelberg

Contents

ATOMIC PHYSICS 4

EDWARD UHLER CONDON, A PERSONAL RECOLLECTION

Lewis M. Branscomb

IBM Corporation

Old Orchard Road, Armonk, New York 10504

Edward Uhler Condon was one of the young American physicists who, in 1926, enjoyed a Fellowship at Göttingen and Munich, in time to participate in the revolution of theoretical physics arising from the new quantum mechanics. Son of a railroad engineer in the western United States, sometime newspaper man in San Francisco, Condon's scientific career combined the abstract with the practical, theory and observation in a unique way. The most fruitful period of his work in theoretical physics occurred between 1926 and 1936, encompassing collaborations with many of the most distinguished scientists of that period. Others in today's symposium will discuss these contributions in more detail. I would like to recall his impact on science after he began to combine the academic and administrative life, including major positions in industry (Westinghouse and Corning Glass) as well as in government (Directorship of the National Bureau of Standards).

When Condon went to Pittsburgh as Associate Director of Research for the Westinghouse Electric Corporation in 1937, he set about strengthening the basic research and preparing Westinghouse for the capability to move into the nuclear field. His success in attracting highly competent young scientists to an industrial laboratory, while focusing the total research effort in a direction of enormous future economic importance, set a precedent that has since been followed by industrial research managers in many enlightened companies. It was not long, of course, when World War II caused his effort to focus in microwave technology in support of the radar work at the M.I.T.

1

Radiation Laboratory. He also became involved in the nuclear weapons program, serving as Associate Director of the Los Alamos Scientific Laboratory, and later headed the Theoretical Physics Division of the Berkeley Radiation Laboratory.

In this connection, Condon became a close associate of the substantial British group, including Sir Mark Oliphant and Sir Harrie Massie. The research they did in electromagnetic separation of isotopes resulted in major advances in the science of mass spectroscopy. This was later reflected not only in Westinghouse products but also in the major contributions to atomic physics that resulted from mass spectrometry after the war.

Of all the American scientists whose concern about the use of atomic weapons at Hiroshima and Nagasaki led them to devote themselves to the cause of peace, none was more dedicated nor more effective than Condon. His efforts as Science Adviser to the Select Committee on Atomic Energy of the United States Senate after the war had much to do with the decision by the Congress to place government responsibility for atomic energy in a civilian rather than a military agency. The price he paid for his relentless pursuit of this goal is another story, taking us away from his scientific contributions.

His responsibilities as adviser to the Senate were concurrent with his tenure as Director of the National Bureau of Standards. Just as he had done at Westinghouse, he brought the Bureau of Standards in a very few years into the world of modern physics and chemistry. All of the Bureau's contributions in atomic collision physics, quantum chemistry and low temperature physics stem from his enthusiasms and the people he attracted to NBS. The program in atomic spectra, which predated his arrival, received the kind of theoretical underpinnings it required. But Condon did not restrict his interests at NBS to the areas of modern physics he knew best.

Probably his most important contribution was his recognition in the late 1940's that the field of building technology was probably the most important neglected area of technology in the country. Condon created the Building Research Division and set it on the path that now enables NBS to claim for its Center on Building Technology a leading position among such laboratories in the world.

Condon also anticipated by about 20 years the need for government to put research to work in behalf of the consumer. Indeed, he

*moved so effectively in this direction that his successor reaped a
whirlwind over battery additive ADX2 after the elections of 1953.*

*Condon also created the Central Radio Propagation Laboratory of the
NBS, locating it in Boulder, Colorado. The ionospheric research of
this laboratory became a very fruitful test bed for the applications of
atomic physics during the 1950's and 1960's. In two respects the
ionosphere has been critical to the development of atomic and molecular physics. First, it provides a laboratory for ion chemistry in
which reactions continue to completion without the influence of walls.
Secondly, the upper regions of the atmosphere are a rich showcase
of phenomena based on non-equilibrium energy distributions for both
photons and electrons. Although one could hardly call the structure
of the atmosphere simple, more progress has been made in basic
understanding of this environment than in the glow discharge, even
though one captures the latter in one's laboratory.*

*The Central Radio Propagation Laboratory subsequently evolved into
two institutions, the Office of Telecommunications of the Department
of Commerce and the Research Laboratories for the National Oceanic
and Atmospheric Administration.*

*In 1951, Condon left the Bureau of Standards to become Director of
Research and Development in the Corning Glass Works. There he
was to apply his knowledge of solid state theory to the many glass
and ceramic projects of Corning, to develop materials for use in the
heat shields of spacecraft, and to help Corning move into the
emerging field of electronics.*

*In 1956, he returned to the university world as Chairman of the
Department of Physics at Washington University in St. Louis. Moving
in 1963 to Boulder as Fellow of the Joint Institute for Laboratory
Astrophysics and Professor of Physics, he once again returned to his
original love, the theory of atomic spectra. Together with Hallis
Odabasi, a graduate student who subsequently worked with Condon
as post-doctoral research associate, Condon began the task of a new
and up-to-date version of his treatise with George Shortley on the
Theory of Atomic Spectra. A major portion of this work was complete before his recent death. If present hopes to bring this work
into publishable form by 1976 are borne out, it would be a fitting
tribute to the 50th anniversary of the Ph.D. degree and first publication of Edward Uhler Condon, a scientist whose life has touched
many fields and many people.*

E. U. CONDON - THE PHYSICIST AND THE INDIVIDUAL

Transcript of a talk given by

I. I. Rabi

Columbia University, New York

Having heard Dr. Branscomb's talk[*] I think we will turn the program around, as he gave Condon's importance for physics, and I will talk more about Condon the individual, and the times in which he lived, and something about the scientific scene in the United States at that time and some of his influence on that and through that on the world.

First let me say that I find it very astonishing that I should be here in Heidelberg at this big conference talking to you in English. When I first came to visit Germany in 1927 such a thought would have been ridiculous, incomprehensible, unbelievable. In the first place Germany and the German-speaking world were the center of physics at that time. If you wanted to learn physics you learned German, and very few Germans knew any English -- it was a very esoteric kind of language. And in particular they did not know many Americans and there was no need to know many Americans at that time.

Fifty years ago, about 1924, physics seemed to have come to an end in frustration. There had been no great experimental discovery for a few years and the quantum theory in the form of the correspondence principle of Bohr and his students had after its magnificent career come to the end of the line. The two-electron spectrum presented a clear paradox and obstructed

---
[*]see p.1

5

future progress along the lines of the classical correspondence principle. It is true that towards the end
of 1924 deBroglie had published his brilliant paper,
but with all due apologies very few physicists in
Germany or in the English-speaking world read Annales
de Physique. Practically nobody read his paper
except for provincial graduate students like myself
in far-away New York. The reason I read that paper was
that I read everything because we were so far away
from the center.

The final blow of this period came in 1925. If you
look up the Zeitschrift für Physik there is a long paper, not very long but 8 or 10 pages, by Bohr. This
paper did not contain a single equation. It spoke very
broadly of things like "Unmechanischer Zwang" and
things of this sort -- all very mysterious, as Bohr
knew how to be when he wanted to be mysterious. From
that point on one seemed to take off on wings into me-
taphysics. I remember listening to a paper by a distinguished American scientist who had a theory that
the mass of the electron was in one place, its charge
in another, and its name somewhere else. This period
of frustration fortunately lasted a very short time.

The end of 1925 and the year 1926 saw the second
great revolution of ideas in this century: the birth
and the formulation of quantum mechanics. The first
one, and perhaps as great or greater a revolution, had
occured in 1905: the work of a single individual --
Albert Einstein. Though an enormous development of the
Bohr theory and the correspondence principle took place
in the intervening years and many people had worked on
it, it had come to an end. Now in 1926 a whole new
world opened.

So I take this occasion of the Condon Memorial
Session to make a few personal remarks which may help
to communicate some of the quality of the time when
Condon and I and quantum mechanics were young. Sometimes it seems that events conspire to thwart one's
desires and hopes, and at other times it is happily
just the other way around. In my own case after many
false steps I became a graduate student of physics at
Columbia University in the period from 1923 to 1927.
I call this the Golden Age of deBroglie, Schrödinger,
Dirac, Heisenberg, and Pauli. In this period physics
emerged from inspired groping in darkest shadows into
the new universe of clarity.

Condon was also one of the fortunate students to
come into the world of physics at that time. One could
talk endlessly about these wonderful few years when
young physicists walked around with a shining light
in their eyes.

In the mid-1920's every American physics student
who could get fellowship support came to Germany or to
German-speaking Europe, the land of Einstein, Bohr and
the young Turks like James Franck, Otto Stern, Sommerfeld, Debye, Ehrenfest, Born, and a host of others. I
was one of those and Condon was too. He came a bit
earlier, in 1926. I arrived in the summer of 1927.

I had just completed my Ph. D. at Columbia and
set off for Europe on a minute stipend, first to
Zürich with Schrödinger, Debye, Scherrer, Linus
Pauling. It was Stratton who recommended a pension
in München to be with Sommerfeld. It was the Pension
Elvira, which unfortunately had disappeared from
the face of the earth. I think of this with tremendous nostalgia because one could have room and board
and all meals for six Marks a day, just exactly my
size. I came to this pension and when I came down to
breakfast in the morning there was one of the most
German looking men I have ever seen. He had the German
hair-do of that time: "ein Millimeter". Anyway, there
was this man, Condon, reading the American Mercury,
which was a literary critical magazine of that time.
We struck up a warm friendship immediately.

Now a few words about Condon before he came to
Europe. As mentioned by Branscomb, he had has a varied
career, although very young. He started to be an experimental physicist and he went to the stockroom and
got a lot of glassware. On the way up to his room he
stumbled and all the glass fell down and broke. He decided experimental physics was not for him. And thus
began a career in theoretical physics which was in many
ways unique. He attended the lectures by Birge there at
the University of California. Birge mentioned some rules for molecular transistions which interested and im-
pressed Condon very much. He went home and quantified
them, and came back and showed the paper and it looked
very good. He had the original idea of submitting
it as a dissertation (it was the Franck-Condon-Prin-
ciple) for his doctor's degree although he had only
been a graduate student for a few months. Of course,
there was tremendous opposition; "you can't do a
dissertation in a weekend". But that was what he

did. But there were people there at Berkeley, like
Loeb, who said, "Why can't you do a dissertation in a
weekend, or a day for that matter. It's the quality
of the dissertation that counts, and this should be
accepted." Being very aggressive he won; so Condon
got his doctor's degree in one year and got a fellowship, a pretty good fellowship, a National Research
Council fellowship, and he came to Göttingen and then
München. There he found Oppenheimer and H.P. Robertson,
and others. Actually we came there as if on a pilgrim-
age. Most of the people of my age that I knew in physics
I first met in Germany, like Condon, Robertson, Loomis,
Stratton, Oppenheimer and so on. So there he came with
his Ph. D. and his Franck-Condon principle.

In the year 1926 he went to Sommerfeld in München.
This is where I met him. There were Condon and Robertson and I, and we had a wonderful time there. I recall
that we went to a fair in München. As we went along, the
smell of beer in the air became thicker and thicker. There
was a brass band guzzling the beer in at one end and blow-
ing it out through the trumpets at the other end. The late
professor Robertson, the famous relativist, and Condon got
together and played the "Peas, Porridge Hot .." game and
pretty soon we had a circle of people around us watching,
and I passed the hat and got a few Pfennigs. Altogether
we had a wonderful time. Although we were not German
students, we lived a little bit of the life, to a point
where somehow or other, we never succeeded in finishing
a meal in a restaurant before we were thrown out, especially with Robertson, who was rather boisterous. We
were rather sensitive about the whole thing because
German is a difficult language. It was not so difficult for me because I had studied chemistry first and
even undergraduates in chemistry at that time had to
read German because the real literature was in German.
We had to look up things in Beilstein and so on. At
one time, Condon, being very adventurous, bought a
Lederhosen outfit. So there he was very beautiful in
this thing going along with the short pants and the
straps. And he explained to the maid that he would not
have much use back home in America. But she said,
"Doch, doch, abends zum Spazieren!"

Jessie Dumond had written Sommerfeld about visi-
ting him. Sommerfeld had never seen him, neither had
we. Sommerfeld asked us to meet Dumond at the railroad
station and Condon went in this outfit. I went along
because his German was poor and he was afraid that if
people talked to him and he answered in his language,

with this outfit on, he would be lynched!

We came along and we recognized Dumond immediately because it was easy to recognize an American then and I suppose it is now too, because of his luggage and his shirt. Well, Dumond had married a French wife, and when she looked at Condon she was very turned off. We got together that evening but she was too "sick" to join us. It took longer for the first world war to end than the second world war, it took a very long time. In 1927 I was in Zürich, and there was the first meeting where French and German physicists came together on an informal basis.

Now just a little bit more about those days. I would like to tell about it because it may give a lot of encouragement to people in less developed countries. We felt that we came from a less developed country scientifically in 1927. For example, the Physical Review was taken by the library in Göttingen but was made accessible only at the end of the year. You can see it was not a very exciting journal even though I published my dissertation in it. And we felt this very keenly. Here was the United States, a vast and rich country but on a rather less than modest level in its contribution to physics, at least per capita. And we resolved that we would change the situation. And I think we did. By 1937 the Physical Review was a leading journal in the world.

Condon contributed greatly to this sucess because of his early journalistic career. This included writing books of having books written and, particulary, editing which he did brilliantly through the following years. As told already by Dr. Branscomb, he had had an upbringing where his family went from place to place in the United States. So he never had any deep roots in any one place. He was able to make friends, to pick up friends and acquaintances very easily.

When he left Germany he went to Bell Telephone Laboratories. Later he recalled to me that on the way home to London he overheard an early transatlantic telephone conversation and he was struck by the extraordinary range difference between the high level of technology that developed the transatlantic telephone connection and the trivialities that go over it.

        Condon left Bell Telephone Laboratories and became
professor of physics at Princeton and then at the University of Minnesota for a short time, then came back to
Princeton. At Princeton at that time, both the Physics
Department and the Institute were at a very high level
mathematically. There he gave these wonderful courses and
papers on relativity. One day Condon announced an elementary course on relativity. All of these high powe-
red people came to his lecture to find out what relativity was really about. He had this very remarkable
gift of making complicated things simple and clear.

        He then went to Westinghouse. He was of an age,
of course, where he knew all the good people in the
field who were young. He had a great capability for
finding talented students and interesting them. At
Westinghouse he brought a group of brilliant young
people together. Of course it was not so difficult
then. This was in the 1930's and a job was simply not
to be had. The general idea that one gets a doctor's
degree in physics and immediately moves to a job was
not the case then; there were no jobs. I know when I
got my degree at Columbia I did not expect a job to go
with it. But I was interested in physics and I did
want to do it anyhow. Under those circumstances he
could pick a very good group of people, many from the
then advanced subject of nuclear physics. At that time
the young nuclear physicists were just as arrogant as
the young people now in high energy physics. Dr. Ste-
pian, a brilliant electrical engineer at Westinghouse,
came around to converse with them about some subjects
other than nuclear physics, only to be rebuffed. He
said he couldn't understand why a man should be so
proud of knowing so little physics. This group was very
successful and some of the leaders of the Westinghouse
Company came from this group.

        When the war came and we all became engaged in
some war activity, Condon was always called on to take
a very important position. We were together at the
Radiation Laboratory in Cambridge, Massachusetts in
the development of radar. This very successful laboratory was organized under Lee DuBridge. It began with
about 30 people in 1940 and ended with about 4000 in
1945. A very unique experience, having a laboratory of
this size and spending this enormous amount of money,
having the very vital connections with the army and
navy and with the British and with people of other con-
tries. Well, nobody in the government group of the la-

boratory had ever had any administrative experience of
that sort. When it came to making the budget for the
following year we always did very well but we were
afraid to give away the secret. We simply plotted an
exponential curve. That always did it, because if you
have more and more physicists you have more and more
ideas and this goes exponentially. It's as simple as
that. Condon was a great help to us because of his connections with the Westinghouse Company.

Later on, Los Alamos was organized. It was natural
to bring in Condon and he wrote the first primer on
how to build an atomic bomb. It was a good thing to
have in those days a book about how to build an atomic
bomb; it explained things to the young people who came
in. He didn't stay very long in Los Alamos because he
and the General Groves were orthogonal. He went back
to Westinghouse. This battle with Groves was to haunt
him one way or another for the rest of his life.

Wherever the action was thick, there was Condon.
He performed a very great job as a consultant to a
U.S. Senate special committee. At the end of the war
there was this big question of what to do with atomic
energy. The law which was proposed was the May-Johnson
bill, to put the whole thing under the control of the
military. And a small group of scientists outside the
government got into action to oppose the bill in the
most formidable manner to the surprise of all politi-
cians. They organized speeches, they set up an office
in Washington, they disturbed all kinds of documents,
and in spite of the strong backing of the military and
the government itself for the May-Johnson bill, this
group won out. A new bill put the control over nuclear
energy under a civilian administration. Simultaneously
a small group of scientists was formed to advise the
government about nuclear energy.

I would like to turn to Condon's later career,
his unfortunate difficulties with the United States
government. He was a man of independent mind, and
could not be pushed around by people in the government.
He suffered very greatly from this. There were times
when he could not get a passport to go abroad. Anybody
who could get a look at Condon or listen to one sen-
tence of Condon's speech would have had no doubts about
his basic and essential Americanism. But he was attacked
and although he had a very natural and cheerful manner,
this troubled his life in many respects. These battles

took a great fraction of his time and were a drain on
his emotions for many, many years. At the same time he
did not stop in his various efforts in teaching and in
writing. He was a great idealist, an idealist who could
laugh at himself with some bitterness.

When we came together we talked over the old times
in Germany when we were young. The gaiety and spirit of
fun with which we approached physics at that time remained with us mostly for the rest of our lives.
Physics was great fun, a wonderful thing. But beyond
that it did for us, and I think for that group, my age
group, something that I sense is somewhat less important nowadays. I don't believe any of us took physics
in a sense of a profession, it was not a trade, it was
a way of life, it was a form of interest, it was an
orientation, a philosophical orientation, not articulate
but with a certain feeling that there was something
exciting about this whole quest of mankind to extend
its understanding, to make the universe more understandable, to clarify man's place in the universe, to develop tools to extend his power like the radiotelescope,
to help mankind in understanding itself. I think this
was a basic attitude and assumption at that time. Perhaps the great success of physics and the large sums
of money that were going into it have to some degree
-- I hope not too much -- turned physicists into pro-
ressionals rather than what used to be called "natural
philosophers".

Condon got his Ph. D. in a year and took the rest
of his life in completing his education and thus remained young and fresh in spirit and in intellect until the very end.

PERSPECTIVES ON "THE THEORY OF ATOMIC SPECTRA"

B. R. Judd

Physics Department, The Johns Hopkins University

Baltimore, Maryland

"The Theory of Atomic Spectra" by E. U. Condon and G. H. Shortley[1] has dominated atomic spectroscopy over the last forty years in a way that few books in other fields have done. It is not difficult to see how this has come about. In 1935, when the book was published, it was possible to give a reasonably complete account of the subject without making concessions to the reader. The frontiers of the field are described in a lucid and compelling style. It is wholly accurate. Most important of all, the theory is firmly based on quantum mechanics. Condon's wide experience in different areas of physics is apparent here. In fact, his publication list[2] reveals only six articles on atomic spectra that were written before 1935 - and even fewer afterwards. But although his work in other fields, such as the enunciation of the Franck-Condon principle, almost certainly made a greater impact on the development of physics, his book with Shortley has probably attracted more readers than any of his other writings.

The reasons for the appeal of "The Theory of Atomic Spectra" become better appreciated if viewed in the perspective of the 1930's. The journal "Nature" reviewed[3] it at the same time as White's book[4] "Introduction to Atomic Spectra." The review was highly complimentary to both. The reviewer felt that possibly the best way to gain an understanding of atomic spectra was to absorb the rules of the vector model of the atom, described, for example, in White's book. For a deeper interpretation in terms of quantum mechanics, the reader could turn to Condon and Shortley. The simplicity of the vector model is, of course, deceptive. Its promulgation has done much to retard atomic spectroscopy. Consider, for example, Fig. 12.5 of White. It represents a two-

13

electron atomic system in the LS and jj coupling schemes. Although it is pictorially effective, it is useless for performing calculations. Suppose we ask the simplest possible question: What is the overlap between the two states represented by the diagrams? The vector model is unable to supply the answer for the simple reason that there is no way to interpret the geometry of the diagrams in an algebraic way. Condon and Shortley's book is free from the baleful influence of the vector model. If we turn to their Table 1[12], we find the overlap to be $(45)^{-\frac{1}{2}}$. Today this would be recognized as a tabulation of

$$[(2j_1 + 1)(2j_2 + 1)(2S + 1)(2L + 1)]^{\frac{1}{2}} X$$

where X is a 9-j symbol. If we use the tables of Howell[5] for evaluating X, we find, in Howell's notation,

$$X = (32^+1^+, \ 321, \ 10^+0^+) = (22680)^{-\frac{1}{2}}$$

for $s_1 = s_2 = \frac{1}{2}$, $S = 1$, $\ell_1 = 1$, $\ell_2 = 2$, $L = J = 3$, $j_1 = 3/2$, and $j_2 = 5/2$. Putting in the factor prefacing X, we recover $(45)^{-\frac{1}{2}}$ for the overlap.

The transformation from LS to jj coupling is but one example of a number of transformations studied by Condon and Shortley. They show how the theory of angular momentum can be used to relate any state of an atomic system to a basic expansion in Slater determinants. Once in this primitive form, the matrix elements of any operator can be calculated. Part of the success of "The Theory of Atomic Spectra" is due to the confidence that the reader gets when he realizes that he has the key to solve, at least in principle, any problem in atomic spectroscopy.

The basic theory of angular momentum is set out in Chapter 3. Racah told the writer that he considered this chapter one of the greatest he had ever encountered. It was certainly the main source of inspiration for Racah's work on spherical tensors. [6] As such, it can be considered to have played a crucial role in nuclear as well as atomic shell theory. Not the least of its virtues is the care with which phases are defined. The physicist approaching the atomic f shell has 16384 independent phase choices to decide on. The remarkable degree of unanimity one finds in the work of different spectroscopists stems in large measure from the original selection imposed on the subject by Condon and Shortley's book. Of course, some phases were not systematically analyzed; for example, those of Table 1[12] are arbitrary. But at least some important ground rules were set out.

Another interesting aspect of the influence of "The Theory of Atomic Spectra" on subsequent work is the way in which it

anticipated the use of group theory. The famous and rather com-
placent remark in the introduction that the authors "manage to
get along without it" did much to release the practising spectros-
copist from the thrall of Wigner and Weyl. However, with hind-
sight we can see that the group theory is there, but in a concealed
form. Nowhere is its presence felt more strongly than in the
analysis of the separation of the two ^{2}D terms of d^3. Spelling
out the procedure in greater detail than had the original analysis,[7]
Condon and Shortley noted that angular-momentum theory alone
could not fix the coefficients a, b,...,f in the determinantal expansion

$$|^2D, \ M_S = \tfrac{1}{2}, \ M_L = 2\rangle$$
$$= a(2^+1^+ - 1^-) + b(2^+ - 1^+1^-) + c(2^-1^+ - 1^+) + d(2^+0^+0^-)$$
$$+ e(2^+ - 2^+2^-) + f(1^+1^-0^+).$$

They therefore suggested that one linear combination be chosen at
random (subject, of course, to the constraints $L = M_L = 2$,
$S = M_S = \tfrac{1}{2}$), the other being then fixed by the condition of orthogonality. Their choice was

$$\tfrac{1}{2}[(2^+1^+ - 1^-) - (2^+0^+0^-) + (2^+ - 1^+1^-) - (2^+ - 2^+2^-)]$$

and

$$(84)^{-\tfrac{1}{2}} \, [3(2^+1^+ - 1^-) - 3(2^+0^+0^-) - (2^+ - 1^+1^-) + 5(2^+ - 2^+2^-)$$
$$- 4(2^-1^+ - 1^+) - (24)^{\tfrac{1}{2}}(1^+1^-0^+)].$$

The first linear combination can be constructed by taking the ^{1}S
state of d^2, namely,

$$(5)^{-\tfrac{1}{2}} \, [-(2^+ - 2^-) + (1^+ - 1^-) - (0^+0^-) + (- 1^+1^-) - (- 2^+2^-)],$$

and slipping the state 2^+ of a single d electron within each pair of
parentheses. This is how Racah[8] generated states of a common
seniority; the first linear combination thus possesses a seniority
v of 1 (because 2^+ is a one-electron state), while the second corresponds to v = 3. Later, it was recognized[9] that seniority could
be interpreted in group-theoretical terms. The two linear combinations correspond to two different irreducible representations
of R_5, the rotation group of the five-dimensional space of a single
d orbital. They also correspond to two different irreducible representations of Sp_{10}, the symplectic group of the ten-dimensional
space of a single d spin-orbital. Thus Condon and Shortley's in-
tuitive sense of what states are the most appropriate led them to
choose precisely the ones that exhibit the highest degree of

intrinsic symmetry. The choice is not at all obvious because the
two linear combinations are not eigenfunctions of the Coulomb
interaction.

It must be said that a book as monumental as "The Theory of
Atomic Spectra" tends to inhibit the free development of new ideas.
A tendency to defer to the printed text is difficult to resist. A
diagram which dominated for twenty years the initial attempts to
understand the spectroscopy of the rare earths is Fig. 5[7], which
represents the energies of the terms of LaII $4f^2$. Even though the
terms are perturbed by configuration interaction and also correspond to a 4f eigenfunction that is much more extended than would
be expected for the typical rare-earth atom, all early analyses of
such rare-earth spectra as PrIV $4f^2$ were discussed in relation to
the lanthanum spectrum. A strange numerical accident complicated matters. The perturbed lanthanum configuration, when analyzed in terms of the Slater integrals F_k, leads to an exception-
ally small value of F_6. This fortuitous result gave rise to the im-
pression that F_6 might be small for PrIV. This idea was rein-
forced by Tables 1[6] and 2[6] of Condon and Shortley. The denominator 7361·64 appearing there is 736164/100, but was widely interpreted as 7361 x 64 by physicists who were not familiar with the
typography of Cambridge University Press, which followed the
English convention for decimal points. The spurious factor of 64
supported the small value of F_6. This error effectively prevented
any real understanding of the atomic 4f shell until 1955. The various attempts to come to grips with PrIV $4f^2$ are summarized in
Fig. 12 of the Handbuch der Physik article of Fick and Joos. [10]
As a result of the artificially small value of F_6, the energy of $^1I_6$
is consistently too low. This level is now known to lie quite close
to the ^3P multiplet. [11] It is ironic that a straightforward calculation, in which hydrogenic radial eigenfunctions are used, gives an
excellent representation of the true situation.

On another level, the pre-eminence of Condon and Shortley's
book has inhibited a proper appreciation of the developments of
Racah. Given enough time, all problems in atomic spectroscopy
can be solved by an application of the methods described in "The
Theory of Atomic Spectra." The ready availability of modern
computers has made comparative inefficiency of the mathematical
techniques inconsequential. The book has served a generation of
spectroscopists well: one might say, too well. By providing accurate and convenient formulas for term energies, transition probabilities, and the like, it has given spectroscopists a rock to build
on. At the same time, it has constrained their outlook. It is difficult to see how any work that achieves great fame can avoid this
weakness. In any case, it can be legitimately argued that the
fault lies with the readers rather than the writers. The book, now
in paperback and selling at a price only marginally higher than

that of the original 1935 edition, will undoubtedly enjoy many more years of useful life.

REFERENCES

1. E. U. Condon and G. H. Shortley, "The Theory of Atomic Spectra," Cambridge University Press (1935).

2. W. E. Britten and H. Odabasi (Eds.), "Topics in Modern Physics," Colorado University Press (1971).

3. R. H. F (...), Nature, 138, 525 (1936).

4. H. E. White, "Introduction to Atomic Spectra," McGraw-Hill, New York (1934).

5. K. M. Howell, "Tables of 9_j-Symbols," University of Southampton Research Report 59-2 (1959).

6. G. Racah, Phys. Rev. 62, 438 (1942).

7. C. W. Ufford and G. H. Shortley, Phys. Rev. 42, 167 (1932).

8. G. Racah, Phys. Rev. 63, 367 (1943). See also J. C. Slater, "Quantum Theory of Atomic Structure," Vol. II, McGraw-Hill, New York (1960), p. 159.

9. G. Racah, Phys. Rev. 76, 1352 (1949).

10. E. Fick and G. Joos, Handbuch der Physik, Vol. 28 (Ed. S. Flügge), Spectroscopy II, Springer-Verlag, Berlin (1957), p. 269.

11. G. H. Dieke and R. Sarup, J. Chem. Phys. 29, 741 (1958); K. H. Hellwege, G. Hess, and H. G. Kahle, Zeits. f. Phys. 159, 333 (1960).

NEW DEVELOPMENTS IN LEVEL CROSSING SPECTROSCOPY

N.I. Kalitejewski and M. Tschaika

University of Leningrad

I. HIGH RESOLUTION OPTICAL AND LEVEL CROSSING SPECTROSCOPY

The level crossing method permits to measure small energy splittings of the atomic states which cannot be resolved by classical spectroscopy. This is due to the fact that the resolution is not determined by the Doppler width but by the natural line width. On the other hand: Because the level crossing method does not involve the Doppler width it cannot be applied to the investigation of isotope shifts which play an important part in precission measurements of optical line structures, nor can it be used to study effects which influence the Doppler shape. So the range of application of the two methods - level crossing and high resolution optical spectroscopy - is clearly defined and it is evident that both should be improved and perfectioned.

In this paper we report on experiments that have been performed in recent years at the Laboratory for Coherent Optics at Leningrad University. First we will report on investigations concerning the development of the level crossing method. In addition to the magnetic field of the conventional method an electric field was applied simultaneously.[1,2,3,4,5,6,7] This permits the measurement of the Stark parameter β_i in addition to the hyperfine constants A and B and the lifetime τ of the level under investigation. The Stark parameter β_i describes the splitting of the magnetic sublevels in an electric field. A detailed investigation of the statistical and

systematic errors involved shows that all those constants can be determined with high accuracy.[8] Calculations have been performed to estimate the influence of
reabsorption[9] and of depolarizing collisions which limit
the accuracy in the determination of the natural life
time. Relations between the decay contants Γ_1 and Γ_2 of
circular and linear polarization are deduced and checked
experimentally. A theory of relaxation via atomic collisions is developed for levels with hyperfine structure.

To give examples for this kind of investigations
we will concentrate on a number of experiments on alkalis which have been done to determine their atomic
constants in magnetic and electric fields. First these
methods were applied to Cesium[1,2,3], more recently to
Rb-isotopes in a more refined modification[4,5]. Fig. 1
shows the experimental apparatus schematically.

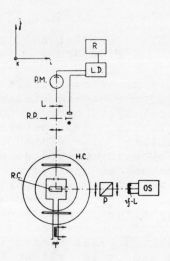

Fig. 1: Scheme of apparatus for observing
level crossings in Rubidium

In this work resonance vessels have been used with inner
electrodes for production of the electric field. A high
homogeneity of the electric field was achieved by carefully adjusting the electrode plates (0.1 % parallel)
and by reducing effects of the plate edges. The resonance vessel was situated in the center of Helmholtz
coils. The coils were calibrated by magnetic resonance
of optically pumped Cesium atoms. Additional coils compensated disturbing laboratory fields. The axis of irradiation $\vec{t}$, observation $\vec{f}$, and external fields were per-

pendicular to each other. Bell Bloom lamps were used
as light sources. The atoms were excited with light
linearly polarized along $\vec{\jmath}$ so that magnetic sublebels
with $\Delta m = \pm 2$ were excited coherently. In this geometry
the signals have a Lorentzian shape. In order to obtain
sufficient pressure of the Rubidium vapour the vessel
was mounted in a heated chamber. The fluorescence light
passed a rotating polaroid filter, thus a signal proportional to the difference of the two polarizations
was given to the lock-in amplifier. Because in this experiment the information regarding the atomic constants
was obtained from line shapes the linearity of registration had to be checked. The nonlinearity was below 1 %.

Depolarizing curves were taken at constant electric
field at 40 to 50 values of the magnetic field in a
field region of 0 to 20 Oe. In this region the Hanle
signal and the first three level crossings were observed. The time constant of registration was 1.5 to
2 min. After each signal the zero of the recorder was
checked, that is zero field intensity of the fluorescence was measured. In order to eliminate effects of
the instability of the fluorescence intensity on the
line shape the signal height was registered regularly
and the whole curve normalized to it. The constants
A, B, τ and β_i were obtained by fitting experimental and
theoretical depolarization curves. The theoretical cur-
ves were computed by using the density matrix formalism.

Let us examine more closely how the information on
the atomic states is obtained from the level crossing
signals. In most older works only the magnetic fields
of the crossing points (see Fig. 2 as example) were
used to fit A, B and the β_i-values, and τ was taken
from the width of the signals. Here, however, those
constants are used as parameters to fit the whole of
the depolarization curves. First by some variations
the regions of the curves are found which are most sensitive to a change in one or the other of the parameters. In this way A, B and τ can be determined with
sufficient accuracy. Finally from the best fit the
Stark parameter is obtained (Fig. 3). This kind of analysis permits us to obtain information on the atomic
constants even in case of non resolved signal structures. It gives higher accuracy compared to the case
when only certain parts of the curves are fitted.

In this context we want to make the following
point: In the incoming light two hyperfine components
of different intensity are present (the ground state

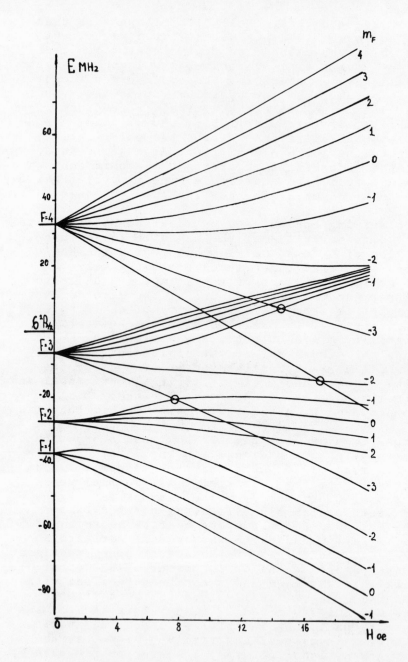

Fig. 2: Zeeman sublevels of the $6^2P_{3/2}$ state of ^{85}Rb.
The open circles indicate level crossings of
sublevels with $\Delta m_F = \pm 2$

$5^2S_{1/2}$ of ^{85}Rb is split by 3036 MHz). Therefore a parameter γ is introduced which describes the intensity
ratio of the components. It is taken into account when
the density matrix of the excited state is calculated.
The parameter affects the width of the Hanle signal
and thus is relevant to the determination of the decay
constants (Fig. 4).

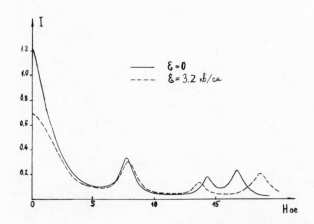

Fig. 3: Level crossing signals of the Rb $6^2P_{3/2}$
        state
        ———— magnetic field only
        ---- electric field present

The experimental results could only be explained by
taking into account this parameter. For instance could
the Hanle effect be observed in connexion with the
transition J = 1/2 → J = 1/2 under irridiation with
linearly polarized light. This is due to the fact that
the intensity ratio of the hyperfine components differs
from the statistical one due to different absorption in
the resonance vessel (Fig. 5). For ^{85}Rb the following
constants have been obtained:

A = (8.16 ± 0.01) MHz τ = (97 ± 3) ns

B = (8.20 ± 0.10) MHz β_i = (0.260 ± 0.007) MHz/$(\frac{kV}{cm})^2$

Taking into account the systematic errors doubles the
limits of error.

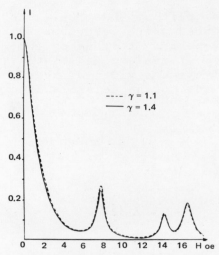

Fig. 4: Influence of parameter γ on the
shape of the ^{85}Rb $6^2P_{3/2}$ signals

Of particular interest is the observation of the fluorescence light when $\vec{E}$ and $\vec{H}$ are perpendicular to each
other. In this case so called anticrossings[6] are observed. For the anticrossings the linewidth increases
with the electric field (Fig. 6).

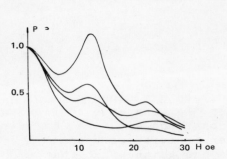

Fig. 5: Influence of the
spectral composition of the
exciting light ($3^2P_{3/2}$ Na)

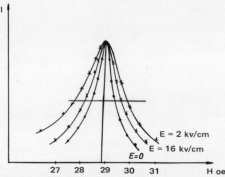

Fig. 6: Broadening of the
anticrossings

This broadening can be explained quantitavely and the
measurements permit to obtain the parameter β at lower
electric field as in the case of $\vec{E}$ parallel $\vec{H}$, with
smaller accuracy, however. Measurements of this kind
have been performed on Cesium.[7]

    In the following we will discuss to what extent
the level crossing method can be used to determine lifetimes and depolarizing collision cross-sections for

laser levels. First experiments to measure the depolarization of spontanous radiation of gas lasers in longitudinal magnetic fields have been described independently in three papers.[13,14,15]. The widths of the level crossing signals were determined under constant
conditions of the medium and constant laser power. However, in order to develop on the basis of these measurements a method to determine lifetimes and depolarizing
cross-sections in the oscillation regime one encounters
certain difficulties: The density of the atoms in the
excited state is not large. To obtain a sufficiently
large interference signal one has to switch to high
intensities of the radiating field. Thus the laser field
acting on the gaseous medium is fairly high and the
"weak field approximation" which underlies the theory
of these effects is no longer applicable.

     No broad line excitation covering the whole spec-
tral region under investigation is used here, but a
narrow quasi monochromatic line of high intensity. As
a consequence complicated nonlinear processes take place.
They result in the phenomenon, for instance, that the
level crossing signal in the spontanous radiation contains information on the state connected to the one
under investigation by stimulated transitions. For this
reason measurements at various laser power levels have
to be done and then to be extrapolated to power zero
in order to obtain τ_0.

For a qualitative description of the signals perturbation theory up to fourth order has to be applied. In
most cases this led to reasonable agreement between
theory and experiments, certain phenomena, however, remained unexplained. A certain role in this context plays
the transfer of coherence from other levels that are
connected to the investigated one by spontanous ransi-
tions. For instance possible radiation cascades have to
be taken into account when observing the lower working
level of the laser. In our laboratory the method was
applied to investigate some He-Ne-lasers which worked
at 0.63μm, 1.15μm and 3.39μm (Fig. 7)[17]. The observed
region of the discharge was exposed to an axial magnetic field produced by Helmholtz coils. The spontanous
emmission which was polarized perpendicularly to the
direction of the magnetic field was observed through
the wall of the discharge tube. A monochromator selected the desired spectral line. By modulating the laser
oscillation with low frequency the weak level crossing
signals could be monitored by a lock in amplifier. From

the widths of the signals, extrapolated to zero pressure
and zero laser power, lifetimes and depolarization cross-
sections could be determined. In Fig. 8 the levelcrossing signal of the $3p_4$ state of Neon is shown as an example.

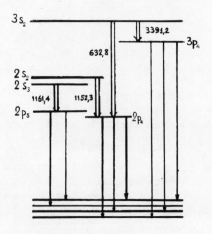

Fig. 7: Level scheme of Neon

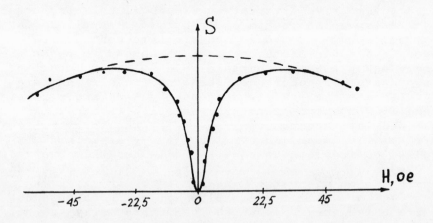

Fig. 8: Levelcrossing signal from the level $3p_4$ of Neon

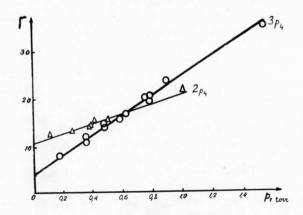

Fig. 9: Extrapolation of the widths of level
crossing signals to zero pressure for
the Neon levels 2p₄ and 3p₄

Fig. 9 shows the extrapolation of the signal widths to
zero pressure for the level crossings of two laser levels (3p₄ and 2p₄ of Neon). Table I shows our results.

Table I

Level	Γ_{rad} /MHz	$\delta(2)Ne^*\text{-}Ne$ /$10^{-16}cm^2$	$\delta(2)Ne^*\text{-}He$ /$10^{-16}cm^2$	λ /µm
$2p_4$	$11,4 \pm 0,8$	74 ± 5	72 ± 5	$0,63;1,15$
$3p_4$	$3,0 \pm 0,5$	90 ± 20	100 ± 10	$3,39$
$3s_2$	$8,3 \pm 0,3$	58 ± 5	24 ± 5	$0,63;1,15$

Recently the level crossing method was applied to
the investigation of pulsed lasers. To do this a number of experimental problems had to be solved. The apparatus consisted of a Cd-laser, an optical and a system for the registration of the signals. The cadmium
laser had a number of puls oscillations in the afterglow. (1.912μm, 1.398μm and 1.433μm). The oscillations
lasted for about 50 to 70μs. The pulse front was retarded relative to the current puls by about 40 to
50μs. In order to observe only a single oscillation
the active cell contained a quartz prism. The spontanous radiation was observed through the side windows,
two polarizations vertical to each other were registered. In the cell which was filled with Cd-vapour and
a buffer gas a discharge was maintained at a constant
level. The intensity of the interference signal was
about 5.10^{-5} to 5.10^{-3} compared to the total observed
light intensity. The duty cycle of the laser was 40.
Signal averaging techniques as well as computer analysis were used in order to handle signals of that
small size. For results see table II. The experimental
results were compared to calculations based on fourth
order perturbation theory (which includes nonlinear
effects) as in the case of continous lasers. This is
particularly important for pulsed lasers because of
the short and intense light pulses. So all the effects
mentioned above show up. For instance is the mixture
of two interacting levels demonstrated in Fig. 10 were
the results for the $5^3D_3 \rightarrow {}^3P_2$ transistion are shown.
Calculation of the signal shape up to fourth order in
the field shows that the curve can be considered as a
superposition of two Lorentz curves, the widths of the
curves being given by the Hertz coherence times of the
upper and lower levels of the 1.648μm laser transistion.
This decomposition into two components allows the determination of the lifetimes and depolarizing cross-
sections for the Cd levels 5^3D_3 and $4^3F_{2\ 3\ 4}$. Agreement with the values known from literature'is good
which indicates that the estimates of reabsorption
effects and the extrapolations are appropriate.

Some years ago E.B. Aleksandrov discovered a new
effect[17] which can be considered at fundamental to the
investigation of atomic systems and which is closely
related to level crossing. The method consists in observation of certain interference effects which follow
a coherent excitation of the atomic system in the kind
of quantun beats. The author examined the conditions
that give rise to this effect and make possible its
observation, various methods of modulation in the ex-

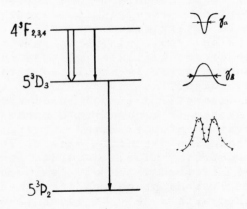

Fig. 10: Occurence of level crossings
in the 5^3D_3 Cadmium level

Table II

Level	Γ_{rad}/MHz	$\delta(2)$ Cd*-Ne /10^{-16}cm^2	λ/μm
6^1D_2	35 ± 10	270 ± 30	1,912
6^3S_1	8 ± 1	< 20	1,433
5^3D_3	9,2 ± 1,5	210 ± 30	1,648
4^3F_{234}	110 ± 60	100 ± 60	1,648

citing light and of the frequency under investigation
(Switching on of a magnetic field) were used. A detailed review of this interesting work has been published
recently. We will examine this method somewhat more
closely in view of our processes in gaseous lasers:
The possibility to observe these phenomena in spontanous emission has been pointed out in ref.[18] The beats
develop when the laser oscillates in two modes. In
this case the output power is modulated with the beat
frequency of the modes. The interaction of the atomic
system with the laser field is modulated with the same
frequency. The system is brought in a magnetic field
and if the seperation of the Zeeman sublevels and the
beat frequency coincide resonance occurs. The beat
signal has a Lorentzian shape. The width is determined
by the effective lifetime which equals the natural
lifetime when collisions processes are absent. The
method was checked with a He - Ne laser $(\lambda=0.63\mu m)$[19].
In order to decrease the seperation of the modes the
resonator had a length of 5 m. To have a more compact
form it was Z-shaped. (Fig. 11). On the photocathode
of a photomultiplier SEV-38 the beat frequency was
mixed with the frequency of a heterodyne, then amplified and registered. The beat signal was recorded as
a function of the magnetic field. The field was produced by Helmholtz coils. (Fig. 12)

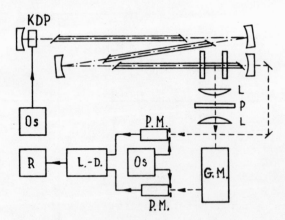

Fig. 11: Scheme of the apparatus for obser-
         vation of beats in the spontaneous
         radiation of the He-Ne-laser

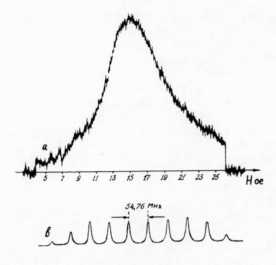

Fig. 12: Beat signals in the spon-
            taneous radiation of the
            He-Ne-laser

In order to improve the signal to noise ratio the axial
modes of the laser were synchronized via an electrical
modulation (KDP-crystal). The value for the lifetime τ
of the lower level $2p_4$ obtained by this method agrees
well with values from other measurements (see Table I).
In the theoretical analysis of the results the fact
was taken into account that the recorded signal is determined by the beats of the spontanous radiation that
develop at $H = H_{res}$ as well as by the modulation of the
occupation numbers of the laser levels under consideration which do not depend on the magnetic field. According to ref.[20] the lifetime of the level can be determined from the phase retardation of the modulation of
the occupation numbers compared to the modulation of
the excitation. By this method the value $\tau = 1/\Gamma_0 =$
(14.7 ± 2)ns was measured for the $2p_4$ state while
$\tau_{coh} = (1/\Gamma_2 = 12.5\pm0.4)$ns were measured by the beat
method. It should be remarked that also in this case
a careful fit of experimental to theoretical results
was made by varying $1/\Gamma_0$ and $1/\Gamma_2$. It follows from
table I that the results agree very well with the values for τ_0 and τ_2 obtained from level crossing experiments.

     Experiments on the alignment in gaseous laser dis-
charges will be dealt with later in this paper.

Let us now consider some new varieties of methods of optical spectroscopy which have been applied to study line broadening and line shifts. At low pressures of the buffer gas a line broadening of the spectral lines can hardly be detected. In the last years, however, new ways were found to determine the apparatus function via registration of the line shape of a single mode laser. So the true spectral energy distribution can be calculated by computer analysis and the influence of collisions can be determined. However, the noise leads to serve difficulties in the solution of this type of "reserved" problems as has been described somewhat in[21].

A useful new way to attack these problems is offered by the method of "magnetic scanning" recently developed in our laboratory: The light absoption of vapours in a magnetic field is measured[22]. In this way a component of the absorption line of ^{87}Rb can be made coincident with a hyperfine component of the ^{85}Rb emission line[24]. From the size of the magnetic field the line shifts in the presence of various buffer gases (He,Ne,Ar,Kr,N$_2$) have been determined. Presently experiments are in progress[25] to obtain from the comparison of measured and calculated signals not only the shifts but also the line broadening due to collisions for Rb atoms. First experiments indicate that a knowledge of the excitation profile is required. For this reason the use of fluorescence of vapours seemed suitable. The particular difficulties are partly compensated for by the fact that contrary to a spectral lamp the line shape of fluorescence can be considered as a Doppler profile with sufficient accuracy when observed perpendicularly to excitation. In addition to that the same isotope can be used in the absorption cell and in the resonance vessel. The absorption lines then are added up, each corresponding to a certain Zeeman component with hyperfine structure. The line widths of those "Voigt curves" are equal but positions ν_i and relative intensities are different and depend on the magnetic field. It should be noted here that in order to calculate ν_i and $k_i(\nu_i)$ the mixing of the hyperfine components by the magnetic field in the upper as well as in the lower state has to be taken into account. The matrix elements and the dipole transistion prohabilities are calculated with an accuracy of about 40 MHz. The dependence of the signal from the magnetic field is a function of some parameters which are introduced in the calculation of $k_i(\nu_i)$. Fig. 13 shows the calculated signals and it can be seen that the curves a=1.0, 0.9, 0.8 differ well from each other. Here

$$a = \sqrt{\ln 2}\, \Delta\gamma_{Voigt} / \Delta\gamma_{Doppler}$$

is the "broadening parameter" which determines the overall Voigt shape of the absorption line.These results give
rise to expectations that from the extended calculations
and comparison with the magnetic scanning experiment not
only the shifts but also the collision broadening can be
determined with errors not larger than $0.1\Delta\gamma_{Doppler}$
At the end of this section it should be remarked that
those measurements are also interesting in that they permit to continue our comparison of broadening and depolarizing collision cross-sections.[26]

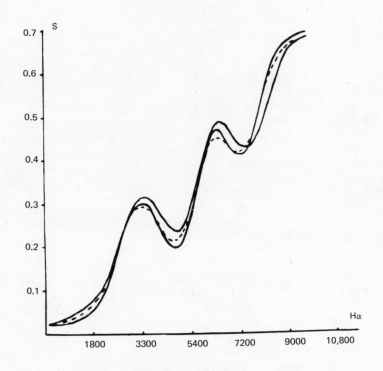

Fig. 13: Transmission of the π-component of
        the D_1-line of [87]Rb through a [87]Rb
        absorption cell as a function of
        the magnetic field

II. ALIGNMENT OF EXCITED ATOMS IN A
DC - GAS DISCHARGE.

Investigating the emission of a spectral line in
Ne excited by a dc - gas discharge, we discovered in
1968 a Hanle signal [26]. The strength and sign of this
signal depended both on the direction of the applied
magnetic field and on the region of the gas discharge
observed with the monochromator. For example the signal has opposite sign in the center and at the edge
of the gas discharge tube (Fig. 1). From the observed
signals the distribution of the alignment over the
tube volume can be determined. This distribution is
shown in Fig. 2 with the length of the dipole representing the intensity of the dipole emission. In the
center of the tube the alignment is uniaxial with its
axis parallel to that of the discharge. Close to the
zero signal region the alignment is biaxial (with the
Hanle signal being observable for any direction of the
magnetic field). The second axis coincides with the
radius of the tube.

Two conditions have to be fulfilled before an
atomic state can be aligned:
1. The angular momentum has to be larger than 1/2.
2. The process leading to this state must be anisotropic, or the original state must have had an alignment. In the gas discharge in the main three processes
can produce excited atoms:collisions between the particles, absorption of light, and the decay of higher
excited states. All three processes can give rise to
aligned atomic states, but we ,as well as the authors
of a later independent paper[27],believe that alignment
is mainly due to the generation of excited states by
absorption of light.

The contribution of the reabsorption of light to
the generation of the alignment was shown in a special
experiment. In both sides of the discharge tube
two reversing prisms were positioned (oposite to the
region from which the radiation could be observed)
which reflected the visible light back into the tube.
The prisms increased the portion of light in the direction perpendicular to the observation. In this case
the signals due to the radial alignment increased at
the edge of the tube and decreased in the center. These
results simultaneously confirm the model of the alignment distribution deduced from other experiments.

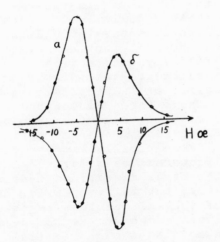

Fig. 1   Hanle signals in the center (a)
         and at the edge (b) of the dis-
         charge tube.

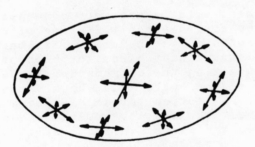

Fig. 2   Distribution of the alignment
         over the volume of the discharge
         tube.

The transfer of alignment from higher states has also been observed. The experimental measurement of the alignment signals in the gas discharge is very simple. The radiation of the discharge is observed through the wall of the tube. With a common monochromator or a filter one spectral line is separated and the intensity is recorded with a photodetector as a function of the applied magnetic field. The direction of the magnetic field is either parallel to that of the observation or to the discharge axis, or is perpendicular to both of these directions. These three directions are the most favourables ones. A signal can of course be observed also with other directions of the magnetic field H. With the magnetic field parallel to the direction of observation the light must be polarized under any circumstances; for the other directions a polarizer is not absolutely necessary, but is favourable. Because of the low intensity I of the emitted line several variants of measurements are used. In one case the derivative $\partial I/\partial H$ is measured for a modulated magnetic field. A second scheme involves the use of a dielectric polarizer to separate the output beam of the monochromator into two beams polarized perpendicularly to each other, and the substraction of the photocurrents associated with the two beams.

The alignment signals are very convenient to determine the time of coherence (lifetime of the alignment) which depends on the natural lifetime, on depolarizing collisions, and on the reabsorption of the radiation. By investigating the alignment signals for different pressures of the buffer gas, both the natural lifetime and the cross section for depolarizing collisions can be determined. The results of measurements for a series of states are listed in Table III.

In the favourable cases without substantial complications (these will be discussed later) and with the function $I = I (H)$ being Lorentzian, the coherence time is obtained directly from the width of the signal. The extrapolation to zero pressure gives the natural lifetime and the slope of the extrapolation line gives the cross sections for depolarizing collisions. In this way the data for the He states $3,4,5\,^1D_2$ were obtained.

Table III

Element	Atomic state	τ / ns	$\sigma^2/10^{-16}\,cm^2$
He	3^1D_2	17 ± 2	70 ± 10
	4^1D_2	33 ± 7	150 ± 20
	5^1D_2	80 ± 40	270 ± 40
	3^1P_1	1.8 ± 0.1	
	4^1P_1	4.0 ± 0.4	
Ar	$3p^54p\,^1P_1$	32 ± 3	
	$3p^54s\,^1P_1$	1.8 ± 0.4	
Ne	$2p^53s\,^3P_1$	18 ± 1	
	$2p^53s\,^1P_1$	2.5 ± 0.4	

In the most cases, however, we have to deal with
more complicated alignment signals because of reabsorption and transfer of alignment through transitions from
higher excited states. This transfer can be observed in
a particularly distinct way for Ar lines from the transition 2p - 1s (lines with λ = 706, 738, 801, and 842 nm;
for the conditions: dc - gas discharge with a current
of 50 mA, a pressure of 0,03 - 0,2 Torr, and a diameter
of the discharge tube of 6 mm). The signal from these
lines is a curve with two extrema (the derivative $\partial I/\partial H$
is recorded and represents the sum of two Hanle signals
with different width. For all lines the relative height
of the signals changes with the pressure. For pressures
smaller than 0.02 Torr the broad component which we attribute to the usual Hanle effect vanishes completely, for
an example, see Fig. 3.

For a series of lines for which the change in
signal form can not be followed so distinctly, the signal width does not depend linearly on the pressure p.
For small pressures the p-dependence of particular lines
deviates to one side or the other from a straight line.
We believe that the reason for the change of the signal
width comes from the influence of the depolarizing collisions for high pressures and from the influence of the
cascade-Hanle-effect for small pressures. There is no
inconsistency in the assumption that an observable align-

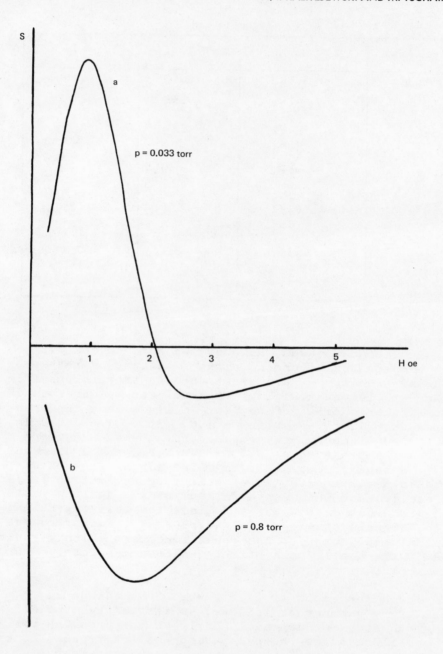

Fig. 3   Alignment signals for the Ar line
706.7 nm consisting of two Hanle curves.
Curve a: 0.033 Torr Ar,
Curve b: 0.8 Torr Ar.

ment exists even for small pressures, since the states
are coupled with the ground state and radiation of the
transistion is strongly reabsorbed.

Additionally the alignment of the state depends
strongly on the relative intensities of the lines which
couple the state with lower ones,and on the relative
population densities of these since the absorption of
the light by different lower states leads to an alignment with different signs.

The influence of reabsorption of the signal width
which is linked to a partial conservation of coherence
by the reabsorption manifests itself when one compares
the signals representing the destruction of the alignment parallel to the axis with that along the radius.
In Fig. 2 the distribution of the alignment over the
tube volume was shown. We should remark here that the
conservation of the alignment due to reabsorption can
be taken into consideration for the dipoles of each
direction individually. The effect of reabsorption con-
sist of partial transfer of the alignment and its direction from the radiating atoms to the absorbing ones. In
the beginning we consider the conservation of the axial
alignment which is characterized by the difference of
magnitude of the dipole from the mean value. Independent of the location of the radiating and absorbing
volume element, the resulting alignment will be added
to that already present. Thus the lifetime of the alignment is increased. Such an alignment is destroyed by a
magnetic field perpendicular to the discharge axis. The
signal belonging to this alignment can be observed in
the perpendicular magnetic field from the radiation polarized parallel to the discharge axis. This signal is
decreased by reabsorption.

It is more complicated to explain the conservation
of the radial alignment by reabsorption of the radiation.
The difficulty of this case is connected with the axis
of the alignment not being parallel in different parts
of the tube. Thus the reabsorption of radiation from
and by different volume elements can either increase
or decrease the already present alignment.

After developing this model of reabsorption we also expected a difference in the experimental signal
width of the axial and radial alignment. For the investigation of these signals not only a change of magnetic
field direction but also different methods of measurement become necessary. It should be mentioned that the
signal from the radial alignment is much smaller than

from the axial one. For the Ar 794.8 nm line both signals could be investigated. The extrapolated curves for
the signal width are shown in Fig. 4.

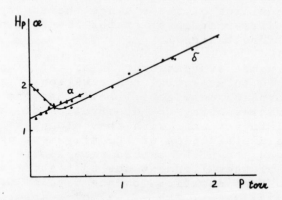

Fig. 4:   Dependence of the signal-
           width of the Ar 794.8 nm
           line on the gas pressure:
           (a) for the radial and (b)
           for the axial alignment.

The signal widths differ significantly. The signal width
of the radial alignment depends linearly on the pressure
and the extrapolation to zero pressures yields the correct value for the natural lifetime[28].Thus it could be
shown that the reabsorption differently influences the
alignments in the different directions and that the signals from the radial alignment are less deformed by reabsorption than those from the axial one. The signals,
e.g. for the He lines 501.6 nm (3^1P_1 - 2) and 396.5 nm
(4^1P_1 - 2^1S_0)[29,30]are strongly deformed by reabsorption.
The states from which the lines originate are coupled
through resonant transistions to the ground state and
for these of course complete reabsorption takes place
even for the lowest pressures for which signals can be
observed. Therefore the dependence of the signal width
on pressure is linear. The "hook" which is characteristic for reabsorption of radiation could not be observed
close to zero pressure and the natural lifetimes can not
be obtained by extrapolation to zero pressure. The linear

extrapolation of the curves to zero pressure yields coherence times of τ = 5.3 ns and 12.5 ns. But just the
fact of complete reabsorption made it possible to determine the lifetimes from the measured coherence times.
The narrowing of the signal width for complete reabsorption in a transition from the states in consideration
is given by the following expression:

$$\Gamma_{reab} = \Gamma_2 - \Delta\Gamma = \Gamma_2 - 2.1 \cdot d_r \Gamma_0 \, (2J_1+1) \cdot \begin{Bmatrix} 1 & 1 & 2 \\ J_1 & J_1 & J_0 \end{Bmatrix}^2$$

with d_r = resonance transition probability,

$\quad\quad d_o$ = transition probability of the observed
$\quad\quad\quad\quad$ transition, and

$\quad\quad d_r + d_o = 1$.

With this expression the values of τ for the He I states
3^1P_1 and 4^1P_1 in Table III are determined.

For the heavier noble gases alignment signals are
also observed. The energy levels of these atoms lie closer together. Therefore for the same pressure an alignment due to reabsorption in a larger number of states
results and the transfer of coherence from spontaneous
emission plays a much greater role than for the lighter
noble gases. For a series of lines, signals are observed
which can be divided into two or three components of
different width (587.1 nm, 446.4 nm, 760.2 nm, 826.3 nm,
and 556.2 nm)[31].

In addition the odd isotopes in a natural abundant
gas have hyperfine structure. Since the Landé g-factors
of hyperfine components of the levels are ordinarily unequal the signal has a complicated form and the analysis
is difficult. This can be avoided by using a separated
even isotope. Nevertheless the field of application for
the above described simple experimental technique to
measure atomic constants is rather limited.

We call all the observed and described signals
macroscopic alignment signals as distinguished from
the "hidden" alignment which we will deal with in the following[32]. We consider an isotopic source of light with a
Doppler profile. In this case a moving atom absorbs more
of the light propagating perpendicular to its motion than
of that propagating parallel because of the Doppler detu-

ning of the latter. The excitation of atoms with light
propagating perpendicular to the direction of motion
produces an alignment of the atoms moving in a particular direction. The axis of the alignment coincides with
the direction of motion. The alignment vanishes by integrating over all directions of atomic motion, therefore
we called it "hidden" alignment. But this alignment causes a change in the distribution of the absorption and the
spontaneous emission. An ensemble of atoms which is iso-
tropically excited with a light having a Doppler profile
into a state which cannot have an alignment, e.g. with
J = 0, emits and absorbs light with a narrower profile
than the absorption line from the ground state. The
"hidden" alignment changes the profile in as much as it
can become narrower or broader depending on the details
of the transition. An example of the change of the line
because of the alignment is shown in Fig. 5 for a J = 1
→ J = 0 transistion.

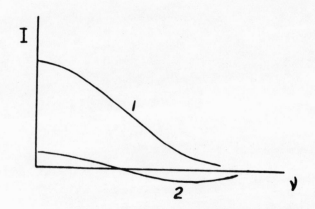

Fig. 5   Admixture of interference (2) to
          the spectral profile of the emission
          line (1). Curve (2) is enlarged eight
          times compared to curve (1).

The admixture causing the change is partly destroyed
by applying a magnetic field. This is independend of
the frequency. The admixture is described by the term:

$$f \ (H) \ \sim \ \frac{1}{1 \ + \ \omega^2/\Gamma^2} \ + \ \frac{1}{1 \ + \ 4 \ \omega^2/\Gamma^2} \ .$$

A change in the profile of the emission line leads to
a change in the reabsorption and therefore to a change
in the population densities in the intermediate state.
Both give rise to a change in the population densities
of higher states and also the state with $J = 0$. Thus the
line intensities from the $J = 0$ state and the π -compo-
nents of other lines must depend on the hidden alignment and therefore on the magnetic field. These signals
were observed already in the first experiments but only
two years ago did we succeed in explaining them.

Since the signals from the hidden alignment are determined by the characteristics of the intermediate level, these can be also measured. For instance, the natural lifetime of the state $2p^5 3s \ \ ^3P_1(^1S_4)$, see TableIII,
was measured from the Ne 607.4 nm line coming from the
$2p^5 3p \ \ ^3P_0 \ (2P_3)$ state (by means of extrapolation to
zero pressure).

This example is interesting since it shows the possibility of determining the relaxation time of a state
with spontaneous emission in the UV via the radiation
in the visible spectrum by just observing the radiation
of the gas discharge. From a technical point of view,
however, the experiment is simple.

REFERENCES

1. H. Kallas, G. Markova, G. Chvostenko, M. Tschaika,
   Opt. i Spektr., 19, 303, 1965.
2. G. Markova, G. Chvostenko, M. Tschaika, Opt. i
   Sepktr., 23, 835, 1967.
3. G. Chvostenko, M. Tschaika, Opt. i Spektr., 25, 450,
   1968.
4. L. A. Wolikova, W. N. Grigorjeva, G. I. Chvostenko,
   M. P. Tschaika, Opt. i Spektr., 30, 170, 1971.
5. W. N. Grigorjeva, G. I. Chvostenko, M. P. Tschaika,
   Opt. i Spektr., 34, 1224, 1973.
6. G. I. Chvostenko, Opt. i Spektr. 26, 63, 1969.
7. G. I. Chvostenko, G. I. Chutorschikov, M. P. Tschaika,
   Opt. i Spektr., 36, 814, 1974.
8. A. L. Maschinski, M. P. Tschaika, Opt. i Spektr.,
   28, 1093, 1970.
9. M. P. Tschaika, Opt. i Spektr., in press.
10. W. N. Rebane, T. K. Rebane, Opt. i Spektr., 33, 405,
    1972.
11. W. N. Rebane, T. K. Rebane, W. A. Tscherenkovski,
    Opt. i Spektr., 33, 616, 1972.
12. E. L. Altmann, A. L. Maschinski, M. P. Tschaika,
    Opt. i Spektr., 26, 126, 1969.
13. T. I. Krupenikova, M. P. Tschaika, Opt. i Spektr.
    20, 1087, 1966.
14. T. Hänsch, P. Toschek, Phys. Lett., 20, 273, 1966.
15. B. Decomps, M. Dumont, Compt.Rend. Acad. Sci., 262,
    1004, 1966.
16. E. Kotlikov, G. Todorov, M. Tschaika, Opt. i Spektr.,
    30, 185, 1971.
    G. Todorov, M. Tschaika, Opt. i Spektr. 23, 326, 1967.
17. E. B. Aleksandrov, Opt. i Spektr., 14, 436, 1963.
18. M. Tschaika, Sammelbd. "Physik der Gaslaser", LGU,
    S. 117.
19. E. I. Iwanov, M. P. Tschaika, Opt. i Spektr., 29,
    124, 625, 1970.
20. W. Demtröder, Z. Physik 166, 42, 1962.
21. S. L. Isotova, N. G. Preobraschenski, M. S. Frisch,
    IV. International Conference on Atomic Physics,
    Heidelberg, 1974, Abstracts, p.221.
22. Ju. W. Jewdokimov, N. I. Kalitejevski, M. P. Tschaika,
    Opt. i Spektr., 27, 186, 1969.
23. W. W. Gerschun, N. N. Jacobson, W. I. Chutorschikov,
    Opt. i Spektr., 31, 866, 1971.
24. O. M. Risch, M. P. Tschaika, Opt. i Spektr.,in press.
25. Ju. W. Jewdokimov, N. I. Kalitejevski, Opt. i Spektr.,
    31, 656, 1971.

26. H. Kallas, M. Tschaika, Opt. i Spektr.,27, 694, 1969.
27. C. G. Carrington, A. Corney, Opt. Commun.,1, 115,1969.
28. S. Kasanzev, A. Kießling, M. Tschaika, Opt. i Spectr. 36,1030,1974
29. S. Kasanzev, W. Markov, Opt. i Spektr., 36, 613, 1974.
    S. Kasanzev, A. Kießling, M. Tschaika, Opt. i Spektr., 34, 1227, 1973.
    S. Kasanzev, W. Markov, M. Tschaika, Opt. i Spektr., 34, 854, 1973.
    S. Kasanzev, M. Tschaika, Opt. i Spektr., 31, 510, 1971.
    S. Kasanzev, A. Kießling, W. Markov, M. Tschaika, Vestnik LGU, in press.
30. C. G. Carrington, J. Phys. B 5, 1572, 1972.
31. S. Kasanzev, W. Markov, M. Tschaika, Federal Union Symposion:"Investigation of Spatially In-homogeneous Plasmas from Spectral Line Shapes", Petrosavo-dsk,May 1974.
32. M. Tschaika, Opt. i Spektr., 30, 822, 1971, 31, 513, 1971, 31, 670, 1971.

CORRELATIONS OF EXCITED ELECTRONS [*]

U. Fano and C. D. Lin

Department of Physics, University of Chicago

Chicago, Illinois 60637, U.S.A.

This progress report descends from the report on doubly excited states of atoms presented at the first of these Conferences. [1] It develops the trend introduced at Belgrade last year[2] toward a descriptive treatment of electron correlations by summarizing the results on doubly excited He and H⁻ obtained by one of us. [3] It will also outline, in the initial section, the background problems of excitation dynamics which have lent special interest to the theory of doubly excited states from its very beginning. [4] The progress achieved thus far encourages us to articulate such broader problems, the more so as we are now beginning to discern how our approach to double excitation may eventually encompass the Wannier theory of threshold ionization and the transition to the Born and impulse approximations which are characteristic of higher-energy collisions. Remarks on this aspect will be made in a concluding section.

1. The Problem of Excitation

Bohr remarked at the outset of quantum physics that the stability and the well defined identity of each chemical species rests on the existence of stationary states and on the energetic separation of the ground state from excited states. Yet all forms of reactivity, whether in a chemical laboratory or in astrophysical or biological settings, involve transformations of ground states and thus proceed by overturning, at least locally, the stability stressed by Bohr. Their pathway often passes through intermediate stages of electronic excitation. It is then a main task of atomic physics to trace these pathways in the configuration space of the atomic electrons.

We circumscribe here this broad task firstly by confining ourselves to excitations from one shell to another. (Excitations within the ground state shell are actually more likely but involve more complex patterns of electron correlation, whose study we are not yet attempting.) Secondly, we need not consider excitations of single electrons by photons inasmuch as this process can be treated as a weak perturbation exerted by an external field on a ground state electron; the escape of this electron from its initial shell, subsequent to energy absorption, is rather well understood and has been dealt with extensively. Similarly, excitation by fast charged particles reduces either to a long-range weak perturbation — akin to a radiation field — or to an impulsive localized action, both of which are also well understood. Our field thus narrows to photoexcitation of two electrons and to excitation by collision of slow electrons or of slow ions or atoms. We restrict it further to electronic processes noting only that, whereas excitation by slow heavy particles and by electrons used to be regarded as quite different, a striking similarity between them will emerge in this paper. Finally, we shall deal explicitly only with two-electron atoms, namely, He and its isoelectronic sequence (H^-, Li^+, ...), for simplicity. However our study is intended to be relevant to the excitation of many-electron atoms, with adaptations required by the replacement of the point nucleus of He by an ionic core.

The processes of direct interest, two-electron photoexcitation and single excitation by impact of a slow electron, are regarded here as two aspects of a single process, in which a complex consisting of two excited electrons in a Coulomb field is formed through alternative input channels as shown by the diagrams

$$
h\nu + H^- \rightleftharpoons \overset{\textstyle\quad}{\underset{\displaystyle \updownarrow}{H^{-**}}} \rightleftharpoons H^* + e \; , \qquad h\nu + He \rightleftharpoons \overset{\displaystyle h\nu + He^* \atop \updownarrow}{\underset{\displaystyle \updownarrow}{He^{**}}} \rightleftharpoons He^{+*} + e \qquad (1)
$$

$$
e + H \hspace{8cm} e + He^+
$$

where a single (double) "*" indicates single (double) excitation. The equivalence of the two processes lies in the fact that, for collision energies of the order of the excitation threshold, the incident electron's velocity becomes comparable to that of the excited electron after the initial stage of inelastic collision. Thus the two electrons remain correlated until the complex dissociates again. The study of this complex and of its eventual dissociation through any of the channels radiating from it — including a return to the entrance channel — constitutes our objective.

To illustrate this goal we give here a list of qualitative questions, some of which we can answer rather confidently, others only quite tentatively.

a) Why are the cross section ratios for inelastic vs. elastic collision and double vs. single photoexcitation small, generally of order $\lesssim 10^{-1}$, even though the interaction of any two electrons within the ground state shell is far from weak?

b) To what extent can one treat excitation to lower shells (say, with n=2 or 3) without allowing in detail for possible excitation to higher shells?

(No such allowance is made, e.g., in close-coupling calculations.) A related question often asked by experimentalists is: When the energy available in a collision rises above the threshold of a new process, does the total excitation probability increase or does the new process occur at the expense of the probability of lower excitations?

c) To calculate the probability of excitation of an atom to its n-th state by a slow electron collision, does one have to take into account correlations up to the radius of this state, or are the essential interactions confined within a smaller volume?

d) Does collisional ionization near threshold, or double photoionization, emerge as a limit of the processes of excitation to lower levels?

e) Do individual electrons within a doubly excited complex ever attain a high orbital momentum? Do they retain this momentum after the complex breaks up?

f) At energies $\gtrsim 100$ eV, the reactions (1) are known to proceed directly, e.g., by an impulsive momentum transfer, without passing recognizably through an intermediate doubly excited complex of the kind we want to study. What criterion identifies the minimum energy required for a direct reaction, that is, the upper energy limit at which complex formation is relevant?

We aim in the following at least at indicating the types of calculation that should answer these questions.

2. Mapping in Configuration Space

To examine the mechanisms of excitation we should construct approximate eigenfunctions of the two-electron Hamiltonian and use them to map the probability distribution $|\Psi(\vec{r}_1, \vec{r}_2)|^2$ in the six-dimensional configuration space of the electron pair. Significant electron correlations are represented, in a mapping, by a narrow spread of this probability distribution in one or more of its variables.[2] (By "variable" we mean here either a position coordinate or its conjugate momentum or an equivalent quantum number; the identification of suitable variables is itself an important stepping stone.) The mapping should provide both a qualitative description of excitation and guidance for laying out effective calculations of observable parameters.

More precisely, we shall not construct eigenfunctions of exact stationary states directly but, rather, those of quasi-stationary states which are weakly coupled to one another. The existence of such weakly coupled states was a main result of early experiments on double excitation, soon confirmed and extended by direct numerical solutions of the Schroedinger equation. Discrete states belonging to a Rydberg series, and the continuum states beyond the series limit, were found to have common properties; in current language such a group of states is said to constitute a channel. Reference 1 reviewed these facts and stressed that calculations had not yet provided any clue to the weakness of coupling or to the characteristics of the various channels. It also pointed out that weakness of coupling implies the existence of some unknown approximate constant of the motion which characterizes each channel. Such a constant will serve, when identified, not only as a channel label but also as one of the variables of our problem, quasi-independent of the other five.

In a narrow sense, the correlation of the electron positions r_1 and r_2 within an atom is fully represented by the probability distribution of two vari-ables only, which identify the shape of the triangle in Fig. 1. Other variables, which do not represent correlations, are the three Euler angles which define the orientation of this triangle in space. (Two of these are usually replaced by their conjugate momenta, or rather by the orbital quantum numbers L and M which are constants of the motion; the third one does not similarly separate out, because the triangle constitutes an asymmetric rotor.) A sixth variable should represent the size, or scale, of the triangle in Fig. 1. The angular cor-relation of the electron positions is represented by a plot of $|\Psi|^2$ against the angle θ_{12} shown in Fig. 1. The radial correlation is represented by an analo-gous plot against a second variable that specifies the relative magnitude of the vectors in Fig. 1. For this purpose we choose the angle $\alpha = \arctan(r_2/r_1)$ shown in Fig. 2. The point that represents the pair of variables (r_1, r_2) in this figure is the projection of the 6-dimensional vector $(\vec{r}_1, \vec{r}_2)$ onto the plane (r_1, r_2). The distance $R = (r_1^2 + r_2^2)^{\frac{1}{2}}$, also shown in Fig. 2, repre-sents then the scale of the triangle in Fig. 1; it will serve as a measure of the size of the atom. Recall, finally, that correlations also result from the iden-tity of electrons depending on their mutual spin orientation. These so-called "exchange correlations" are represented by symmetries of the mapping on Fig. 2 under reflection on the $\alpha = 45°$ axis.

For purposes of orientation we indicate here how different types of ex-citation yield qualitatively different mappings of the distribution of $|\Psi|^2$ onto

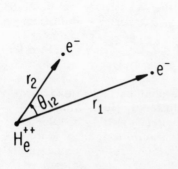

Fig. 1

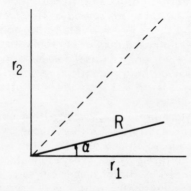

Fig. 2

the coordinate plane of Fig. 2. Excitation in general implies an increase of the size of an atomic system and hence expansion of the $|\Psi|^2$ distribution toward large values of R as well as the occurrence of nodal surfaces of this distribution. Excitation of a single electron implies expansion in the range of $r_1 \gg r_2$ (or $r_2 \gg r_1$), that is, toward large R with $\alpha \sim 0$ (or $\alpha \sim \frac{1}{2}\pi$); in the example of Fig. 3a the radial correlation confines $|\Psi|^2$ to a narrow range of α and one nodal surface occurs with $R \sim$ constant. Single ionization implies that this expansion extends to $R = \infty$. Double excitation evenly shared between the two electrons implies, instead, that the distances r_1 and r_2 of the electrons remain comparable, whereby the $|\Psi|^2$ distribution spreads over the whole range of α, from 0 to $\frac{1}{2}\pi$; in the example of Fig. 3b the distribution is marked by two nodal surfaces running from $R = 0$ to $R = \infty$. Double excitation of He in which one electron is highly excited while the He^+ ion has only a low radial excitation is represented by a pattern similar to that of Fig. 3b but extending further towards large R near the $\alpha = 0$ axis and with a succession of nodal surfaces $R \sim$ constant. In the event of double ionization the distribution spreads over the whole range of α all the way to $R = \infty$; at large R each value of α corresponds then to a specific partition of escape energy between the two electrons.

These considerations suggest that wave functions of each channel are characterized by a particular distribution of $|\Psi|^2$ in the radial correlation variable α, and presumably also in θ_{12} and other direction variables. Different states of the same channel correspond to different levels of excitation; hence their eigenfunctions differ mainly in their extension toward larger R, reaching all the way to $R = \infty$ for states of the continuum. Accordingly, each of these wave functions should be approximately factorable, into one factor Φ, which represents the correlations and depends on α and other angular variables, and a factor F which depends on R and is analogous to the radial wave functions of the usual one-electron problems.

As we shall see, the correlations — represented by the factor Φ — are not constant for each channel but depend themselves on R. This is not surprising since, for example, angular correlations disappear when one of the electrons escapes from the atom leaving the other one behind. The physical situation thus resembles that of diatomic molecules, whose wave functions split approximately into electronic and vibrational factors; the electronic wave function does depend on the internuclear distance R, but the vibrational motion couples the different electronic states only weakly. In both cases the weakness of coupling rests actually on the slow response of one motion to the other. In the case of molecular vibrations the slowness of response is attributed to the large mass and slow velocity of nuclei; however the coupling remains remarkably weak even in atomic collisions with nonnegligible nuclear velocity. We suspect that in both cases the weak response of electronic wavefunctions to changes of the parameter R rests ultimately on the long range of Coulomb forces.

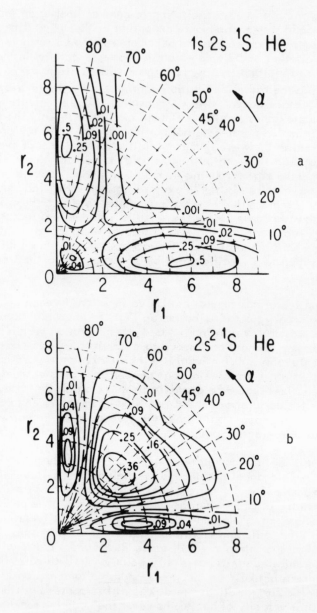

Fig. 3. Contour plots of approximate probability distribution $|\Psi(\vec{r}_1,\vec{r}_2)|^2$ for He. a) 1s2s ^{1}S , b) 2s^2 ^{1}S. (Angular correlations represented by admixture of p^2, d^2, etc. disregarded.)
——— equidensity contours
—·—·— nodal lines

3. Channel Expansion

The key evidence on double excitations, reviewed in ref. 1 and sum-marized in Sec. 2, points to the existence of <u>weakly coupled and characteris-tically different channels</u> of excitation. We seek to formulate this concept mathematically by reducing the six-dimensional Schroedinger equation to a system of weakly coupled, single variable radial equations, one for each channel.

An approach followed since the mid-60's is to study expansions of the Schroedinger equation for the electron pair in polar coordinates of the six-dimension space $(\vec{r}_1, \vec{r}_2)$ (hyperspherical coordinates). These coordinates con-sist of the size parameter R shown in Fig. 2 (hyperspherical radius), of the angle α pertaining to radial correlations, and of any choice of four additional angles which identify the directions $(\hat{r}_1, \hat{r}_2)$ of
Macek[5,6] and to ref. 1, the schematic form of the equation, in a.u., is

$$\left[\frac{d^2}{dR^2} - \frac{\Lambda^2 + 15/4}{R^2} + \frac{C(\alpha, \theta_{12})}{R} + 2E \right] (R^{5/2} \Psi) = 0 . \tag{2}$$

Here Λ^2 indicates the generalized squared angular momentum in the configu-ration space, being a combination of the squared orbital momenta, $\vec{\ell}_1^2$ and $\vec{\ell}_2^2$, of the two electrons and of a squared mock-angular momentum conjugate to the coordinate α. The function

$$C(\alpha, \theta_{12}) = \frac{2Z}{\cos\alpha} + \frac{2Z}{\sin\alpha} - \frac{2}{(1 - \sin2\alpha\cos\theta_{12})^{\frac{1}{2}}} , \tag{3}$$

a function of the shape of the triangle in Fig. 1, represents an effective atomic number such that $-C/R$ equals the Coulomb potential energy of the whole sys-tem. (The factor $R^{5/2}$ has been set off to eliminate terms in d/dR.) An essen-tial feature of this form of the equation lies in the analogy of its structure to the radial equation for the H atom: The d^2/dR^2 term represents the kinetic energy of the (hyperspherical) radial motion, $(\Lambda^2 + 15/4)/R^2$, represents a mock-centrifugal potential energy and $-C/R$ the potential energy. The rela-tive importance of the three terms <u>varies simply and smoothly</u>, in proportion to R^{-2} and R^{-1}. The complication of the problem lies in the fact that Λ^2 and C do not commute. Incidentally, the structure (2) holds for any atomic, or molec-ular, system with internal Coulomb forces; one needs only to adjust the defini-tions of Λ^2 and C and the power 5/2 of R.

In the limit of $R \sim 0$, the mock-centrifugal term of Eq. (2), $(\Lambda^2 + 15/4)/R^2$, predominates.[7] Here, then, the equation is locally separable, with solutions of the form $\Phi_\mu(\Omega) F_\mu(R)$, where $\Phi_\mu(\Omega)$ is an eigenfunction of Λ^2 and Ω indicates the set of five angular coordinates. The eigenvalues of Λ^2 are $\nu(\nu+4)$, with $\nu = 0, 1, 2, \ldots$, and are degenerate for $\nu > 0$. The expression of Λ^2 in terms of the orbital momenta, $\vec{\ell}_1$ and $\vec{\ell}_2$, of the two electrons

$$\Lambda^2 = -\frac{1}{\sin^2\alpha\cos^2\alpha}\frac{d}{d\alpha}\left(\sin^2\alpha\cos^2\alpha\frac{d}{d\alpha}\right)+\frac{\vec{\ell}_1^{\,2}}{\cos^2\alpha}+\frac{\vec{\ell}_2^{\,2}}{\sin^2\alpha}\tag{4}$$

leads to eigenfunctions $u_{\ell_1\ell_2 m}^{SLM}(\Omega)$ antisymmetric under permutation of the two electrons,[5] whose quantum numbers ℓ_1, ℓ_2, S, L, and M have their usual meaning, while m indicates the number of nodal hypersurfaces $\alpha=$const, that is, the degree of excitation of radial correlations. The corresponding quantum number ν is given by $\nu=\ell_1+\ell_2+2m$.[8] As R increases, the potential energy term of Eq. (2) becomes increasingly important, though this happens gradually and at larger and larger values of R for increasing values of ν. Figure 4 shows a relief map of this term on the (α,θ_{12}) coordinates; note how narrow is the base of the positive energy spike, which represents the electron-electron repulsion, and how the nuclear attraction tends to draw the system into the ditch along $\alpha=0$ (or in its mirror image at $\alpha=\frac{1}{2}\pi$).

Equation (2) is transformed into a system of coupled ordinary equations by expanding Ψ at each value of R into a complete system of functions of the five angular variables, $\Phi_\mu(R;\Omega)$. Substituting

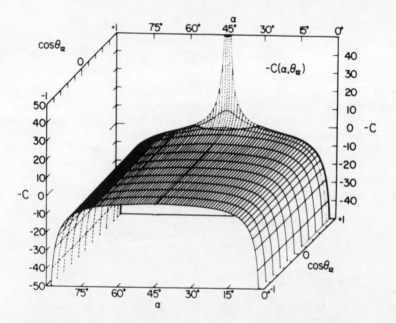

Fig. 4. Relief plot of $-C(\alpha,\theta_{12})$ with Z=1. The ordinates represent a potential surface in Rydberg units at R=1 bohr. (Courtesy, C.E. Theososiou)

$$R^{5/2} \sin\alpha \cos\alpha \, \Psi = \Sigma_\mu \Phi_\mu(R;\Omega) F_\mu(R) \tag{5}$$

in Eq. (2) and projecting onto Φ_μ gives

$$(\frac{d^2}{dR^2} + \frac{1/4}{R^2} + 2E) F_\mu(R) - \Sigma_{\mu'} [U_{\mu\mu'}(R) - W_{\mu\mu'}(R)] F_{\mu'}(R) = 0 \tag{6}$$

where

$$U_{\mu\mu'}(R) = \frac{1}{R^2} (\Phi_\mu | - \frac{d^2}{d\alpha^2} + \frac{\vec{\ell}_1^2}{\cos^2\alpha} + \frac{\vec{\ell}_2^2}{\sin^2\alpha} - RC | \Phi_{\mu'}) \tag{6'}$$

and

$$W_{\mu\mu'}(R) = 2(\Phi_\mu | \frac{\partial \Phi_{\mu'}}{\partial R}) \frac{d}{dR} + (\Phi_\mu | \frac{\partial^2 \Phi_{\mu'}}{\partial R^2}) . \tag{6''}$$

We interpret the existence of weakly coupled channels as implying the existence of functions Φ_μ such that the terms of Eq. (6) with $\mu' \neq \mu$ can be treated
as small perturbations.

Macek[6] approached the search for such functions by determining eigenfunctions Φ_μ of the operator U in (6'), following thus the adiabatic approximation of molecular physics. Each eigenvalue $U_\mu(R)$ represents then an effective potential for the radial motion in the μ-th channel. Macek's numerical
procedure, however, prevented any analysis of the eigenfunctions Φ_μ and restricted evaluation of the matrix $W_{\mu\mu'}$ to a minimal level. Macek proceeded
nevertheless by disregarding all off-diagonal elements of $W_{\mu\mu'}$ and calculating series of discrete eigenvalues for several of the equations thus decoupled.
In each case he obtained a Rydberg series of eigenvalues in remarkable agreement with one of the series that had emerged from the analysis of earlier data.
Thereby each of these series was linked, at least semiempirically, to one of the
optical potential wells $U_\mu(R)$. The weak processes of double excitation and
autoionization were tentatively interpreted as interchannel transitions occurring at "avoided crossings" between different potential curves. However, the
validity of this most suggestive analogy — which would merge the treatments of
electronic and atomic collisions — was not really tested by Macek.

This program has now been reactivated by one of us, as reported in
ref. 3. The eigenfunctions $u_{\ell_1 \ell_2 m}^{SLM}(\Omega)$ of the operator (4) served as a basis
for constructing the functions Φ_μ of Eq. (6). An early remark was that in this
basis the matrix elements of the operator C are larger by one-half order of
magnitude when diagonal in the quantum numbers (ℓ_1, ℓ_2) than otherwise. The
reason is that off-diagonal elements involve a transfer of orbital momentum between the electrons and thus arise only from a portion of the last term of Eq.
(3). This circumstance suggested the construction of functions Φ_μ with fixed
values of (ℓ_1, ℓ_2), as well as fixed "good" quantum numbers (S, L, M); this is

done by superposing only functions $u_{\ell_1\ell_2m}^{SLM}$ with different m. Using such functions, the matrix U is not made quite diagonal but its diagonal elements $U_\mu(R)$, are generally close to those calculated by Macek. To construct the adiabatic potential curves of Macek, one needs then to diagonalize only a small matrix of $\underline{U}(R)$. The greater simplicity in the construction of these Φ_μ and U_μ permitted a much more extensive and rewarding analysis of their properties, outlined in the next section. It also permitted pilot calculations of the off-diagonal elements of the matrix $W_{\mu\mu'}$ of Eq. (6") and of its effect on the coupled Eqs. (6). As we shall see, these elements are not large; however their coupling effects cannot be worked out readily by perturbation techniques. Further efforts will thus be required to decouple the Eqs. (6) to the extent anticipated on the basis of experimental and computational evidence.

In planning such efforts one may wish to leave flexible the choice of basis functions Φ_μ. One also needs experience on solving coupled systems with the velocity-dependent interaction of Eq. (6). This explicit dependence on velocity can be by-passed by transforming the system (6) into a nonlinear but first-order system through the Phase Amplitude Methods[9] which are finding increasing and varied application in atomic and molecular collisions.[10] We give an example of such a transformation in the Appendix.

4. Description of Channels

Each excitation channel is characterized by its potential function $U_\mu(R)$ and by its pattern of electron correlations represented by the function $\Phi_\mu(R;\Omega)$. Both functions are easily identified in the limiting cases of $R\to 0$ and $R\to\infty$ which we call "condensed atom" and "dissociated atom" in analogy to the "united atom" and "separated atoms" limits of the molecular analog. As in the molecular analog, interest lies in constructing a correlation diagram which connects the two limits of each channel and in describing the evolution of the functions U_μ and Φ_μ as R varies. Again as in the analog, this goal is sought by labeling each channel, at least initially, by quasi-good quantum numbers. We proceed then at fixed ℓ_1,ℓ_2, L and M, as indicated in Sec. 3. The Φ_μ to be considered will be explicit functions of α, so that the dependence of $|\Phi_\mu(R;\Omega)|^2$ on α represents the radial correlations directly. Angular correlations, on the other hand, will be represented only indirectly, through the orbital momentum equation $\vec{L}=\vec{\ell}_1+\vec{\ell}_2$; the value of $\frac{1}{2}(\ell_1+\ell_2-L)$ can be shown to represent the tightness of this correlation. More specifically, $\Phi_\mu(R;\Omega)$ factors for $\ell_1=\ell_2$ in the form $g_\mu(R;\alpha)\,\mathcal{Y}_{\ell_1\ell_2LM}(\hat{r}_1,\hat{r}_2)$, for $\ell_1\neq\ell_2$ it is instead a symmetrized combination of this product and of $g_\mu(R;\frac{1}{2}\pi-\alpha)$ $\mathcal{Y}_{\ell_2\ell_1LM}(\hat{r}_1,\hat{r}_2)$. For simplicity, we regard nevertheless $|g_\mu(R;\alpha)|^2$ as representative of the pattern of radial correlation of each channel μ.

In the condensed atom limit, $R\sim 0$, and for (ℓ_1,ℓ_2) fixed each channel function Φ_μ coincides with one of the eigenfunctions $u_{\ell_1\ell_2m}^{SLM}$ of Λ^2 to within a factor $\sin\alpha\cos\alpha$. We take this quantum number m as a channel label, i.e., we set $\mu \equiv m$. The function $g_m(R;\alpha)$ has m nodes in this limit, and in fact for all R. The squared function $|g_m(R;\alpha)|^2$, characterizing the probability distribution of the relative distances (r_1, r_2) of the two electrons, has then $m+1$ peaks. As one does for the probability distributions of single electrons, we shall regard m as measuring the degree of excitation of the radial correlation.

The value of U_m in the condensed atom limit is then a diagonal element of the matrix $\underline{U}$ in the basis $u^{SLM}_{\ell_1\ell_2 m}$,

$$U_m(R) = [(\ell_1 + \ell_2 + 2m + 2)^2 - R(\ell_1\ell_2 m \,|\, C \,|\, \ell_1\ell_2 m)]/R^2 \text{ at } R \sim 0; \quad (7)$$

this value is positive as the first term predominates. At large R, on the contrary, the nuclear attraction predominates and confines the distribution $|g_m(\alpha)|^2$ within the "ditches" shown in Fig. 4 at $\alpha \sim 0$ and at $\alpha \sim \frac{1}{2}\pi$. For $\alpha \ll 1$, where one electron is far closer to the nucleus than the other, the operator U defined in Eq. (6') reduces to the radial hydrogen Hamiltonian of one electron, $-d^2/d(R\alpha)^2 + \ell_2^2/(R\alpha)^2 - 2Z/R\alpha$ plus the screened potential acting on the other electron, $-2(Z-1)/R$, to order $O(R^{-1})$ and $O(\alpha)$.[6] We have then

$$U_m(R) \xrightarrow[R \to \infty]{} - \frac{Z^2}{n^2} - \frac{2(Z-1)}{R} \,, \quad g_m(R;\alpha) \xrightarrow[R \to \infty]{} P_{n\ell_2}(R\alpha)\,, \quad (8)$$

where $P_{n\ell_2}$ is a hydrogenic wavefunction with principal quantum number n. The correspondence between the values of m and n in Eq. (8) should emerge from the calculation at intermediate R. The symbol ℓ_2 indicates in Eq. (8) either element of the pair (ℓ_1, ℓ_2). Indeed the value of U_m at large R is <u>degenerate</u> under permutation of the quantum numbers (ℓ_1, ℓ_2); for this reason Φ_m may have to be taken as a linear combination of two $P_{n\ell_2}$ with the same n and alternative ℓ_2, as discussed in ref. 3. In this dissociated atom limit, the complete wave function resolves into a symmetrized product of functions of $\vec{r}_1$ and $\vec{r}_2$; to this extent the electrons are no longer correlated.

Proceeding away from each of the limits, of small and large R, the functions $\Phi_m(R;\Omega)$ are determined as the linear superpositions $\sum_{m'} a^m_{m'}(R) u^{SLM}_{\ell_1\ell_2 m'}(\alpha)$ which diagonalize the submatrix $(\ell_1\ell_2 m \,|\, U \,|\, \ell_1\ell_2 m')$. The $U_m(R)$ are the corresponding eigenvalues, examples of which are plotted in Fig. 5. Remarkably, these eigenvalues remain close to one or the other of their limiting values, (7) or (8), except in a rather narrow intermediate range. This behavior is evidenced at low R by the near-linearity of the plots of $R^2 U_m(R)$ in Fig. 6.

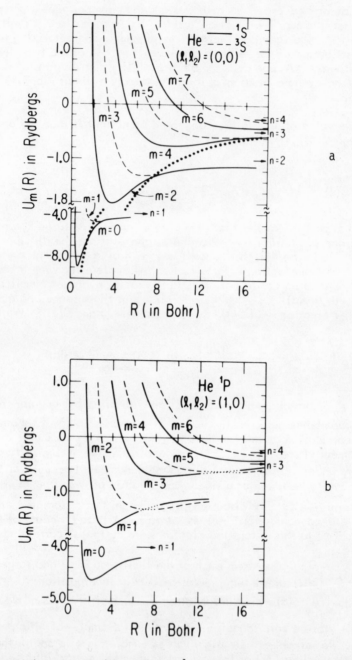

Fig. 5. Potential curves $U_m(R)$ for S and ^1P channels of He (from ref. 3).
   a) ——— s^2 ^1S ; ---- s^2 ^3S .... Eq. (9)
   b) ——— sp $^1P+$; ---- sp $^1P-$

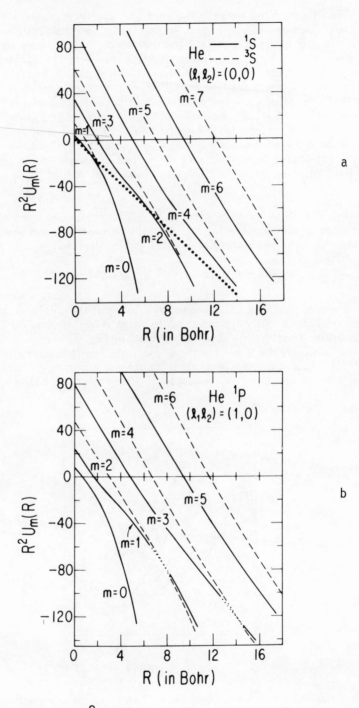

Fig. 6. Plots of $R^2 U_m(R)$; data and notation as in Fig. 5.

The intermediate range is the seat of the effects of greatest relevance to our study. In the plots of Fig. 6, in particular, one notices the occurrence of two distinct families of curves. The dashed curves turn very smoothly, almost imperceptibly, from their linear trend at low R to the quadratic trend at large R which corresponds to the constant limit of U_m in Eq. (8). For the full lines, instead, the transition from linear to quadratic trend is marked by bumps, characteristic of curve "repulsion" at "avoided crossings." This second family consists of ^1S — and generally of singlets with $\ell_1 = \ell_2$ — and of the so-called "+" channels[4] with $\ell_1 \neq \ell_2$ and any (L, S, π). The dashed-line family consists instead of the triplets with $\ell_1 = \ell_2$ and of the "–" channels with $\ell_1 \neq \ell_2$. For each (ℓ_1, ℓ_2) with $\ell_1 \neq \ell_2$, two channels converge to the same $R \to \infty$ limit (8), except for the lowest values of n; one of these is + and one is – .

The transition from the small-R to the large-R behavior takes place for each channel when its potential $U_m(R)$ drops below the broad saddle point which separates the two potential ditches in Fig. 4, i.e., when

$$U_m(R) \sim - C(\tfrac{1}{4}\pi, \pi)/R = - \sqrt{2}\,(4Z-1)/R \ . \tag{9}$$

Figures 5 and 6 show that (9) is indeed the locus of the avoided crossings. This connection of the transition to the saddle point is highlighted by the evolution of the channel functions $g_m(R;\alpha)$, illustrated in Fig. 7. For U_m values well above the value (9) the wave functions oscillate freely from $\alpha=0$ to $\alpha=\tfrac{1}{2}\pi$. As the critical value is approached and then passed, $g_m(\alpha)$ first bulges at $\alpha=45°$ and then rapidly drops to very low values, its oscillations being now confined

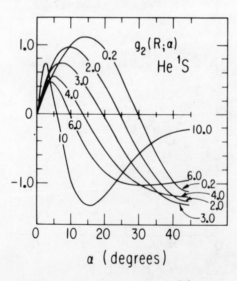

Fig. 7. Channel functions $g_2(R;\alpha)$ for helium $s^2\,^1S$ for various R. Curves extend symmetrically to the range $45° \leqslant \alpha \leqslant 90$.

to the two potential ditches separated by a barrier. The flatness of the potential saddle contributes to the rapidity of this evolution of the functions $\Phi_m(R;\Omega)$. Rapid evolution implies in turn large values of the projection matrix elements $(\Phi_m|\partial\Phi_{m'}/\partial R)$ which determine the channel couplings in Eqs. (6) or (18-19); we return to this point below. Note that the examples shown in Fig. 7 pertain only to the simple case of a singlet with $\ell_1 = \ell_2$, in which radial and angular correlations factor out. So does, strictly speaking, much of the discussion in this paragraph. The corresponding analysis for $\ell_1 \neq \ell_2$, developed in ref. 3, shows an analogous behavior for the + channels. For triplet channels with $\ell_1 = \ell_2, g_m(R;a)$ has a node at $a = 45°$ and its values never even bulge for a near 45° but decrease progressively as R increases beyond the transition region. For – channels with $\ell_1 \neq \ell_2$, the $g_m(R;a)$ have no symmetry under reflection at $a = 45°$ but have nevertheless a node in the vicinity of this point; their general transition behavior resembles that of triplet channels with $\ell_1 = \ell_2$. This remarkable behavior confirms and extends the initial characterization of + and – channels in ref. 4.

The eigenvalues $U_m(R)$ of the matrix $(\ell_1\ell_2m | U | \ell_1\ell_2m')$ — which represent, as we know, the potentials of alternative channels with the same (ℓ_1,ℓ_2) — are not degenerate except in the limit of large R, where pairs of U_m converge to the same value of the limit (8) in the case of $\ell_1 \neq \ell_2$. Converging pairs are also nearly degenerate at large but finite values of R; this complication, involving long range dipole interaction between the two electrons, is discussed in ref. 3. Otherwise the plots of successive curves $U_m(R)$ remain in the same order as they start from the condensed atom limit, R = 0. Accordingly the quantum number m, characteristic of the condensed atom limit, serves as a convenient label at all R.

We proceed now to consider simultaneously channels with different pairs of quantum numbers $(\ell_1 \ell_2)$, though with the same values of the good quantum numbers S, L, M and of the parity π. Off-diagonal elements of the matrix U, which are smaller than the diagonal elements but far from negligible, play an important role where two or more channels with different $(\ell_1\ell_2)$ are degenerate or nearly degenerate. To consider this effect, we begin by analyzing the pattern of degeneracies (or quasidegeneracies) of the $U_\mu(R)$ with different $(\ell_1\ell_2)$ and equal L, which emerge as crossings (or quasi-crossings) when their plots are superimposed. In Fig. 8, we notice two systematic types of degeneracy: a) At large R, all the channels approaching the same limit (8) are nearly degenerate; the matrix elements that couple them arise mainly from the familiar long range dipole component of the electron interaction. b) At moderately small R the $U_\mu(R)$ values of channels with lower values of $\ell_1 + \ell_2$ and higher m decrease more rapidly than those with higher $\ell_1 + \ell_2$ and lower m. The reason is that a radial correlation with higher m requires proportionately higher excitation energy at low R than angular correlation with a higher value of $\frac{1}{2}(\ell_1 + \ell_2 - L)$, since the eigenvalues $v(v+4)$ of Λ^2 are given by $v = \ell_1 + \ell_2 + 2m$. This situation tends to invert with increasing R, thus leading to curve crossings. An additional type of degeneracy occurs at R = 0, where the same value of v can be partitioned differently into ℓ_1, ℓ_2 and 2m. Reference 3 gives examples and details.

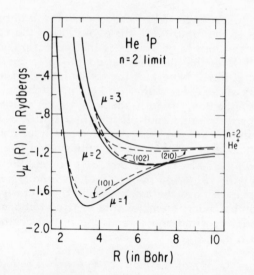

Fig. 8. Potential curves for helium ^1P channels converging to He^+ (n=2).
----- (sp+), (sp-), (pd) channels disregarding interaction off-diagonal in (ℓ_1, ℓ_2).
——— $U_\mu(R)$, eigenvalues of complete matrix U.

        In the regions of degeneracy, the coupling between channels labeled
by different (ℓ_1, ℓ_2) is sufficiently strong to make this classification unrealistic, at least locally. Accordingly the excitation channels with which we are
really concerned, namely, those identified experimentally as weakly coupled,
are only partially characterized by a ($\ell_1 \ell_2$) label. The approach we have
followed is thus viewed as an initial basis for describing the weakly coupled
channels, to be elaborated upon as required for specific purposes. This approach has also proven effective in ref. 3 to abbreviate numerical calculations.

        Ultimately we should regard the weakly coupled channels as identified
successfully when the coupling terms defined by Eqs. (6") or (19) and constructed with the functions $\Phi_\mu(R;\Omega)$ of these channels have an effect as low as
is observed in the experiments or in the calculations reviewed in ref. 1. For
purposes of orientation, ref. 3 has evaluated some sets of the matrix elements
($\Phi_\mu |d\Phi_{\mu'}/dR$) which appear in Eqs. (6") and (19). The results indicate that
the coupling terms, while not large, are not as small as to produce a satisfactory interpretation of the observed weakness of coupling.

5. Mechanisms of Excitation

        We return now to considering our entire problem with the intent of pro-
viding answers to the questions raised in Sec. 1. The excited two-electron
complexes appear to be formed initially in the lowest-energy channel for what-

ever orbital momentum L is relevant. Specifically, photoexcitation from the
^1S ground state of He or H^- will lead to states of the ^1P channel with $\mu=0$
– which is primarily ($\ell_1=1$, $\ell_2=0$, $m=0$) – because other channels have such
a large U_μ at low R that their radial functions F_μ would not overlap the
ground state at energies within our range of interest. The same states of the
^1P channel with $\mu=0$ arise from the combination of an incident p electron
with either He^+ or H in its ground state. Formation of a doubly excited complex occurs, therefore, indirectly, by transition from a singly excited state of
the lowest ($\mu=0$) channel to higher channels.

This transition should occur at hyperspherical radii where the interchannel coupling is strongest, e.g., at $R=2$ or 3 a.u. for transitions from the
$\mu=0$ to the $\mu=1$ $^1P^o$ channel of He. That $\mu=0$ is most strongly coupled to
$\mu=1$ than to other channels was anticipated in ref. 6 on qualitative grounds
and is now illustrated by data on coupling matrix elements. Reference 3 also
shows how the coupling strengths ($\mu \rightarrow \mu'$) are inversely related to the energy
differences $U_{\mu'} - U_\mu$, other circumstances being equal. However, we still cannot demonstrate that the coupling of the $\mu=0$ channel is as weak as observed.
Nor can we decide yet whether in He the far weaker coupling of the $\mu=0$ $^1P^o$
channel to the $\mu=2$ (sp-) and $\mu=3$ (pd) channels results mostly from direct interaction or indirectly through the $\mu=1$ (sp+) channel.

We proceed then on the assumption that transitions occur primarily at
avoided crossings of pairs of potential curves, $U_\mu(R)$ and $U_{\mu'}(R)$, as they do in
the molecular analog. Section 4 has stressed the occurrence of a characteristic
sequence of avoided crossings between successive curves of the + family. (For
simplicity the "+" label will be applied also to singlets with $\ell_1=\ell_2$.) The
main locus of these avoided crossings is represented by Eq. (9) and extends to
increasingly large radii R with increasing μ. We suggest that excitation proceeds mainly through this family of crossings, in slow electron collisions and
in double photoexcitation. This suggestion is consistent with the limited sample of data in ref. 3, but requires far more extensive verification.

Avoided crossings are also characterized by rapid variations of the
channel functions $\Phi_\mu(R;\Omega)$, with the resulting large values of the coupling ma-
trix elements $(\Phi_{\mu'}|\partial\Phi_\mu/\partial R)$. The variation of channel functions, illustrated
initially in Fig. 7, is demonstrated with greater emphasis by two-dimensional
mappings of channel functions $|g_m(R;\alpha)|^2$ proportional to the probability den-
sity $|\Psi(\vec{r}_1,\vec{r}_2)|^2$. Figure 9 shows that, as R increases through the range of
avoided crossing between $m=0$ and $m=2$, the density $|g_0(R,\alpha)|^2$ drops
sharply in the region of the saddle point of $C(\alpha,\theta_{12})$, that is, at $\alpha \sim 45°$ and
$\theta_{12} \sim 180°$, while the density $|g_2(R,\alpha)|^2$ of the next channel experiences a
compensating increase. Transition to the higher channel has then the effect of
maintaining a higher density $|\Psi|^2$ in the range of correlation corresponding to
the saddle point, while the complex expands toward larger R. This permanence
of a probability distribution parallels the familiar phenomenon in atom-atom
collisions, where the electronic wave function remains unchanged if the system
follows a diabatic curve, skipping from one adiabatic curve to another at an
avoided crossing. The same remark also ties our discussion to the Wannier
theory of threshold ionization, a connection to be developed further below.

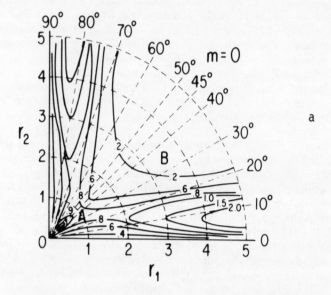

a

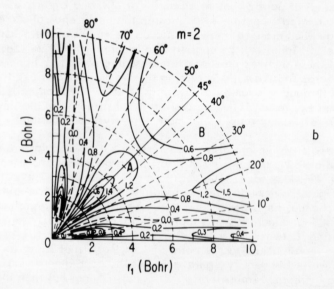

b

Fig. 9. Contour plots of $|g_m(R;\alpha)|^2$ for helium ^1S channels; a) m=0 , b) m=2. (Plots of Fig. 3 represent $|g_m|^2$ multiplied by appropriate radial functions.)

For purposes of orientation we outline the development of double photo-excitation in He in a schematized model where the channel coupling is re-garded as wholly concentrated at the avoided crossings of $^1P+$ channels identi-fied by Eq. (9). As mentioned above, the direct result of photoabsorption is a transition to the single-excitation, $\mu=0$, $^1P^o$ channel. The relevant part of the final state wave function consists of a wave propagating in this channel from the region $R \lesssim 2$ outward towards $R \to \infty$. At the first avoided crossing, at $R \sim 3$, the coupling acts as an optical beam splitter transferring a fraction of the amplitude to the $\mu=1$, double-excitation channel. This action repeats at successive avoided crossings, at larger and larger values of R, where suc-cessively smaller fractions of the amplitude are transferred to higher and higher channels of the $+$ family. The squared amplitude of the wave propagating to $R \to \infty$ in the various channels determines the probability of photoionization com-bined with excitation of He^+ to its various states.

This framework permits us to sharpen some of the questions raised in Sec. 1. Thus the probability of double photoexcitation, or of single excitation by electron collision, is now seen to be limited by the strength of coupling be-tween the $\mu=0$ channel and all higher channels. To question b) we can say that excitation to higher channels will indeed proceed at the expense of the probability of excitation to one of the lower channels under two conditions whose applicability can be tested: 1) channel coupling does in fact act chain-wise, splitting off a fraction of a wave's amplitude into a higher channel as postulated in our schematization, 2) neither the coupling nor simple propaga-tion in any channel reflects any appreciable fraction of the wave backwards toward lower R. Reflections can be attributed, broadly speaking, to depar-tures from a semiclassical (i.e., geometrical optics, or WKB) approximation. Their presence causes readjustments of all excitation probabilities and prevents the calculation of lower excitation without reference to higher excitations.

Reflection certainly occurs upon fractional transfer of the wave func-tion to a "closed channel," i.e., to a channel whose asymptotic potential $U_\mu (R=\infty)$ exceeds the energy of the state, which is represented by $2E$ in Eq. (6). In this event the wave function is trapped within the potential well rep-resented by plotting $U_\mu(R)$ as in Fig. 5. Reflection is then localized semi-classically at turning points $R_{t\mu}$ identified by

$$U_\mu(R_{t\mu}) + 2E - \tfrac{1}{4}/R_{t\mu}^{\,2} = 0 . \tag{10}$$

Multiple reflections, back and forth between these turning points, damp the amplitude of the wave in the closed channel and minimize its reaction on lower channels, unless an integral number of half-waves fits between the turn-ing points thus generating a resonance. Resonance amplifies, instead, the ef-fect of reflection which propagates then back to the lower channels.

As we know, the process of excitation to higher and higher channels does not spread uniformly through all channels with the same L, S, and parity. Rather, excitation will proceed predominantly among channels with the same

(ℓ_1, ℓ_2) classification and also, at constant (ℓ_1, ℓ_2), within the same + or –
family. This tentative conclusion rests on the analysis of Sec. 4 and is borne
out by the available experimental evidence. While the lowest channels are
weakly coupled, various evidence indicates that the coupling increases between successively higher channels of the same classification. Analytically
the coupling matrix elements $(\Phi_\mu \mid \partial \Phi_{\mu'} / \partial R)$ are inversely related to the energy difference $U_{\mu'}(R) - U_\mu(R)$,[3] which decreases with increasing μ at constant $\mu' - \mu$. Experimentally, collisional excitation of He^+ or H to their $n=2$
levels is unlikely, but higher excitations are not much less likely. Similarly
double photoexcitation of He to a channel converging to the $n=3$ limit of He^+
is not much less intense than it is for the corresponding $n=2$ channel;[11] discrete levels of the $n=3$ channel are broader, though, indicating likelihood of
autoionization to the $n=2$ rather than to the $n=1$ channel. Again in electron
collision processes resonances generally stand out far more sharply in the spectra of inelastic rather than of elastic scattering;[12] this indicates that a discrete doubly excited level of a complex is far more likely to decay into the
next lower accessible channel rather than into the lowest one.

    We come now to consider the highest excitations of a two-electron
complex, near to and beyond the energy threshold for escape of both electrons.
Our previous discussion suggests that the coupling among higher channels of
double excitation increases in strength, in the range of avoided crossing, to a
point where some new type of channel expansion should be sought. One different channel function, which we call $\Phi_W(R;\Omega)$, has in fact been provided
by Rau[13] and Peterkop's[14] wave mechanical version of Wannier's work[15] on
the threshold problem. The construction of Φ_W relies on approximate separa-
bility of the Schroedinger equation (2) in the neighborhood of the flat saddle
point of the potential (3). Residual terms of the equation, arising from
$\partial \Phi_W / \partial R$, vanish to $O(R^{-3/2})$ as $R \to \infty$. This reduces the coupling of the channel W at large R. The analytical form of Φ_W,

$$\Phi_W(R;\alpha, \theta_{12}) = \exp\{i f(R)(\tfrac{1}{4}\pi - \alpha)^2 - g(R)(\pi - \theta_{12})^2\} \tag{11}$$

characterizes the correlations in the W channel. Both $f(R)$ and $g(R)$ are proportional to

$$R[2E + C(\tfrac{1}{4}\pi, \pi)/R]^{\frac{1}{2}} = [2ER^2 + \sqrt{2}(4Z-1)R]^{\frac{1}{2}}. \tag{12}$$

These expressions show the angular correlation to sharpen as R increases, re-
stricting the two electrons to nearly opposite directions ($\hat{r}_2 \sim -\hat{r}_1$, i.e., $\theta_{12} \sim \pi$);
the mean value of $[\tfrac{1}{2}(\ell_1 + \ell_2 - L)]^2$ increases in proportion to the coefficient
g, thus showing the increasing role of coupling to high-(ℓ_1, ℓ_2) channels.[16]
The radial correlation has also a maximum at $\alpha = 45°$ (i.e., at $r_2 \sim r_1$), which
sharpens with increasing R; however, the imaginary unit in the exponent of Eq.
(11) represents a propagation of the complete wave function away from the
point of maximum radial correlation, thus indicating that this correlation may
dissolve progressively. Fuller analysis of the wave functions in refs. 13 and 14

bears out this interpretation and shows a maximum of the probability
$|\Psi(\hat{r}_1, \hat{r}_2)|^2$ to propagate outward, toward R = ∞, along the α = 45° line of
Fig. 9. This maximum may be viewed as the envelope resulting from the superposition of the maxima in Figs. 9a) and 9b) and of analogous maxima pertain-
ing to the distribution of $|\Phi_\mu(R;\Omega)|^2$ for higher and higher values of μ within
a + family.

For energies 2E < 0, i.e., below the threshold for escape of both electrons, there is a value of R at which the coefficient (12) vanishes and the
WKB approximation which underlies Eq. (11) fails altogether. This value will
be indicated by $R_{tW} = C(\frac{1}{4}\pi, \pi)/(-2E)$ because it represents the classical
turning point of radial motion along the "Wannier line" (α = $\frac{1}{4}$π, θ_{12} = π). According to Eqs. (10) and (9), R_{tW} coincides with the turning point $R_{t\mu}$ of a
channel μ for which $U_\mu(R)$ is just dropping below the saddle point of Fig. 4
(to the extent that the $\frac{1}{4}/R_{tW}^2$ term of Eq. (9) is negligible). If, then, high
excitations of the complex give rise to outward propagation along the Wannier
line, with the channel function $\Phi_W(R;\alpha,\theta_{12})$, resonances might be expected
at the energies 2E for which an integral number of half wavelengths fits along
the Wannier line up to the turning point R_{tW}. The possibility of such resonances seems worth noticing even though they have not been observed and even
though they might in fact be obscured altogether by the progressive decay of
excitation from the Wannier channel back into the channels μ. At this stage,
the relevance of the Wannier channel to excitation of the complex near to and
beyond the threshold for double escape seems firmly established, especially
following recent and detailed experimental verification.[17] However, the
problem of the formation, and presumable decay, of the channel function Φ_W
as a superposition of channel functions Φ_μ remains untouched.[18]

We forego any discussion of the Wannier law concerning the energy
dependence of the cross sections for excitations very near the threshold for
double escape, for which the basic theory is given in refs. 13-15, the recent
experiment in ref. 17 and additional fragmentary understanding in ref. 16.
However, one point of the initial theory seems to acquire much broader significance in the context of our survey. This concerns the existence, at any
energy 2E > 0 above the threshold, of a "far zone" within which all the
Coulomb forces become unimportant as compared to the kinetic energies of the
electrons, and hence no longer control the electron correlations. The inner
boundary of the far zone is set at a critical radius R_c at which the two terms
in the brackets of Eq. (12) are equal,

$$R_c = C(\frac{1}{4}\pi, \pi)/2E = \sqrt{2}(4Z-1)/2E \text{ a.u. .} \tag{13}$$

(Note that R_c equals the turning point radius R_{tW} pertaining to a negative
energy - |2E|.) While the concept of a far zone was introduced with refer-
ence to threshold phenomena, we appreicate now its relevance to processes at
energies far above threshold. It implies that the electron correlations, which
characterize the mechanics of our excited complex, no longer form at radii
R > R_c, however small R_c may become as the excitation energy increases.
Indeed the very concept of complex formation remains relevant to the escape

of electrons in the far zone only by determining boundary values of the wave function at the inner limit $R = R_c$. The upper energy limit at which complex formation is relevant appears thus to be reached at the level for which the critical radius R_c has shrunk down to the ground state radius of the atom or negative ion of interest.

As this last remark provides a tentative answer to question f) of Sec. 1, we may conclude with summary comments on the other questions posed there. The weakness of excitation seems to derive from the wide energy separation of the two lowest channels. Excitation to higher shells seems merely to reduce the intensity of lower excitation provided reflections are negligible (see above, in this section). Excitation to the m-th channel requires treatment of correlations up to the radius R which satisfies Eq. (9), or up to R_c, whichever is smaller. The Wannier picture of ionization near threshold does begin to emerge as the limit of excitation to lower channels. High-ℓ excitations seem likely to occur but only within a narrow energy band centered at the ionization threshold.[16] We have stressed in this section the role of the saddle point of $C(\alpha, \theta_{12})$ in promoting the complex from one channel to a higher one; this promotion appears to take place at the avoided crossings, whose locus is defined by Eq. (9), and to be accompanied by the retention of a local maximum of $|\Psi|^2$ on the saddle point as in the case of the maximum A in Fig. 9 a) and b). This concept may extend to many-electron processes, in the sense that saddle points of the generalized function C may have a catalytic role, as the seats of electron transitions from one channel to another.

Appendix

We outline here a procedure to reduce the system of coupled equations (6) to a first order system. Call $U_\mu(R)$ the diagonal elements of the matrix U in some chosen basis Φ_μ, and assume that the radial equations

$$[\frac{d^2}{dR^2} + \frac{1}{4R^2} - U_\mu(R) + 2E] u_\mu(R) = 0 ,$$
(14)

have been solved, disregarding the coupling terms, as was done by Macek.[6] However, for each of these equations a pair of independent solutions, $f_\mu(R)$ and $g_\mu(R)$, should be obtained, with Wronskian $W(f_\mu, g_\mu) = 1$ and with f_μ regular at $R = 0$. Consider now an expansion of the complete wave function of the form

$$R^{5/2} \sin\alpha \cos\alpha \, \Psi = \Sigma_\mu \, \Phi_\mu(R;\Omega) \, F_\mu(R, \delta_\mu) a_\mu(R)$$
(15)

where

$$F_\mu(R, \delta_\mu) = f_\mu(R) \cos\delta_\mu(R) - g_\mu(R) \sin\delta_\mu(R) ,$$
(16)

and where $a_\mu(R)$ and $\delta_\mu(R)$ are a new pair of dependent variables defined by (15) and (16), with the further restriction

$$\frac{\partial}{\partial R} R^{5/2} \sin\alpha\cos\alpha\,\Psi = \Sigma_\mu \Phi_\mu(R;\Omega) \frac{\partial F_\mu(R,\delta_\mu)}{\partial R} a_\mu(R) \ . \tag{17}$$

Substitution into Eq. (2) reduces it now to the coupled system

$$a_\mu(R) \frac{d\delta_\mu}{dR} = \Sigma_{\mu'} M_{\mu\mu'} a_{\mu'} \ , \quad \frac{da_\mu}{dR} = \Sigma_{\mu'} \frac{\partial M_{\mu\mu'}}{\partial \delta_\mu} a_\mu(R) \ , \tag{18}$$

whose matrix,

$$M_{\mu\mu'}(R,\delta_\mu,\delta_{\mu'}) = - F_\mu \left[U_{\mu\mu'}(R) - U_\mu(R)\delta_{\mu\mu'} \right] F_{\mu'}$$

$$+ (\Phi_\mu | \frac{\partial \Phi_{\mu'}}{\partial R}) W(F_\mu, F_{\mu'}) \ , \tag{19}$$

vanishes for $\mu' = \mu$. The system (18) represents thus explicitly the influence of channel coupling alone upon the phases and amplitudes of the channel wave functions. Its numerical solution should describe the operation of channel interactions. While the procedures used thus far for constructing channel functions Φ_μ aim, in essence, at cancelling the first term on the right of Eq. (19), the coupling might actually be minimized by some prescription that minimizes the whole expression (19) rather than one of its terms at the expense of the other.

References

* Work supported in part by the U.S. Atomic Energy Commission, Contract No. COO-1674-96 and in part by the Advanced Research Projects Agency.

1. U. Fano, "Doubly Excited States of Atoms", in Atomic Physics (Plenum Press, 1969) p. 209.

2. U. Fano and C.D. Lin, "Correlations in He**, e+He and in Related Systems", in The Physics of Electronic and Atomic Collisions, ed. by B.C. Čobić and M.V. Kurepa (Institute of Physics, Belgrade, Yugoslavia, 1973) p. 229.

3. C.D. Lin, "Correlations of Excited Electrons. I. The Study of Channels" submitted to the Phys. Rev. A in June 1974, and to be published.

4. J.W. Cooper, U. Fano, and F. Prats, Phys. Rev. Letters $\underline{10}$, 518 (1963).

5. J.H. Macek, Phys. Rev. $\underline{160}$, 170 (1967).

6. J.H. Macek, J. Phys. B $\underline{2}$, 831 (1968).

7. Note that this term is positive definite in a space with >3 dimensions.

8. The symmetry of eigenfunctions of Λ^2 for $N>2$ electrons coupled to spin S has been studied recently by D.L. Knirk, J. Chem. Phys. $\underline{60}$, 66 (1974).

9. See, e.g., F. Calogero, Variable Phase Approach to Potential Scattering (Academic Press, New York, 1967); V. V. Babikov, Metod phasovikh Functsii v kvantovoi mekhanikhe (Nauka, Moscow, 1968); J.L. Dehmer and U. Fano, Phys. Rev. A $\underline{2}$, 304 (1970).

10. See, e.g., J. B. Delos, W. R. Thorson, and S. K. Knudson, Phys. Rev. A $\underline{6}$, 709 (1972).

11. R. P. Madden and K. Codling, Astrophys. J. $\underline{141}$, 364 (1965); P. Dhez and D. L. Ederer, J. Phys. B $\underline{6}$, 259 (1973).

12. Compare, e.g., Fig. 2 of C. E. Kuyatt, J. A. Simpson, and S. R. Mielczarek, Phys. Rev. $\underline{138}$, A386 (1965) with Fig. 3 of G. E. Chamberlain, Phys. Rev. $\underline{155}$, 46 (1966).

13. A. R. P. Rau, Phys. Rev. A $\underline{4}$, 207 (1971).

14. R. Peterkop, J. Phys. B $\underline{4}$, 513 (1971).

15. G. Wannier, Phys. Rev. $\underline{90}$, 817 (1953).

16. U. Fano, J. Phys. B $\underline{x}$, xxx (1974).

17. S. Cvejanovic and F. H. Read, J. Phys. B, to be published, (1974).

18. An analogous problem occurs in the motion of a single highly excited electron under the combined influence of a Coulomb and a magnetic field, in which case the Schroedinger equation involves two nonseparable variables only. Resonances observed near the ionization threshold [W. R. G. Garton and F. S. Tomkins, Astrophys. J. $\underline{158}$, 839 (1969)] have been interpreted in terms of excitation of a Wannier-like channel [A. F. Starace, J. Phys. B $\underline{6}$, 76 (1973)] . A treatment of this problem along the lines of the present paper and of ref. 3 is now being developed by A. F. Starace.

THE SEARCH FOR VIOLATION OF P OR T INVARIANCE IN ATOMS OR MOLECULES

P.G.H. Sandars

Clarendon Laboratory, Oxford University

1. INTRODUCTION

My task in this talk is to review current work in the search for any violation of parity (P) or time-reversal invariance (T) in atoms or molecules. Because the field is at present in a phase of rapid development this means that I shall spend much of my time discussing experiments which are not yet completed and proposed experiments which are still under construction. Some aspects of past work are described in two previous talks by Feinburg[1] and Commins[2] at these conferences.

The subject matter of my talk breaks naturally into two parts: the search for P violation with simultaneous violation of T and the search for P violation not accompanied by T violation. Interestingly, there does not seem to be any simple way in which T violation unaccompanied by P violation can occur in atomic systems[3].

The search for the two different types of violation have a number of features in common but also a very significant difference. In both cases the motivation comes from high energy physics and the basic structure of the weak interactions is involved. In both cases one can make predictions based on high energy physics results which suggest that there may be observable effects in atomic systems. Clearly such effects will be very small or they would have already been observed so the

71

relevant experiments must not only be of high sensitivity, but in order to separate the violating interaction from the much larger normal interaction one must look for effects whose existence indicates explicitly a violation of the symmetry concerned.

There is a very significant difference here between P and T violation and P violation without T violation. One can show on very general grounds that the former can give rise to energy and frequency shifts when an atom or molecule is placed in an external electric field while the latter cannot. On the other hand it can produce observable effects when radiation is absorbed or emitted, in particular a dependence on the circular polarization of the radiation which would be zero in the absence of a parity violating interaction. The most general test for P and T consists in the search for an effect of the form

$$\nu(E, B) \neq \nu(-E, B)$$

where ν is a transition frequency of an atom or molecule in externally applied electric and magnetic fields E and B. The corresponding test for a P violation unaccompanied by T violation is the search for circular polarization

$$P_c = \frac{I_+ - I_-}{I_+ + I_-}$$

where I_+ and I_- represent intensities with right and left handed circular polarizations.

In section 2 we discuss the search for P and T violation with particular reference to experiments designed to look for electric dipole moments, though we also include other possible sorts of violations. We include a discussion of Ramsey's work on the neutron because the methods used are very similar to those in the atomic and molecular beam experiments. In section 3 we discuss a number of proposed experiments which look as though they have sufficient sensitivity to allow detection of the recently predicted weak P violating effects caused by neutral currents.

2.   P AND T VIOLATIONS

2.1   The Present Status of T Violation

The current interest in searching for evidence of T violation in atomic systems stems largely from the important discovery by Christenson et al in 1964[4] that the long lived K^O meson decays into two pions and into three pions as illustrated below

$$K^O_L \rightarrow 2\pi$$
$$K^O_L \rightarrow 3\pi$$

The significance of this is that the 2π and 3π final states have opposite CP quantum numbers and hence the simultaneous existence of the two decay modes indicates a violation of CP invariance. This in turn implies through the CPT theorem that T must also be violated. More recently Casella[5] has argued that certain phase relations required by T invariance are not obeyed in the K^O system and that one can therefore demonstrate T violation without requiring the CPT theorem. While this argument has been questioned[6], there seems little doubt that T invariance is violated in the K^O system. There is, however, as yet no evidence of T violation outside the K^O system; it is clearly important to look for such an effect with the maximum attainable sensitivity.

2.2   Theoretical Predictions on Electric Dipole Moments

Perhaps the most fascinating feature of the CP violating K^O decay is the wide range[7] of interactions which can be invoked to explain it. These range from milli-strong interactions some 10^{-3} times the strong interaction, through the weak interactions, to a possible superweak interaction 10^{-6} times weaker than the weak interaction. All these proposed interactions violate T and, since P is known to be violated in the ordinary weak interaction, they all can, in principle, give rise to an electric dipole moment (edm) which requires simultaneous P and T violation. The various orders of magnitude which result are set out in table 1 for convenience of reference.

Interaction	Symmetries violated	Strength	EDM (e x cm)
Millistrong	C, T	10^{-3}	10^{-23}
Electromagnetic	C, T	10^{-2}	10^{-21}
Weak	P, T	10^{-7}	10^{-20}
Milliweak	P, T	10^{-10}	10^{-23}
Superweak	P, T	10^{-15}	10^{-29}

Table 1.   Nucleon edm orders of magnitude.

In figure 1 we also show graphically the large number of predictions of the nuclear edms which have been made since the discovery of CP violation.

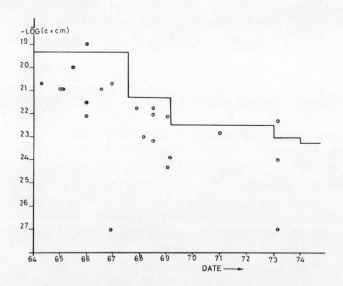

Figure 1.   Theoretical predictions and experimental limits for the neutron edm.

We can draw two conclusions: first, that a number of models predict values for the edm which are of measurable magnitude. Second, the range of values predicted by the different models suggests that edm experiments form a useful means for deciding between them.

A number of the models also predict a non-zero edm for the electron though generally at least $\frac{m}{M} \sim 10^{-3}$ smaller than for the neutron.

2.3   Experiments on the edm of the Neutron

Over a period of years Ramsey and his co-workers have carried out an important series of experiments to look for an edm on the neutron. Full details can be found in the review article by Golub and Pendlebury[8] and only the briefest sketch will be given here. Basically, the Ramsey method is to use a beam resonance technique as illustrated schematically in figure 2. The resonance is observed in parallel electric and magnetic field and search is made for any change in the resonance frequency on reversing E or B with respect to the other.

With the usual Hamiltonian:

$$H = -\mu \,\underline{\sigma}.\underline{B} - D\,\underline{\sigma}.\underline{E}$$

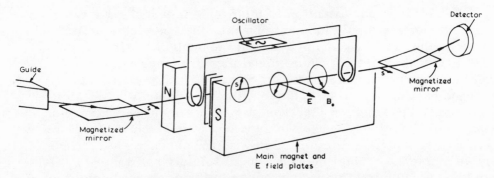

Figure 2.   Schematic diagram of neutron edm experiment
(taken from reference 8 with permission).

the frequency change on reversing E or B is given by

$$\delta \nu = 4DE$$

In order to obtain maximum sensitivity the rf oscillator is set on the point of maximum slope on the side of the resonance so that any change in frequency is translated into a change of intensity. One can readily see that, in the absence of spurious effects, the ultimate sensitivity is determined by the resonance line width and by the signal to noise with which it is observed. Ramsey[9] has achieved a narrow line-width of order 50 H_z by using a long resonance region and slow neutrons, and a high signal to noise by carrying out the experiment at the reactor with the highest available slow neutron flux, and by counting for a long time. The results of Ramsey's experiments are indicated by the solid line in figure 1. The interaction between experiment and theory is clearly indicated by the declining values of the predictions under the pressure of the experimental results.

As a result of this edm work it is now generally held that the T violation in the K^o decay cannot be due either to a milli-strong or to the electromagnetic interaction. It is likely that the present round of experiments which may lead to a sensitivity of order 1×10^{-24} excm will see an effect if the violation lies in the weak interactions. If nothing is seen, there must be the strong supposition[10] that the T violation lies in the suggested superweak interaction.

2.4   Electric Dipole Moments on Charged Particles

We now turn our attention to the problem of looking for possible electric dipole moments on the other particles of atomic physics, the electron, the proton and the nucleus. The difficulty here is that they are all charged and there is a very general theorem, most fully expounded by Schiff[11], to the effect that a charged particle which is in an equilibrium quantum mechanical state under the action of electrostatic forces must experience an electric field which averages to zero and which cannot therefore give rise to a non-zero interaction with the particle's electric dipole moment. If this theorem was absolutely valid there would be no atomic effects first order in the electric dipole moments. Luckily, it has only restricted validity and as Schiff has shown it breaks down (i) when relativistic spin dependent effects are present or (ii) when the particle involved has a finite size and structure. The former effect applied to an electron in an atom might be expected to yield an effect of order $Z^2\alpha^2$. Both effects occur for the proton in hydrogen and for the nucleus in a heavier atom and are of relative order $\dfrac{m}{M}\,\alpha^2$ and $\left(\dfrac{r_n}{a_o}\right)^2$ respectively.

Symmetry considerations suggest that the residual effect will take the form of an interaction between the edm of the particle concerned and an effective electric field along the direction of the externally applied field, e.g. $H_{eff} = -D \cdot E_{eff}$

Schiff[11] and the author[12] have calculated the value of E_{eff} for the proton and the electron in the ground state of H with the results shown in table 2.

Clearly, in hydrogen the edm is effectively shielded from an applied field, and precision experiments would be difficult. However, as the author[13] has pointed out, the size of E_{eff} increases rapidly with Z, and for high Z can be appreciable. Indeed, as is shown in table 2, one can even get a situation for the electron where $E_{eff} > E$ so that the effective interaction with the electron edm is actually greater than it would be with a free

Particle	'Host'	Effective field (in v/cm)
Electron	H	10
	Xe, Cs	5×10^5
	BaF	2×10^9
Proton	H	5×10^{-3}
	Tl	5×10^1
	TlF	1×10^4
Neutron	Free	5×10^4
	PbF	2×10^4

Table 2.   Effective electric fields. (All values assume an applied electric field of 5×10^4 v/cm.)

electron in the same field. This result forms the basis for the
experiments to be described in the next section.

Even for high Z, E_{eff} for the proton bound into a nucleus is
still appreciably less than the applied field. A breakthrough
here was the realisation[14] that one could make use of the high
electric fields in polar molecules which can themselves be
polarised by relatively low externally applied fields. Detailed
calculations of the effective electric fields give the results in
table 2. The relatively high effective field at the nucleus in TlF
is the basis of the experiment described in section 2.6. The very
high value[15] for the effective field on the electron in BaF is the
basis for an experiment currently under consideration in my
laboratory.

2.5   The Electric Dipole Moment on the Electron

The experimental method is very similar to that in the
neutron experiment except that one uses atomic beam resonance
and E_{eff} replaces E. Details of the experiments and their
difficulties have been published[16] and we simply quote the
results in table 3 below, together with some limits obtained in
other ways.   It is hoped that the BaF experiment currently
under development may have a sensitivity two or three orders of
magnitude higher than the best atomic experiment. If so, an
electron edm in the range $e \times 10^{-24}$cm - $ex\ 10^{-27}$cm will become
accessible.

1958	Lamb shift analysis	Salpeter, Feinberg	2.10^{-13}
1962	g-value of electron	Crane	4.10^{-16}
1964	Cs atomic beam	Sandars, Lipworth	2.10^{-21}
1968	Reversible Cs beam	Sandars	1.10^{-22}
1968	Cs-Na comparison	Lipworth	3.10^{-24}
1970	Metastable xeonon beam	Sandars	2.10^{-24}

Table 3.   Experimental limits on electron edm.
            (All values in e x cm)

2.6   The Electric Dipole Moment on the Proton

Here, the proton is the odd $S_{1/2}$ nucleon in the Tl nucleus which is bound in a TlF molecule. The transition used is the nuclear resonance $M_{Tl} = \frac{1}{2} \leftrightarrow -\frac{1}{2}$ in the $J = 1$, $M_J = 1$ rotational state. Since electric deflection is used a triple resonance scheme[17] is required to observe the transition. As before, the experiment consists in reversing the direction of the electric field and looking for a shift in the resonance frequency $\delta\nu = 4DE_{eff}$ which would indicate the presence of an edm. The result of a pilot experiment has been published[17] and is listed in table 4. In order to improve on the sensitivity we have constructed a 'monster' beam machine 20 m long with a beam 8 cm high and 0.5 cm wide. The observed wide Ramsey resonance pattern is illustrated in figure 3. With this new apparatus, which is now running, we hope to be able to reduce our present limit on the proton edm by two or three orders of magnitude.

2.7   Odd Nuclear Multipole Moments

Thus far we have discussed P and T violation in terms of edms, but if P and T is violated a nucleus can, provided its spin is high enough, have a magnetic quadrupole moment or electric octupole moment. Now we have seen that edm effects are suppressed by a factor depending on $\frac{m}{M}\alpha^2$ or $\left(\frac{r_n}{a_o}\right)^2$ which are just the factors by which the magnetic quadrupole and electric octupole interactions are smaller than the electric dipole interaction. Thus for a given size of P and T violation one would expect observable effects of similar magnitude arising via these higher moments as through the suppressed edm interaction. The experiments are

1959	Hydrogen energy levels	Sternheimer	1.10^{-13}
1960	Precession of free protons	Manning	3.10^{-15}
1969	TlF molecular beam	Sandars	9.10^{-21}

Table 4.   Experimental limits on proton edm.
         (All values in e x cm)

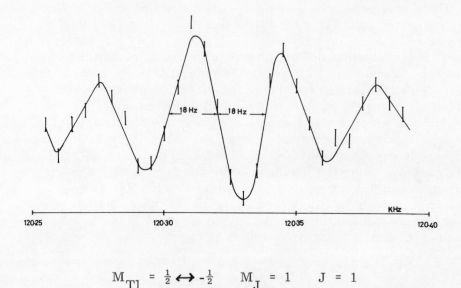

$$M_{Tl} = \tfrac{1}{2} \longleftrightarrow -\tfrac{1}{2} \qquad M_J = 1 \qquad J = 1$$

Figure 3.   Tl nuclear resonance in TlF using a Ramsey double
loop of 10 m separation.

essentially the same, one can show that both types of effect give
rise to a frequency shift when an applied electric field is
reversed with respect to a magnetic field.   The main difference
is that the edm effect is present for $I = \tfrac{1}{2}$ while the magnetic
quadrupole requires $I > \tfrac{1}{2}$ and the electric octupole requires
$I > 1$.   Thus the TlF experiment has no contribution from the
higher moments.   Indeed, the only experiment to date which was
sensitive to them was the Lipworth experiment on Cs.[16]
Interpretation of the result in terms of calculations[13] carried out
to find E_{eff} gives the following limit on the magnetic quadrupole
moment of the Cs nucleus:

$$M_2 (Cs^{133}) \lesssim 1 \times 10^{-4} \, \mu_n \, r_n$$

where μ_n is the nuclear magneton and r_n the nuclear radius.
A corresponding limit on the electric octupole moment is
currently being calculated. While these results are not themselves significant, it may well be that an experiment designed
specifically to look for one of the higher moments may turn out
to be the most sensitive test for P and T violation in the nucleus.

2.8   P and T Violating Short Range Interaction

The possible existence of P violating short range weak
interactions in atoms which is discussed in the next section
together with the known violation of T in the weak interactions
suggest that we interpret our present experimental results to set
limits on any P and T violating short range interactions. One
can show that there are only two such interactions possible
between an electron and a nucleon, a scalar and a tensor interaction:

$$H_S = C_s \frac{G}{\sqrt{2}} \sum_{e,n} \left(i \gamma_0 \gamma_5 \right)_e \delta(\underline{r}_{en})$$

$$H_T = C_T \frac{G}{\sqrt{2}} \sum_{e,n} i \underline{\sigma}_n \cdot \left(\gamma_0 \underline{\alpha} \right)_e \delta(\underline{r}_{en})$$

One can readily show that detection of H_S requires non-zero
electron spin whereas H_T requires non-zero nuclear spin, thus
the electron edm experiments will yield a limit on C_S and the TlF
experiment a limit on C_T. Using the wave-functions developed in
the calculation of E_{eff} we can readily evaluate the expectation
values of H_S and H_T; comparison with experiment then yields the

Interaction	System	Group	Limit
Scalar	Tl	Lipworth	2×10^{-4}
Tensor	TlF	Sandars	3×10^{-6}

Table 5.   Experimental limits on short range P and T violating
electron-nucleon interaction.
(All results in terms of the weak interaction coupling constant G.)

limits on C_S and C_T set out in table 5. The limits are
surprisingly small: our experiments are sensitive to effects
which are many orders of magnitude weaker than the weak interactions. Furthermore, future experiments are likely to be as
much as 1000 times more sensitive. It is not unlikely that the
information which these experiments can yield on short range
interactions may be as significant as the limits which we can set
on edms.

3.   PARITY VIOLATION AND WEAK NEUTRAL CURRENTS

3.1   Neutral Currents

Conventional weak interaction theory is expressed in terms
of charged currents as illustrated in figure 4. This implies that
one can have a nucleon-nucleon interaction as shown but not an

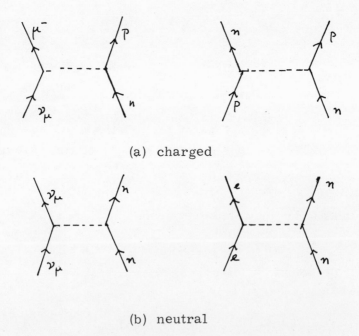

(a) charged

(b) neutral

Figure 4. Charged and neutral weak currents.

equivalent electron-nucleon or electron-electron interaction since charged current theory requires that in an interaction an electron disappear and a neutrino appear, but in an atom the number of electrons must remain constant.

The degree of belief in charged currents has always been rather stronger than the experimental evidence on which it was based. Recently a number of theoreticians[18] have constructed models of the weak interactions which require the existence of neutral currents. Stimulated by the theorists a number of high energy experiments have been carried out in the past year at CERN[19], N.A.L.[20] and Argonne[21] to look for interactions of the type illustrated in figure 5 in which a muon neutrino interacts but does not produce a muon, staying as a neutrino. Such events have been found and there is now strong, though perhaps not yet conclusive, evidence that weak currents exist. These results indicate that the neutral current interaction is similar in strength to the charged current one in agreement with the theoretical predictions of Weinberg's model[18].

3.2 The Magnitude of Weak Neutral Current Effects in Atoms

If neutral currents exist then it is likely that there will be parity violating interactions of order G between electrons and nucleons and between pairs of electrons, and the form of these interactions is specifically predicted in some of the models referred to above. Bouchiat and Bouchiat in an important paper[22] have discussed the magnitude of such interactions in atoms and muonic atoms; Feinberg[23] has made a similar study of the muonic case.

Basically there are two interactions to be considered which are in Feinberg's notation

$$H_1 = C_1 \frac{G}{\sqrt{2}} \bar{\psi}_e \gamma_\alpha \gamma_5 \psi_e \bar{\psi}_n \gamma_\alpha \psi_n$$

$$H_2 = C_2 \frac{G}{\sqrt{2}} \bar{\psi}_e \gamma_\alpha \psi_e \bar{\psi}_n \gamma_\alpha \gamma_5 \psi_n$$

where C_1 and C_2 are dimensionless constants whose values indicate the strength of the interaction. Now we are interested in those parts of these interactions which have electron operators of

odd parity. It is also convenient to write the interactions in
Dirac form more familiar to atomic physicists. We find:

$$H_1 = \sum_{e,n} C_1 \frac{G}{\sqrt{2}} (\gamma_5)_e \, \delta(\underline{r}_{en})$$

$$H_2 = \sum_{e,n} C_2 \frac{G}{\sqrt{2}} \, \underline{\sigma}_n \cdot \underline{\alpha}_e \, \delta(\underline{r}_{en})$$

where we have omitted functions which distinguish between
protons and neutrons and allow for a possible iso-spin dependence
of the interaction.

There is an important feature of these interactions which
should be mentioned at this stage. H_1 is independent of nucleon
spin and the interaction will be additive for all the nucleons in an
atom. H_2 on the other hand will only have a contribution from
the unpaired nucleon and will vanish for I = o nuclei. This means
that we must expect H_1 to give an effect which is larger than that
produced by H_2 by a factor of order 100 for heavy atoms. This
enhancement is crucial in the experiments to be described in the
next section.

It is useful at this stage to consider the likely magnitude of
these parity violating interactions in atoms. Because they are
odd parity they will only have matrix elements between single
particle states of opposite parity. But they are also short range
and require electron density at the origin. It follows that the
only significant elements will be between s and p states.
Bouchiat and Bouchiat have used a most ingenious extension of
the Fermi-Segre method to obtain an approximate expression for
these matrix elements in terms of observable quantities

$$\langle ns_{1/2} | H_1 | n'p_{1/2} \rangle \approx \frac{G}{2\pi} Z^2 \alpha \, F(Z,N) \left(\varepsilon_{ns} \, \varepsilon_{n'p} \right)^{3/4} K_r$$

where G the weak interaction constant $G = 2.18 \times 10^{-14}$ atomic
units, α is the fine structure constant and F (Z, N) depends on the
iso-spin dependence of H_1 and the nuclear charge and mass
numbers. ε_{ns} and $\varepsilon_{n'p}$ are the eigen-values of the two states
and K_r is a relativistic correction factor which can be
appreciable for high Z. In most models F (Z, N) $\sim$ Z. Putting in
the various quantities we find for Tl as an example

$$\langle 7s_{1/2} | H_1 | 6p_{1/2} \rangle \approx 2 \times 10^{-11} \text{ atomic units} \approx 100 \text{ kHz}$$

Thus the interaction is not negligible by the standards of
precision atomic experiments.

Bouchiat and Bouchiat have refined the formula above so
that it can be used for reliable atomic calculations.
Alternatively, one can use the perturbed relativistic wavefunction, which has been developed by the author for dealing with
the closely related P and T violating effects[15].

3.3   The Detection of Atomic Parity Violations

As was mentioned in the introduction, the parity violating
interaction cannot give rise to an observable energy or frequency
shift in an atom or molecule. While the proof of this is trivial in
the case of an isolated atom or molecule which must have
definite parity a detailed time-reversal treatment is required
when there is an externally applied electric field. On the other
hand the parity violating interaction can effect transition
probabilities and intensities. In particular, the admixture of
states of opposite parity allows interference between electric and
magnetic multipoles of the same order, though in atoms only the
interference between electric and magnetic dipole moments is
large enough to be of interest. The most important result of this
interference is that the transition matrix element is different for
right and left handed circular polarisations; the parity violating
interaction has introduced a 'handedness' into the system. Very
roughly the circular polarization produced by the parity violation
is given by T^{pv}/T, the ratio of the induced transition element to
the allowed one. Clearly, this ratio is increased if the original
element T is small and this enhancement is the basis of a number
of the experiments to be described. T is normally chosen to be
a magnetic transition or even a forbidden magnetic transition so
that a small parity violating admixture of an electric dipole
element will produce a much amplified value of the T^{pv}/T ratio.

3.4   Circular Polarization in Emission

A few years ago Poppe carried out a very neat experiment in
which he looked for any evidence of circular polarization P_c in

the $1_{S_o} \rightarrow 3_{P_1}$ (4618 Å) magnetic dipole line in Pb. He observed
only the same degree of circular polarization as in the allowed
transitions and set a limit $P_c \sim 5 \times 10^{-4}$. Bouchiat and Bouchiat[22]
have estimated P_c using their approximate formula for the matrix
elements of the parity violating interaction. They obtain a value
of P_c more than three orders of magnitude smaller than Poppe's
limit. Thus this experiment is not of sufficient sensitivity to
show up neutral current interactions of order G. Nor is it
likely that it can be improved to the required degree; it would
seem virtually impossible to control conditions in an optical
source to ensure no spurious circular polarization at the 10^{-6},
10^{-7} level.

A situation where this difficulty is avoided and which may
possible be experimentally feasible is the $2s_{\frac{1}{2}} \rightarrow 1s_{\frac{1}{2}}$ transition in
high Z hydrogenic ions. One might expect a rather high
admixture of $2p_{\frac{1}{2}}$ state because the $2s_{\frac{1}{2}}$ and the $2p_{\frac{1}{2}}$ are separated
only by the Lamb shift. Feinberg[23] has carried out the requisite
calculations and finds that the admixture of $P_{\frac{1}{2}}$ ranges from
2×10^{-11} to 6×10^{-9} over the range from Z=1 to Z=50. While
this admixture seems very small, the $2p_{\frac{1}{2}} \rightarrow 2s_{\frac{1}{2}}$ is a fast allowed
dipole transition while the $2s_{\frac{1}{2}} \rightarrow 1s_{\frac{1}{2}}$ is a doubly forbidden
transition being allowed as a magnetic dipole transition only when
relativistic terms of order $(Z\alpha)^2$ are included. The resulting
circular polarization ranges from 4×10^{-3} for Z=1 to 7×10^{-7}
for Z=50. These results would seem to suggest that low Z atoms
would be the most favourable. Unfortunately, for low Z ions the
alternative mode of decay via two photons from the $2s_{\frac{1}{2}}$ state
dominates the one photon decay; the branching ratio is
approximately $3 \times 10^{-7}Z^4$. This becomes of order unity for
Z = 30 where the circular polarization is 1.6×10^{-6}. A measurement in this region would be very hard but could conceivably be
possible.

A related question which has been examined in detail by
Feinberg[23] and by Bouchiat and Bouchiat[22] is the feasibility of
observing parity violating effects in the $2s_{\frac{1}{2}} \rightarrow 1s_{\frac{1}{2}}$ transition in
muonic atoms. This would be important because there are
models[24] which predict a muon interaction but not an electron one
and vice-versa. It would, however, be an extraordinarily
difficult one. The first problem is that in the cascade process
following capture relatively few of the muons finish up in the $2s_{\frac{1}{2}}$
state. Although the circular polarization of the $2s_{\frac{1}{2}} \rightarrow 1s_{\frac{1}{2}}$ 1δ
transition is higher than for the hydrogenic atom, ranging from

6×10^{-3} for $Z=6$ to 2×10^{-5} for $Z=82$ there are also more
alternative decay modes so that the branching ratio for the
transition is worse. The region $Z < 10$ is impossible because
of competition from a radiative 1γ decay in which the electrons
finish up in an odd parity state thus conserving overall parity.
In the region between $Z=20$ and $Z=30$ auger type processes are
important and lead to a branching ratio of order $10^{-5} - 10^{-6}$.
The situation looks a little better for $Z > 20$, for example for
Ti Bouchiat and Bouchiat[22] predict a circular polarization
$P_c \sim 5 \times 10^{-4}$ and a branching ratio of 5×10^{-4}. They estimate
that detection of this effect would require of the order of 10^{15}
stopped muons, a figure which might possibly become available
in the future. However, Bernaben, Ericson and Jarlskog[25]
point out that the detection of circular polarization is difficult and
that it would be better to look for either correlation between the
photon direction and the initial muon polarization or between the
photon direction and the direction of the electron in the final
decay of the muon. Finally, Simons[26] has suggested that higher
sensitivity might be achieved by looking for asymmetry in the
$3d_{3/2} \to 1s_{\frac{1}{2}}$ transition for example. Although the expected
values for the asymmetry are smaller than for the $2s_{\frac{1}{2}}$ decay the
counting statistics should be more favourable.

3.5   Optical Rotation and Circular Dichroism

A number of years ago Bradley and Wall set a limit on the
circular dichroism close to the magnetic dipole transition
$3\Sigma \to 1\Sigma$ (7600 Å) in molecular oxygen. Because of the low Z,
their experiment was many orders of magnitude too insensitive
to see any effect.

However, calculations by the author suggest that close to a
suitable transition in a very heavy element one might expect an
optical rotation or a circular dichroism of order 10^{-6}. (One can
show that it follows from dispersion relations that the two are
normally of comparable magnitude.) Effects of this order of
magnitude should be observable provided adequate signal to noise
is available. Unfortunately, nature is against us and the only
allowed magnetic dipole transition from a ground state in the
visible region where one can use tunable lasers is in Bi and this
is heavily masked by an allowed electric dipole band system from
Bi_2. The best remaining possibilities appear to be the 6p

$^2P_{\frac{1}{2}} \rightarrow {}^2P_{3/2}$ (12, 833 Å) in Tl or the $6p^2\ {}^3P_1 \rightarrow {}^1S_0$ (4624 Å) in Pb. The former is in a wave-length range where light sources are poor and detection is inefficient; the latter has as a starting level a metastable level which has a thermal population relative to the ground level of only 10^{-2}. Detailed design studies are now under way to see if either experiment is feasible. An analogous experiment on neutron polarization has been proposed by Stodolsky[21a].

3.6   Optical Fluorescence

Bouchiat and Bouchiat[22] are now carrying out a very ingenious experiment to look for any circular polarization dependence in the $6s_{\frac{1}{2}} \rightarrow 7s_{\frac{1}{2}}$ (5395 Å) transition in Cs using the method illustrated schematically in figure 5. The transition is very highly forbidden, being allowed only when relativistic effects are included. Its transition matrix is some $10^{-4}\mu_B$, so that any small admixture of an allowed electric dipole element is greatly amplified. Calculation suggests that using a tunable laser of intensity 3 x 10^{17} photons/sec. there should be about 10^9 transitions per sec. and a circular polarization dependence of order $P_c = 10^{-4}$. Thus this experiment should certainly have adequate sensitivity to see the effect of the nuclear spin independent neutral current interaction if it is of the expected size.

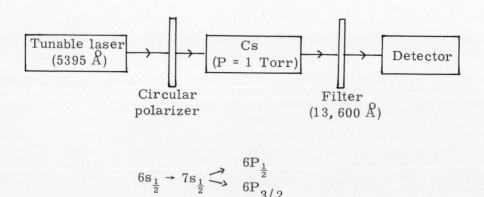

Figure 5.   The experiment of Bouchiat and Bouchiat to look for
the circular polarization dependence of the
$6s_{\frac{1}{2}} \rightarrow 7s_{\frac{1}{2}}$ transition in Cs.

Bouchiat and Bouchiat[22] have also tentatively proposed a similar experiment in Pb which would be sensitive only to the spin-dependent interaction. As they point out this experiment is hardly possible with currently available technology. Not only is the interaction to be looked for about 100 times smaller but the experiment itself is intrinsically less sensitive than the Cs one by at least one or two orders of magnitude.

3.7   Atomic Beam Electric/Magnetic Resonance

Stimulated by the difficulty of measuring the spin-dependent term by optical means and the importance of doing so to determine the structural form of the neutral current interaction, the author has recently put forward a proposal for an atomic beam experiment which should be of sufficient sensitivity to allow the detection and measurement of the spin dependent inter-action if it is of the expected magnitude.

Basically our proposal, which is illustrated schematically in figure 6, is to apply to an atom in the resonance region of an atomic beam machine parallel electric and magnetic fields of the same frequency but 90° out of phase. One would then look for a parity violating interference between the two fields by reversing the phase of one with respect to the other and detecting any consequent change in resonance intensity.

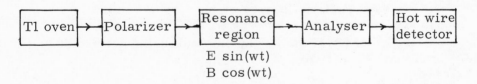

Figure 6.   Schematic diagram of proposed atomic beam electric/magnetic resonance experiment.

One can show that only the spin-dependent term will
contribute and that even this gives zero unless the transition
involves more than one hyperfine state. An obvious difficulty is
that such a transition is normally in the 1000 MHz range. At
these frequencies an oscillating high electric field has
associated with it via Maxwell's equations a magnetic field
which would be sufficiently big to saturate the resonance and
completely mask the small parity violating term. To avoid this
we propose to use a transition at high DC magnetic field which is
between two levels close to the point where they cross so that
the transition frequency can be made very low.

The choice of the optimum atomic transition is a complicated
one which is currently being investigated.

4.   CONCLUSION

In this talk I have attempted to review as widely as time
permits the current state of play in atomic physics experiments
on P and T violation. I hope that I have succeeded in showing
that in both the search for P and T violations and the search for
P without T violation atomic physics experiments are sufficiently
sensitive to provide useful information on the fundamental
structure of the weak interactions.

1.   G. Feinberg, Atomic Physics, (Plenum Press, 1969) p.1.

2.   E.D. Commins, Atomic Physics 2, (Plenum Press, 1971)
        p.25.

3.   see however F.R. Calaprice et. al. Phys.Rev. 190D;
        579 (1974) and R.L. Steinberg et. al. Phys.Rev. Letters
        33; 41 (1974) for   decay experiments using atoms and
        neutrons respectively.

4.   J.H. Christenson, J.W. Cronin, V.L. Fitch and
        R. Turley, Phys.Rev. Letters 13; 138 (1964); see also
        The CP Puzzle: Strange Decays of the Neutral Kaon,
        P.K. Kabir (Academic Press, 1968).

5.   R.C. Casella, Phys.Rev. Letters 21; 1128 (1968) and
        22; 554 (1969).

6.   B.G. Kenny and R.G. Sacks Enrico Fermi Institute pre-
      print no. 72/53 (unpublished) 1972.

7.   see for example the review by Golub and Pendlebury,
      ref. 8 below.

8.   R. Golub and J.M. Pendlebury,Contemporary Physics 13;
      579 (1972).

9.   W.B. Dress, P.D. Miller, N.F. Ramsey, Phys.Rev. 7D;
      3147 (1973).

10.   L. Wolfenstein, Cern pre-print T4 1837 (1974).

11.   L.I. Schiff, Phys.Rev. 132; 2194 (1963).

12.   P.G.H. Sandars, Journal of Physics B1; 511 (1968).

13.   P.G.H. Sandars, Physics Letters 14; 194 (1965).

14.   P.G.H. Sandars, Phys.Rev.Letters 19; 1396 (1967).

15.   P.G.H. Sandars, to be published.

16.   P.G.H. Sandars and E. Lipworth, Phys.Rev.Letters 13;
      529 (1964).
      M.C. Weisskopf et.al. Phys.Rev.Letters 21; 1645 (1968).
      M.A. Player and P.G.H. Sanders, J.Phys.B. 3;
      1620 (1970).

17.   G.E. Harrison, P.G.H. Sanders, S.J. Wright, Phys.Rev.
      Letters 22; 1263 (1969).

18.   A. Salam and J. Ward, Phys.Letters 13; 168 (1964).
      S. Weinberg, Phys.Rev.Letters 19; 1264 (1967).

19.   F.J. Hasert et.al. Phys.Letters 46B; 138 (1973).

20.   B. Aupert et.al. Phys.Rev.Letters 32; 1454 (1974) and
      references contained therein.

21.   S.J. Barish et.al. Argonne National Laboratory pre-print
      (1974) unpublished.

21a. L. Stodolsky, Phys. Letters B 50; 352 (1974).

22.   C.C. Bouchiat and M.A. Bouchiat, Phys. Letters 48B;
      111 (1974) and M.A. Bouchiat and C.C. Bouchiat, pre-
      print (1974) to be published.

23.   G.E. Feinberg and M.Y. Chen, pre-print (1974) to be
      published.

24.   A. Pais, H. Georgi, to be published.

25.   J. Bernaben, T.E.O. Ericson and C. Jarlskog, Phys.
      Letters B 50; 467 (1974).

26.   L.M. Simons, private communication (1964).

RYDBERG CONSTANT[*]

Theo W. Hänsch[†]

Department of Physics, Stanford University

Stanford, California 94305

I.   INTRODUCTION

Recent advances in high resolution spectroscopy with tunable lasers have made it possible to determine a new value of the Rydberg constant with an almost tenfold improvement in accuracy over recent other experiments. Our research group at Stanford University, i.e. M. H. Nayfeh, S. A. Lee, S. M. Curry, I. S. Shahin, and myself, has recently completed a measurement of the absolute wavelength of the optically resolved $3D_{5/2} - 2P_{3/2}$ component of the red Balmer line H_α of atomic hydrogen and D_α of deuterium.[1] Doppler broadening was eliminated by saturation spectroscopy[2] with a pulsed tunable dye laser. An iodine stabilized He-Ne laser served as wavelength standard.[3] The same measurements provide a new precise value for the $H_\alpha - D_\alpha$ isotope shift. In addition, the splittings between the stronger fine structure components in the optical spectrum were determined to within a few MHz.

The Rydberg constant, an important cornerstone in the evaluation of the other fundamental constants, is related to the wavelength in the spectra of one-electron atoms via Bohr's formula, corrected for finite nuclear mass, Dirac fine structure and Lamb shifts.[4]

---

[*]Portions of this work supported by the National Science Foundation under Grant GP-28415, by the U.S. Office of Naval Research under Contract ONR-0071, and by a Precision Measurement Grant from the National Bureau of Standards.

[†]Alfred P. Sloan Fellow, 1973-75.

The red Balmer line of hydrogen is perhaps the most extensively studied of all spectral lines. But precise wavelength measurements always encountered a peculiar problem, because this line consists of seven closely spaced fine structure components, which always appeared blurred and masked by Doppler broadening to the random motion of the light atoms. This problem is only partly alleviated, when Doppler broadening is reduced by operation at cryogenic temperatures and by use of the heavier isotopes deuterium and tritium. The fine structure intervals are well known from radiofrequency spectroscopy, but the relative line intensities emitted by a gas discharge follow no simple rule, and the mathematical line fitting of the observed blend of unresolved components always involved troublesome uncertainties.

G. W. Series pointed out in 1970, after a careful analysis, that the accuracy of the accepted Rydberg value (1 part in 10^7) may well have been optimistically overstated.[5] On the other hand it became increasingly clear that future consistency tests and adjustments of the fundamental constants will soon call for an accuracy of the Rydberg to within a few parts in 10^8. B. N. Taylor pointed out that an uncertainty of a few parts in 10^7 would severely limit the interpretation of several future precision experiment in physics.[6]

II.   RECENT CONVENTIONAL MEASUREMENTS OF THE RYDBERG

Responding to this situation, several research groups have recently made new efforts to determine a precise Rydberg value by measuring the wavelength of gas discharge emission lines with modern versions of conventional high resolution spectroscopy. B. P. Kibble and coworkers[7] at the National Physics Laboratory, England, reported in 1973 on a wavelength measurement of the red Balmer line of deuterium and tritium, comparing it directly with the present standard length, a spectral line of Kr^{86} at 606 μm. By cooling an rf discharge tube with liquid helium, they obtained an emission line profile of D_α, as shown in the center part of Fig. 1. The theoretical fine structure components with their relative transition probabilities are shown for comparison. The wavelengths were determined with the help of a deconvolution and fitting procedure. A comparison with numerical calculations by G. Erickson (1974) yielded a new Rydberg value $R_\infty = 109737.326(8)$ cm^{-1}, with a somewhat smaller uncertainty than the accepted value $R_\infty = 109737.312(11)$ cm^{-1}.

E. G. Kessler, Jr., at the National Bureau of Standards in Washington, D. C. reported on another new Rydberg measurement in 1973.[8] He measured the wavelength of the n = 3-4 line in ionized helium in terms of a well known Hg line. The fine structure of this line is considerably larger than that of hydrogen and can be optically resolved. But the ionic line is weak and easily subject to systematic shifts, which limit the obtainable accuracy. Kessler's Rydberg value, $R_\infty = 109737.3208(85)$ cm^{-1} agrees with Kibble's result within the quoted standard deviations.

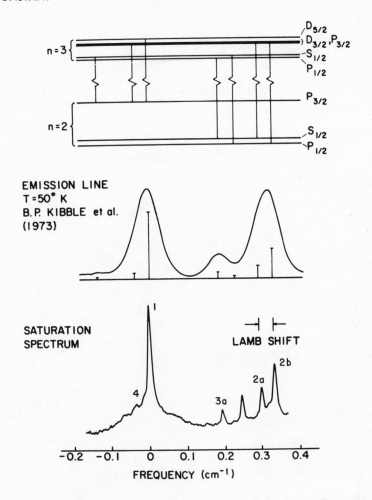

Fig. 1 Deuterium Balmer Line D_α
top: energy levels with fine structure transitions;
center: emission line profile of a cooled deuterium
gas discharge, and theoretical fine structure lines with
relative transition probabilities.[7] bottom: saturation
spectrum with optically resolved Lamb shift.[1]

III.   MEASUREMENT OF THE RYDBERG CONSTANT BY LASER
SATURATION SPECTROSCOPY OF H_α AND D_α

The feasibility of a dramatically improved Rydberg measurement
became apparent, when we succeeded at Stanford in resolving single

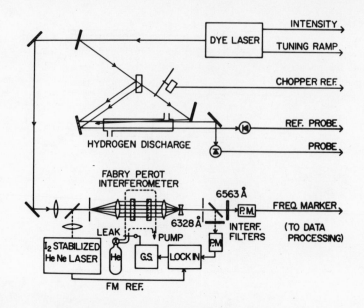

Fig. 2   Scheme of laser saturation spectrometer and
wavelength comporator for measurement of
the Rydberg constant.[1]

fine structure components of H_α in laser saturated absorption.[2]   The
recent new accurate Rydberg measurement is based on these earlier
qualitative experiments.

(a) Laser Saturation Spectrometer

A simplified optical scheme of the Stanford experiment is given
in Fig. 2.   The saturation spectrometer, shown in the top half, is
essentially the same as before.[2]   Hydrogen or deuterium is generated
electrolytically and continuously pumped through a Wood type dc
discharge tube, where it is excited to the atomic n = 2 state (1 m
long folded pyrex tube, 8 mm diameter, 0.1-1 torr, cold Al cathode,
8-20 mA, electronic switch for afterglow measurements).   The
saturated absorption is observed in a 15 cm long center section of
the positive column (typical absorption at resonance 20-50%).   The
light source is a nitrogen laser pumped dye laser[9] with an external
confocal filter interferometer (71 pulses per sec, 30 MHz line width
FWHM, 8 nsec pulse length).   It provides two weak probe beams (1 mm
diameter, 4 mm apart, symmetric with respect to the tube axis, peak
power < 5 mW) and a counter-propagating alternately chopped saturating beam, which overlaps one of the probes in the discharge (crossing angle 2 mrad).   At exact resonance the two counter propagating
waves are interacting with the same atoms, those with essentially

zero axial velocity, and the saturating beam can bleach a path for
its probe, producing a saturation signal. The use of the second
reference probe in a differential detection scheme reduces the
noise due to randomly fluctuating laser amplitudes, which exhibit
an exponential pulse height distribution after the external
interferometer.[10]

(b) Wavelength Measurement

To measure absolute wavelengths, part of the dye laser light
is sent through a fixed Fabry-Perot interferometer (bottom), whose
spacing is accurately known in terms of the He-Ne laser standard
wavelength. While the saturation spectrum is produced by slowly
scanning the laser frequency across the line profile, the resulting
transmission maxima are recorded simultaneously as wavelength
markers.

The iodine stabilized He-Ne laser is similar to that described
by Schweitzer et al.[3] Its frequency is electronically locked to
the n hyperfine component of the coincident $I_2{}^{129}$ absorption line.
This wavelength has been measured in terms of the Kr^{86} standard to
within 1.4 parts in 10^9, limited in precision only by the definition
of the present standard length. The He-Ne laser beam and part of
the dye laser beam are focused through a common pinhole (140 µm
diameter) and sent through a collimating quartz lens (f = 105 cm)
perpendicularly into the plane Fabry-Perot interferometer. Feedback
into the laser cavity is minimized by an optical isolator, an
attenuator, and by random phase modulation of the back-reflected
light with a vibrating mirror. The transmitted light passes through
a Galilean telescope (20x), and a second pinhole (1 mm dia) in its
image plane eliminates any ghost fringes. The two laser beams are
finally split by filters and monitored by separate photomultipliers.

The interferometer is mounted in an air-tight chamber with high
quality windows for pressure tuning. The Ag-coated plates are flat
to within λ/200 over the utilized area (1 cm dia) and are separated
by a spring loaded Invar spacer of up to 64 mm length. Such a large
spacing greatly facilitates an accurate wavelength comparison.
Earlier interferometric measurements of gas discharge emission lines
required much shorter spacers to avoid multiple interferometer
orders within the line profile.

The spacer length can be measured mechanically to within a few
µm. For a more precise determination, to within one order number,
we compared the measured relative positions of the Fabry-Perot
fringes of up to eight known spectral lines (Cd and Hg) with
numerical computations. During the experiment, the interferometer
is kept in exact resonance (± 1 MHz) with the He-Ne laser by manually
controlling the pressure of the tuning gas. A sensitive indicator
is the zero crossing of the dispersion shaped differentiated Fabry-

Perot resonance observed with a lock-in detector, utilizing the
frequency modulation of the Ne-He laser (2 MHz sweep).

The strong $2P_{3/2} - 3D_{5/2}$ component (see Fig. 1) was chosen for
the absolute wavelength measurement because it exhibits the smallest
hyperfine splitting. The effect of tuning nonlinearities is
minimized by selecting the order of the pressure tuned interferometer
so that a marker resonance appears in near coincidence with this
line, as shown in Fig. 3. The accuracy of earlier wavelength
measurements is sufficient to uniquely identify the order number of
this marker, and the precise wavelength can be determined from the
relative separation of the marker and the fine structure line.

(c) Data Processing

Six critical experimental parameters, as indicated in Fig. 2,
are stored for each laser pulse in digital form on a magnetic disk
memory for later processing with a small laboratory computer. About
6×25000 numbers are accumulated in this way as raw data during a

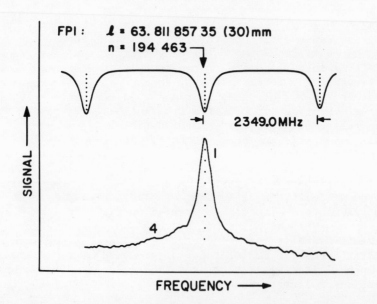

Fig. 3 Section of saturation spectrum of Balmer line D_α, with
simultaneously recorded transmission maxima of Fabry-Perot
interferometer (length 1). The wavelength of the
component $2P_{3/2} - 3D_{5/2}$ (1) is determined by
measuring its distance from the near-
coincident marker resonance
(interferometer order n).

typical run of 6 min duration. The saturation signal is computed, without model-dependent normalization, by taking the intensity difference e I_p - I_r of probe and reference probe in the presence of the saturating beam, and subtracting a correction due to imperfect balancing of the two probes. The correction $<(I_p° - I_r°)/I°>$ I is calculated from the neighboring parameter values without saturating beam (°). The saturation spectrum is obtained by sorting these signals according to tuning ramp, i.e. laser frequency, into an 800-element array. A corresponding array is calculated for the normalized wavelength marker signal. These spectra are smoothed without introducing systematic line shifts, by convolution with a Gaussian or Lorentzian profile. The spectra shown in Fig. 1 and Fig. 3 were obtained by averaging over about 10 individual runs, after correcting the frequency scale so that the centers of gravity of the marker resonances coincide.

Both the center of gravity and the center of area were calculated for that part of the fine structure line which reaches about half miximum. The line profile is sufficiently symmetric so that these two values are in good agreement, and the cut-off level ensures that any line shifts due to the weak neighboring $2P_{3/2}$ - $3D_{3/2}$ (4) component and its intermediate cross-over line remain negligible. This simple procedure has the advantage of not depending on any line profile models.

(d) Systematic Errors

A series of wavelength measurements was performed for hydrogen and deuterium. The optical geometry was frequently optimized in between runs to randomize any alignment errors. The refractive index dispersion of the tuning gas (N_2 or He) was taken into account in the evaluation. Most data were taken with a spacer of 63.8 mm length. It was confirmed, however, that a shorter 31.8 mm spacer gives the same results to within one standard deviation. A correction of 8 parts in 10^9 (for the longer spacer) was subtracted from the wavelengths obtained with the Ag etalon coatings used originally (transmission T = 5%) to account for a wavelength dependent phase change in reflection.[7] Later, the etalon plates were stripped and recoated with Al (T = 11%), which has a negligible phase dispersion in the red, and good agreement was found with the earlier corrected data.

Measurements in light hydrogen require an additional correction due to unresolved hyperfine splitting because saturation spectroscopy at low intensity weights the line components with the square of their oscillator strengths. An upward wavelength shift of 3.6 parts in 10^9 is expected for the $2P_{3/2}$ - $3D_{5/2}$ components, if cross over lines are absent. If we assume the presence of additional unresolved cross over lines with a magnitude equal to the geometric means of their parents, then we expect a shift of 4.2 parts in 10^9 for quadratic response (and of 2.1 parts in 10^9 for linear response). In the final analysis, we applied a correction of 4.2 parts in 10^9 to the hydrogen data. The correction for deuterium is negligible.

Several other sources of possible systematic errors have been carefully studied. To investigate possible Stark shifts we measured the wavelength both with running discharge and in the afterglow, 0.5 μsec after electronically stalling the discharge. No systematic line shifts could be found within standard deviations of 2 parts in 10^8. A shift of -4.5 parts in 10^9 is expected for a dc field of 10 V/cm in the positive column of the discharge.[11] We have experimental evidence that the Stark effect due to the microfields of free electrons and ions in the afterglow is smaller than that. A field of 10 V/cm is sufficient for Stark mixing of the closely spaced hydrogen levels $3P_{3/2}$ and $3D_{3/2}$,[12] and a corresponding cross-over signal has been observed halfway between the lines $2S_{1/2} - 3P_{3/2}$ (2a) and $2P_{1/2} - 3D_{3/2}$ (2b), if the discharge is running. This cross-over signal is essentially absent in the afterglow measurements. A Stark shift correction of 4.5 parts in 10^9 was applied to wavelengths measured with running discharge.

For possible pressure shifts we established an experimental upper limit of 13 parts in 10^9 per torr, or 2 parts in 10^9 at the typical pressure of 0.15 torr, by comparing wavelength measurements at hydrogen pressures of 0.15, 0.3, and 0.44 torr. Intensity dependent light shifts on the order of 0.5 parts in 10^{10} per mW/mm^2 are theoretically expected due to the weak neighboring fine structure lines $2P_{3/2} - 3D_{3/2}$ (4) and $2P_{3/2} - 3S_{1/2}$.[13] Sorting of the saturation spectra according to laser pulse height into eight ranges from 2 to 180 mW/mm^2 peak power gave an experimental upper limit for such shifts of 2 parts in 10^{10} per mW/mm^2. Table 1 gives a summary of the various systematic corrections, applied to the measured wavelengths, and the uncertainties associated with these corrections.

(e) Results

The absence of unknown systematic errors was further confirmed by optical measurements of the intervals between different fine structure components of H_α. Closely spaced frequency markers were generated with a confocal calibration interferometer of 200 MHz free spectral range. For the $2P_{1/2} - 3D_{3/2}/2S_{1/2} - 3P_{3/2}$ (2b - 2a) interval, measured in the afterglow, we obtained 1052.7 (1.7) MHz, and for the $2P_{1/2} - 3D_{3/2}/2P_{3/2} - 3D_{5/2}$ (2b - 1) interval 9884.5 (2.9) MHz. Both values agree within their error limits with theoretical values, and seem to give an experimental confirmation for the expected small $3D_{3/2} - 3P_{3/2}$ Lamb shift (5.3 MHz).[12]

In the final data analysis we averaged the wavelengths measured under various experimental conditions, so that the standard deviation gives some measure for residual systematic errors. For the $2P_{3/2} -$

Source	Correction (parts in 10^9)	Uncertainty (parts in 10^9)
Phase Dispersion in Etalon Coatings	-8	4
Refractive Index of Tuning Gas	-11...-16	2
He-Ne Laser Wavelength Standard	0	4
Unresolved HFS and Cross Over Lines	-4.2 (H_α)	4 (H_α)
Stark Shifts	+4.5	3
Intensity Shifts	0	2
Pressure Shifts	0	2

Table 1 Systematic corrections applied to
measured wavelengths of H_α and D_α

	H_α	D_α
Fine Structure Interval $3P_{1/2}-3D_{3/2}$ / $2S_{1/2}-3P_{3/2}$	1052.7(1.7) MHz	
$2P_{1/2}-3D_{3/2}$ / $2P_{3/2}-3D_{5/2}$	9884.5(2.9) MHz	
Vacuum Wavelength^{-1} $2P_{3/2}-3D_{5/2}$	15233.07021(9) cm^{-1}	15237.21538(8) cm^{-1}
Isotope Shift	4.14517(12) cm^{-1}	
Rydberg Constant	109737.3130(6) cm^{-1}	109737.3150(6) cm^{-1}
	109737.3143(10) cm^{-1}	

Table 2 Laser Saturation Spectroscopy of
H_α and D_α Summary of Results

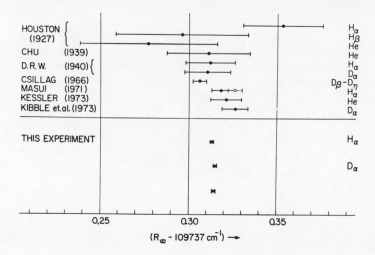

Fig. 4 Results of conventional Rydberg measurements (top)
and new Rydberg values (H_α, D_α, and combined average),
obtained by laser saturation spectroscopy (bottom).

$3D_{5/2}$ component of H_α we determine an inverse vacuum wavelength of
15233.07021(9) cm^{-1} from 17 individual spectra, where the uncertainty is one standard deviation. From the corresponding D_α line
we obtain 15237.21538(8) cm^{-1} from 33 spectra. The resulting H_α –
D_α isotope shift is 4.14517(12) cm^{-1}.

To determine the Rydberg constant, we fitted the measured
wavenumbers with calculated wavenumbers, which are proportional to
an assumed Rydberg value. We used calculations by G. Erickson,[14]
which are similar to those by Garcia and Mack,[15] but take the 1973
adjustments of the fundamental constants into account. In this way
we obtain R_∞ = 109737.3130(6) cm^{-1} for H_α and R_∞ = 109 737.3150(6)
cm^{-1} for D_α. If we ignore the small discrepancy between the two
isotopes and combine all individual measurements, we arrive at
R_∞ = 109 737.3143(4) cm^{-1}: with a standard deviation of 4 parts in
10^9. Taking the uncertainties of the various systematic corrections
into account, we arrive at a larger rms error of 0.001 cm^{-1} or 1
part in 10^8.

The numberical results of this experiment are summarized in
Table 2. Figure 4 gives a comparison of the new Rydberg values
with previously reported measurements of the Rydberg constant. The
new result is only slightly outside the error limits of Kibble's
recent measurement[7] and agrees with Kessler's result.[8]

Future refinements of the present experiment can be expected, when the discharge is replaced by a beam of metastable hydrogen atoms. Observation of atomic beam fluorescence, possibly using a cw dye laser, may be an attractive alternative to saturation spectroscopy. A very accurate Rydberg value can also be expected, when the new technique of two-photon spectroscopy without Doppler broadening[16] is applied to the 1S - 2S transition in hydrogen.

ACKNOWLEDGEMENTS

We are indebted to Professor A. L. Schawlow for his continuing stimulating interest in this research. And we thank Professor G. W. Series for stimulating discussions and valuable advice. We are also grateful to Dr. B. Kibble for many helpful suggestions and for lending us a Cd spectral lamp, and last, but not least, to Professor G. Erickson for making his computer calculations available to us prior to publication.

REFERENCES

[1] T. W. Hänsch, M. H. Nayfeh, S. A. Lee, S. M. Curry, and I. S. Shahin, Phys. Rev. Letters 32, 1336 (1974).

[2] T. W. Hänsch, I. S. Shahin, and A. L. Schawlow, Nature 253, 63 (1972).

[3] W. G. Schweitzer, Jr., E. G. Kessler, Jr., R. D. Deslattes, H. P. Layer, and J. R. Whetstone, Appl. Opt. 12, 2827 (1973).

[4] G. W. Series, Contemp. Phys. 14, 49 (1974).

[5] G. W. Series, *Proceedings of the International Conference on Precision Measurements and Fundamental Constants*, NBS Gaithersburg, August 1970. NBS Spec. Pub. 343, U.S. Dept. of Commerce, pp 73-82.

[6] B. N. Taylor, private communication.

[7] B. P. Kibble, W. R. C. Rowley, R. E. Shawyer, and G. W. Series, J. Phys. B, 6, 1079 (1973).

[8] E. G. Kessler, Phys. Rev. 7A, 408 (1973).

[9] T. W. Hänsch, Appl. Opt. 11, 895 (1972).

[10] S. M. Curry, R. Cubeddu, and T. W. Hänsch, Appl. Phys. 1, 153 (1973).

[11] J. A. Blackman and G. W. Series, J. Phys. B, 6, 1090 (1973).

[12] H. A. Bethe and E. E. Salpeter, in *Encyclopedia of Physics*, Vol. XXXV, Atoms 1, S. Flügge, ed., Springer-Verlag, Berlin (1957).

[13] M. Mizushima, Phys. Rev. 133, A414 (1964).

[14] G. Erickson, to be published.

[15] J. D. Garcia and J. E. Mack, J. Opt. Soc. Am. 55, 654 (1965).

[16] T. W. Hänsch, K. C. Harvey, G. Meisel, and A. L. Schawlow, Opt. Comm., 11, 50 (1974).

A NEW VALUE OF THE FINE STRUCTURE CONSTANT FROM HELIUM FINE STRUCTURE*

Michael L. Lewis

J.W. Gibbs Laboratory, Yale University

New Haven, Connecticut, U.S.A.   06520

For almost half a century, the triplet states of helium have been investigated both theoretically and experimentally by atomic physicists. From the early calculations by Heisenberg[1] to the present, the determination of the 2^3P fine structure splittings have demonstrated the validity of the theory of the interaction of electrons with the electromagnetic field. If one assumes that quantum electrodynamics properly describes the electron-electron interaction in helium, one can use the measured fine structure intervals and the calculated values to precisely determine the fine structure constant α.

Figure 1 shows the fine structure of the 2^3P state of helium and the other low-lying helium states. The $2^3P_0 - 2^3P_1$ splitting is about 29 616 MHz and the $2^3P_1 - 2^3P_2$ splitting is about 2 291 MHz.

In 1926 Heisenberg calculated the fine structure of helium using a semi-classical spin-spin interaction. (See Tables 1 and 2.) Optical experiments by Houston,[2] Hansen,[3] and Meggers[4] from 1927 to 1935 yielded different values. Gaunt[5] used a two-particle Dirac equation but he neglected retardation.

At the same time Breit[6] derived the equation bearing his name with a relativistic retarded interaction

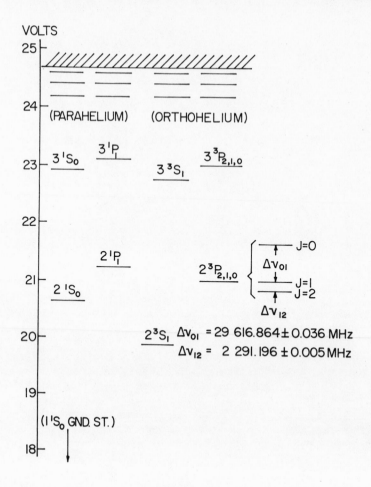

Figure 1. Energy levels of helium.

$$\left(H_1 + H_2 + \frac{e^2}{r_{12}} + B\right)\Psi = E\Psi$$

$$H_i = \beta_i m^2 - e\phi(\vec{r}_i) + \vec{\alpha}_i \cdot [\vec{p}_i + e\vec{A}(\vec{r}_i)] \tag{1}$$

$$B = -\frac{e^2}{2r_{12}}\left[\vec{\alpha}_1 \cdot \vec{\alpha}_2 + \frac{(\vec{\alpha}_1 \cdot \vec{r}_{12})(\vec{\alpha}_2 \cdot \vec{r}_{12})}{r_{12}^2}\right]$$

Table 1. Early 2^3P Helium Fine Structure Results

	Theory or experiment	ν_{01}	ν_{12}	
W. Heisenberg (1926)	T	0.56 cm^{-1}	0.056 cm^{-1}	
W. Houston (1927)	E	0.992 cm^{-1}	0.071 cm^{-1}	optical
G. Hansen (1927,1929)	E	0.990 cm^{-1}	0.077 cm^{-1}	optical
J. Gaunt (1929)	T	0.73 cm^{-1}	0.035 cm^{-1}	
G. Breit (1929, 1932)	T	0.97 cm^{-1}	0.14 cm^{-1}	
H.A. Bethe (1933)	T	0.84 cm^{-1}	-0.05 cm^{-1}	
W.F. Meggers (1935)	E	0.99 cm^{-1}	0.08 cm^{-1}	optical
G. Araki (1937)	T	0.94 cm^{-1}	0.068 cm^{-1}	
J. Brochard, R. Chabbal, H. Chantrel, and P. Jacquinot (1952)	E	0.988 cm^{-1}	0.0765 cm^{-1}	optical
I. Wieder and W.E. Lamb (1957)	E		2 291.72±0.36 MHz	optical microwave
G. Araki, M. Ohta, and K. Mano (1959)	T	0.997 457 11 cm^{-1}	0.075 974 45 cm^{-1}	
F.D. Colegrove, P.A. Franken, R.R. Lewis, and R.H. Sands (1959)	E	29 650±280 MHz	2 291.56±0.09 MHz	optical level crossing
J. Traub and H.M. Foley (1959)	T	0.985 198 cm^{-1}	0.083 352 cm^{-1}	

Table 2.   Recent 2^3P Helium Fine Structure Results

	Theory or experiment	ν_{01}	ν_{12}	
C.L. Pekeris, B. Schiff, and H. Lifson (1962)	T	0.987 905 cm^{-1}	0.076 271 cm^{-1}	
C. Schwartz (1964)	T	0.987 837 cm^{-1}	0.076 530 2 cm^{-1}	
J. Lifsitz and R.H. Sands (1965)	E	29 616.76±0.40 MHz	2 291.200±0.022 MHz	optical level crossing
B. Schiff, C.L. Pekeris, and H. Lifson (1965)	T	0.987 998 cm^{-1}	0.076 372 9 cm^{-1}	
C.E. Johnson, F.M.J. Pichanick, A. Kponou, S.A. Lewis, and V.W. Hughes (1968)	E	29 616.99±0.13 MHz		
F.M.J. Pichanick, R.D. Swift, C.E. Johnson, V.W. Hughes (1968)	E		2 291.195±0.007 MHz	atomic beam optical microwave
S.A. Lewis, F.M.J. Pichanick and V.W. Hughes (1970)	E		2 291.196±0.005 MHz	atomic beam optical microwave
A. Kponou, V.W. Hughes, C.E. Johnson, S.A. Lewis and F.M.J. Pichanick (1971)	E	29 616.864±0.036 MHz		atomic beam optical microwave
J. Daley, M. Douglas, L. Hambro and N.M. Kroll (1972)	T	29 616.83±0.18 MHz	2 291.36±.36 MHz	

H_i is the Dirac Hamiltonian for the ith particle, $\dfrac{e^2}{r}$
is the Coulomb interaction, B is the Breit term and[12]
Ψ is the sixteen component wavefunction. B includes the
effect of the exchange of a transverse photon between
the electrons. Breit[7] later showed that B is to be used
only in first order yielding energies accurate to α^2 Ry.
Breit reduced his equation the the Pauli approximation
involving 4-component wavefunctions (equation (2) includes the contact interaction term not given by Breit):

$$(\mathcal{H}_0 + \mathcal{H}_1 + \mathcal{H}_2 + \mathcal{H}_3 + \mathcal{H}_4 + \mathcal{H}_5 + \mathcal{H}_6)\ U = WU$$

$$\mathcal{H}_0 = -eV + \frac{P_1^2}{2m} + \frac{P_2^2}{2m}$$

$$\mathcal{H}_1 = -\frac{1}{8m^3c^2}(P_1^4 + P_2^4)$$

$$\mathcal{H}_2 = -\frac{e^2}{2(mc)^2}\frac{1}{r_{12}}\left[(\vec{P}_1 \cdot \vec{P}_2) + \frac{\vec{r}_{12} \cdot (\vec{r}_{12} \cdot \vec{P}_1)\vec{P}_2}{r_{12}^2}\right]$$

$$\mathcal{H}_3 = \frac{\mu_0}{mc}\left\{[\vec{E}_1 \times \vec{P}_1 + \frac{2e}{r_{12}^3}\vec{r}_{12} \times \vec{P}_2] \cdot \vec{s}_1 + [\vec{E}_2 \times \vec{P}_2 + \frac{2e}{r_{12}^3}\vec{r}_{21} \times \vec{P}_1] \cdot \vec{s}_2\right\}$$

$$\mathcal{H}_4 = \frac{ie\hbar}{(2mc)^2}(\vec{P}_1 \cdot \vec{E}_1 + \vec{P}_2 \cdot \vec{E}_2) \tag{2}$$

$$\mathcal{H}_5 = 4\mu_0^2\left\{-\frac{8\pi}{3}\vec{s}_1 \cdot \vec{s}_2 \delta^{(3)}(\vec{r}_{12}) + \frac{1}{r_{12}^3}[\vec{s}_1 \cdot \vec{s}_2 - \frac{3(\vec{s}_1 \cdot \vec{r}_{12})(\vec{s}_2 \cdot \vec{r}_{12})}{r_{12}^2}]\right\}$$

$$\mathcal{H}_6 = 2\mu_0[\vec{H}_1 \cdot \vec{s}_1 + \vec{H}_2 \cdot \vec{s}_2] + \frac{e}{mc}[\vec{A}_1 \cdot \vec{P}_1 + \vec{A}_2 \cdot \vec{P}_2]$$

$$V = \frac{Ze}{r_1} + \frac{Ze}{r_2} - \frac{e}{r_{12}} + \phi(r_1) + \phi(r_2)$$

$\mathcal{H}_3$ and $\mathcal{H}_5$ contribute to the fine structure splitting. $\mathcal{H}_3$ contains
spin-orbit and spin-other-orbit interactions while $\mathcal{H}_5$ includes the
spin-spin terms. Breit calculated the fine structure using a 2
term wavefunction with a $\cos\theta_{12}$ correlation term where θ_{12} is the
angle between the electrons. His result for $E(2^3P_0) - E(2^3P_1) = \nu_{01}$

was accurate to 2%. For $E(2^3P_1) - E(2^3P_2) = \nu_{12}$ the results are less accurate because ν_{12} depends on the difference between two approximately equal numbers.

In 1933 Bethe[8] computed the fine structure using a wavefunction that was not antisymmetric. Araki[9] included polarization terms of all orders of the 1s electron by the 2p electron and considered the effect of the 2^1P level on the 2^3P states. The singlet state lowers the 2^3P_1 level by about 5 MHz. The optical experiment by Brochard[10] and colleagues yielded results accurate to .02%. Wieder and Lamb[11] used an optical microwave technique to determine ν_{12}. Colgrove[12] measured ν_{12} by an optical level crossing experiment. Araki, Ohta and Mano[13] included the polarization of the s orbital, the admixture of the 2^1P state and the anomalous magnetic moment of the electron. Traub and Foley[14] used an 18-term Hylleraas-type wavefunction to calculate the fine structure.

Pekeris, Schiff and Lifson[15,16] first used a wavefunction with 220 terms and then with 560 terms and obtained accuracies of about 1 ppm in the fine structure. In 1964 Charles Schwartz[17] and Vernon Hughes embarked on a program to determine the fine structure constant from theoretical calculations and experimental measurements of the fine structure. The energy levels in helium can be expressed by a power series in α.

$$E_J = E_0 + \alpha^2 <H_2>_J + \alpha^4 <H_2 \frac{1}{E_0 - H_0} H_2>_J + \alpha^4 <H_4>_J \qquad (3)$$

$$+ \ldots$$

H_2 is the Breit interaction, taken in first and second order. H_4^2 is a higher-order spin dependent operator derived from the covariant two-particle Bethe-Salpeter equation. A reduction of this equation to nonrelativistic form has been carried out by Douglas and Kroll.[18]

There are a number of terms not included in equation 3. The anomalous magnetic moment of the electron must be considered yielding terms of order α^3. The reduced mass must be correctly inserted in the problem. The motion of the nucleus is accounted for by the use of the operator $\vec{p}_1 \cdot \vec{p}_2/M$ where M is the nuclear mass. This operator contributes in second-order when mixed with a spin-dependent Breit operator and gives energies of order $\alpha^2(m/M)$.

Schwartz computed the matrix elements of the fine structure operators to order α^2 Ry to better than 1 ppm. He used the

variational principle to minimize the energy eigenvalue of the
Schrodinger equation for helium as more terms were added to the
Hylleraas-like wavefunction. Schwartz first used the wave function

$$\Psi = (\frac{1-P_{12}}{4\pi\sqrt{2}}) \sum_{\ell,m,n=0}^{\ell+m+n\leq\omega} C_{\ell m n} \; \vec{r}_1 r_1^m r_2^n r_{12}^\ell \; e^{-(\kappa\sigma/2)r_1} \; e^{-(\kappa/2)r_2} \quad (4)$$

where P_{12} interchanges coordinates r_1 and r_2. The P state is
designated by the inclusion of $\vec{r}_1$. The screening constants
κ and σ where extrapolated from the results of Traub and Foley.[14]
Schwartz used up to 286 terms in this expansion to compute the
energy eigenvalue and the matrix elements of the spin-orbit term

$$H_{SO} = (\frac{e\hbar}{2mc})^2 \; (\frac{\vec{\sigma}_1+\vec{\sigma}_2}{2}) \cdot [Z(\frac{\vec{r}_1 \times \vec{p}_1}{r_1^3} + \frac{\vec{r}_2 \times \vec{p}_2}{r_2^3})$$

$$- \frac{3(\vec{r}_1-\vec{r}_2) \times (\vec{p}_1-\vec{p}_2)}{r_{12}^3}] \qquad (5)$$

and the spin-spin term

$$H_{SS} = (\frac{e\hbar}{2mc})^2 \; \frac{1}{r_{12}^3} \; (\vec{\sigma}_1\cdot\vec{\sigma}_2 - \frac{3\vec{\sigma}_1\cdot\vec{r}_{12}\vec{\sigma}_2\cdot\vec{r}_{12}}{r_{12}^2}) \qquad (6)$$

The first order calculation of the fine structure intervals are
usually given in terms of two constants C (from spin-orbit) and D
(from spin-spin):

$$\nu_{01} = C + 5D \qquad (7)$$

$$\nu_{12} = -2C + 2D$$

The accuracy Schwartz obtained by this method was about 100 ppm
for C and D.

The precision of this calculation could be improved by using a great many more terms. However Schwartz chose to include in the basis set terms that would add flexibility as r_1 and r_2 approach zero in accordance with the work of Fock.[19] He replaced the $C_{\ell mn}$ in equation 4 with

$$C_{\ell mn} + D_{\ell mn} \, (r_1 + r_2)^{1/2}$$

where $D_{000}=0$ and computed wavefunctions with up to 439 terms. The extrapolated results give the fine structure intervals to better than 1 ppm.

The experimental measurement of the fine structure was active-ly pursued at Yale with an atomic beam optical microwave method (Figure 2). Helium atoms in the $2\,^3S_1$ state were excited optically to the $2\,^3P$ states. Magnetic dipole resonance transitions between the $2\,^3P_J$ Zeeman levels were observed by the population of the $2\,^3S$ states after decay. The results give ν_{10} to 1.2 ppm and ν_{12} to 2.2 ppm. These are the most precise measurements of the fine structure available. If all the theoretical terms were calculated to this accuracy, it would enable us to determine α to 0.6 ppm.

The expectation values of the higher-order operator H_4 were computed by Daley.[20] Also included are corrections of order $\alpha^2(m/M)$ due to the recoil motion of the nucleus.[21]

The contributions in second-order to relative order α^4 and $\alpha^2(m/M)$ have been computed by Hambro,[22] and by Lewis and Serafino[23] using the method of Dalgarno and Lewis.[24] An inhomogeneous Schrodinger equation is solved variationally for the correction to the wavefunction for each operator. The second-order energies are then given by integrals. The symmetry of the intermediate states may be ^3P, ^1P, ^3D, ^1D and ^3F. Hambro calculated the second-order energies with ^3P, ^1P and ^3D symmetry that contribute to the ν_{01} splitting. His results, however, were not sufficiently accurate-there was poor convergence of some of the second order energies as the number of terms in the wavefunction was increased. Some of the ^3P second order energies were calculated with wavefunctions with inverse powers in the expansion (3). Hambro's results[25] are given in Table 3. Almost the entire uncertainty in the fine structure calculations is due to the second order results.

Lewis and Serafino have computed the second order energies using wavefunctions with up to 286 terms. In addition to the ^3P, ^1P and ^3D intermediate states, we computed the ^1D and ^3F

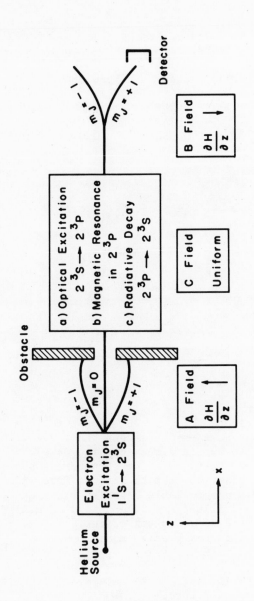

Figure 2. Atomic beam optical microwave apparatus for measuring helium fine structure.

Table 3.   Second-Order Energies Calculated by Hambro

	ν_{01} (MHz)	ν_{12} (MHz)
^3P	5.060 ± 0.18	-0.360 ± 0.36
^1P	6.505 ± 0.06	-6.505 ± 0.06
^3D	0.027 ± 0.002	0.072 ± 0.003

Table 4.   Second-Order Energies

	ν_{01} (MHz)	ν_{12} (MHz)
^3P	5.075 ± 0.11	-0.439 ± 0.21

energies that contribute to ν_{12}. Our result for the ^3P state
appears in Table 4. Some of our second order energies are in disagreement with Hambro but in agreement with Daley.[20] The total
theoretical contribution to the fine structure appears in Table 5.
The fine structure constant from the helium fine structure is
$\alpha^{-1} = 137.035\ 98(2.0\ ppm)$. Table 6 presents the other determinations of α. Figure 3 illustrates the values of α listed in Table 6.
Note that although the ac Josephson effect provides a highly accurate
value of e/h, the different precise measurements of γ_p, the proton
gyromagnetic ratio, yield two values of α that differ by 4 ppm.[26]

    Further work is in progress to determine α to better than 1 ppm
from the helium fine structure.

*Research supported in part by the Air Force Office of Scientific
Research, AFSC, under contract No. F44620-70-C-0091.

Table 5. Theoretical contributions to fine-structure intervals (MHz). The values[a] of α^{-1}, c, R_∞ and (m/M_{He}) are 137.036 04 (11) (0.82 ppm), 2.997 924 58 (12) x 10^{10} cm sec^{-1} (0.004 ppm), 109 737.317 7 cm^{-1} (0.075 ppm), 1.370 934 x 10^{-4} respectively. Thus $\frac{1}{2} \alpha^2 c R_\infty$ = 87.594 22 GHz (1.6 ppm).

Interval	ν_{01}	ν_{12}
$\alpha^4 mc^2$	29 564.557±0.006 (0.21 ppm)	2 317.202±0.0018 (0.76 ppm)
$\alpha^5 mc^2$	54.708	-22.548
$\frac{m}{M_{He}} \alpha^4 mc^2$	-10.707 ± 0.000 44 (0.015 ppm)	1.952±0.000 88 (0.39 ppm)
Second order	11.61±0.11 (4 ppm)	-6.87±0.21 (92 ppm)
$\alpha^6 mc^2$	-3.331±0.0039 (0.13 ppm)	1.542±0.0068 (3.0 ppm)
ν_{theory}	29 616.834±0.11 (4 ppm)	2 291.28±0.21 (92 ppm)
ν_{expt}	29 616.864±0.036[b] (1.2 ppm)	2 291.196±0.005[c] (2.2 ppm)
$\nu_{theory} - \nu_{expt}$	-0.03±0.11 (1 ppm)	0.08±0.21 (35 ppm)

[a]E.R. Cohen and B.N. Taylor, J. Phys. Chem. Ref. Data 2, 663 (1973).

[b]A. Kponou, V.W. Hughes, C.E. Johnson, S.A. Lewis, F.M.J. Pichanick, Phys. Rev. Letters 26, 1613 (1971).

[c]S.A. Lewis, F.M.J. Pichanick and V.W. Hughes, Phys. Rev. 2, 86 (1970).

Table 6. Values of the Fine Structure Constant α

Source	Value of α^{-1} (1 standard deviation error)
Helium fine structure	
$2^3P_1 \rightarrow 2^3P_0$	137.035 98(27) (2.0 ppm)
$2^3P_1 \rightarrow 2^3P_2$	137.038 0(63) (46 ppm)
Hydrogen fine structure	
$2^2P_{3/2} \rightarrow 2^2P_{1/2}$	137.035 44(54) (3.9 ppm)
Hydrogen hyperfine structure	
$1^2S_{1/2}$, F=1 $\rightarrow$ F=0	137.035 97(22) (1.6 ppm)
Hydrogen Lamb shift	137.034 16(20) (1.5 ppm)
Muonium hyperfine structure	
$1^2S_{1/2}$, F=1 $\rightarrow$ F=0	137.036 34(21) (1.5 ppm)
Electron anomalous moment	137.035 63(42) (3.1 ppm)
Josephson effect	137.035 91(14) (1.0 ppm)
	137.036 45(18) (1.3 ppm)
1973 Adjustment by E.R. Cohen and B.N. Taylor, J. Phys. Chem. Ref. Data 2, 663 (1973)	137.036 04(11) (0.82 ppm)

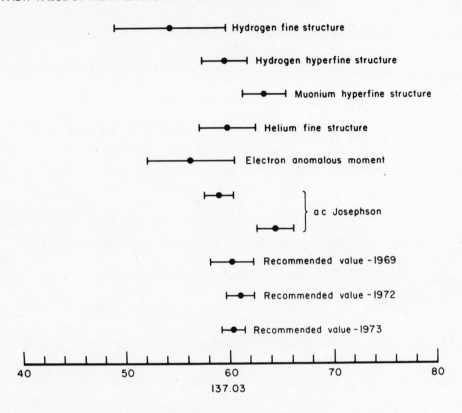

Figure 3. Values of the fine structure constant α.

REFERENCES

1. W. Heisenberg, Z. Phys. 39, 499 (1926).

2. W.V. Houston, Proc. Nat. Acad. Sci. 13, 91 (1927).

3. G. Hansen, Nature (London) 119, 237 (1927); Verh. Deut. Phys. Ges. 10, 5 (1929).

4. W.F. Meggers, J. Res. Nat. Bur. Stand. 14, 487 (1935).

5. J.A. Gaunt, Phil. Trans. Roy. Soc. London, A 228, 151 (1929).

6. G. Breit, Phys. Rev. 34, 553 (1929); Phys. Rev. 36, 383 (1930).

7.   G. Breit, Phys. Rev. 39, 616 (1932).

8.   H.A. Bethe, Hanbuch der Physik, Vol. 24/1, Springer-Verlag, Berlin, 1933.

9.   G. Araki, Proc. Phys. Math. Soc. Japan 19, 128 (1937).

10. J. Brochard, R. Chabbal, H. Chantrel and P. Jacquinot, J. Phys. Radium 13, 433 (1952).

11. I. Wieder and W.E. Lamb, Jr., Phys. Rev. 107, 125 (1957).

12. F.D. Colgrove, P.A. Franken, R.R. Lewis and R.H. Sands, Phys. Rev. Letters 3, 420 (1959).

13. G. Araki, M. Ohta, and K. Mano, Phys. Rev. 116, 651 (1959).

14. J. Traub and H.M. Foley, Phys. Rev. 116, 914 (1959).

15. C. Pekeris, B. Schiff and H. Lifson, Phys. Rev. 126, 1057 (1962).

16. B. Schiff, C.L. Pekeris and H. Lifson, Phys. Rev. 137, A1672 (1965).

17. C. Schwartz, Phys. Rev. 134, A1181 (1964).

18. M. Douglas and N.M. Kroll, Annals of Physics 82, 89 (1974).

19. V.A. Fock, Kgl. Norske Videnskab. Selskabs. Forh 31, 138 (1958).

20. J. Daley, thesis, University of California, Berkeley.

21. M. Douglas, Phys. Rev. 6, 1929 (1972).

22. L. Hambro, Phys. Rev. 5, 2027 (1972); Phys. Rev. 6, 865 (1972); Phys. Rev. 7, 479 (1973).

23. M.L. Lewis and P.H. Serafino, Bull. Am. Phys. Soc. 18, 1510 (1973).

24. A. Dalgarno and J.T. Lewis, Proc. Roy. Soc. London, A 233, 70 (1956).

25. J. Daley, M. Douglas, L. Hambro, and N.M. Kroll, Phys. Rev. Letters 29, 12 (1972).

26. E. Richard Cohen and B.N. Taylor, J. Phys. Chem. Ref. Data 2, 663 (1973).

RECENT FINE STRUCTURE MEASUREMENTS IN HYDROGEN-LIKE ATOMS*

Francis M. Pipkin

Harvard University

Lyman Lab. of Physics, Cambridge, MA.   02138

INTRODUCTION

In spite of the advances elsewhere in physics, measurements of the hydrogen fine structure still give a very significant test of quantum electrodynamics and an important source for the determination of the fine structure constant.[1]   Until very recently the precision of the measurements was limited by the natural line width due to the lifetime of the states. Recent advances in technique make it possible to overcome this limitation and give promise of increasing the precision of the measurements by an order of magnitude. In this review, I will summarize the present state of our knowledge of the hydrogen fine structure, describe the recent advances in technique which enable one to reduce the line width below the natural line width, and describe methods on the horizon which give promise of even greater precision.

THEORETICAL BACKGROUND

Figure 1 shows the energy levels for a hydrogen like atom in the absence of any hyperfine interaction. The chief interest centers on the $^2S_{1/2} \to {}^2P_{1/2}$ Lamb shift interval ($\mathcal{L}$) and the $^2P_{1/2} \to {}^2P_{3/2}$ fine structure interval (ΔE). ΔE is usually determined from independent measurements of $\mathcal{L}$ and the $^2S_{1/2} \to {}^2P_{3/2}$ fine structure interval ($\Delta E - \mathcal{L}$).

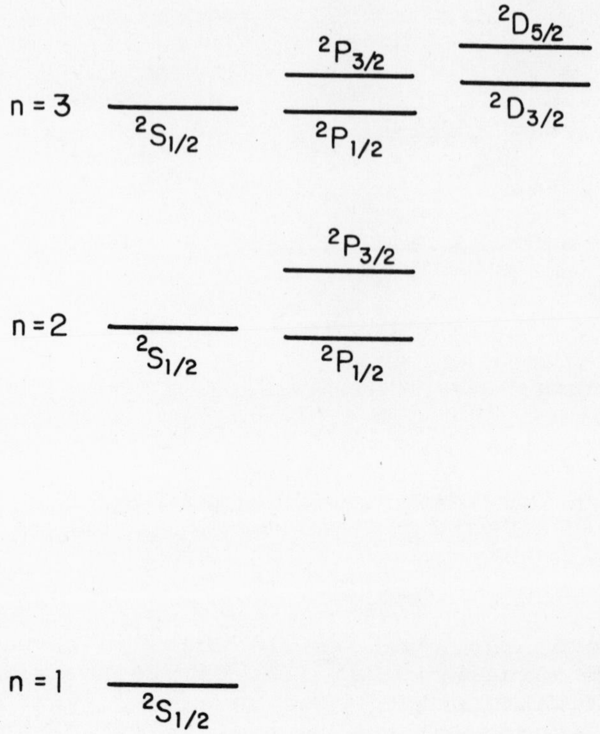

$$
\begin{array}{llll}
 & & \dfrac{}{^2P_{3/2}} & \dfrac{}{^2D_{5/2}} \\[2mm]
n=3 \quad \dfrac{}{^2S_{1/2}} & & \dfrac{}{^2P_{1/2}} & \dfrac{}{^2D_{3/2}}
\end{array}
$$

$$
\begin{array}{lll}
 & & \dfrac{}{^2P_{3/2}} \\[3mm]
n=2 \quad \dfrac{}{^2S_{1/2}} & & \dfrac{}{^2P_{1/2}}
\end{array}
$$

$$
n=1 \quad \dfrac{}{^2S_{1/2}}
$$

Figure 1 -- The lower three energy level complexes for a hydrogen like atom. This figure is not drawn to scale.

The fine structure splitting, ΔE, is due to the interaction of the magnetic moment of the electron with the magnetic field produced through the motion of the electron in the electric field from the nucleus. The theoretical value for the n=2, $^2P_{3/2} \rightarrow {}^2P_{1/2}$ interval for a hydrogen like atom is[1-3]

$$
\Delta E = \frac{R_\infty (Z\alpha)^2}{16}\left\{ \left[1 + \frac{5}{8}(Z\alpha)^2 \right]\left(1 + \left(\frac{m}{M}\right)\right)^{-1} - \left(\frac{m}{M}\right)^2 \cdot \left(1 + \frac{m}{M}\right)^{-3} \right.
$$

$$
\left. + \; 2\,a_e \left(1 + \frac{m}{M}\right)^{-2} + \frac{\alpha}{\pi}(Z\alpha)^2 \log Z\alpha \right\}
$$

Here R_∞ is the Rydberg, Z is the charge on the nucleus, α is the fine structure constant, m is the mass of the electron, M is the mass of the nucleus, a_e is the anomalous part of the electron magnetic moment. This splitting can be accurately calculated and does not depend strongly on the quantum electrodynamic correction to the hydrogenic energy levels. The a_e term is the first term of radiative origin; it contributes roughly 0.1%. The last term, which is also a radiative correction, contributes only about 1ppm.[4-5] Bounds to the next uncalculated terms, $O[(\alpha/\pi)(Z\alpha)^2]$, have been estimated by Erickson.[6]

On the other hand the Lamb shift, $\mathscr{L}$, is due entirely to the QED corrections to the theory. The largest contribution to $\mathscr{L}$ comes from the electron self energy and can be viewed as a smearing out of the electron charge due to the emission and reabsorption of virtual photons. Figure 2 summarizes the different theoretical components of $\mathscr{L}$.[7]

Determination of the Fine Structure Constant. At present the most precise value for α comes from a combination of several measurements -- namely, the Rydberg,[8] the value of 2e/h determined using the AC Josephson effect,[9] the magnetic moment of the proton in Bohr magnetons[10],the velocity of light,[11] and the absolute measurement of the proton g factor.[12] Figure 3 summarizes the relevant constants and the calculation used to determine α. The weakest link in this determination of α is the absolute value for the gyromagnetic ratio of the proton. This measurement requires the construction of a precision solenoid whose field can be calculated from the measured current and the dimensions of solenoid. The U.S. Bureau of Standards now has underway a new program to construct an accurate solenoid and to make an absolute determination of the proton g factor to a few parts in 10^7.

On the other hand if one could make a sufficiently precise measurement of α, one could use the equation in Fig. 3 to determine γ_p and thus through the standard solenoid define the unit of current. This technique would essentially refer all magnetic fields to the standard magnetic field encountered in the n=2 state of hydrogen. This possibility gives one of the prime motivations for using the hydrogen fine structure to make a high precision measurement of α. Various other measurements such as the hydrogen fine structure,[8,13] the hydrogen hyperfine structure,[8] the muonium hyperfine structure,[14-17] the helium fine structure,[18,19] and the g-2 value for the electron[20-22]

Order	Description	Numerical Value (MHz)
$\alpha(Z\alpha)^4m$	2nd order self energy	1009.920
$\alpha(Z\alpha)^4m$	2nd order magnetic moment	67.720
$\alpha(Z\alpha)^4m$	2nd order vac. polarization	-27.084 ± 0.00557
$\alpha(Z\alpha)^5m$	2nd order binding	7.140
$\alpha(Z\alpha)^6m$	4th order binding and higher order	-0.372±0.00491
	Higher order $Z\alpha$ and α uncertainty	±0.00568
$\alpha^2(Z\alpha)^4m$	4th order self energy	0.444
$\alpha^2(Z\alpha)^4m$	4th order magnetic moment	-0.102
$\alpha^2(Z\alpha)^4m$	4th order vac. polarization	-0.239
	Red. mass. uncertainty	±0.00341
$\alpha(Z\alpha)^4\frac{Zm}{M}m$	Recoil corrections	0.359
$\alpha(Z\alpha)^4\left(\frac{mR_p}{e}\right)^2 m$	Proton size	0.125±0.00634
	Proton structure uncertainty	±0.00063
	Total	1057.911±0.012

Figure 2 -- Compilation of the contributions to the Lamb shift in the n=2 state of hydrogen. For these calculations α^{-1} was assumed to be 137.0608(26).

$$\alpha^{-1} = \left(\frac{1}{4R_\infty} \; \frac{1}{\gamma_p} \; \frac{c\Omega_{ABS}}{\Omega_{NBS}} \; \frac{\mu_p}{\mu_B} \; \frac{2e}{h} \right)^{1/2}$$

$R_\infty = 1.097\ 373\ 12(11) \times 10^5\ \text{cm}^{-1}$ (0.1 ppm)

$\gamma_p = 2.675\ 196\ \ 5(82) \times 10^4\ \text{rad sec}^{-1}\text{G}^{-1}$ (3.1 ppm)

$c = 2.997\ 924\ 562(11) \times 10^4\ \text{cm/sec}$ (0.003 ppm)

$\dfrac{\Omega_{ABS}}{\Omega_{NBS}} = 1 + (0.36 \pm 0.7)\ \text{ppm}$ (0.7 ppm)

$\dfrac{\mu_p}{\mu_B} = 1.521\ 032\ 64(46) \times 10^{-3}$ (0.3 ppm)

$\dfrac{2e}{h} = 4.835\ 937\ 18\ (60)\ \text{MHz}/\mu V_{NBS69}$ (0.12 ppm)

$\alpha^{-1} = 137.036\ 10(22)$ (1.6 ppm)

Figure 3 -- The equation used to determine α and the constants that enter this equation.

have been suggested as sources for high precision values of α. Figure 4 summarizes the values of α determined from these sources.

Source	α^{-1}
Hydrogen Fine Structure $2\,^2P_{1/2} \longrightarrow 2\,^2P_{3/2}$	137.035 4 (6) (4.4 ppm)
Hydrogen Hyperfine Structure $1\,^2S_{1/2},\, F=0 \longrightarrow F=1$	137.035 91(34) (2.5 ppm)
Muonium Hyperfine Structure $1\,^2S_{1/2},\, F=0 \longrightarrow F=1$	137.036 31(32) (2.4 ppm)
Helium Fine Structure $2\,^3P_1 \longrightarrow 2\,^3P_0$	137.035 95(41) (3.0 ppm)
Electron g-2	137.035 93(41) (3.0 ppm)

Figure 4 -- The values of the fine structure constant determined from other sources.

MEASUREMENTS OF HYDROGENIC FINE STRUCTURE

Four distinct techniques have been used to measure the fine structure of hydrogen like atoms. They are radiofrequency or level crossing spectroscopy on a thermal atomic beam,[23] radiofrequency or level crossing spectroscopy on excited atoms produced by electron bombardment in a bottle,[24] electric field quenching of a fast ion beam[25] and radiofrequency spectroscopy on a fast atomic beam[26]. For most of these measurements the line width is determined by the natural lifetime of the states involved. The natural line width divided by the frequency of the fine structure transition does not vary appreciably with n or Z. For the $^2S_{1/2} \to\, ^2P_{1/2}$ Lamb shift transition, it is 1 part in 10; for the $^2S_{1/2} \to\, ^2P_{3/2}$ fine structure transition it is 1 part in 100. Thus a measurement of α to a few parts per million requires the determination of the line center

Atom	n	Experimental Value (MHz)	Theoretical Value (MHz)	Ref
^1H	2	1057.90(6)	1057.911(12)	31
		1057.86(6)		32
^1H	3	314.819(48)	314.896(3)	33
^1H	4	133.18(59)	133.084(1)	34
		133.53(78)		24
^1D	2	1059.28(6)	1059.271(25)	35
$^4He^+$	2	14045.4(12)	14044.765(613)	36
		14045.2(18)		37
$^4He^+$	3	4183.17(54)	4184.42(18)	38
$^4He^+$	4	1776.0(75)	1769.088(76)	39
		1769.4(12)		40
$^4He^+$	5	905.2(10)	906.2	41
$^4He^+$	6	526.7(25)	524.2	41
$^6Li^{2+}$	2	63031.0(3270)	62762(9)	42
		62765.0(210)		
$^{12}C^{5+}$	2	780100.0(80000)	783680(250)	25

Figure 5 -- A summary of the reported measurements of the Lamb shift of hydrogen like atoms.

to a few parts in 10^5. Figure 5 summarizes the reported measurements of the Lamb shift of hydrogen like atoms.

Only the measurements on hydrogen and helium are sufficiently precise to give an interesting test of quantum electrodynamics. Figure 6 shows the comparison between theory and experiment for the Lamb shift in hydrogen. For the construction of this figure the theoretical value of the fine structure interval ΔE has been used to convert the measurements of $\Delta E - \mathcal{L}$ to an equivalent value of $\mathcal{L}$. The overall agreement between theory and experiment is not good.

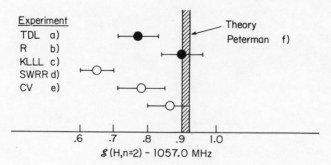

a) S. Triebwasser, E.S. Dayhoff, W.E. Lamb Jr., Phys. Rev. 89, 98 (1953)

b) R.T. Robiscoe, T.W. Shyn, Phys. Rev. Letters 24, 559 (1970); R.T. Robiscoe, Phys. Rev. 168, 4 (1968)

c) S.L. Kaufman, W.E. Lamb Jr., K.R. Lea, M. Leventhal, Phys. Rev. Letters 22 507 (1969) ΔE-S=9911.377(26)

d) T.W. Shyn, W.L. Williams, R.T. Robiscoe, T. Rebane, Phys. Rev. Letters 22, 1273 (1969) ΔE-S=9911.213(58)

e) B.L. Cosens and T.V. Vorburger, Phys. Rev. Letters 23 1273 (1969) ΔE-S=9911.173(42)

f) A. Peterman, Physics Letters 38B 330 (1972)

Figure 6 -- A comparison between theory and experiment for the Lamb shift in hydrogen.

One of the noteable new measurements of the Lamb shift in higher Z hydrogenic atoms is the measurement by Leventhal and Havey of the Lamb shift in the n=2 state of $^6Li^{2+}$.[27] Figure 7 shows a schematic drawing of their apparatus. They used a pulsed electron beam to produce ions in the n=2, $^2S_{1/2}$ metastable state of $^6Li^{2+}$, used an electric field to trap the metastable ions, and then observed the resonances as a function of magnetic field with a fixed radiofrequency field. The final result which was an average of 76 experimental runs gave $\mathscr{S}$ = 62765±21 MHz. This value is in good agreement with Erickson's recently calculated value of $\mathscr{S}$ =62762±9 MHz. The experiment is unfortunately not sufficiently precise to make a significant test of QED.

SEPARATED OSCILLATORY FIELD FINE STRUCTURE MEASUREMENT

In a paper presented to this conference, S. Lundeen and the present author[28] have reported preliminary results of a measurement of the Lamb shift in the n=2 state of hydrogen by a fast beam, radiofrequency method.

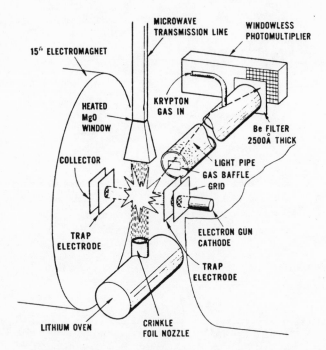

Figure 7 -- Schematic diagram of the apparatus used by Leventhal and Havey for the measurement of the Lamb shift in the n=2 state of $^6Li^{2+}$

Figure 8 shows a schematic diagram of the apparatus. The fast hydrogen atoms were produced from a 50 or 100 keV proton beam through charge capture collisions in a nitrogen filled charge exchange cell. The atoms then passed through a series of radiofrequency fields prior to reaching the observation region where there was a detector which monitored the Lyman α photons emitted by the atoms passing in front of the detector. The radio-frequency signals were observed by comparing the count rates when the rf fields in the interaction region were on and off. The resonances were swept over by changing the radiofrequency field. The first rf field and the rf field in the observation region were used to make the apparatus sensitive to only one hyperfine state.

Figure 9a shows in detail the $^2S_{1/2}$ and $^2P_{1/2}$ energy levels in the n=2 state of hydrogen when account is taken of the hyperfine structure. Figure 9b shows the expected line profile if there is no hyperfine state selection. In order to make the apparatus sensitive to

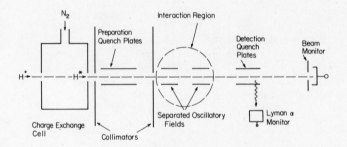

Figure 8 -- A schematic diagram of the apparatus used
for the fast beam measurement of the Lamb shift in the
n=2 state of hydrogen.

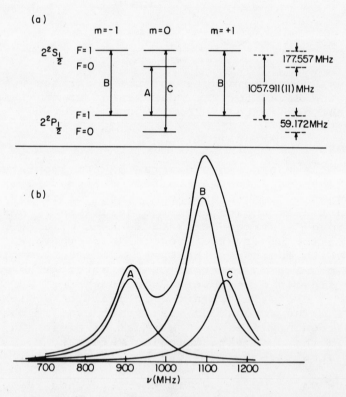

Figure 9 -- (a) The $^2S_{1/2}$ and $^2P_{1/2}$ levels in the n=2
state of hydrogen taking into account the hyperfine
interaction. (b) The expected line profile if there is
no hyperfine state selection.

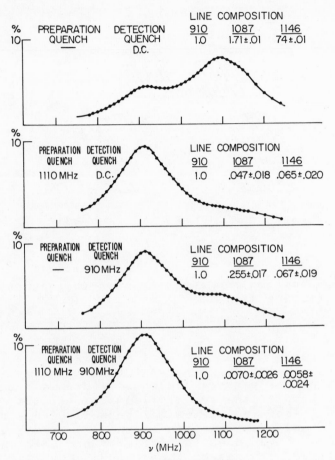

Figure 10 -- Resonance curves showing the effect of the hyperfine quenching fields prior to the interaction region and in the detection region.

only the $^2S_{1/2}$, F=0 $\to$ $^2P_{1/2}$, F=1 hyperfine transition, a continuous rf field at the frequency of the $^2S_{1/2}$, F=1 $\to$ $^2P_{1/2}$, F=0 transition is used prior to the spectroscopy field and a continuous rf field at the frequency of the $^2S_{1/2}$, F=0 $\to$ $^2P_{1/2}$, F=1 transition is used in the observation region. Figure 10 shows the effect of each of these fields applied singly and together on the observed rf transition.

By separating the two spectroscopy fields and taking as the signal the difference in the counting rates when

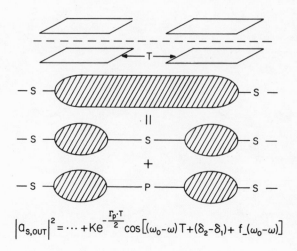

$$\left|a_{s,OUT}\right|^2 = \cdots + K e^{-\frac{\Gamma_p \cdot T}{2}} \cos\left[(\omega_0-\omega)T + (\delta_2-\delta_1) + f_-(\omega_0-\omega)\right]$$

Figure 11 - Schematic representation of the formation
of the interference signal.

the relative phase of the two rf fields is 0^0 and 180^0,
one can obtain a line which shows the characteristic
Ramsey interference pattern and whose width is determined
by the separation of the two rf fields rather than the
natural lifetime of the states. Figure 11 give a schematic
representation of the formation of the interference signal. One in essence uses the interference between the
S and P amplitudes to select those atoms that live longer
than the natural lifetime and reach the second field.
The signal decreases exponentially with the separation
of the two fields, the line width decreases linearly.
Since the signals are good due to the high intensity of
the proton beam, it is feasible to significantly decrease
the line width and still make a precision measurement.
Figure 12 shows the calculated signal for the 0^0 and
180^0 components and how each contributes to the interference signal. Figure 13 shows a typical observed line
profile. Figure 14 shows a comparison between the line
profile observed in this experiment and that observed
in earlier Lamb shift measurements. Figure 15 shows the
comparison between the earlier results, the preliminary
results from this experiment and the theoretical value
for the Lamb shift. It is clear that this technique
will make it possible to substantially increase the
precision of the measurements and to make a new test

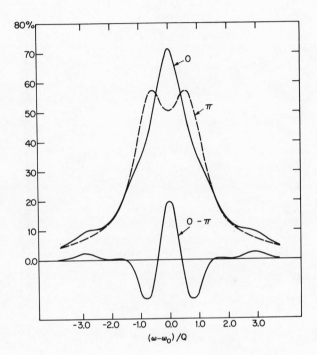

Figure 12 -- The calculated line shape for the relative phase of the rf fields at 0^0 and 180^0 and the interference signal.

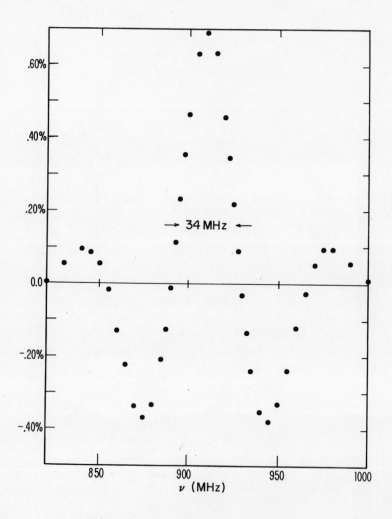

Figure 13 -- Typical line profile for the $^2S_{1/2}$, F=0 → $^2P_{1/2}$, F=1 fine structure transition.

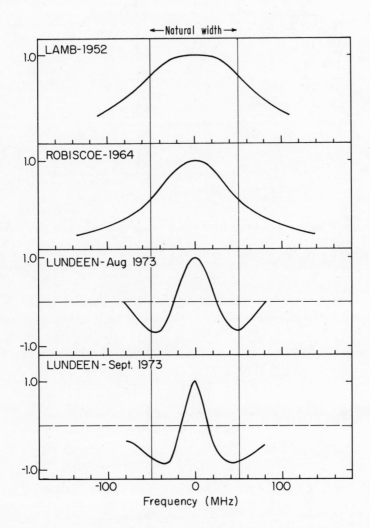

Figure 14 -- A comparison of the line profiles observed using the separated oscillatory field technique and those observed in earlier Lamb shift measurements.

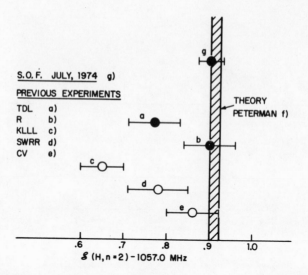

Figure 15 -- A comparison between the preliminary fast
beam measurement of the Lamb shift, the earlier thermal
beam measurements and the theoretical value.

of QED. This same technique can also be used on the
n=2, $^2S_{1/2} \rightarrow {}^2P_{3/2}$ transition and thus it should also
give to more precise value of the fine structure constant.

MULTIPLE QUANTUM FINE STRUCTURE SPECTROSCOPY

More recently a multiple quantum transition technique has been suggested as a means of making a more precise measurement of α.[29] Figure 16 shows the energy
levels in the n=3 state of hydrogen. A two quantum
transition can be used to drive atoms from the $^2S_{1/2}$
state to the $^2D_{5/2}$ state. Since the lifetimes of these
states are longer than that for the P state, the line
width will be smaller. For the n=3 state the fractional
line width for the $^2S_{1/2} \rightarrow {}^2D_{5/2}$ transition is a factor
of 4 smaller than that for the $^2S_{1/2} \rightarrow {}^2P_{3/2}$ transition.

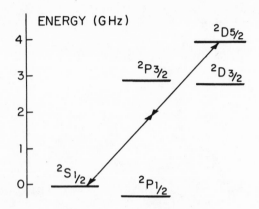

Figure 16 -- The energy levels in the n=3 state of hydrogen with the assumption there is no hyperfine
structure.

The apparatus described earlier for the fast beam measurements of $\mathcal{L}$, n=2 was also used to study this double
quantum transition. Figure 17 shows the observed spectrum when the rf quenching technique is used to select
only one hyperfine component. The agreement between
the measurements and the calculated shape is satisfactory
and suggests that one can use this technique to make a
precision measurement of the hydorgen fine structure.
It remains to be seen how serious are the shifts due to
the required large amplitude of the radiofrequency field.

The multiple quantum technique become particularily
attractive when it is applied to the higher excited
states of helium. As n increases the lifetime of the S
state becomes equal to that of the highest angular
momentum state and thus the fractional line width decreases. As an example for the n=6 state of He^+, the
natural line width for the multiple quantum transition
$6^2S_{1/2} \rightarrow 6^2H_{11/2}$ is 40 kHz and the frequency is 2165 MHz.
This gives a fractional line width of 2 in 10^5. Again
it remains to be seen how serious are the shifts due to
the large rf fields required.

The availability of tunable lasers with high
stability gives rise to still another method for the
precise measurement of the Lamb shift.[30] The multiple
quantum technique can be used to make transitions from

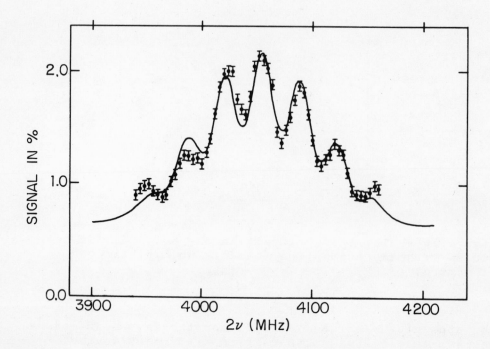

Figure 17 -- The line profile of the $^2S_{1/2} \rightarrow {}^2D_{5/2}$
double quantum resonance with selection of the F=0,
$^2S_{1/2}$ state. The solid curve is a computer fit using
two-level theory.

the ground state to the n=2, $^2S_{1/2}$ state or the n=3,
$^2S_{1/2}$ state. The natural width of these transitions
would be 0.4 Hz and 1 MHz respectively. The combination
of the two frequencies together with the theoretical
dependence of the Lamb shift on the n state could then
be used to determine with high precision both the Rydberg
and the Lamb shift in the ground state of hydrogen. This
is an exciting possibility. There are, however, many
technical problems to overcome in order to obtain lasers
of the required frequency and stability.

REFERENCES

*The preparation of this manuscript was supported in part by NSF Grant GP-22787.

1. For two relatively recent reviews of the status of QED experiment and theory see (a) V. W. Hughes, Atomic Physics 3, edited by S. J. Smith and G. K. Walters (Plenum, New York, 1973) p. 1. (b) B. E. Lautrup, A. Peterman and E. deRafael, Physics Reports 3, 193 (1972).

2. H. Grotch and D. R. Yennie, Rev. Mod. Phys. 41, 350 (1969).

3. W. A. Barker and F. N. Glover, Phys. Rev. 99, 317 (1955).

4. A. J. Layzer, Phys. Rev. Letters 4, 580 (1960).

5. H. M. Fried and D. R. Yennie, Phys. Rev. Letters 4, 53 (1960).

6. S. Brodsky and R. Parsons, Phys. Rev. 163, 134 (1967).

7. This table was taken from reference 1b.

8. B. N. Taylor, W. H. Parker, and D. N. Langenberg, Rev. Mod. Phys. 41, 375 (1969).

9. T. F. Finnegan, A. Denenstein and D. N. Langenberg, Phys. Rev. B4, 1487 (1971).

10. P. F. Winkler, D. Kleppner, T. Myint and F. G. Walther, Phys. Rev. A5, 83 (1972).

11. K. M. Evenson, J. S. Wells, F. R. Petersen, B. L. Danielson, G. W. Day, R. L. Barger and J. L. Hall, Phys. Rev. Lett. 29, 1346 (1972).

12. N. Langenberg and B. N. Taylor, Precision Measurement and Fundamental Constants, NBS Special Pub. 343, U. S. Govt. Printing Office, (1971).

13. J. C. Baird, J. Brandenberger, K. I. Gondaira and H. Metcalf, Phys. Rev. A5, 564 (1972).

14. R. D. Ehrlich, H. Hofer, A. Magnon, D. Y. Stowell R. A. Swanson and V. L. Telegdi, Phys. Rev. A5,

2357 (1972).

15. R. DeVoe, P. M. McIntyer, A. Magnon, D. Y. Stowell, R. A. Swanson, and V. L. Telegdi, Phys. Rev. Lett. 25, 1779 (1970)

16. D. Favart, P. M. McIntyre, D. Y. Stowell, V. L. Telegdi, and R. DeVoe, Phys. Rev. A8, 1195 (1973).

17. T. Crane, D. Casperson, P. Crane, P. Egan, V. W. Hughes, R. Stambaugh, P. A. Thompson and G. zu Putlitz, Phys. Rev. Lett. 27, 474 (1971).

18. F. M. J. Pichanick, R. D. Swift, C. E. Johnson, and V. W. Hughes, Phys. Rev. 169, 55 (1968).

19. A. Kponov, V. W. Hughes, C. E. Johnson, S. A. Lewis and F. M. J. Pichanick, Phys. Rev. Lett. 26, 1613 (1971).

20. J. C. Wesley and A. Rich, Phys. Rev. D4, 1341 (1971).

21. A. Rich and J. C. Wesley, Rev. Mod. Phys. 44, 250 (1972).

22. S. Granger and C. W. Ford, Phys. Rev. Lett. 28, 1479 (1972).

23. For a general treatment see R. T. Robiscoe, Lamb shift Experiments in Cargese Lectures in Physics edited by M. Levy (Gordon and Breach, New York, 1968) p. 3.

24. For a general treatment see R. A. Brown and F. M. Pipkin, Annals of Physics 80, 479 (1973).

25. See for example, H. W. Kugel, M. Leventhal and D. E. Murnick, Phys. Rev. A6, 1306 (1972).

26. C. W. Fabjan and F. M. Pipkin, Phys. Rev. A6, 556 (1972).

27. M. Leventhal and P. E. Havey, Phys. Rev. Lett. 32, 808 (1974).

28. S. Lundeen, B. Clark, and F. M. Pipkin, Abstract of 4th International Conference on Atomic Physics, p. 57, Heidelberg, 1974.

29.   P. B. Kramer, S. R. Lundeen, B. O. Clark and F. M. Pipkin, Phys. Rev. Lett. 32, 635 (1974).

30.   D. E. Roberts and E. N. Fortson, Phys. Rev. Lett. 31, 1539 (1973).

31.   R. Robiscoe and T. Shyn, Phys. Rev. Lett. 24, 559 (1970).

32.   S. Triebwasser, E. S. Dayhoff and W. E. Lamb Jr., Phys. Rev. 89, 98 (1953).

33.   C. W. Fabjan and F. M. Pipkin, Phys. Rev. A6, 556 (1972).

34.   C. W. Fabjan, F. M. Pipkin and M. Silverman, Phys. Rev. Lett 26, 347 (1971).

35.   B. Cosens, Phys. Rev. 173, 49 (1968).

36.   M. Narashimham, Thesis, University of Colorado (1969) unpublished.

37.   E. Lipworth and R. Novick, Phys. Rev. 108, 1434 (1957).

38.   O. Mader, M. Leventhal and W. E. Lamb Jr., Phys. Rev. A3, 1832 (1971).

39.   L. Hatfield and R. Hughes, Phys. Rev. 156, 102 (1967).

40.   R. Jacobs, K. Lea and W. E. Lamb Jr., Bull. Am. Phys. Soc. 14 , 525 (1969).

41.   A. Eibofner, Phys. Lett. 47A, 399 (1974).

42.   C. Fan, M. Garcia-Munoz and I. Sellin, Phys. Rev. 161, 6 (1967).

PRELIMINARY EXPERIMENTAL RESULTS ON THE ENERGY DIFFERENCE BETWEEN THE $2S_{1/2}$ AND $2P_{3/2}$ LEVELS OF THE $(\mu^4He)^+$ SYSTEM [*]

A. Bertin[**], G. Carboni[**], J. Duclos[***], U. Gastaldi[+],
G. Gorini[+], G. Neri[+], J. Picard[***], O. Pitzurra[+],
A. Placci[**], E. Polacco[+], G. Torelli[+], A. Vitale[++],
E. Zavattini[**], and K. Ziock[+++]

[**]CERN, Geneva, Switzerland

[***]Centre d' Etudes Nucléaires de Saclay, France

[+]Istituto di Fisica dell' Università, and INFN, Pisa,
Italy

[++]Istituto di Fisica dell' Università, and INFN, Bolo-
gna, Italy

[+++]University of Virginia, Charlottesville, Va., U.S.A.

1. INTRODUCTION

Since 1972 a pioneer experimental set-up was built,
and successively used at CERN,[1] in order to perform a precise measurement of the energy difference S_1 between the

---

[*]Presented by E. Zavattini.

$2S_{1/2}$ and $2P_{3/2}$ levels [2-4] of the helium muonic ion $(\mu^4He)^+$. We report here on the present status of the experiment, giving preliminary results on S_1.

The experiment has been performed sending electromagnetic radiation of proper wavelength on muonic helium metastable systems previously formed in a target of pure helium at 40 atm and 293 °K. The radiation was supplied by a high-power tunable dye-laser device.[1] The $2S_{1/2} \to 2P_{3/2}$ transition, when present, was identified by looking at the subsequent de-excitation 8.2 keV X-rays released by the $(\mu He)^+$ system in the $2P_{3/2} \to 1S_{1/2}$ transition (see Fig. 1), which proceeds at a rate of 2×10^{12} sec^{-1}.

The energy difference S_1 depends on the r.m.s. charge radius $(\langle r^2 \rangle)^{1/2}$ of the ^{4}He nucleus as is shown in Fig. 2 in wavelength units. If λ_1 is the wavelength corresponding to S_1, one can then set for convenience

$$\lambda_1 = (8000 - D_{np} + ((\langle r^2 \rangle)^{1/2} - 1.570) \times 2282.8) \overset{o}{A} \qquad (1)$$

where D_{np} is the nuclear polarizability contribution. Taking for D_{np} the value recently calculated by Bernabeu and Jarlskog, [3] and for $(\langle r^2 \rangle)^{1/2}$ the weighted average $\overline{(\langle r^2 \rangle)^{1/2}}$ of the presently available experimental results, [4-7] (see Fig. 3) one gets from Equation (1) – which stands on the validity of Q.E.D. – the expected theoretical value

$$\lambda_1^{th} = (8167 \pm 57) \overset{o}{A} \qquad (2)$$

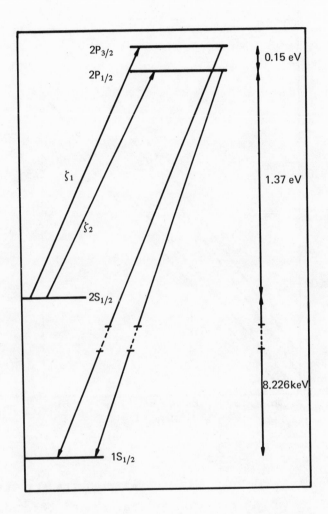

Fig. 1.— Scheme of the lowest levels of the $(\mu He)^+$ muonic ion.

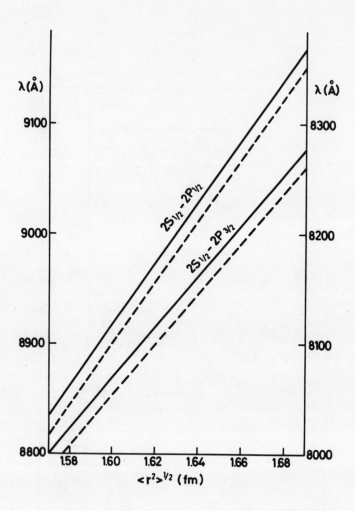

Fig. 2.- Theoretical values of the wavelengths for the $2S_{1/2} \rightarrow$ $2P_{3/2}$ (right scale of ordinates) and $2S_{1/2} \rightarrow 2P_{1/2}$ separation (left scale of ordinates) as a function of the r.m.s. radius of the ^{4}He nucleus. The solid (dashed) line corresponds to the prediction without (with) the nuclear polarizability contribution. (Figure taken from reference 3).

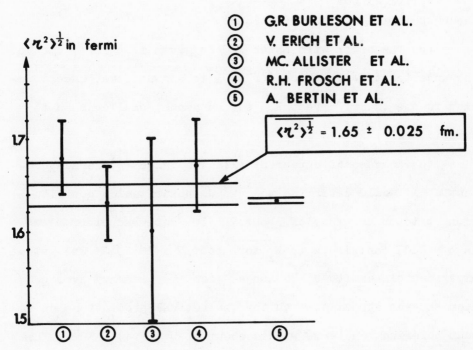

Fig. 3.- Experimental results on the r.m.s. radius of the ^{4}He nucleus as obtained from different authors (1-4) and from the present experiment assuming validity of Q.E.D. (5)

which is in the near-infrared region. The natural linewidth [2]
of λ_1 is $d\lambda_1 \cong 8 \overset{o}{A}$.

As far as regards the experimental determination of λ_1,
we wish to underline the following points:

(a) The value given by Equation (2) is mainly due to the
vacuum polarization contribution to S_1.

(b) The error with which λ_1^{th} is given is chiefly due to
the uncertainty with which $(\langle r^2 \rangle)^{1/2}$ is known; other uncertainties in the theory affect λ_1^{th} for an amount smaller than the
linewidth $d\lambda_1$. [2,3]

(c) In case of agreement with the value (2), a measurement of λ_1 would determine the vacuum polarization contribution up to an accuracy of about 7×10^{-3} (second order terms
in $\alpha = 1/137$ contribute to λ_1 for about 7×10^{-3}). [2] The latest
experimental results by Robiscoe [8] on the ordinary hydrogen
atom show an agreement with the theoretical predictions on vacuum polarization terms within about $(1.0 \pm 0.4) \times 10^{-2}$. Therefore, even with the considerable uncertainty pointed out in
(b), one sees that a measurement of λ_1 in the $(\mu He)^+$ system
would give a check on the vacuum polarization contribution with
an accuracy comparable to the one obtained from the experimental results on the ordinary (ep) system.

The present measurements were performed at the 600 MeV
CERN Synchrocyclotron (SC) between July 1972 and June 1973,
when the SC was temporarily closed to start an improvement

program.

Given the tight machine schedule, it was planned to search systematically the S$_1$ transition before the SC shutdown only within about one half the wavelength range allowed by the extreme values of Equation (2), starting from 8100 Å upwards.

In alternance with these measurements, some other runs were performed, using part of the experimental apparatus to determine - as a function of the helium pressure P - the fraction ε_{2S}(P) of muons stopped in helium which form the $(\mu He)^+_{2S}$ state, and the lifetime τ_{2S}(P) of such metastable state. The results of these measurements [9] indicate that at the pressure of 50 atm ε_{2S} is about 5 x 10^{-2}, whereas τ_{2S} is about 1.4 μsec.

Such second result shows that - even at a pressure as high as 50 atm - density-dependent effects (such as external Auger effect or Stark-mixing collisions) contribute in a small way to decrease the lifetime of the $(\mu He)^+_{2S}$ state with respect to the zero-density value (1.8 μsec, see Table I). This fact is essential for the feasibility of the present experiment, as will be understood later on.

2. EXPERIMENT

As was previously mentioned, the experiment consists in sending infrared radiation of a chosen wavelength on $(\mu He)^+_{2S}$

systems, and in looking for that wavelength for which 8.2 keV X-rays appear in coincidence with the laser pulse, due to the de-excitation of the $2P_{3/2}$ state attained through the infra-red radiation absorption.

A schematic view of the apparatus used is shown in Fig. 4. The negative muon beam was supplied by the muon channel of the CERN SC. A negative muon stopping within the target was defined by the coincidence-anticoincidence signal MUSTOP = $(1,3,4,-(2+5+\sum A_i))$ (where (-) means "not") between the prompt pulses coming from the various counters in Fig. 4.

The A_i counters were NaI(Tl) detectors, which were also requested to detect the 8.2 keV X-rays in coincidence with the laser pulse. The calibration of the energy response of the A_i detectors was obtained on line, by looking at the "prompt" K-lines X-rays emitted in coincidence with the MUSTOP signal (see Fig. 5), which for 62% are due to the K_α line (8.2 keV).[15]

Besides the different counters, the target contained also a 38 mm in diam., 140 mm long and 12 μm thick aluminum cylinder, which limited the useful volume V of helium for stopping muons. The cylinder was internally covered with 150 Å of gold in order to reflect as much as possible the infrared radiation within the volume V. The thickness of the cylinder was such to allow the 8.2 keV X-rays to reach the A_i detectors.

The rate of muons actually stopping within the volume V of helium gas was about 50 % of the total MUSTOP counts.

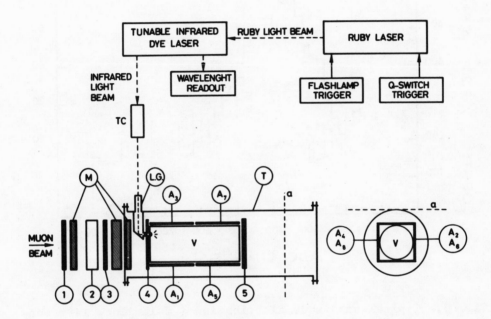

Fig. 4.- Very simplified view of the present experimental apparatus. M = CH_2 moderators; 1,3,4,5 = plastic scintillators; 2 = anticoincidence Cerenkov counter; T = invar steel tank; V = useful volume for stopping muons; A_1 - A_8 = NaI(Tl) counters; L.G. = Light guide used to inject the infrared radiation into the target; TC = optical telescope.

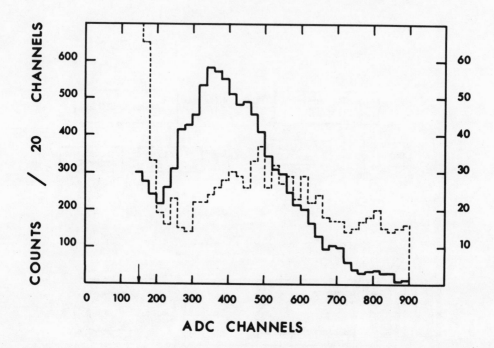

Fig. 5.- Typical amplitude distribution (full line, left scale of ordinates) supplied by the NaI(Tl) detectors for
the K-lines X-rays coming from the prompt cascade process of the $(\mu He)^+$ ion (typical energy: 8.2 keV). The
data were taken by the present apparatus working at
50 atm helium pressure. Dotted line (right scale of
ordinates) indicates the background distribution, ob
tained for the same amount of total incoming muons
when the target was kept under vacuum. The arrow on
the abscissae indicates the pedestal of the used ana-
log-to-digital converters (ADC).

The infrared radiation of variable wavelength was supplied by a dye laser excited by a Q-switched ruby laser.[10]
The chief characteristics of the laser devices are summarized
in Table II. The ruby laser was a standard one.[1,11] The dye
solution was chosen in order to get output radiation within
a 200 Å interval around 8100 Å.[12]   The solvent was chosen with
the aim of getting the maximum conversion efficiency (25-30%)
between the input energy (from the ruby laser) and the ouput
infrared radiation.

To tune the infrared radiation from the dye laser on a
given wavelength, we used a diffraction grating, following the
technique by Soffer and McFarland.[13]

The wavelength of the radiation selected was measured
by a digitalized spectrometer, having a linear dispersion of
4 Å/mm. The absolute scale of the device was calibrated by
standard spectroscopic lamps. In this way, the absolute value
of λ was known within $\pm$ 2 Å.

The Q-switch operation was performed by a Pockels cell,
which was triggered by a fast MUSTOP signal: in this way, the
infrared radiation was let into the target's useful volume V
at a time when a $(\mu He)^+_{2S}$ system might have been present.

It has to be noticed, however, that the time necessary
to pump up the ruby rod above the critical inversion was about
2 msec, which is much too long if compared with the lifetime
τ_{2S} of the $(\mu He)^+_{2S}$ system. For this reason, it was impossi-

ble to trigger the pumping flashlamp directly by the MUSTOP
signal.

On the other hand, for the ruby lasers used it was verified that the critical inversion lasted for a time interval
δ of about 1 msec. Therefore, a fairly long time interval
was available to obtain effective stimulated emission, which
could be exploited, within δ , by closing the Q-switch of the
ruby laser.

For this reason, the proton beam of the SC was bunched [14]
in a burst few msec long, with the same repetition rate as the
ruby laser; and the start signal to the flashlamp exciting
the ruby rod was supplied at a time τ_F in advance with respect
to the useful muon bunch entering the target. The muons stopping
within the target were accepted only within the gate $\Delta\tau$, which
was centred on the peak value of the particle bunch; and the
electronics was adjusted in such a way that δ was overlapping
with the first part of $\Delta\tau$. The stopping muons accepted after
δ were used for background studies (see Fig. 6).

If during δ a MUSTOP signal was coming, the latter closed the fast Q-switch (at a time τ_P in Fig. 6), starting the
effective lasing operation. The delay with which the infrared
light entered the target with respect to the MUSTOP signal was
of about 900 nanosec, i.e. sensitively smaller than τ_{2S}.

A small probe was inserted in the target cavity defining
the volume V, in order to monitor the amount of energy entering

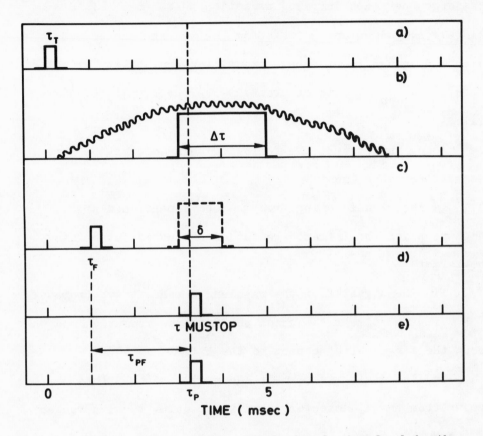

Fig. 6.– Timing scheme of the chief signals involved in the
synchronism operation between muon beam and laser
trigger. (a) time τ_T at which the flipping target
crosses the proton beam; (b) muon bunch time distribution observed thereafter; the $\Delta\tau$ gate, during which
the electronics are able to count muons, is also shown
in the figure; (c) flashlamp trigger time τ_F for the
ruby laser, and useful lasing interval δ ; (d) time
MUSTOP of a possible first muon, falling within δ and
$\Delta\tau$, accepted by the electronics; (e) trigger time
τ_P to the Q-switch of the ruby laser; (the typical duration of the laser radiation pulse is 20 nanosec).

the target per each infrared radiation pulse. Each of these
pulses conveyed an energy which on the average was of 0.3 J,
which was enough to produce a transition probability from the
$2S_{1/2}$ to the $2P_{3/2}$ state of about 20–30% per formed $(\mu He)^+_{2S}$
state.

Among the others, the following main information were
recorded for each event:

i) The signal coming from the above-mentioned probe,
which measured the infrared radiation energy injected into
the target.

ii) The signal from the wavelength readout spectrometer.

iii) The time t_L at which the infrared radiation had entered the target with respect to the MUSTOP signal.

iv) The digitalized amplitude of the signals (one or more)
coming from the A_i detectors, and their delay t_X with respect
to the MUSTOP time.

v) The delay t_e of the muon decay electron – when this
was present – with respect to the MUSTOP time.

In this way, the energy and wavelength of the laser radiation were continuously monitored during the experiment. Furthermore, the amplitudes from the A_i detectors, as well as the
time information t_L, t_X, and t_e were available for successive
off-line selections.

As a final remark, we should mention that the special
synchronization procedure illustrated in Fig. 6 – which was

essential to couple the laser pulse with the muon beam - com-
pelled us to work with an effective duty cycle which was about
100 times worse than the one obtained for ordinary machine o-
peration. Since tha NaI(Tl) detectors are highly background-
sensitive ones, such restriction represented the main draw-
back of the chosen experimental technique.

3. RESULTS

The data were taken in three running periods, starting
from July 1972 until June 1973. A wavelength range about 60 Å
wide was explored, starting from 8100 Å upwards. This region
was swept for three times, in such a way that it was eventually scanned in steps of 4 Å maximum size.

The results we are going to present were obtained by a-
nalyzing the data through the following steps:

i) Only the events for which the NaI(Tl) detectors showed one signal having an amplitude in the range 3 - 15 keV
were considered.

ii) Among these, the events for which the amount of infrared radiation energy injected into the target was at least
0.1 J were accepted.

iii) Only those events for which the delay t_X of the NaI(Tl)
signal with respect to the MUSTOP time was included in a proper
time interval were retained; such time interval was about 900
nanosec wide, centred on the average time $\overline{t_L}$ at which the in-

frared light entered the target (i.e. on the time at which the delayed "good" 8.2 keV X-rays were expected to appear).

iv) For each value of the wavelength $\bar{\lambda}$, the runs performed on a wavelength region 10 Å wide around $\bar{\lambda}$ were singled out; and the time distribution (A) of the corresponding events versus $\bar{t} = t_X - t_L$ was constructed. The same time distribution (B) for the events collected in all the remaining runs ($\lambda \neq \bar{\lambda} \pm 5$ Å) was subtracted from the spectrum A after proper normalization.

As a result of this analysis, we found that the subtracted spectrum (A – B) was in general identically zero, except for the value $\bar{\lambda} = 8120$ Å, for which an abundance of events appeared in proximity of the value $\bar{t} = 22$ channels; such value corresponds to the expected one when an X-rays is emitted in coincidence with the laser pulse.

Within the statistics available – which corresponds to about 2/3 of the total number of collected events – this was verified independently for all the three runs which were performed over a time period of about one year.

The time distribution (A – B) obtained for $\bar{\lambda} = 8120$ Å is shown in Fig. 7. The signal-to-background ratio in the peak region (Channels 20 – 30) was about 1:2.

A subset of the events shown in Fig. 7 – corresponding to the expected reduction due to the efficiency for detecting electrons – are also asssociated to a muon decay electron de-

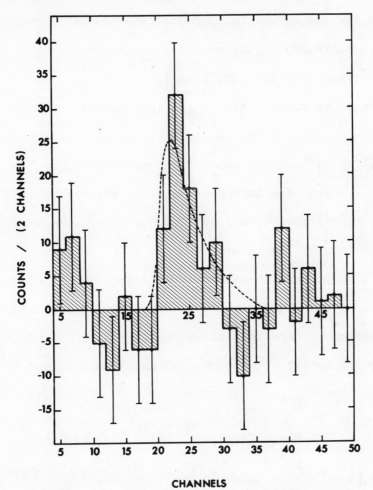

Fig. 7.— Background-subtracted time distribution (A-B spectrum, see text) of the delayed X-rays observed in the runs performed with the dye laser tuned within the region $\lambda = 8120 \pm 7$ A. Abscissae: 20 nanosec/ch. The dashed curve represents the observed time distribution – normalized to the area of the events between channels 20 and 30 – of the prompt 8.2 keV X-rays observed in coincidence with the MUSTOP signal, located where the $2P_{3/2} \rightarrow 1S_{1/2}$ de-excitation 8.2 keV X-rays are expected to appear.

tected by the apparatus. Furthermore, the events in the peak
are also associated, on the average, to higher values of the
infrared energy sent into the target.

The analysis of the data is now in progress. The number
of events expected for the set of analysed data at the resonance value of λ, on the base of the measured values for
ε_{2S}, τ_{2S}, and of the infrared radiation energy injected
per pulse in the target is about 70. The events observed in
the peak region of Fig. 7 are 78 ± 20.

We interpretate these events as due to the $2P \rightarrow 1S$ de-
excitation of the $(\mu He)^+_{2P}$ state obtained from the initial
$(\mu He)^+_{2S}$ system through the laser-induced $2S_{1/2} \rightarrow 2P_{3/2}$ transition. The particular value to which Fig. 7 refers is

$$\lambda_1^{exp} = (8120 \pm 7) \overset{o}{A} . \qquad (3)$$

From a comparison between the values (2) and (3), one gets

$$\lambda_1^{exp} - \lambda_1^{th} = - (47 \pm 60) \overset{o}{A}. \qquad (4)$$

The agreement between λ_1^{th} and λ_1^{exp} shown by Equation (4)
gives then a verification of the vacuum polarization contribu-
tion to the energy difference S_1 up to an accuracy of 7×10^{-3}.

On the other hand, if one assumes the exact validity of
the present Q.E.D. calculations,[2,3] the present result (3) can
be used to extract a value of the r.m.s. charge radius $(\langle r^2 \rangle)^{1/2}$
of the ^{4}He nucleus. If this is done, one gets

$$(\langle r^2 \rangle)^{1/2} = (1.630 \pm 0.003) \text{ fm} \qquad (5)$$

which is the value (5) shown in Fig. 3.

TABLE I

Disappearance channels for the $(\mu^4 He)^+_{2S}$ system and corresponding rates

Process	Rate symbol	Rate (sec^{-1})
Muon decay	λ_o	4.54×10^5
Muon capture	λ_c	45 ± 5 [a]
Stark-mixing collision [b]	$\lambda_{St}(P)$ [c]	$\leqq 3 \times 10^2 \times P$ [d]
External Auger effect	$\lambda_A(P)$ [c]	$(2.7 \pm 1.6) \times 10^3 \times P$ [e]
Spontaneous M_1 transition [b]	λ_{M_1}	0.53 [f]
Two-quantum decay	λ_{2X} [g]	1.06×10^5 [h]
Total disappearance rate: at zero density	$\lambda_{2S,tot}(0 \text{ atm})$	5.6×10^5
at pressure P	$\lambda_{2S,tot}(0 \text{ atm}) +$ $\lambda_{St}(P) + \lambda_A(P)$	

a) Experimental value (see ref. 18) corrected for the case that the nuclear capture proceeds from the 2S state of the μHe atom.

b) These processes yield the emission of an 8.2 keV X-ray.

c) P is the helium pressure in atm.

d) See ref. 15.

e) See ref. 9.

f) See ref. 16.

g) This process occurs through the emission of two X-rays; the sum of their energies is 8.2 keV.

h) See ref. 17.

TABLE II

Characteristics of the dye laser used in the present apparatus [a]

Dye	HITC [b]
Solvent	DMSO [c]
Molar concentration (moles/litre)	5×10^{-5}
Average ruby pumping energy (J)	1.2
Average pulsed infrared output energy (mJ)	300
Infrared power at 8150 A	15
Radiation pulse duration (nanosec)	20
Bandwidth of radiation (A)	6

a) Optimal values, referring to the use of the Apollo ruby Laser.

b) 1,3,3,1',3',3'- Hexamethyil -2,2'- indotricarbocyanine iodide.

c) Dimethylsulphoxide.

REFERENCES

1. A. Bertin, G. Carboni, A. Placci, E. Zavattini, U. Gastaldi, G. Gorini, G. Neri, O. Pitzurra, E. Polacco, G. Torelli, G. Stefanini, A. Vitale, J. Duclos, and J. Picard, to be published on Il Nuovo Cimento.

2. E. Campani, Lettere al Nuovo Cimento 4, 982 (1970). This author does not take into account the contribution due to the nuclear polarizability of the ^{4}He nucleus.

3. J. Bernabeu and C. Jarlskog, CERN TH 1796 (1974), Nucl. Physics, Vol B 75, 59 (1974).

4. V. Erich, H. Frank, D. Haas and H. Prange, Z. Phys. 209, 208 (1968).

5. R.W. Mc Allister and R. Hofstadter, Phys. Rev. 102, 851 (1956).

6. G.R. Burleson and H.W. Kendall, Nucl. Phys. 19, 68 (1960).

7. R.H. Frosch, J.S. McCarty, R.E. Rand, and M.R. Yearian, Phys. Rev. 160, 874 (1967).

8. R.T. Robiscoe, Phys. Rev. 168, 4 (1968).

9. A. Bertin, G. Carboni, A. Placci, E. Zavattini, J. Duclos, J. Picard, U. Gastaldi, G. Gorini, G. Neri, O. Pitzurra, E. Polacco, G. Torelli, and A. Vitale, paper submitted to the present Conference (presented by G. Carboni).
A. Bertin, G. Carboni, G. Gorini, O. Pitzurra, G. Torelli, A. Vitale, and E. Zavattini, to be published on Phys. Rev. Letters.

10. P.P. Sorokin, J.R. Lankard, E.C. Hammond, and V.L. Moruzzi, IBM J. of Res. and Dev. 11, 130 (1967).

11. For the first set of measurements, a modified version of a Bradley ruby laser (model 351) was used; lately, an A-

pollo laser (supplied by Apollo Lasers Inc., Los Angeles, California, U.S.A.) was purchased. The wavelength of the radiation supplied by the ruby lasers is 6943 A.

12. Y. Miyazoe and M. Maeda, Appl. Phys. Letters 12, 206 (1968).

13. B.H. Soffer and McFarland, Appl, Phys. Letters 10, 266 (1967).

14. J.J. Domingo, K. Gase, U. Gastaldi, E.G. Michaelis, G. Torelli and E. Zavattini, CERN/MSC/72/4 (1974), unpublished.

15. A. Placci, E. Polacco, E. Zavattini, K. Ziock, G. Carboni, U. Gastaldi, G. Gorini, G. Neri, and G. Torelli, Il Nuovo Cimento 1 A, 445 (1971).

16. R.W. Schmieder and R. Marrus, Phys. Rev. Letters 25, 1692 (1970).

17. J. Shapiro and G. Breit, Phys. Rev. 113, 179 (1959).

18. R. Bizzarri et al., Il Nuovo Cimento 33, 1497 (1964); M.M. Blok et al., Nuovo Cimento 55, 501 (1968).

PARTICULAR PROPERTIES OF HADRONIC (EXOTIC) ATOMS

G. Backenstoss

University of Basel, Switzerland
and
CERN, Geneva, Switzerland

INTRODUCTION

We first should like to make a few introducing remarks concerning the definition of exotic atoms, and the experimental techniques to produce and observe them and to measure their properties. We will then proceed to discuss particularly those properties of the exotic atoms which constitute the main differences between both the ordinary electronic and the exotic atoms.

We finally try to demonstrate that the investigation of the particular effects in exotic atoms provides us tools with the help of which we are able to study problems in physics which range from solid-state physics and chemistry on the one side to atomic physics, nuclear physics, and elementary particle physics on the other.

An "ordinary" atom consists of a positive nucleus which interacts electromagnetically with negative electrons in bound atomic states. Already in 1947 Wheeler[1] suggested that also other negative particles may be captured by the nucleus to form an "exotic" atom. The particle in question may be either a lepton, e.g. a muon, or a hadron, e.g. a π or K meson or baryons such as antiprotons or Σ hyperons. The properties of such exotic atoms have recently been reviewed by Burhop[2,3]; and in particular those of muonic atoms by Devons and Duerdoth[4] and Wu and Wilets[5], those of pionic atoms by Backenstoss[6] and those of strange hadronic atoms by Backenstoss and Zakrzewski[7] whose articles contain references to earlier work in this field.

GENERAL PICTURE

The atomic levels between which the observed X-ray transitions occur originate from the interaction of the nuclear Coulomb field with the negative particles and are basically the same as the levels of the ordinary electronic atoms. Hence, already the simple Bohr formulae lead to a good understanding of their properties:

$$E_n = - \frac{\mu_h c^2}{2} \left(\frac{Z\alpha}{n}\right)^2 \tag{1a}$$

$$r_n = \frac{\hbar^2}{\mu_h e^2} \frac{n^2}{Z} , \tag{1b}$$

where n is the principal quantum number, and r_n is the radius of the corresponding Bohr orbit; μ_h is the reduced mass of the hadron, Z the atomic number of the nucleus, and α the fine structure constant. From Eqs. (1a) and (1b) it follows that $E_n = E_{ne}(\mu_h/\mu_e)$, $r_n = r_{ne}(\mu_e/\mu_h)$, where quantities denoted by "e" refer to an electron in an ordinary atom. Since even for the lightest hadron of Table 1, $\mu_h/\mu_e \approx 273$, this has the important consequence that the energy levels in an exotic atom lie much deeper than in an ordinary one for the same value of n, while the Bohr radii are smaller in the same proportion. Thus the hadron is affected by the screening effects of atomic electrons only in states with very large n-values, whereas for states with smaller n-values these effects become negligible. An exotic atom can, therefore, be treated as a hydrogen-like atom of an atomic number Z. Very little is known about the initial state into which the hadron is captured. However, one might assume that capture takes place where the greatest overlap between electron and hadron wave functions occurs. Then with a hadron in place of a K-shell electron, an exotic atom would be formed with $n = (\mu_h/\mu_e)^{\frac{1}{2}}$. The values of this quantum number, together with those of the energy of the ground state and of the corresponding Bohr radius, obtained from the simple Bohr model, are given in Table 2 for different exotic atoms.

For the properties of hadronic atoms discussed later, the energy levels must be calculated more exactly, which is done by means of the Klein-Gordon or Dirac equations for a boson or a fermion, respectively. For a point charge nucleus, the solutions have the form

$$E_{n,j} = - \frac{\mu_h c^2}{2} \left(\frac{\alpha Z}{n}\right)^2 \left[1 + \left(\frac{\alpha Z}{n}\right)^2 \left(\frac{n}{j + \frac{1}{2}} - \frac{3}{4}\right) - \cdots\right] , \tag{2}$$

Table 1

Properties of Stable Negative Hadrons

Hadron	Baryon number B	Spin parity J^P	Isospin I	Strangeness S	Mass m_h (MeV)	Mean life τ (sec)
π^-	0	0^-	1	0	139.569	2.6×10^{-8}
K^-	0	0^-	$\frac{1}{2}$	-1	493.691	1.2×10^{-8}
$\bar{p}$	1	$\frac{1}{2}^+$	$\frac{1}{2}$	0	938.259	∞
Σ^-	1	$\frac{1}{2}^+$	1	-1	1197.34	1.5×10^{-10}
(Ξ^-)	1	$\frac{1}{2}^+$	$\frac{1}{2}$	-2	1321.29	1.7×10^{-10}

Table 2

Properties of Exotic Atoms in the Simple Bohr Model

Hadronic atom	Bohr energy*) $E_B = E_0(m_h/m_e)Z^2$ (keV)	Bohr radius*) $a_B = (m_e/m_h)(a_0/Z)$ (fm)	Principal quantum number of the atomic capture state
π^-	$3.7\,Z^2$	$193.5/Z$	17
K^-	$13.1\,Z^2$	$54.7/Z$	31
$\bar{p}$	$24.9\,Z^2$	$28.8/Z$	43
Σ^-	$31.8\,Z^2$	$22.6/Z$	48
(Ξ^-)	$35.2\,Z^2$	$20.5/Z$	57

*) Values are given for an infinitely heavy nucleus; for an ordinary hydrogen atom $E_0 = 13.6$ eV; $a_0 = 5.3 \times 10^4$ fm.

where $j = \ell$ for a spin-zero boson while $j = \ell \pm 1/2$ for a spin-half fermion, ℓ being the angular momentum quantum number. Thus, for π and K^- mesons, the energy levels of an exotic atom are singlets, whereas for an antiproton or a hyperon they are doublets.

Several electromagnetic corrections have to be applied to the solutions (2), arising from the finite size of the nuclear charge ΔE_{FS}; from radiative corrections, of which vacuum polarization ΔE_{VP} is dominating; and from electron screening ΔE_S. The energy value of a level in an exotic atom, with all electromagnetic effects taken into account, can finally be written as

$$E_{em} \approx E_{KG,D} + \Delta E_{FS} + \Delta E_{VP} + \Delta E_S , \qquad (3)$$

where in all cases the dominating term is $E_{KG,D}$, the solution (2) of the Klein-Gordon or Dirac equations, respectively. Actually the first two terms of the right-hand side of Eq. (3) and the first-order part of the third term are calculated simultaneously by solving numerically the wave equation with the corresponding potentials.

All information obtained on exotic atoms originates from the observation of X-ray transitions. Since the atom is formed in a highly excited state with $n \gtrsim (m_h/m_e)^{\frac{1}{2}}$, these X-rays are naturally present. De-excitation of all exotic atoms occurs via electromagnetic transitions between energy levels with the transfer of energy to an electron by the Auger effect or with its release in the form of X-rays. Calculations of these cascades have been performed by several authors[8,9]. The first process prevails in the earlier stages of the cascade while the second one dominates for states with smaller n, i.e. larger energies. The particular selection rules for these processes, $\Delta\ell = \pm 1$ with $\Delta\ell = -1$ favoured, Δn as small as possible for the Auger transition, and Δn as large as compatible with $\Delta\ell = -1$ for the X-ray emission, tend to drive the hadron towards states of maximal ℓ-values, $\ell = n-1$. Once a state (n, $\ell = n-1$), a "circular orbit", has been reached, the only electromagnetic transitions that can occur are of the type (n, n-1) $\rightarrow$ (n-1, n-2) $\rightarrow$ $\rightarrow$ (n-2, n-3), etc. For states with small n-values, radiative E1 transitions dominate, until the hadron reaches a state for which nuclear absorption becomes a competing process which rapidly takes over.

Experimental Techniques

The experimental technique to measure the X-rays is in principle simple. If negative particles come at rest in material they cannot avoid being captured by the nuclei provided their lifetime is sufficiently long. Since the particles in question are produced from accelerators with energies of typically hundreds of MeV and atomic

capture takes place nearly at rest, the particle lifetime should be
large not only compared with the time the atomic cascade process
lasts but also with the slowing-down time of the particles. Since
this already may be as large as $\sim 10^{-10}$ sec, one sees from Table 2
that, for example, Σ^- particles have lifetimes that are too short
for Σ^- atoms to be produced by stopping Σ particles of a hyperon
beam. An important part of the experiments on hadronic atoms is to
obtain and identify the hadrons stopping in a target in such a way
that the X-rays produced simultaneously can be measured with best
possible precision. Without entering into details, it should be
said that measurements with heavier hadrons are much more difficult
than for pions, since protons of 20 GeV hitting a target produce
three orders of magnitude more pions than kaons or antiprotons, and
particle separation as well as electronic selection between the
particles by means of Čerenkov counters and pulse-height discrimination of the counters of the beam telescope is necessary. In typical
experiments, about 300–600 K$^-$ mesons and about half as many antiprotons could be stopped per second in a target as compared with about
5×10^4 π^- mesons.

For Σ^- hyperons the situation is even more difficult. Because
of its short lifetime, a hyperon beam would be useless for producing
Σ hyperonic atoms, since all Σ^- would decay before being captured by
atoms. However, Σ^- hyperons are produced by strongly interacting
K mesons by the reaction

$$K^- + N \rightarrow \Sigma(\text{or } \Lambda) + \pi . \tag{4}$$

About (8–10)% Σ^- are produced per stopped K$^-$ with an energy of about
20 MeV. These Σ^- are rapidly stopped in the neighbourhood of their
origin and thus form Σ hyperonic atoms. Therefore Σ hyperonic X-ray
spectra are observed simultaneously with the kaonic spectra as demonstrated in Fig. 1.

The identification of the exotic X-ray transitions is mainly
done on the basis of their energies as calculated according to
Eqs. (2) and (3). The measurement of the energies is performed by
calibrating with radioactive γ-ray standards or known pionic X-ray
transitions. Pionic X-rays also visible in Fig. 1 are present, not
as a consequence of an incomplete rejection of the telescope trigger
but because of the pions produced by reaction (4). If one aims at
high-precision energy measurements, precautions must be taken particularly when using radioactive γ-rays which arrive randomly and thus
differently than particles from a pulsed accelerator[10]). Furthermore, the identification of the X-ray transition is facilitated by
the typical pattern provided by the X-ray series as known from ordinary X-ray spectroscopy. Hence the identification can always be
based on a particular sequence of lines.

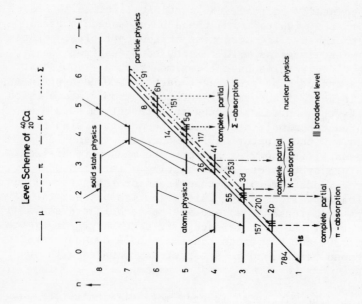

Fig. 2 Scheme of the atomic cascade for different particles and the related fields in physics

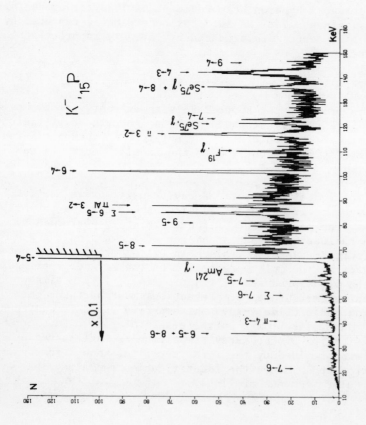

Fig. 1 K-mesonic spectrum in $_{15}$P from 0-200 keV. The transitions shown are K-mesonic transitions, except where indicated (π, Σ). The broadening of the K 4-3 transition is clearly seen [from Backenstoss et al., Phys. Letters 38 B, 181 (1972)].

RESULTS

The results obtainable from hadronic atoms are manyfold. Since
the atom consists essentially of two components, namely the nucleus
and the orbiting hadron -- the electrons being far outside can be
neglected in most circumstances -- one might expect to gain knowledge
about both the nuclei and the hadrons which as we will show extends
beyond the previously established facts. Furthermore, we may hope
to learn details about the interaction between hadrons and nuclei.
Besides the electromagnetic interaction which is of course responsible for the bare existence of the atoms, hadrons experience the
strong interaction which is, however, of a very short range compared
with the long-range $\sim 1/r$ Coulomb field or even the field of higher
multipoles. This is important since it enables us to select atomic
states where the well-known electromagnetic interaction prevails and
those where additionally the strong interaction sets in. This occurs
only when the hadronic wave functions begin to overlap the nucleus
which takes place with increasing particle mass and increasing Z at
increasing principal quantum numbers n, as can easily be concluded
from Table 2. In fact the strong interaction sets in so rapidly that
its influence can be observed at most for two levels of one atom
where it acts as a perturbation to the electromagnetic interaction.
Then it is so overwhelming that the concept of the atom is no longer
valid, and the particle quickly undergoes nuclear absorption. This
is demonstrated in Fig. 2. As the hadron cascades down from the
highly excited atomic states towards the ground state, the energies
vary from eV to MeV. Accordingly, effects of the molecular binding
and solid-state phyiscs have been observed[11] which must be caused in
the highly excited levels. Further down, the atomic cascade can be
studied; and also in this region, free from the not too well-known
strong interacting fields, particle properties have been explored.
It should be pointed out that very strong electric fields of up to
10^{18} V/cm are present. The last two levels before the particle is
absorbed by the nucleus, yield the information about the strong
interaction typically for hadronic atoms.

Atomic Capture Process

Although the atomic capture process has been discussed as early
as 1947[12] before experimental evidence for the existence of muonic
atoms was proved, not too many details about it are known. It is
generally more assumed than proved that capture takes place at a
certain n state with a population of the ℓ substates in the simplest
approximation proportional to the statistical weight $(2\ell + 1)$. There
are simple arguments that capture occurs when the particle approaches
the orbit of the K-shell electrons having the highest electron density by ejecting one K-electron and being then captured in a state
$n_h = \sqrt{m_h/m_e}$. This picture is certainly much too rough, particularly

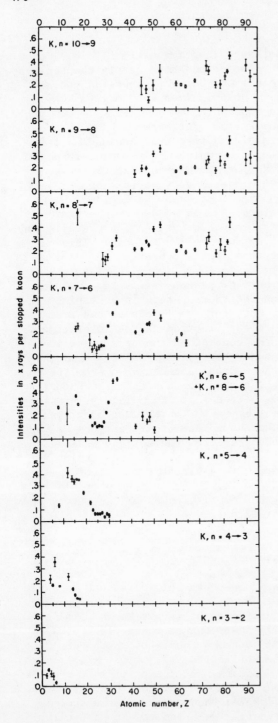

Fig. 3   Intensities of different K X-ray transitions per stopping K⁻ mesons (Ref. 18)

since one has observed transition, for example of muons from states
$n > n_h$ but it is nevertheless convenient since only for $n < n_h$
the problem is hydrogen-like.   There are also calculations through
the electron cloud and attempts to calculate the initial distribution of the captured particles with classical or semiclassical methods[13-17].   The experimental test of these models is complicated
by the fact that at present it can be made only indirectly.   Assuming an initial distribution over the ℓ-states at a fixed n-value,
the cascade is calculated with the known transition probabilities
for electric dipole radiative- and Auger-transitions.   Since at the
higher states of the cascade Auger transitions dominate which cannot
be observed, the comparison between prediction and experiment must
take place at the lower states where radiative transitions occur.
It is convenient to parametrize the distribution over ℓ by

$$P_\ell \sim (2\ell + 1)\, e^{\alpha \ell} \qquad\qquad\qquad (4)$$

for a fixed n where α is a free parameter.   It has been shown[16]
that this distribution is only weakly varying with n in the upper
part of the cascade, and hence the experimentally observable X-ray
intensities do not depend strongly on the value chosen for n.
During the cascade there is a tendency to drive the particles towards orbits of the circular type ($\ell = n - 1$).   It is obvious that
levels with $\ell < n - 1$ will be more frequently populated if the initial distribution will be less peaked at large ℓ-values ($\alpha < 0$).
Hence relative intensity measurements between transitions of the
circular type ($n, \ell = n - 1 \rightarrow n - 1, \ell = n - 2$) and those from non-
circular levels are sensitive tests where non-circular levels will
favourably be depopulated by transitions with Δn as large as compatible with $\Delta \ell = -1$.

Since the atomic capture takes place far outside the range of
the strong interaction and hence one would expect that the same information can be obtained from muonic atoms there are nevertheless
a few points where hadrons may contribute additional information.
Because of their strong nuclear absorption they will be lost during
their cascade towards the ground state as soon as they reach any
s-level.   The total number of X-rays observed in the lower transitions will therefore depend on the probability for the hadrons to
avoid s-levels, which again depends on the initial ℓ-distribution.
Recently a great number of absolute kaonic X-ray intensities per
stopping K^- meson have been published by Wiegand and Godfrey[18]
which are shown in Fig. 3.   The most striking feature is the strong
variation of the X-ray intensities by as much as a factor of 5 with
the atomic number Z which has nothing to do with the strong absorption.   Under these circumstances an agreement claimed for a smoothed
average[17] seems to be not too meaningful.   It is interesting to

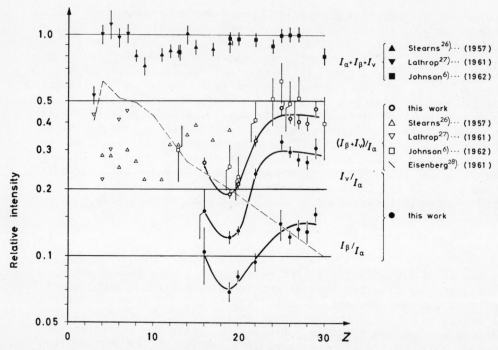

Fig. 4   Intensity ratios I_β/I_α and I_γ/I_α of the K_α, K_β and higher
K_ν lines in muonic atoms versus atomic number Z (Ref. 20)

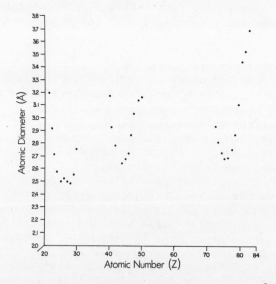

Fig. 5   Internuclear distances (atomic diameter in Å) versus atomic
number Z (Ref. 21)

note that similar fluctuations had been seen also for pions[19] and
even earlier for muons[20] always occurring at the same Z-values.
The measurements on muons performed with NaI detectors and based on
relative intensity measurements between $\Delta n = 1$ and $\Delta n = 2$ (K_α and K_β)
transitions is shown in Fig. 4. This long-unresolved phenomenon may
have found an explanation in a recent preprint by Condo[21] who compares the curves of Fig. 3 with the intranuclear distances in the
metallic lattices showing exactly the same behaviour (Fig. 5). At
close distances to the next nucleus the high ℓ-values may be truncated as classically obvious, and hence a shift towards smaller ℓ-
values may be caused which might explain the intensity drop in hadronic atoms as well as the relative increase of the K_β radiation.

Relative intensity measurements on kaonic atoms for transitions
with $\Delta n = 1$ until $\Delta n = 5$ have also been performed recently[22]. Here
the cascade parameter α has been fitted to the measured intensities
of up to 11 lines for the elements C (graphite) P (red) and S.
The refilling time for electron holes has also been varied as a
parameter, the depletion of electrons affecting of course the Auger
transition probabilities. For C, $\alpha = -0.011 \pm 0.01$ and fast refilling (i.e. electrons always sufficiently present) yielded the best
fit. For the isolators P and S, α was found to be equal to $+0.10 \pm$
± 0.01 and $+0.14 \pm 0.01$, respectively, and slow refilling in both
cases. These differences are certainly significant. Neither of
these results confirm the predictions of Ref. 17 of an enhancement
at $\ell \approx n/2$ leading to $\alpha < 0$ and are in line with earlier measurement
on muonic and pionic atoms where the X-ray intensities are different
whether they are observed in different chemical compounds or in different physical conditions[23].

In light atoms, particularly H and He, the question of Stark
mixing had been discussed since a long time[24]. The probability
that s-states will be reached is increased, and consequently the
X-ray intensities observed would decrease. Recently, measurements
on muonic and pionic liquid He have been performed[25]. Without being able to repeat here in detail the arguments given, the results
may be summarized. If one assumes a statistical ℓ-distribution,
one needs a moderate amount of Stark mixing (sliding transitions),
otherwise 10 times more pions could decay during the cascade than
those measured. The muonic spectra cannot, however, be explained
in this way, but better with a bell-shaped ℓ-distribution[13] and no
sliding transitions. An external Auger rate being smaller than the
internal one improves the fit. The same assumptions for the pionic
cascade could also explain the pion decays and the total X-ray yield
but not the intensity of the 3p-1s line which was measured to be 25%
larger than that of the 2p-1s line. In any case the Stark mixing
must be moderate otherwise the total yield could not have been as
observed.

174

G. BACKENSTOSS

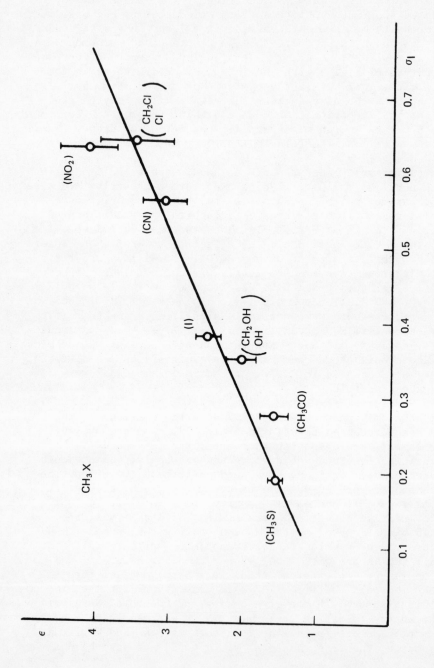

Fig. 6   Measurement of a correlation between the π capture rate in H and the reactivity of the radical CH₃ with various substituents;   see text (Ref. 23).

If more than a single type of atom is involved, the fractional
capture rates may be of interest. The frequent failure of the
"Fermi-Teller law", which was derived only for alloys and predicts
capture rates in the ratio of their atomic number Z, led to other
approaches. Here the mesomolecular concept, where mesons follow
the orbits of the molecular valence electrons before being captured
in atomic orbits, has been promoted by Russian authors [reviewed by
Ponomarev[23)]. Only one application may be mentioned here because
it is restricted to pions. If pions are absorbed by protons, the
only two processes are the two defined by the Panofsky ratio:

$$\pi^- + p \rightarrow n + \pi^0$$
$$ \hookrightarrow 2\gamma \qquad\qquad (5)$$

$$\pi^- + p \rightarrow n + \gamma \;,$$

resulting in γ-rays with energies above 70 MeV. If pions are captured by nuclei, even as light as deuterium, they are predominantly
absorbed by two nucleons, resulting in emission of nucleons without
production of high-energy γ-radiation. Hence the detection of those
γ-rays may serve as signature for π absorption on H, the X-rays of
which would otherwise be undetectable.

It has been shown by this method that in a hydrogen compound
$(Z_m H_n)$, the probability for pion capture in H is

$$W(Z_m H_n) \approx a \, n \, Z^{-3}/m \;,$$

where a is a coefficient of the order of unity. The Z^{-3} dependence
could be satisfactorily explained by the mesomolecular concept. The
exact value of a, however, depends on the chemical bound. In Fig. 6
measurement on various compounds of the radical CH_3 are shown. The
quantity $\varepsilon = a/a_0$ is plotted versus the quantity σ_I which characterizes the reactivity of the radical CH_3 under the effect of various
substituents[23)].

A last indication how the capture process manifests itself in
the transition intensities is now described. In a high-pressure
vessel filled with 1000 atm of gaseous H, $10^{-3} - 10^{-5}$ A gas was admix-
ed. The muons are captured by the H and almost quantitatively transferred to the A atoms[26)]. Whereas in pure A, $I(3p - 1s)/I(2p - 1s) <$
< 0.03, the same ratio reaches 0.19 ± 0.04 for 2×10^{-3} A content.
Probably lower ℓ-states are reached by transferring the muon from H.

Particle Properties

As already mentioned before, particle properties, as far as
they are not related to the strong interaction, may be best investigated outside the range of the strong interaction. The most obvious

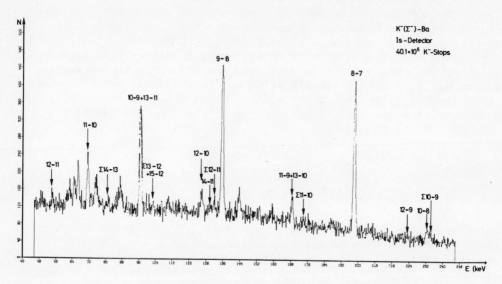

Fig. 7 K(Σ) X-ray spectrum of barium (Ref. 28)

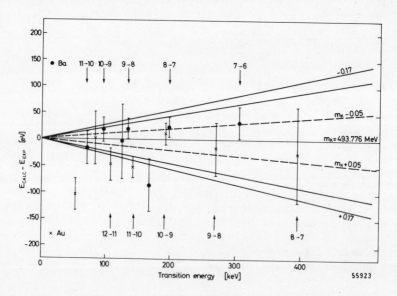

Fig. 8 Determination of the K⁻ mass from X-ray transitions in Ba and Au (Ref. 28)

quantities to be thus measured in exotic atoms are the masses and magnetic moments of the hadrons. But also the presence of an intrinsic structure of the hadrons, such as their polarizabilities, should affect the atomic energy levels.

Indeed the best values of the masses of the negative pion and kaon obtained so far, originate from measurements of the corresponding atomic levels. The same is true for the antiproton if one does not assume its mass to be equal to that of the proton. The purely electromagnetic energy levels can be calculated by solving the Dirac equation for fermions and by the Klein-Gordon equation for bosons with a finite-size Coulomb potential and a potential describing the vacuum polarization in first order, including corrections for the higher-order terms and the electron screening, with a precision considerably higher than the experimental errors of the level energies. Hence the particle masses and their errors are essentially determined from precise absolute energy measurements of suitable transition energies. Thus the pion mass has been determined from measurements on the 5g-4f transition in πI and the 6h-5g transition in πTl to[27] (139.563 ± 0.012) MeV and (139,572 ± 0.009) MeV, respectively, yielding

$$m_{\pi} = (139.569 \pm 0.008) \text{ MeV} .$$

Since the electron screening which causes corrections of -14 eV and -75 eV on πI and πTl, respectively, depends on the probabilities of finding the electron levels (in particular the K shell) occupied, a detailed knowledge of the pionic cascade is required. It turns out that for the 5g level in πI where the screening effect is small, the chance for a K-shell vacancy is large, whereas for the 6h level in Tl the probability for a full K shell is large. Nevertheless, it should be pointed out that the agreement of the two above-mentioned measurements implies that the screening problem could be dealt with sufficiently well with respect to the other errors.

The K⁻ mass has been similarly determined from the kaonic spectra of Ba and Au. In Fig. 7 five main transitions of Ba between 12-11 and 8-7 can be seen. In Fig. 8 the deviation of the measured transitions of Ba and Au from the calculated values assuming a certain K⁻ mass are plotted as function of the transition energy. Thus we obtain the energy and error of the K⁻ mass[28]:

$$m_K = (493.691 \pm 0.040) \text{ MeV} .$$

Although no particular effort has been made, the mass of the antiproton could be already measured at the first observation of antiprotonic X-rays[29] to agree within about 5×10^{-4} with that of the proton.

The magnetic moments of the fermions should be measurable from the fine structure splitting of the levels. Here one has to investigate the heavy elements, the fine structure splitting being proportional to Z^4. From the 11-10 transitions in Pb and U the size of the magnetic moment of the antiproton was derived[30]:

$$\mu_{\bar{p}} = (-2.83 \pm 0.10)\mu_N$$

in agreement with CPT predictions.

The magnetic moment of the Σ^- hyperon is predicted by SU(3). However, symmetry-breaking may be appreciable, so that an experimental determination is very desirable. As mentioned above, these measurements are very difficult owing to a lack of statistics. Nevertheless, measurements on the 12-11 transitions of Σ hyperonic Pb and U have been performed which indicate a broadening of the transition[31] owing to the fine structure splitting from which the authors claim to have derived a magnetic moment

$$\mu_{\Sigma^-} = (-1.48 \pm 0.37)\mu_N \ ,$$

as compared to SU(3) predictions of $-0.9 \ \mu_N$, a deviation which one should not take too seriously in view of the great experimental difficulties.

If a hadron is polarizable, the polarization of it in a strong electric field may lead to a shift of the energy levels[32]. The shift of the level (n,ℓ) is given by[33,34]

$$(\Delta E)_{n,\ell} = -\frac{e^2}{2}(\alpha_N + Z^2\alpha_h) \langle r^{-4} \rangle_{n,\ell} \ ,$$

where α_N is the nuclear and α_h the hadron polarizability. The nucleus and the hadron polarize each other with their fields, where the hadron gains from the fact that it moves in the field with charge Z. To measure those effects, one should look for levels for which $\langle r^{-4} \rangle$ is large, just before the strong interaction sets in. In fact the 7-6 transition in K^- Ba (Fig. 7) and the 8-7 transitions in K^- Au had not been used for the above-mentioned K-mass determination but were considered as a test for the polarization shift. We do not find a significant deviation of the measured energies from the ones calculated without the kaon polarizability shift and derive an upper limit of the polarizability of the K^- meson[28] of $\alpha_K < 0.020$ fm^3.

Interactions

The interactions relevant to exotic atoms, the electromagnetic interactions and additionally the strong interaction in hadronic atoms, may be studied in exotic atoms. Though the electromagnetic

interaction is known and understood to an astonishing level of accuracy, there may still be occasions where tests of the electromagnetic interactions of the orbiting particle in the strong Coulomb
field of the nucleus may touch the limits of our present knowledge.
As an example, the measurements of the vacuum polarization in muonic
atoms is mentioned (this was dealt with in detail by Dr. Engfer at
this Conference) and a test on a possible violation of parity in
purely electromagnetic interactions. The strong interaction between
the different hadrons and the nucleus or the nucleons in the nuclei
is much less well known. Particularly the data at zero energy are
scarce, and thus data from hadronic atoms may complement the scattering data at higher energies.

Parity violating effects. In the light of increasing evidence
for the existence of neutral currents in the weak interaction experienced by high-energy neutrinos where only hadrons are produced[35],
or of the purely leptonic process[36]

$$\bar{\nu}_\mu + e^- \rightarrow \bar{\nu}_\mu + e^- \, ,$$

the possibility of unifying electromagnetic and weak interactions
is being extensively discussed. In this connection the limit at
which parity is conserved in purely electromagnetic interactions
is of great importance. It has recently been pointed out[37] that
with the help of intense polarized photon beams the excitation of
certain states in electronic atoms could serve as a measurement
for parity-violating effects.

Compared with electronic atoms, muonic atoms possess some advantageous properties. They are hydrogen-like systems where levels
of the same n and different parity $(-1)^\ell$ are nearly degenerate and
hence very sensitive to parity mixing. Furthermore, the matrix
element that is decisive for the magnitude of the parity mixing,

$$\eta = \langle \psi_f | V^{vp} | \psi_i \rangle / (E_i - E_f) \, ,$$

where V^{vp} is the parity violating potential, is much larger for
muonic states if one assumes that V^{vp} has essentially nuclear extension. Recently, suggestions have been made to use parity mixing
of the $2s_{\frac{1}{2}}$-$2p_{\frac{1}{2}}$ states in muonic atoms[38] where the M1 (E1) mixing
should induce a one-photon transition 2s-1s. Also the use of the
same mechanism in very light muonic atoms has been discussed[39],
where a particularly high anisotropy effect could be extracted. At
least in the light nuclei it would be difficult to exclude an ordinary transition to the nearly degenerate level by Stark effect or
collisions. Therefore, an interference term must be observed in the
form of a circular polarization or an asymmetry of the X-ray emission
with respect to the muon spin measured by the asymmetry of the decay
electrons. There is also a strong background due to the two-photon

transition of the 2s-1s transition, estimated to be about 10^5 times
more frequent than the one-photon transition[39] for light nuclei.
It is clear that the proposed measurements are very difficult. Therefore, one might consider another possibility described in a contribution to this conference[40]. There E2 (E1) mixing is discussed. E2
transitions have been observed in muonic atoms[41]. Here the transitions are fast and hence the danger of disturbing effects is reduced.
A sizeable fraction of the longitudinal muon polarization is still
conserved in the starting level of the transition in question. Therefore it is only necessary to measure the X-ray anisotropy with respect
to the beam direction.

   Strong Interactions. All hadrons experience strong interactions
with the nucleons and hence with the atomic nuclei. Although the
strength of the interaction is of the same order, there are considerable differences between the different hadrons with respect to the
strength and type of the interaction. These interactions are known
to a different degree from high-energy scattering experiments. At
low energies they are known in the best cases from phase-shift analysis or from more or less accurate extrapolations. The hadrons in
the bound atomic states have almost zero energy, and hence measurements of the strong interaction effects on hadronic atoms may complement the scattering measurements. Since, however, the hadrons interact in the first place with the nucleus, one important question must
be raised: Do nucleons in nuclei behave like nucleons? This question is of course intimately connected with our understanding of nuclei and nuclear matter, and hence any experiment which sheds light
on it is of great importance.

   There are essentially three observable effects due to the strong
interaction which sets in when the hadron is so far de-excited that
it comes into the reach of the strong forces. Already one or two
transitions further down the strong interaction overwhelms the electromagnetic interaction so strongly that the entire concept which
leads to atomic states must be revised. Therefore, strong interaction effects are only observable where they contribute a small perturbation to the Coulomb field. The influence of strong interactions
shows up in a shift of energy levels ΔE_N, a broadening of the levels
Γ_a, and a decrease of the observed X-ray yield Y. The last two are
due to nuclear absorption from the lower and from the upper level of
an X-ray transition, respectively.

   The strong interaction energy shift ΔE_N is determined experimentally as

$$\varepsilon = \Delta E_N = E_{meas} - E_{em} , \qquad\qquad (6)$$

the difference between the values of the energy measured E_{meas}, and
that calculated with the inclusion of all electromagnetic corrections
E_{em} of Eq. (3).

The level width due to strong interaction Γ_a is related to the rate of the nuclear absorption W_a:

$$\Gamma_a = \hbar \, W_a \, . \tag{7}$$

Since, as mentioned above, the strong interaction sets in very rapidly, Γ_a of the lower level is 2-3 orders of magnitude larger than that of the upper level of a transition $(n, \ell) \to (n-1, \ell-1)$. Hence the Lorentzian width determined from the observed line shape of the X-ray transition after unfolding it from the instrumental one, assumed to be Gaussian, measures Γ_a of the lower level $(n-1, \ell-1)$.

Shifts and widths of energy levels. Shifts and widths of energy levels have been observed for pions [6], kaons, and antiprotons. Some typical values are presented in Table 3. The recently obtained high precision in pionic atoms[25] was achieved by the use of muonic X-ray lines as the standard for the energy and the line shape produced by the spectrometer. Muonic lines in light nuclei can be calculated to a precision of a fraction of an electronvolt. Since the muonic lines are produced exactly under the same experimental conditions, then by using this method one avoids the systematic errors that are unevitable if one uses radioactive γ-ray standards (see Fig. 9).

The first measurements of the kaonic level shift ε and line width Γ, from direct observation of the X-ray line, have been obtained at CERN for the 4f-3d transitions in sulphur[42]. The broadening of the line beyond the instrumental widths due to the Lorentzian widths can easily be seen for the kaonic 4-3 transition in Fig. 1 Also in antiprotonic atoms a few absorption widths Γ could be measured[43], whereas the shifts ε are only for $\bar{p}0$ slightly significantly different from zero[44].

No such measurements have yet been obtained for Σ atoms. The reason is not only the fact that the intensities of Σ lines are more than 10 times weaker than for kaonic lines, but that owing to the heavier mass the strong interaction effects set in at larger (n, ℓ) values. Hence the increase of the effects from (n, ℓ) to $(n-1, \ell-1)$ is smaller than for kaons, which means that, for a given X-ray yield, the shifts and widths become smaller with increasing particle mass.

Intensity attenuation. The width of the upper level of a transition, in general over 2-3 orders of magnitude smaller than that of the lower level, can be determined by measuring the yield Y of an X-ray transition defined as

$$Y = \frac{W_x \, P}{W_x + W_A + W_a} \, , \tag{8}$$

Table 3

Experimental and Calculated Energy Shifts and Level Widths in Hadronic Atoms

Transition	Energy (keV)	ε_{low} (eV)		Γ_{low} (eV)		Γ_{upp} (eV)	
		exp	calc	exp	calc	exp	calc
π ^{4}He 2-1	10.6987 ± 0.002	− 75 ± 2	−91	45.2 ± 3.0	52.7	–	–
π ^{12}C 2-1	93.19 ± 0.12	−5950 ± 120	−5960	3250 ± 150	2880	1.02 ± 0.29	0.98
π S 3-2	133.06 ± 0.10	+516 ± 100	+460	790 ± 150	360	0.25 ± 0.15	0.10
K C 3-2	62.73 ± 0.08	−590 ± 80	−597	1730 ± 150	1745	0.98 ± 0.19	0.75
K S 4-3	161.56 ± 0.06	−550 ± 60	−457	2330 ± 200	2170	3.25 ± 0.41	3.79
	$\bar{A} = (+0.56 + i\ 0.69)$ fm						
$\bar{p}$ O 4-3	73.465 ± 0.064[a]	−97 ± 64	−137	648 ± 150	640	0.66 ± 0.15[a]	0.73
$\bar{p}$ Cl 5-4	158.95 ± 0.27					$8.0^{+2.2}_{-1.9}$	7.35
	$\bar{A} = (+2.9 + i\ 1.5)$ fm						
Σ C 4-3	50.65 ± 0.04					0.031 ± 0.012	
Σ Ca 6-5	151.7 ± 0.3					0.40 ± 0.22	

a) Preliminary data[44]

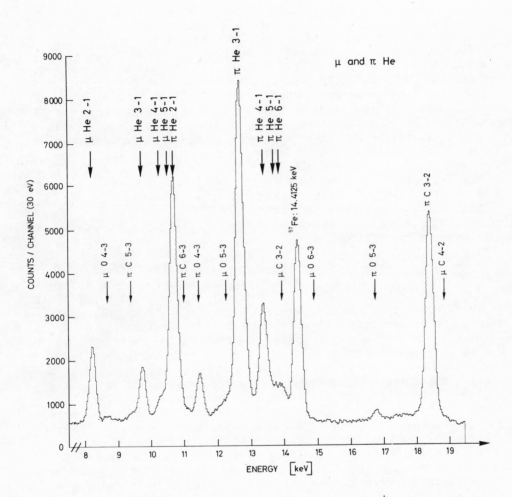

Fig. 9 Muonic and pionic X-ray transitions in ^{4}He (Ref. 25)

where W_X, W_A, and W_a are the X-ray, Auger, and nuclear absorption rates from the upper level, and P is the population of this level. If Y is measured and P, W_X, and W_A are known, one can obtain from this relation the nuclear absorption rate W_a, and hence the width of the upper level:

$$\Gamma_a = \Gamma_X(P/Y - 1) - \Gamma_A . \tag{9}$$

Whereas the termination of the X-ray series is the most striking feature of hadronic atoms as compared with leptonic atoms, the exact determination of absorption widths from intensity measurements remains the most difficult experimental part. Either the population P in Eq. (8) has to be known, for which a detailed knowledge of the atomic cascade is required, or P must be determined by measuring the transitions feeding this level. The advantage of the latter method [Eq. (9)] where the quantity P/Y involves only the measurement of relative intensities, is described in a recent paper[22]. Some of the data thus obtained and similar results for Σ^- atoms are included in Table 3. For kaons it means that three strong interaction effects are available for one nucleus, which provides constraints for the possible theoretical models. For Σ atoms, on the other hand, these measurements constitute the first and only observations so far of strong interaction effects.

Comparison with theory. The strong interaction effects discussed above can be calculated provided the hadron-nucleus interaction is described by an optical potential. The complex potential is introduced simultaneously with the extended Coulomb potential and the vacuum polarization potential in the corresponding wave equation. The situation for pions and the heavier hadrons, e.g. kaons, is rather different. The interaction of the pions is considerably weaker than that of the other hadrons. This may be the reason why the multiple scattering approach devised by Ericson and Ericson[45] to derive the potential parameters from the elementary π-nucleon interaction works remarkably well[6]. On the other hand, the interaction is rather complicated because there exists, besides the s-wave, a strong p-wave interaction related to the $(3/2, 3/2)$ resonance in the πN system, and secondly, pions are for kinematical reasons absorbed predominantly on nucleon pairs. There is not enough time to discuss here all the interesting implications. Except for the recent precise results on very light nuclei[25,46], we have a moderately good picture of the problems[6,47], as can be seen also from the few examples in Table 3.

For kaons the situation seems to be simpler in so far as here only s-wave interaction plays a role and K^- absorption can take place on one nucleon by

$$K^- + N \rightarrow \Sigma(\Lambda) + \pi . \tag{10}$$

But the same approach which was successful for pions, taking into
account the above-mentioned differences, was a failure for kaons.
The optical potential not explicitly quoted for pions because of
its complexity, reads for kaons

$$V_{KN}(r) = \frac{2\pi}{\mu}\left(1 + \frac{m_K}{m_N}\right)\left[A_1 \; \rho_n(r) + \frac{1}{2}(A_0 + A_1)\rho_p(r)\right] \; ,$$

where ρ_n and ρ_p are the nuclear neutron and proton distributions,
and A_i are the KN scattering lengths with isospin i. An experimental
determination[22]) of the complex constant $\bar{A} = \frac{1}{4}(A_0 + 3A_1)$ leads to
the values $\bar{A} = (0.56 + i\; 0.69)$ fm, which means an attractive potential,
whereas from the scattering length one would obtain a repulsive potential with $\bar{A} = (-0.42 + i\; 0.7)$ fm. It is not possible to enter
into details here, but it is clear that in this situation investigations with other hadrons, e.g. the antiproton, are of great importance. Also, here it seems that an attractive potential will fit the
data, as can be seen from Table 3.

From Σ hyperonic atoms only a few upper level widths are known
so far[7,48]) (see Table 3). But here a comparison between kaonic and
Σ hyperonic X-ray intensities tells us something about the interaction of Σ^- inside the nucleus. It is known that by process (10)
about 8% Σ^- per stopped K^- meson are produced in the nuclei of the
photographic emulsion. Hence the $I(\Sigma)/I(K)$ intensity measured and
shown in Fig. 10 enables us to draw conclusions on the probability
that Σ^- escape the nucleus[49]).

OUTLOOK

It was mentioned that questions of strong interactions of hadrons with nuclei and nucleons are intimately connected with problems of the nuclear structure. It is obvious that detailed information on the nuclear structure can only be obtained if the problems
that are still open and not understood will be solved. Nevertheless
I would like to finish by indicating a few points where these studies
may provide new insights. Before the existence of kaonic X-ray spectra it had been suggested[50]) that kaonic atoms may serve as a tool
for studying the nuclear surface. This comes from the fact that the
orbits where nuclear capture takes place increase with the mass of
the particle. Since orbits with $\ell = n-1$ dominate, whose wave function $\psi_\ell(r) \approx r^\ell$, the overlap of ψ_ℓ with the nuclear distribution
will be more and more peaked at the nuclear surface. In Fig. 11 the
situation for K^--S is shown, where the absorption from the (n = 3,
$\ell = 2$) level has its maximum at a nuclear density close to 10% of
its central value. For Σ^--S the absorption would take place around
the 1% density. Therefore investigations with different hadrons

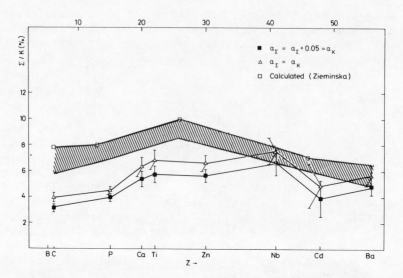

Fig. 10   Ratio of the intensities of the X-ray transitions -- averaged
          over the total cascade -- of Σ atoms and K atoms. The experi-
          mental points are derived under two different assumptions con-
          cerning the cascade parameter α (Ref. 7).

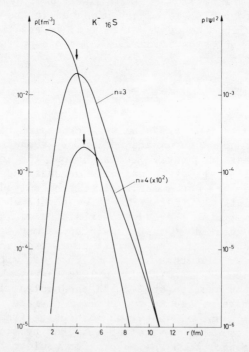

Fig. 11   Nuclear density ρ and overlap $|\psi^2|\rho$ for K^-–S versus
          nuclear radius r

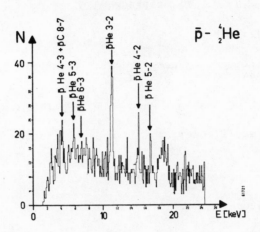

Fig. 12 Antiprotonic spectrum of ^{4}He (Ref. 44)

which also interact differently with protons and neutrons should
give information on proton and neutron distributions at low densities.

The question whether Δ resonances and Y^* resonances observed in
elementary interactions between π^-n and K^-n (n = nucleon), respectively, exist in nuclei is connected to the relation between the interactions of hadrons with free nucleons and those within nuclei.

The pion being absorbed on a correlated pair of nucleons may
be used as a tool to investigate short-range correlations in nuclei.
Here a new aspect of nuclei comes into the picture, the high momenta
behaviour of nuclei, which is not inherent in the independent particle picture as is the shell model.

Finally it would be extremely interesting to learn more about
the basic interaction of hadrons with nucleons at very low energies.
In this respect measurements on the very light nuclei may be promising, if not on H itself. In Fig. 12 we show the $\bar{p}$He spectrum
which could be clearly observed with the 3-2 line near 11 keV, the
4-2 and 5-2 lines also present, and even the 4-3 visible at 4 keV.
This spectrum could serve as a guide to the measurement of the $p\bar{p}$
system with the 2p-1s transition at 9 keV. The observation of such
a line could answer model-independently the question whether the $\bar{p}$p
interaction at low energies is repulsive or attractive; a question
of great importance for astrophysics. However, there may be insur-
mountable difficulties connected with Stark mixing and p-wave absorption which may hide this transition from the eye of the impa-
tient physicist.

REFERENCES

1) J.A. Wheeler, Phys. Rev. $\underline{71}$, 320 (1947).

2) E.H.S. Burhop, High-Energy Physics (Academic Press, Inc.,
    NY, 1969), Vol. 3, p. 109.

3) E.H.S. Burhop, Contemp. Phys. $\underline{11}$, 335 (1970).

4) S. Devons and I. Duerdoth, Advances in Nuclear Phys. $\underline{2}$, 295
    (1968).

5) C.S. Wu and L. Wilets, Annu. Rev. Nuclear Sci. $\underline{19}$, 527 (1969).

6) G. Backenstoss, Annu. Rev. Nuclear Sci. $\underline{20}$, 467 (1970).

7) G. Backenstoss and J. Zakrzewski, Contemporary Physics (1974),
    in press.

8) Y. Eisenberg and D. Kessler, Phys. Rev. $\underline{130}$, 2352 (1963).

9) J.R. Rook, Nuclear Phys. $\underline{B20}$, 14 (1970).

10) G. Backenstoss, H. Daniel, H. Koch, U. Lynen, Ch. von der Malsburg,
    G. Poelz, H.P. Povel, H. Schmitt, K. Springer and L. Tauscher,
    Phys. Letters $\underline{36}$ B, 403 (1971).

11) D. Kessler, H.L. Anderson, M.S. Dixit, H.J. Evans, R.M. McKee,
    C.K. Hargrove, R.D. Barton, E.P. Hincks and J.M. McAndrew,
    Phys. Rev. Letters $\underline{18}$, 1179 (1967).
    H. Daniel, H. Koch, G. Poelz, H. Schmitt, L. Tauscher,
    G. Backenstoss and S. Charalambus, Phys. Letters $\underline{26}$ B, 281
    (1968).
    L. Tauscher, G. Backenstoss, S. Charalambus, H. Daniel, H. Koch,
    G. Poelz and H. Schmitt, Phys. Letters $\underline{27}$ A, 581 (1968).

12) E. Fermi and E. Teller, Phys. Rev. $\underline{72}$, 399 (1947).

13) R.A. Mann and M.E. Rose, Phys. Rev. $\underline{121}$, 293 (1961).

14) A.D. Martin, Nuovo Cimento $\underline{27}$, 1359 (1963).

15) M.Y. Au-Yang and M.L. Cohen, Phys. Rev. $\underline{174}$, 468 (1963).

16) J.R. Rook, Nuclear Phys. $\underline{20}$ B, 14 (1970).

17) M. Leon and R. Seki, Phys. Rev. Letters $\underline{32}$, 132 (1974).

18) C.E. Wiegand and G.L. Godfrey, Phys. Rev. $\underline{9}$ A, 2282 (1974).

19) A.R. Kunselman, Report UCRL-18654 (1969).

20) D. Quitmann, R. Engfer, U. Hegel, P. Brix, G. Backenstoss,
     K. Goebel and B. Stadler, Nuclear Phys. 51, 609 (1964).

21) G.T. Condo, Intensity variations of K-mesonic X rays, submitted
     to Phys. Rev. Letters.

22) G. Backenstoss, J. Egger, H. Koch, H.P. Povel, A. Schwitter and
     L. Tauscher, Nuclear Phys. 73 B, 189 (1974).

23) L.I. Ponomarev, Annu. Rev. Nuclear Sci. 23, 395 (1973).

24) M. Leon and H.A. Bethe, Phys. Rev. 127, 636 (1962).

25) G. Backenstoss, J. Egger, T. von Egidy, R. Hagelberg,
     C.J. Herrlander, H. Koch, H.P. Povel, A. Schwitter and
     L. Tauscher, to be published in Nuclear Phys. A.

26) G. Backenstoss, H. Daniel, K. Jentzsch, H. Koch, H.P. Povel,
     F. Schmeissner, K. Springer and R.L. Stearns, Phys. Letters
     36 B, 422 (1971).

27) G. Backenstoss, H. Daniel, H. Koch, Ch. von der Malsburg,
     G. Poelz, H.P. Povel, H. Schmitt and L. Tauscher, Phys.
     Letters 43 B, 539 (1973).

28) G. Backenstoss, A. Bamberger, I. Bergström, T. Bunaciu,
     J. Egger, R. Hagelberg, S. Hultberg, H. Koch, U. Lynen,
     H.G. Ritter, A. Schwitter and L. Tauscher, Phys. Letters,
     43 B, 431 (1973).

29) A. Bamberger, U. Lynen, H. Piekarz, J. Piekarz, B. Povh,
     H.G. Ritter; G. Backenstoss, T. Bunaciu, J. Egger,
     W.D. Hamilton and H. Koch, Phys. Letters 33 B, 233 (1970).

30) J.D. Fox, P.D. Barnes, R.A. Eisenstein, W.C. Lam, J. Miller,
     R.B. Sutton, D.A. Jenkins, R.J. Powers, M. Eckhause,
     J.R. Kane, B.L. Roberts, M.E. Vislay, R.E. Welsh and
     A.R. Kunselman, Phys. Rev. Letters 29, 193 (1972).

31) J.D. Fox, W.C. Lam, P.D. Barnes, R.A. Eisenstein, J. Miller,
     R.B. Sutton, D.A. Jenkins, M. Eckhause, J.R. Kane,
     B.L. Roberts, R.E. Welsh and A.R. Kunselman, Phys. Rev.
     Letters 31, 1084 (1973); and 32, 1265 (1974).

32) F. Iachello and A. Landé, Phys. Letters 35 B, 205 (1971).

33) T.E.O. Ericson and J. Hüfner, Nuclear Phys. 47 B, 205 (1972).

34) U.E. Schröder, Acta Phys. Austriaca $\underline{26}$, 248 (1972).

35) CERN Gargamelle Collaboration, Phys. Letters $\underline{46}$ B, 138 (1973).

36) CERN Gargamelle Collaboration, Phys. Letters $\underline{46}$ B, 121 (1973).

37) M.A. Bouchiat and C.C. Bouchiat, Phys. Letters $\underline{48}$ B, 111 (1973).

38) G. Feinberg and M.Y. Chen, Columbia University New York Report
    COO-2232B-43.

39) J. Bernabeu, T.E.O. Ericson and C. Jarlskog, CERN Report
    TH-1853 (1974).

40) L.M. Simons, Contribution to this Conference.

41) P. Martin, G.H. Miller, R.E. Welsh, D.A. Jenkins and
    R.J. Powers, Phys. Rev. Letters $\underline{25}$, 1406 (1970).

42) G. Backenstoss, A. Bamberger, J. Egger, W.A. Hamilton,
    H. Koch, U. Lynen, H.G. Ritter and H. Schmitt, Phys.
    Letters $\underline{32}$ B, 399 (1970).

43) P.D. Barnes, S. Dytman, R.A. Eisenstein, W.C. Lam, J. Miller,
    R.B. Sutton, D.A. Jenkins, R.J. Powers, M. Eckhause,
    J.R. Kane, B.L. Roberts, R.E. Welsh, A.R. Kunselman,
    R.P. Redwine and R.E. Segal, Phys. Rev. Letters $\underline{29}$, 113
    (1972).

44) CERN Group, to be published.

45) M. Ericson and T.E.O. Ericson, Ann. Phys. (USA) $\underline{36}$, 323 (1966).

46) G. Backenstoss, I. Bergström, J. Egger, R. Hagelberg,
    C.J. Herrlander, H. Koch, H.P. Povel, R.H. Price,
    A. Schwitter and L. Tauscher, Nuclear Phys. $\underline{B66}$, 125 (1973).

47) M. Krell and T.E.O. Ericson, Nuclear Phys. $\underline{B11}$, 521 (1969).

48) CERN Group, to be published; T. Bunaciu, Thesis, Karlsruhe, 1973.

49) D. Zieminska, Phys. Letters $\underline{37}$ B, 403 (1971).

50) D.H. Wilkinson, Phil. Mag. $\underline{4}$, 215 (1959).

THE VACUUM POLARIZATION CORRECTION IN MUONIC ATOMS

R. Engfer and J.L. Vuilleumier
Laboratorium für Hochenergie der ETH-Zürich
CH-5234 Villigen
E. Borie
SIN, CH-5234 Villigen

1. INTRODUCTION

The quantum electrodynamic theory - a simple, elegant, theory involving only the lepton masses and the coupling constant - has been tested in a large variety of experiments at low and at high energies [BroD 70, Pic 71, AllB 72, Lau P 72]. There are several reasons why experimentalists are searching for a possible breakdown of the present QED: It is difficult to accept that divergent but unphysical expressions appear in the theory. Is the theory still valid at small distances? Since the perturbation theory expands in powers of αZ, discrepancies might occur at high Z.

Nearly all experimental tests are using a system with $Z=1$ and only few data up to $Z=5$ are available [LevM 70]. In addition, the vacuum polarization correction or charge renormalization which can be explained intuitively by Dirac's hole-theory amounts to only a few percent of the total radiative corrections (e.g. 2.7% for the $2s_{1/2} - 2p_{1/2}$ energy in the hydrogen atom) and, therefore, cannot be tested seperately.

In muonic atoms, however, the vacuum polarization turns out to be the major contribution to the radiative corrections. This is due to the large mass ratio of 206 between the muon and the electron. Since the vacuum polarization is an attractive correction, whereas the

vertex correction is repulsive, the total correction in
a muonic atom is, unlike the normal electronic atom,
attractive. The $2s_{1/2}$ - level which is shifted by + 1058
MHz (4.375 µeV) in the electronic Lamb shift is metastable in muonic hydrogen and shifted by - 205 meV
below the $2p_{1/2}$ level due to the vacuum polarization and
by only + 0.66 meV due to the vertex correction. This
property allows for a test of the vacuum polarization
without large competing effects in muonic atoms. Measurements for the $2s_{1/2}$ - $2p_{1/2}$ levels in µ He are reported by the CERN-group [PerC 74]. For a check of the
$(\alpha Z)^n$ expansion in heavy muonic atoms the $2s_{1/2}$ - $2p_{1/2}$
levels cannot be used because of the strongly increasing
finite size effect.

Several experiments have been performed measuring
higher transitions in muonic atoms which are not influenced by finite size effects. In 1970 a CERN group
[Bac C 70] found for the 5g - 4f muonic transition in
lead a 2% agreement of the vacuum polarization correction with Fricke's calculation [Fri 69] up to $\alpha(Z\alpha)^3$
order. A new measurement of the Chicago-Ottawa group
in 1971 [Dix A 71] showed a discrepancy for lead of
130 ± 20 eV in contradiction to the CERN experiment and
to theory. In 1972 Blomquist and others [Blo 72, Sun W
72, Bel 73] found an error of sign and magnitude in the
$\alpha(Z\alpha)^3$ term of ref. [Fri 69] based on Wichmann and
Kroll's calculation [Wic K 56] and a double counting of
a $\alpha^2(Z\alpha)$ term reducing the experimental discrepancy for
lead of ref. [Dix A 71] to about 50 eV. Because of the
discrepancy of 122 eV of the two experiments
an independent measurement with new calibration technique has been performed at CERN; the preliminary analysis [Wal V 72] is in excellent agreement with the
Chicago-Ottawa results. We report on the final analysis
of these measurements [Vui 74] and the present status
of the vacuum polarization correction combining all
experiments and known theoretical calculations.

2. Experiments

2.1 Calibration procedure

The experimental technique for the Chicago-Ottawa experiment is given in ref. [Dix A 71, Mck H 71]. In our CERN experiment [Wal V 72, Vui 74] an additional test for the calibration scheme has been performed besides the conventional technique.

The fact that the configuration of the apparatus differs if muonic X-rays or calibration events are recorded could lead to an electronic shift between the corresponding spectra. To estimate this effect we have measured ^{203}Tl and ^{103}Rh, since for these elements the μ-capture reactions

$$^{203}\text{Tl}(\mu,\ n\nu)^{202}\text{Hg and } ^{103}\text{Rh}(\mu,\ n\nu)^{102}\text{Ru}$$

produce delayed γ rays of 439.5 and 474.8 keV of the excited isotopes ^{202}Hg and ^{102}Ru respectively, which appear in the muonic X-ray spectra. Simultaneously, the same γ lines can be measured in the calibration spectra from the radioactive sources ^{202}Tl and ^{102}Rh, showing no shift within the statistical error of 17eV. Another effect, the directional dependence of energy measurements in Ge(Li) spectrometers [Lic M 74] limits even for planar detectors the absolute accuracy. Extensive studies of this effect are consistent with an error of 16 eV.

2.2 Electron screening

The various contributions to the experimental errors are listed in tables 1-2. Besides the QED corrections the effect of electron screening amounts to one of the largest energy shifts which can be almost independently checked by other muonic transitions in the region of 150-500 keV. The transition 9k - 7i e.g. is shifted by -506 keV due to electron screening and by +225 keV due to QED corrections (see table 1). Such transitions are, therefore, more a check of the electron screening calculation than of the QED corrections.

The experimental energies and transitions of several transitions in muonic rhodium and thallium are given in table 1 and 2. The difference of the calculated and

experimental values for muonic mercury is shown in fig. 1.
The electron screening has been calculated with a self-
consistent Hartree-Fock program of Vogel [Vog 73] assuming
completely filled electron shells for an atom with charge
Z and a muon in its corresponding state. The numerical
accuracy has been estimated to be ± 1 %; however, it
is uncertain how many electrons contribute to the screening of a transition, because preceding low energy
transitions (< 50 keV) are strongly converted to Auger
transitions, ejecting electrons mainly from higher
shells. Since the 1s-electrons contribute about 83 %
of the total screening these effects cannot exceed
about 15 eV for the 5g - 4f transition in μTl. The
measured energies of transitions with an electron
screening of 836 eV (fig. 1) and up to about 2 keV
[Bac E 72] agree with the calculated values in the
mean to 3 %. Therefore, we assume also a 3 % uncertainty for the calculated screening of the
transitions quoted in table 3.

3. Calculation of transition energies and corrections

In tables 1-4 we have summarized energies and all
known corrections which contribute more than about
1/5 of present experimental errors. These calculated
values are based on recent papers as discussed below.
In tables 1-3 only transitions are given which are
not sensitive to the finite size of the nucleus and
therefore could serve as a test of QED or electron
screening corrections.

The charge distribution parameters are most sensi-
tive to the transitions $2p_{1/2} - 1s_{1/2}$, $2p_{3/2} - 1s_{1/2}$,
$2s_{1/2} - 2p_{1/2}$, $2s_{1/2} - 2p_{3/2}$ and to some extent on the
$3d - 2p$, $3p - 1s$, $3p - 2s$. The effect of radiative corrections on these transitions, although smaller than
the effect of finite nuclear size, is not negligible
and must be taken into account. Table 4 shows values
of various QED corrections, as well as the nuclear
polarization correction for 1s, 2s, $2p_{1/2}$ and $2p_{3/2}$ and 3d
levels along with estimates of the uncertainty in these
corrections.

3.1 Vacuum polarization:

Order $\alpha Z\alpha$: The Uehling potential, calculated with
the Coulomb potential for a finite nucleus [Bar B 68],
is added to the Coulomb potential in the Dirac equation,
so that it is included to all orders in perturbation
theory. In any case the difference from first order
perturbation theory is less than 90 eV for the $1s_{1/2}$
state and 7 eV for the 4f state in muonic lead.

Order $\alpha^2 Z\alpha$: The result of Källen and Sabry
[Käl S 55, Bar R 73] is reduced to a potential by
Blomquist [Blo 72], and was folded with the finite
charge distribution e.g. $H_{int} = \int \rho(r') V_{B\ell} |\vec{r} - \vec{r}'| d^3 r'$
so that finite size effects are included.

$\alpha(Z\alpha)^3$: Here we use the expectation value of the
potential V_{13} [Blo 72] folded with the finite charge
distribution (see above), with muon wave functions
for the Dirac equation with finite size. The effect
of finite nuclear size on the $e^+ e^-$ propagators is
not included. This has recently been calculated by
several authors [Gyu 74, Ara 74, Bro C 74, Rin W 73],
mainly for the 5g - 4f transition in lead, where the
correction amounts to 6 eV. Since this has not been
calculated for other nuclei and for most other transitions it has been omitted in tables 1-4. An uncertainty of ± 7 % (roughly the ratio of the 6 eV
correction to the total shift of the 4f level of
90 eV) has been arbitrarily added in table 4. Further
calculations of the finite size corrections for the
1s, 2s and 2p levels would be desirable.

$\alpha(Z\alpha)^5$, $\alpha(Z\alpha)^7$: No finite size effects for fifth
and seventh order other than in the muon wave function are
included.

$\alpha^2(Z\alpha)^2$: The contribution of the order $\alpha^2(Z\alpha)^2$
(analogous to virtual Delbrück scattering) is unknown.

3.2 Lamb shift

The main source of this correction is due to the
virtual emission and reabsorption of photons, with
a contribution also from vacuum polarization due to
$\mu^+\mu^-$ pairs. This is shown graphically by:

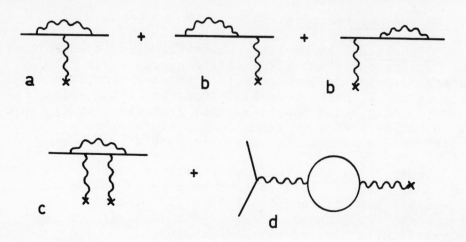

Here graphs (b) contribute mainly to the renormalization
for large virtual photon momenta of graph (a). Graph
(c) is calculated with a nonrelativistic approximation
to the propagators; its main contribution is the Bethe
logarithm [Jau R 55 p. 345-355]. After all divergences
are removed, the resulting operator is taken in the non-
relativistic limit and external lines are replaced with
relativistic Coulomb wave functions. Relativistic corrections, especially to graph (c) are neglected but have
been estimated by Barrett et al. [Bar B 68] to be of
order 30 % of the lowest order result. Evaluation of
the Bethe Logarithm is also uncertain in heavy atoms.
These uncertainties amount to a few hundred eV for the
1s, 2s, 2p states of lead. Improved calculations using
a partial wave expansion of the Coulomb propagator, as
described by Brown et al., [Bro L 59, see also Des J 71]
are in progress.

3.3 Reduced mass

       Corrections due to relativistic reduced mass are
larger and more uncertain for 2p - 1s than for 5g - 4f
transitions (here we speak of corrections over and above
that resulting from use of reduced mass in the Dirac
equation). Equivalent calculations by Calmet et al.
[Cal B 73] and Friar and Negele [Fri N 73] are the
best available. Uncertainly as to the proper form of
the Breit interaction when finite size is included,
is estimated by Friar and Negele [Fri N 73].

A further contribution from two photon exchange, where one of the photons is transverse is of order $(2Zm/M)\Delta E_{Lamb}$ (estimate by Salpeter [Sal 52] and Fulton [Ful 54]). In absence of detailed calculation of this effect for muonic atoms we must regard this quantity as an extra uncertainty.

3.4 Nuclear polarization

Nuclear polarization gives a further uncertain correction of non QED nature. Interpolated values from refs. [Col 69, Ska 70, Che 70, Eri H 73] have been used. The uncertainty can be estimated to be 30 % - 50 %.

4. Present status of vacuum polarization in μ-atoms

The energies of the 5g - 4f transitions in μ Hg and μ ^{203}Tl and the 4f - 3d transitions in μ Rh of our CERN experiment [Wal V 72, Vui 74] agree well with the data of the Chicago-Ottawa group [Dix A 71] whereas the older values of Backenstoss et al. [Bac C 70] disagree by 120 eV for the 5g - 4f transition in μ Pb. A comparison of the two recent experiments with present theoretical calculations is shown in fig. 2 and table 3. Muonic X-rays in the energy region of 150 - 350 keV are in good agreement with theory; the values of the 5g - 4f transitions in Hg, Tl, Pb and 4f - 3d in Ba at about 420 keV, however, deviate by about 50 eV. Assuming that no systematic errors occured in the two independent experiments an energy or Z dependence can be deduced from the deviations of the selected transitions shown in fig. 2. The systematic deviations of the other transitions in fig. 1 with unfortunately large statistical errors might indicate such a systematic experimental error but could also be attributed to an underestimated screening effect.

Before new accurate experiments at the new meson factories will be performed, which would with an absolute accuracy of ± 10 eV at 450 keV confirm or disprove the indicated discrepancy, we can only speculate on possible effects.

(i)   A finite size of the muon sufficient to explain
        the discrepancy [Iac L 71, Rin W 73] would be in
        strong disagreement with electron - proton and
        muon-proton scattering, the (g - 2) experiment,
        and nuclear charge radii deduced from μ-atoms
        and electron scattering [Rin W 73, Dub S 74].

(ii)   It is hard to see, how nonelectromagnetic μ-nucleus
        coupling [Bro R 68, Lee Z 67, Fea 72, Res S 73]
        will have sufficiently long range to significantly
        affect the 5g - 4f transitions. Upper limits for
        the coupling constant deduced from e-p and e-μ
        scattering are too small to produce a shift of
        50 eV.

(iii)   A breakdown of perturbation theory at high Z
        (αZ = 0.6 for lead) should manifest itself in
        heavy ion reactions.

(iv)   A more probable solution would be the inclusion
        of higher order terms. The $\alpha^2(Z\alpha)^2$ term for
        examples which is analogous to a virtual Delbrück
        scattering, is estimated by simple arguments to
        be less than a few eV, but it has to be calculated
        exactly to prove that it does not contribute as
        much as 50 eV.

References

All B 72   V. Alles-Borelli et al., Nuov. Cim. $\underline{7A}$
            (1972) 330, ibid. p. 345

Ara   74   J. Arafune, Phys. Rev. Letters $\underline{32}$ (1974)
            560

Bac C 70   G. Backenstoss et al., Phys. Letters $\underline{31B}$
            (1970) 233, Phys. Letters 43B (1973) $\overline{539}$

Bac E 72   H. Backe et al., Nucl. Phys. $\underline{A189}$ (1972)
            472

Bar   68   R.C. Barrett, Phys. Letters $\underline{28B}$ (1968) 93

Bar B 68   R.C. Barrett et al., Phys. Rev. $\underline{166}$ (1968)
            1589

Bar O 73   R.C. Barrett et al., Phys. Letters $\underline{47B}$
            (1973) 297

Bar R 73   R. Barbieri and E. Remiddi, Nuov. Cim.
            $\underline{13A}$ (1973) 99

Bel   73   T.L. Bell, Phys. Rev. $\underline{A7}$ (1973) 1480

Blo   72   J. Blomquist, Nucl. Phys. $\underline{B48}$ (1972) 95

Bro C 74   L.S. Brown et al., Phys. Rev. Letters $\underline{32}$
            (1974) 562

Bro D 70   S.J. Brodsky and S.D. Drell, Ann. Rev.
            Nucl. Scien. $\underline{20}$ (1970) 147

Bro L 59   G.E. Brown et al., Proc. Roy. Soc.
            $\underline{A251}$ (1959) 92

Bro R 68   S.J. Brodsky and E. de Rafael, Phys. Rev.
            $\underline{168}$ (1968) 1620

Des J 71   A.M. Desiderio and W.R. Johnson, Phys. Rev.
            $\underline{A3}$ (1971) 1264

Dix A 71   M.S. Dixit et al., Phys. Rev. Letters $\underline{27}$
            (1971) 878

Dub S 74   T. Dubler et al., Nucl. Phys. $\underline{A219}$ (1974)
            29

Eng S 74   R. Engfer, H. Schneuwly, J.L. Vuilleumier,
            H.K. Walter and A. Zehnder, Nuclear Data
            Tables, to be published

Eri H 73   T.E.O. Ericson and J. Hüfner, Nucl. Phys.
            $\underline{B57}$ (1973) 604

Fea   72   H.W. Fearing, Phys. Rev. D6 (1972) 203

Fri   69   B. Fricke, Z. Physik 218 (1969) 495

Fri N 73   J.L. Friar and J.W. Negele, Phys. Letters
           46B (1973) 5

Ful M 54   T. Fulton and P.C. Martin, Phys. Rev. 95
           (1954) 811

Gyu   74   M. Gyulassy, Phys. Rev. Letters 32 (1974)
           1393

Jau R 55   J.M. Jauch and F. Rohrlich, The Theory of
           Photons and Electrons, Addison Wesley,
           Cambridge 1955

Käl S 55   G. Källen and A. Sabry, Dan. Mat. Fys. Medd.
           29 (1955) No. 17

Lau P 72   B.E. Lautrup et al., Physics Reports 3C, 4
           (1972) 196

Lee Z 67   T.D. Lee and B. Zumino, Phys. Rev. 163
           (1967) 1667

Lev M 70   M. Leventhal and D.E. Murnick, Phys. Rev.
           Letters 25 (1970) 1237

Lic M 74   P.C. Lichtenberger and I.K. MacKenzie, Nucl.
           Instr. Meth. 116 (1974) 177

McK H 71   R.J. McKee et al., Nucl. Instr. Meth. 92
           (1971) 421

Per C 74   A. Pertin, G. Carboni, A. Placci, E. Zavattini,
           K. Ziock, H. Gestaldi, G. Gorini, G. Neri,
           O. Pitzurra, E. Polacco, G. Torelli, A. Vitale,
           J. Duclos and J. Picard, contributed paper to
           this conference

Pic   71   E. Picasso, Seminar on Electromagnetic Inter-
           actions, Trieste 21-26.6.1971; Daresbury
           Study Week-End on Photon and Lepton Inter-
           actions 1-3.10.1971

Res S 73   L. Resnick et al. Phys. Rev. D8 (1973) 172

Rin W 73   G.A. Rinker and L. Wilets, Phys. Rev. Letters
           31 (1973) 1559, Phys. Rev. D7 (1973) 2629

Sal   52   E.E. Salpeter, Phys. Rev. 87 (1952) 328

Sun W 72   M.K. Sundaresan and P.J.S. Watson, Phys. Rev.
           Letters 29 (1972) 15

Vog   73   P. Vogel, Phys. Rev. A7 (1973) 63, A8 (1973) 2292

Vui   74   J.L.Vuilleumier, thesis Laboratorium für
            Hochenergiephysik, ETH-Zürich, unpublished

Wal V 72   H.K. Walter et al., Phys. Letters 40B (1972)
            197

Wic K 56   E.H. Wichmann and N.M. Kroll, Phys. Rev. 101
            (1956) 843

Table 1 and 2 Experimental and theoretical muonic transition energies in ^{203}Tl and Rh. All energies are in
eV. Errors are given in parentheses. If not otherwise
specified the centre of gravity of the fine structure
components is given. ΔE_{nl}: correction due to nonlinearity of energy calibration, ΔE_{ge}: geometry effect
of Ge(Li) detector, E_{exp}: corrected experimental value,
ΔE_{QED}: quantum electrodynamical corrections, including
nuclear polarization, ΔE_{elsc}: electron screening, the
errors contain computational uncertainties only, E_{theor}:
total theoretical value. The errors of the differences
$E_{theor} - E_{exp}$ are the quadratically added experimental
and theoretical uncertainties.
1) A fast electronic shift has been determined to be
(0 ± 17) eV for all transitions, which has been taken
into account in the final value E_{exp}.

Table 1 Muonic transition energies in ^{203}Tl. For explanation see preceding page.

Transition	Linear calibration	ΔE_{nl}	ΔE_{ge}	E_{exp} [1]	Dirac equation	ΔE_{QED}	ΔE_{elsc}	E_{theor}	$E_{theor}-E_{exp}$
9k -7i	148801 (13)	55 (29)	14 (24)	148870 (43)	149172 (2)	225 (1)	- 506 (5)	148891 (6)	21 (44)
8i -6h	225082 (22)	65 (15)	12 (16)	225159 (36)	225136 (3)	474 (2)	- 420 (4)	225190 (6)	31 (37)
$6h_{11/2}$-$5g_{9/2}$	227401 (4)	63 (15)	12 (16)	227476 (29)	226817 (3)	787 (4)	- 110 (2)	227494 (6)	18 (30)
$6h_{9/2}$-$5g_{7/2}$	229043 (4)	61 (15)	12 (16)	229116 (29)	228432 (3)	807 (4)	- 109 (2)	229130 (6)	14 (30)
9i -6h	285511 (47)	12 (8)	9 (16)	285532 (54)	285765 (4)	543 (2)	- 739 (7)	285569 (8)	37 (55)
$7h_{11/2}$-$5g_{9/2}$	363647 (30)	-17 (12)	8 (16)	363638 (40)	362893 (5)	1064 (3)	- 332 (4)	363625 (7)	-13 (41)
$7h_{9/2}$-$5g_{7/2}$	365633 (26)	-17 (12)	8 (16)	365624 (37)	364879 (5)	1086 (3)	- 332 (4)	365633 (7)	9 (38)
$5g_{9/2}$-$4f_{7/2}$	420742 (13)	-19 (13)	7 (16)	420730 (29)	418834 (5)	2005 (5)	- 80 (2)	420759 (8)	29 (30)
$5g_{7/2}$-$4f_{5/2}$	426850 (10)	-17 (13)	7 (16)	426840 (29)	424842 (5)	2097 (5)	- 78 (2)	426861 (8)	21 (30)

Table 2 Muonic transition energies in Rh. For explanation see preceding page.

Transition	Linear calibration	ΔE_{nl}	ΔE_{ge}	E_{exp}[1]	Dirac equation	ΔE_{QED}	E_{elsc}	E_{theor}	$E_{theor} - E_{exp}$
10i −6h / +10h −6g	101045 (21)	−15 (3)	12 (32)	101042 (42)	101334 (1)	95 (1)	−380 (4)	101049 (5)	7 (43)
7h −5g	111770 (6)	2 (6)	12 (29)	111784 (35)	111711 (1)	183 (1)	−98 (2)	111796 (3)	12 (35)
5g −4f	128971 (3)	22 (10)	14 (26)	129007 (33)	128630 (2)	404 (2)	−20 (1)	129014 (3)	7 (33)
8h −5g	138919 (18)	30 (8)	14 (24)	138963 (34)	138958 (2)	202 (1)	−189 (2)	138971 (3)	8 (34)
6g −4f	198653 (9)	24 (8)	13 (17)	198690 (27)	198260 (2)	515 (2)	−66 (1)	198709 (3)	19 (27)
7g −4f	240738 (26)	−20 (8)	11 (15)	240729 (36)	240268 (3)	566 (2)	−137 (2)	240697 (4)	−32 (36)
8g −4f	267948 (52)	−40 (10)	9 (15)	267917 (58)	267539 (3)	592 (2)	−230 (3)	267901 (5)	−16 (58)
$4f_{7/2} - 3d_{5/2}$	279269 (5)	−46 (12)	9 (15)	279232 (27)	277923 (4)	1313 (4)	−12	279224 (6)	−8 (28)
$4f_{5/2} - 3d_{3/2}$	282343 (6)	−47 (12)	9 (15)	282305 (27)	280965 (4)	1359 (4)	−12	282312 (6)	7 (28)
5f −3d	409108 (30)	−50 (24)	8 (15)	409066 (45)	407433 (5)	1659 (4)	−43 (1)	409049 (7)	−17 (46)

Table 3 Comparison of experimental and theoretical muonic transition energies in various elements. All energies in eV. Errors in parentheses. 1) ref. [Dix A 71], 2) ref. [Vui 74], 3) ref. [Bio 72], 4) Vertex-correction including self-energy and anomalous magnetic moment, and $\mu^+\mu^-$ vacuum polarization [Bar 68], 5) ΔE_{red}: Relativistic correction to reduced mass [Fri N 73, Bar O 73], 6) ΔE_{npol}: Nuclear polarization [Col 67, Che 68, Eri H 73], 7) ΔE_{elsc}: Electron screening [Vog 73].

Z element	Transition	Point Nucleus Energy	Finite Size Effect	Vac.Pol. of order[3] $\alpha(Z\alpha)$	$\alpha^2(Z\alpha)$	357	Lamb shift[4]	ΔE_{red}[5]	ΔE_{npol}[6]	ΔE_{elsc}[7]	E_{theor}	E_{exp}	$E_{theor}-E_{exp}$
$_{20}$Ca[1]	$3d_{3/2}-2p_{1/2}$	157518 (2)	- 78(2)	734(2)	5	-1	1	1	3(2)	- 1	158182 (4)	158173 (18)	9 (19)
	$3d_{5/2}-2p_{3/2}$	156152 (2)	- 28(1)	716(2)	5	-1	1	1	2(1)	- 1	156845 (4)	156830 (16)	15 (17)
$_{22}$Ti[1]	$3d_{3/2}-2p_{1/2}$	191115 (3)	- 154(3)	947(2)	7	-2	2(1)	1	5(3)	- 2	191919 (6)	191921 (18)	- 2 (19)
	$3d_{5/2}-2p_{3/2}$	189108 (3)	- 58(1)	920(2)	6	-1	-3(1)	1	4(2)	- 2	189975 (5)	189967 (17)	8 (18)
$_{26}$Fe[1]	$3d_{3/2}-2p_{1/2}$	268404 (4)	- 439(9)	1473(3)	10(1)	-3	2(2)	2	10(5)	- 2	269457 (12)	269427 (17)	30 (21)
	$3d_{5/2}-2p_{3/2}$	264460 (4)	- 163(4)	1419(3)	10(1)	-3	-5(2)	2	4(2)	- 2	265722 (7)	265705 (16)	17 (18)
$_{38}$Sr[1]	$4f_{5/2}-3d_{3/2}$	199425 (3)	- 4	852(2)	6	-5	2	1	1(1)	- 8	200270 (4)	200254 (20)	16 (21)
	$4f_{7/2}-3d_{5/2}$	197882 (3)	- 2	833(2)	6	-5	-1	1	1(1)	- 8	198707 (4)	198708 (18)	- 1 (19)
$_{45}$Rh[2]	$4f_{5/2}-3d_{3/2}$	280985 (4)	- 20(1)	1350(3)	9(1)	-9(1)	4	1	4(2)	- 12	282312 (6)	282305 (27)	7 (28)
	$4f_{7/2}-3d_{5/2}$	277930 (4)	- 7	1311(3)	9(1)	-9(1)	-3	1	4(2)	- 12	279224 (6)	279232 (27)	- 8 (28)
$_{47}$Ag[1]	$4f_{5/2}-3d_{3/2}$	306971 (4)	- 29(1)	1519(3)	10(1)	-11(1)	5	2	5(3)	- 12	308460 (6)	308428 (19)	32 (20)
	$4f_{7/2}-3d_{5/2}$	303330 (4)	- 11	1470(3)	10(1)	-11(1)	-3	2	4(2)	- 12	304779 (6)	304759 (17)	20 (19)
$_{48}$Cd[1]	$4f_{5/2}-3d_{3/2}$	320424 (4)	- 36(1)	1608(3)	11(1)	-12(1)	5	2	5(3)	- 13	321994 (6)	321973 (18)	21 (19)
	$4f_{7/2}-3d_{5/2}$	316458 (4)	- 14	1555(3)	11(1)	-12(1)	-4	2	5(3)	- 13	317988 (6)	317977 (17)	11 (18)

Table 3 continued

50_{Sn}[1] $4f_{5/2}-3d_{3/2}$	348235 (5)	- 50(1)	1795(3)	12(1)	-14(1)	6	2	6(3)	-13	349979 (7)	349953 (20)	26 (21)
$4f_{7/2}-3d_{5/2}$	343555 (5)	- 19	1731(3)	12(1)	-14(1)	-4	2	5(3)	-13	345255 (7)	345226 (18)	29 (19)
56_{Ba}[1] $4f_{5/2}-3d_{3/2}$	439069 (6)	- 140(3)	2435(4)	17(1)	-22(1)	10 (1)	3	9(4)	-17	441364 (9)	441299 (21)	65 (23)
$4f_{7/2}-3d_{5/2}$	431654 (6)	- 53(1)	2328(4)	16(1)	-22(1)	-7 (1)	3	8(4)	-17	433910 (9)	433829 (19)	81 (21)
$5g_{7/2}-4f_{5/2}$	200544 (3)	0	762(2)	5	-12(1)	2	1	1(1)	-31(1)	201272 (4)	201260 (16)	12 (17)
$5g_{9/2}-4f_{7/2}$	199194 (3)	0	748(2)	5	-12(1)	-1	1	1(1)	-31(1)	199905 (4)	199902 (15)	3 (16)
80_{Hg}[2] $5g_{7/2}-4f_{5/2}$	414183 (5)	- 9(1)	2046(3)	14(1)	-48(2)	8 (1)	2	6(3)	-75(3)	416127 (8)	416089 (28)	38 (29)
$5g_{9/2}-4f_{7/2}$	408465 (5)	- 3	1972(3)	14(1)	-47(2)	-6 (1)	2	6(3)	-77(3)	410326 (8)	410281 (28)	45 (29)
81_{Tl}[2] $5g_{7/2}-4f_{5/2}$	424851 (5)	- 9(1)	2116(3)	14(1)	-50(2)	9 (1)	2	6(3)	-78(3)	426661 (8)	426840 (29)	21 (30)
$5g_{9/2}-4f_{5/2}$	418837 (5)	- 3	2038(3)	14(1)	-49(2)	-6 (1)	2	6(3)	-80(3)	420759 (8)	420730 (29)	29 (30)
82_{Pb}[1] $5g_{7/2}-4f_{5/2}$	435666 (5)	- 10(1)	2190(3)	15(1)	-52(2)	9 (1)	2	6(3)	-81(3)	437745 (8)	437687 (20)	58 (22)
$5g_{9/2}-4f_{7/2}$	429345 (5)	- 4	2106(3)	15(1)	-53(2)	-6 (1)	2	6(3)	-83(3)	431328 (8)	431285 (17)	43 (19)

Table 4 Contribution to binding energies (eV) for 1s, 2s, 2p, 3d states of muonic lead. The contributions to vacuum polarization, the Lamb shift and recoil corrections, as well as the associated uncertainties, are discussed in the text. Uncertainties in the nuclear polarization correction are estimated to be about 30 - 50 %, as discussed in the text.

	$1s_{1/2}$	$2s_{1/2}$	$2p_{1/2}$	$2p_{3/2}$	$3d_{3/2}$	$3d_{5/2}$
VP ($\alpha Z\alpha$)	67176	19365	32372	29873	10527	9868
VP ($\alpha^2 Z\alpha$)	552	149	251	229	76	71
VP ($\alpha(Z\alpha)^3$)	-814±54	-302±20	-463±31	-436±29	-201±13	-192±13
VP ($\alpha(Z\alpha)^5$)	- 99	- 37	- 56	- 53	- 24	- 23
VP ($\alpha(Z\alpha)^7$)	- 18	- 7	- 10	- 10	- 5	- 4
Total VP	66797	19168	32093	29603	10373	9719
Lamb Shift	- 2935	- 698	- 349	- 647	+ 43	- 50
ΔBethe log	± 160	± 141	± 139	± 104	± 3	± 2
ΔHigh.Ord.	± 380	± 90	± 220	± 190	± 30	± 20
Red. Mass	380	87	110	95	15	14
Δ(2γ)	± 264	± 61	± 29	± 54	± 4	± 4
Total QED corrections	64242	18557	31854	29051	10431	9683
	± 492	± 179	± 264	± 225	± 33	± 24
Nuclear Polarization	5077	799	1286	1286	67	67

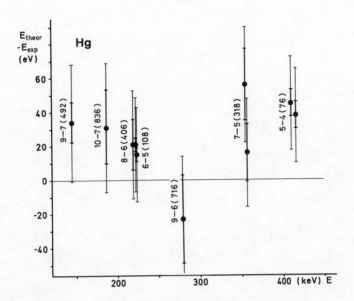

Fig. 1 Differences of calculated and measured energies
        E_{theor} - E_{exp} of X-ray transitions in μ Hg.
        E: energy of the transition
        n-n': charcterize the principal quantum numbers
             of the transition.
        (ΔE_{screen}): The values in brackets are the
                     calculated electron screening of the
                     transition in eV.
        The short thick errors correspond to the statisti-
        cal errors only, the long errors contain the
        statistical and all systematic errors. The two
        fine-structure components $5g_{9/2}$ - $4f_{7/2}$ and
        $5g_{7/2}$ - $4f_{5/2}$ at about 420 keV are the two most
        important transitions for the check of vacuum
        polarization.

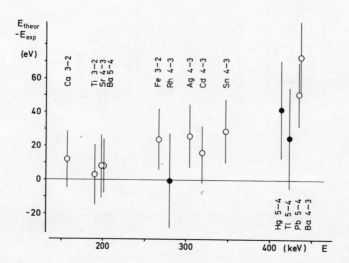

<u>Fig. 2</u> Differences of calculated and measured energies
E_{theor} - E_{exp} of X-ray transitions in several
μ-atoms.
E: Energy of the transition.
n-n': characterize the principal quantum numbers
     of the transitions. In all cases the mean
     value of the two main finestructure compo-
     nents is plotted. The errors contain
     statistical and systematic uncertainties.
     Open points from ref. [Dix A 70], full
     points from ref. [Wal V 72, Vui 74].

THE DECAY OF THE NEUTRAL VACUUM IN OVERCRITICAL FIELDS[x]

Berndt Müller, R.K.Smith, and Walter Greiner[xx]

Institut für Theoretische Physik der Johann

Wolfgang Goethe-Universität, Frankfurt/M.,Ger.

1. INTRODUCTION

The solutions of the Dirac equation are well known.
We learned in school, that, if coupled to the electromagnetic four-potential $A_\mu = \{A, A_0(r)\}$ it reads

$$(\gamma_\mu(p_\mu - \frac{e}{c} A_\mu) + mc^2)\psi = 0 . \tag{1}$$

If A is neglected and the Coulomb potential is that of an
external point charge, i.e. $A_0(r) = \frac{Ze^2}{r}$, the spectrum of
(1) can be obtained analytically and is known as the
Sommerfeld fine-structure formula

$$E = mc^2 \left[1 + \left(\frac{\alpha Z}{N + \sqrt{K^2 - \alpha^2 Z^2}} \right)^2 \right]^{-1/2} \tag{2}$$

$$K = \pm 1, \pm 2,\ldots; \quad N = 0,1,2\ldots$$

It is schematically depicted in Fig. 1 as a function of Z.

What happens to the j=1/2 solutions at this critical point?
The answer is simple: Their radial wave function
$R_{j=1/2}(r) \sim r^{\sqrt{1-(\alpha Z)^2}}$ are, because of the imaginary square
root, non-normalizable for $Z\alpha > 1$. Since normalizability is
an important quantization condition, the solutions disappear
for $Z\alpha > 1$. For modified Coulomb-potentials, as they are

---

[x] This work was supported by the Deutsche Forschungsgemeinschaft, by the Bundesministerium f.Forschung und Technologie,
and by the Gesellschaft f.Schwerionenforschung (GSI).
[xx] Invited speaker at the 4th Int.Conf. on Atomic Physics,
Heidelberg, July 1974.

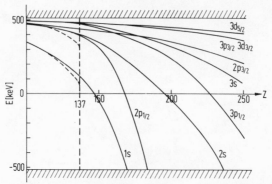

Fig.1: Schematic representation of the solution(2) of the
fine structure formula. At Z=137 the solutions with j=1/2
disappear. The **full** curve indicates the $s_{1/2}$-energy for
modified Coulomb potentials belonging to extended nuclei .
The solutions for higher j are disappearing correspondingly at higher Z-values.

necessary for extended nuclei (see Fig. 2), this problem
can be postponed up to a critical value of the charge
$Z_{cr} \approx 170$. Indeed one obtains for such potentials bound state
solutions up to Z_{cr}, where the bound state <u>dives</u>, i.e.
disappears in the negative energy continuum (see Fig. 1).
It will be an important task for us to clarify the physics
before and after this diving. As we will see, it has fundamental consequences for our understanding of quantum
electrodynamics and the physics of the vacuum states.

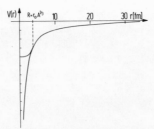

Fig. 2:   Coulomb and modified Coulomb potential.

2.   The Autoionization of Positrons

We are facing here a problem similar to that of ordinary
autoionization of nuclear and atomic physics, where one
or several bound states are imbedded into one or several
continua.   The common autoionization process of nuclei is
illustrated in Fig.3, which shows a bound 2p-2h-state degenerated with a 1p-1h continuum. Both configurations mix

because of residual interactions, which leads to the decay of the bound state.

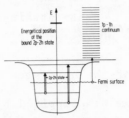

Fig.3: Illustration of the autoionization process in nuclear physics. The 2p-2h configuration (double arrows) is built of bound single particle states and imbedded into the 1p-1h continuum.

In our case the mechanism of imbedding the bound state(s) into the negative energy continua is different from the ordinary autoionization, but the mathematical structure is the same; i.e. bound state(s) ($s_{1/2}, p_{1/2}$... etc.) are imbedded into continua ($s_{1/2}, p_{1/2}$, etc...). We will handle the problem therefore analogously to the former[1,2,5] by using Fano's formalism[3]. This will be useful for the physical understanding; an exact field theoretical treatment will be given later.

Let $D = i\gamma_\nu \frac{\partial}{\partial x_\nu} + m$ be the free electron Dirac operator. We then have to search for bound states of the Hamiltonian $H = D + V(r,Z) = H(Z)$

$$H(Z)|\phi_{nj}\rangle = E_{nj}|\phi_{nj}\rangle \qquad (3)$$

where $V(r,Z) \equiv Z \times U(r,Z)$ is the "rounded off" Coulomb potential of the nucleus. $U(r,Z)$ is depending on Z only via the Z-N-dependent radius of the nuclear charge distribution. Within the range of atomic nuclei considered by us ($170 < Z < 200$) this dependence is very weak and may be omitted for convenience

$$V(r,Z) = Z \times U(r) \quad . \qquad (4)$$

We now use the fact that we know the solution to our problem for $Z = Z_{cr} \approx 170$ and diagonalize $H(Z=Z_{cr}+Z')$ in the basis of eigenstates given by $H(Z_{cr})$. $|\phi\rangle$ be the 1s bound state eigenfunction for $Z = Z_{cr}$, i.e.

$$H(Z_{cr})|\phi\rangle = E_0|\phi\rangle \approx - m_e c^2 |\phi\rangle \qquad (5)$$

and $|\psi_E\rangle$ may denote the S-continuum wave functions to the same Hamiltonian

$$H(Z_{cr})|\psi_E\rangle = E|\psi_E\rangle \quad , \quad |E| > m_e c^2, \langle \psi_{E'}|\psi_{E''}\rangle = \delta(E'-E''). \quad (6)$$

$|\phi\rangle$ and all $|\psi_E\rangle$ serve as a basis for our diagonalization procedure. In doing this we neglect the small contribution from the higher bound ns states (n>1) (see ref. 2a), which are separated by more than 500 keV from the 1s bound state. We need the matrix elements of $H(Z_{cr}+Z')$ in our (restricted) basis

$$\langle\phi| H(Z_{cr}+Z')|\phi\rangle \equiv E_o + \Delta E_o \quad, \tag{7}$$

where

$$\Delta E_o = \langle\phi|Z'U(r)|\phi\rangle = Z'\langle\phi| U(r)|\phi\rangle \quad. \tag{8}$$

$$\langle\psi_E| H(Z_{cr}+Z')|\phi\rangle = \langle\psi_E| H(Z_{cr})|\phi\rangle + \langle\psi_E| Z'U(r)|\phi\rangle=$$

$$= Z'\langle\psi_E|U(r)|\phi\rangle \equiv V_E, \tag{9}$$

and finally

$$\langle\psi_{E''}|H(Z_{cr}+Z')|\psi_{E'}\rangle =$$

$$=\langle\psi_{E''}|H(Z_{cr})|_{E'}\rangle + \langle\psi_{E''}|Z'U(r)|\psi_{E'}\rangle \equiv$$

$$\equiv E'\delta(E''-E') + Z'U_{E''E'} \quad. \tag{10}$$

The matrix elements $U_{E''E'}$ describe the rearrangement of the continuum states ψ_E under the additional potential $U(r)$. For small Z' this, clearly, can be neglected since its influence upon the diving-in bound state is a second order effect. For large Z' the continuum states ψ_E may be prediagonalized (continuum solutions for $Z>Z_{cr}$ are numerically available (see ref. 2)).

Now be $|\psi_E\rangle$ a continuum solution to the Dirac equation for $Z>Z_{cr}$:

$$H(Z_{cr} + Z') |\psi_E\rangle = E|\psi_E\rangle \quad. \tag{11}$$

We may expand $|\psi_E\rangle$ within the approximations stated above into

$$|\psi_E\rangle = a(E)|\phi\rangle + \int_{|E'|>m_e c^2} b_{E'}(E)|\psi_{E'}\rangle dE' \quad. \tag{12}$$

The coefficients $a(E)$ and $b_{E'}(E)$ are readily determined from the solution given by Fano. We are mainly interested in the effect from the bound state $|\phi\rangle$:

$$|a(E)|^2 = \frac{|V_{E'}|^2}{[E - (E_o+\Delta E) - F(E)]^2+\pi^2|V_E|^4} \tag{13}$$

where $F(E)$ is the principal value integral

$$F(E) = P \int_{|E'| > m_e c^2} dE' \frac{|V_{E'}|^2}{E-E'} \tag{14}$$

The continuum functions $|\psi_E\rangle$ are normalized in the usual way

$$\langle \psi_{E'} | \psi_E \rangle = \delta(E'-E) ,$$

so $|a(E)|^2$ gives the probability that an electron bound in $|\phi\rangle$ is promoted to $|\psi_E\rangle$ when an additional charge Z' is "switched on". It can be understood as well as the admixture of the bound state $|\phi\rangle$ to the new continuum state $|\psi_E\rangle$. This admixture has an obvious resonance behaviour. If V_E does not depend too strongly on the energy E, we may neglect $F(E)$ with respect to ΔE_o and set

$$\Gamma = 2\pi |V_E|^2 .$$

Then the bound state amplitude obtains a Breit-Wigner shape

$$|a(E)|^2 = \frac{1}{2\pi} \frac{\Gamma}{\left[E-(E_o+\Delta E_o)\right]^2 + \Gamma^2/4} , \tag{15}$$

with the resonance peaked around $E_o+\Delta E_o$.
Because we have chosen $E_o \approx -m_e c^2$,

$$\Delta E_o = Z' \langle \phi | U(r) | \phi \rangle \equiv - Z' \delta \tag{16}$$

describes the energy shift of the bound 1s-state due to the additional Z'. The width Γ of the resonance is specified as

$$\Gamma = 2\pi |V_E|^2 = 2\pi |Z' \langle \psi_E | U(r)\phi \rangle|^2 \equiv Z'^2 \gamma . \tag{17}$$

Calculations show that

$$\delta \approx 30 \text{ keV}, \quad \gamma \approx 0.05 \text{ keV} . \tag{18}$$

Thus we may explicitly indicate the Z'-dependence of Eq.(15):

$$a(E)^2 = \frac{1}{2\pi} \frac{Z'^2 \gamma}{\left[(E+m_e c^2)+Z'\delta\right]^2 + \frac{1}{4} Z'^4 \gamma^2} \tag{19}$$

Obviously, the bound state $|\phi\rangle$ "dives" into the negative energy continuum for $Z > Z_{cr}$ proportional to $Z' = (Z-Z_{cr})$. At the same time it obtains a width Γ_E in energy distribution proportional to $Z'^2 = (Z-Z_{cr})^2$. This is illustrated in Fig. 4.

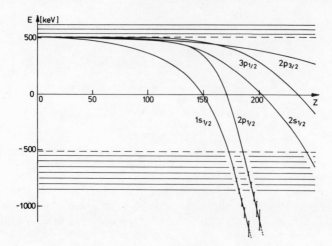

Fig.4: Electronic binding energies for extended
$(R=1.2 \text{ fm} \cdot A^{1/3})$ superheavy nuclei. Fermi charge distributions have been used. The diving points for the
$1s_{1/2}-$, $2p_{1/2}-$ and $2s_{1/2}$ levels are $Z_{cr}=170$, 185 and 245
respectively. The influence of the other electrons are
taken care of by a Thomas-Fermi distribution. The energy spreading of the states is shown by indicating the
spreading width after the diving point (magnified by a
factor 10). As shown in the text it is the same as the
positron decay width of the bound state.

To these mathematical results we give the following
physical interpretation:
If the proton number of a nucleus with $Z<Z_{cr}$ is steadily
increased, the energy of K-shell electrons is decreased
until at $Z=Z_{cr}$ it reaches $-m_e c^2$. During this process
the spatial extension of the K-shell electron charge distribution is also decreasing, i.e. the bound state wave
function becomes more and more localized. When Z grows
beyond Z_{cr} the bound 1s-state ceases to exist. But that
does not mean that the K-shell electron cloud becomes
delocalized. Indeed, according to eq.(12) the bound
state φ is shared by the negative energy continuum states
in a typical resonance type manner over a certain range
of energy by eq.(13) and the negative energy continuum
wave function become - due to the bound state admixture -
strongly distorted around the nucleus. This additional
distortion of the negative energy continuum due to the
bound state can be called <u>real charged vacuum polariza-
tion</u>, because it is caused by a real electron state which
joined the "ordinary vacuum states", i.e. the negative
energy continuum. The charge densities induced by all

the continuum states superpose to form an electron cloud
of K-shell shape. This electron cloud created by the
collective behaviour of all continuum states contains
the charge of two electrons, since the total probability
for finding the 1s-electron state ϕ in any of the continuum states is ($\gamma<<\delta$) :

$$\int_{-\infty}^{-m_e c^2} |a(E)|^2 \, dE = 1 \tag{20}$$

Thus, the K-electron cloud remains localized in r-space.
The surprising fact that it obtains an energy width Γ
though it does not decay is resolved by the fact that
it is composed of many stable states spread in energy,i.e.
the bound state has obtained a <u>spreading width</u> in the negative energy continuum. This can be illustrated in the
following way: Consider the gedanken-experiment of solving the Dirac equation with the cut-off Coulomb potential
inside a finite sphere of radius a. Certain boundary conditions on the sphere have to be fulfilled. In this way
the continuum is discretized (see Fig.5).Fig.5a shows
the situation at $Z=Z_{cr}$, i.e. before diving. After diving
(Fig.5b) the 1s-bound state has joined the lower continuum and is spread over it.

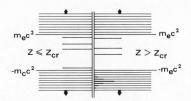

Fig.5: Spreading of the bound state (solid line) over the
negative energy continuum states (weak lines). After
diving i.e. $Z>Z_{cr}$, the negative energy continuum has obtained one more state; the 1s-bound state. It is now
spread, however, over the negative energy continuum states.
a) spectrum before and b) after diving.

If all levels occupied (occupied K-shell), one sees that
the K-shell still exists, but is spread out energetically.
Therefore a γ-absorption line from a 1s-2p-transition
would acquire an additional width, the spreading width.
This is identical with the positron escape width. If one
interprets this behaviour as a kind of strong vacuum polarization, it should be noticed that the vacuum thus defined is doubly charged. The formerly bound state K-electrons have now become a member of the vacuum. See in this

connection also refs.[4,5].
The situation is changed if the bound state $|\phi\rangle$ is unoccu-
pied during the process of "diving in". Then - on grounds
of charge conservation - one of the resulting continuum
states $|\psi_E\rangle$ has to be unoccoupied, i.e. a positron es-
capes. The kinetic energy of the escaping positron is
not sharp, but has a Breit-Wigner type spectrum given by
eq.(15). Of course, the positron-escape-process can be
reversed. If positron scattering from nuclei with $Z > Z_{cr}$
is observed, the scattering cross-section has a reso-
nance at $E = |\Delta E_0|$ with width Γ. The phase shift of the
$s_{1/2}$-positron waves should go through $\pi/2$ at this energy.
This can be used to independently determine the energy
shift ΔE_0 of the 1s-bound state according to the overcritical charge Z'. The probability per unit time for
emission of positrons in the energy interval dE is given
by Fermi's "Golden Rule"

$$P(E)dE = \frac{2\pi}{h}|\langle\phi|H_{int}|\psi_E\rangle|^2\rho(E)dE \qquad (21)$$

In our problem the Ψ_E are normalized to δ-functions and
the probability for finding the bound state at the ener-
gy E is according to (12) given by $\rho(E)dE = |a(E)|^2 dE$,
and $H_{int} = Z'U(r)$. Hence the decay probability of the
ionized K-shell, i.e. the autoionization rate per unit
time, is, using eq.(15)

$$p(E)dE = \cdot \frac{\frac{1}{2\pi}\Gamma_E dE}{[E-(E_0+\Delta E)]^2+\Gamma_E^2/4} \cdot \frac{\Gamma_E}{h} \qquad (22)$$

It can easily be generalized to several "diving states",[2]
This decay must be interpreted as the decay of the normal,
neutral vacuum into a charged vacuum (charge 2) in overcritical fields. Indeed, the Fermi surface has always to
be at least at $E = -mc^2$, it cannot be at $E < -mc^2$, because
otherwise we would immediately have strong contradictions
to the experience. The vacuum state, which is to be identified with $E_F = -mc^2$ obviously is absolutely stable up to
$Z = Z_{cr}$ and becomes instable (spontaneous decay) in overcritical fields. Only the charged vacuum (two positrons
are emitted) is stable in over-critical fields. The vacuum
proceeds to become higher charged as the over-critical
fields (over-critical charge) are further increased.

3. The Change of the Vacuum in Overcritical
Fields - Formal Aspects

Due to the excellent agreement between experiment and
theory quantum electrodynamics (QED) is one of the most

successful parts of modern physics. This has been achieved by a power series expansion of all results in terms
of the parameter α and $Z\alpha$. Unfortunately, little is known
about the convergence of the expansion. In practice, this
was no source of grief, since α and $Z\alpha$ are usually small
parameters. In very heavy atoms, however, we have $Z\alpha=0.7$
and in superheavy atoms we would be confronted with $Z\alpha\approx1$
or even $Z\alpha>1$.

Consequently, it seems to be worthwhile to extend atomic
physics to the case where $Z\alpha>1$. This means that we have
to deal with the unpleasant situation of having neither
weak ($g^2<<1$) nor strong ($g^2>>1$) coupling, but working in
the intermediate region. A first simplification is obtained by the remark that the finestructure constant
$\alpha=1/137$, which couples charged particles to the radiation
field, remains small and these effects may be treated in
perturbation theory. This suggests an expansion

$$B = \sum_{n=o}^{\infty} f_n(Z\alpha)\alpha^n \qquad (23)$$

of observables with exact coefficients $f_n(Z\alpha)$. Conceptually, this corresponds to solving for the exact single
particle solutions in the classical external field of the
nucleus with charge Z and calculating all higher order
effects in perturbation theory.

When dealing with the real electrons alone the most important higher order corrections are conveniently treated by the nonlinear Hartree-Fock equation. It is desirable to generalize these equations so that they include virtual effects as vacuum polarization and self-
energy. We start[4,5] from the operator equation for the
coupled electron and photon fields

$$(\gamma\cdot p-m)\psi = \frac{e}{2}\{\gamma^\mu\psi,A_\mu\} \qquad (24)$$

$$\Box A_\mu = e(j_\mu^{ex} + j_\mu) \qquad (25)$$

where j_μ^{ex} takes into account the classical field of the
nucleus and

$$j = - \frac{1}{2}[\psi,\gamma^\mu\psi] \ . \qquad (26)$$

The second equation can be formally solved for A :

$$A_\mu(x) = A_\mu^{ex}(x) + 2e \int_{-\infty}^{t_x} d^4y \ D_{\nu\mu}^F (x-y)j^\nu(y)+A_\mu^{rad}(x) \qquad (27)$$

In the absence of photons $A_\mu^{rad} = 0$ and we may substi-
tute (27) into (24). After some manipulation this leads
us to the operator equation[6]

$$(\gamma \cdot p - e\gamma \cdot A^{ex} - m)\psi(x) = e^2\int d^4y \quad \gamma^\mu <j^\nu(y)> D^F_{\mu\nu}(\vec{x}-\vec{y};-ty)\psi(x)$$

$$-e^2\int d^4y\gamma^\mu S^F(\vec{x},\vec{y};-t_y)\gamma^\nu\psi(y,t_x+t_y)D^F_{\mu\nu}(\vec{x}-\vec{y};-ty). \qquad (28)$$

In order to derive particle equations we have to take matrix elements between a complete set of particle states.
First we have to specify which states we intend to regard as occupied, i.e. we have to define a fermi surface
F(Fig.6).

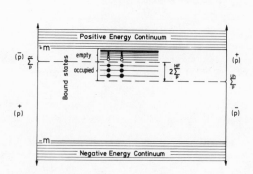

Fig.6: The single particle spectrum and the Fermi sur-
face; definition of the tilde(Σ) and the Uehling sum ($\overset{Uh}{\Sigma}$)

The field operator is

$$\Psi(x) = \underset{p>F}{\Sigma} b_p\phi_p(x)e^{-iE_p t_x} + \underset{p<F}{\Sigma} d^+_p \phi_p(\vec{x})e^{-iE_p t_x} \qquad (29)$$

The Hartree-Fock approximation is introduced by restrict-
ing the basis to a single configuration, i.e. not allowing for any particle-hole excitations. Then the matrix
elements are reduced to those between the vacuum and
single particle states

$$<o| \ldots b^+_q| o> \quad \text{or} \quad <o| d_q \ldots | o> . \qquad (30)$$

The resulting nonlinear equations are [4,5,6]

$$(\gamma_0 E_q - \vec{\gamma}\cdot\vec{p} - e\gamma \cdot A^{ex} - m)\phi_q(\vec{x}) = \qquad (31)$$

$$= \frac{e^2}{4\pi} \gamma_\mu\int d^3y \underset{p}{\Sigma}\overline{\phi}_{\tilde{p}}(\vec{y})\gamma^\mu\phi_p(\vec{y})|\vec{x}-\vec{y}|^{-1}\phi_q(\vec{x}) -$$

$$- \frac{e^2}{4\pi} \gamma^\mu\int d^3y \underset{p}{\Sigma}\phi_p(\vec{y})\gamma^\mu\phi_q(\vec{y}) \frac{F_{pq}(|\vec{x}-\vec{y}|)}{|\vec{x}-\vec{y}|} \phi_p(\vec{x}) .$$

We have now developed the general basis for atomic physics with $Z_\alpha > 1$. In Fig. 4 the binding energies of the lowest lying atomic states are shown as a function of nuclear charge Z. At $Z_{cr} = 172$ the 1s-state is degenerated with the states of the negative energy continuum. It is obvious that a single-particle description will break down at this stage.[2,4,5]

For simplicity we assume all higher bound states as empty. We have two complete bases of single particle states to describe different situations. The states $|q,Z_{cr}+Z'\rangle$ of a supercritical atom without a 1s-state and the just critical state $n,Z_{cr}\rangle$ with a bound 1s-state. They are connected via a unitary transformation:

$$|q,Z_{cr}+Z'\rangle = \sum_n \langle n,Z_{cr}|q,Z_{cr}+Z'\rangle|n,Z_{cr}\rangle \equiv \sum_n a(n,q)|n,Z_{cr}\rangle$$

(32)

The critical 1s-state can be constructed from the supercritical continuum states

$$|1s,Z_{cr}\rangle = \sum_q a^*(1s,q)|q,Z_{cr}+Z'\rangle \quad .$$

(33)

The $a(n,q)$ can be derived explicitly, if one neglects the interaction with higher bound states and among the continuum states. If $Z'U(r)$ is the supercritical part of the Coulomb potential, we have

$$\langle 1s,Z_{cr}|Z'U(r)|1s,Z_{cr}\rangle = E_{1s}+mc^2$$

$$\langle 1s,Z_{cr}|Z'U(r)|E,Z_{cr}\rangle = V_E$$

(34)

$$\langle E,Z_{cr}|Z'U(r)|E',Z_{cr}\rangle = U_{EE'}$$

with the solution

$$a(1s,E) \approx V_E^*[E-E_{1s} - F(E) + i\pi|V_E|^2]^{-1}$$

(35)

$$a(E,E') \approx U_{EE'}(E'-E)^{-1} + \delta(E-E')$$

(36)

where $F(E) = p\int|V_{E'}|^2 (E-E')^{-1} dE'$. This is analogous to our more "anschauliches" treatment in the foregoing section (eqs. 7 ff.).

When two electrons occupy the 1s-states, the state vector can be expanded in both the critical and the supercritical basis:

$$\psi = \sum_{E>F} \widetilde{b}_E|E,Z_{cr}\rangle + \sum_\epsilon \widetilde{d}^+_{1s,\epsilon}|1s,\epsilon,Z_{cr}\rangle + \sum_{E<-m} \widetilde{d}^+_E|E,Z_{cr}\rangle =$$

$$= \sum_{E>-m} b_E|E,Z_{cr}+Z'\rangle + \sum_{E<-m} d^+_E|E,Z_{cr}+Z'\rangle$$

(37)

where ϵ denotes the two spin orientations.

According to (33) the "critical" hole operator $\tilde{d}^+_{1s}$ is connected with the supercritical ones by[5]

$$\tilde{d}^+_{1s} = \int dE' \, d^+_E \, a(1s,E).$$ (38)

The familiar vacuum $|o\rangle$ is defined by

$$|\tilde{o}\rangle = \tilde{d}^+_{1s\uparrow} \, \tilde{d}^+_{1s\downarrow} |o\rangle$$ (39)

We shall show that $|o\rangle$ is unstable for supercritical fields and decays into the twice charged supercritical vacuum defined by the state vector (37). Let us therefore prepare a K-hole in an atom with $Z = Z_{cr} + Z'$. The complete time-dependent state will be $(y(o) \overset{cr}{\equiv} 1)$:

$$|\psi(t)\rangle \equiv y(t) \, \tilde{d}^+_{1s} \, |o\rangle + \sum_{E<-m} W_E(t) \tilde{d}^+_E |o\rangle$$ (40)

Here $\tilde{d}^+_{1s}$ is the "collective" 1s-state defined in (38) and $\tilde{d}^+_{E}$1s are the operators for a positron in supercritical fields. The time dependence of (40) is determined by the Hamiltonian

$$H = : \Psi^+ \, H_D(Z_{cr}+Z') \, \Psi :$$ (41)

and the Schrödinger equation

$$H|\psi(t)\rangle = i \frac{\partial}{\partial t} |\psi(t)\rangle \quad .$$ (42)

Explicit solution yields[5]

$$y(t) = e^{i E_{1s} t} \, e^{-\frac{1}{2} \Gamma t}$$ (43)

with $E_{1s} = E_{1s} + F(E_{1s})$ and $\Gamma = 2\pi |V_{E_{1s}}|^2$,

and finally

$$|W_E(\infty)|^2 = |a(1s,E)|^2$$ (44)

This result shows that the K-vacancy brought into the supercritical atom decays with the resonance width Γ from (33) as decay width. It produces positrons in the continuum states with an energy distribution equal to that of the spread out supercritical 1s-state. Fig.7 shows this spreading for an atom with $Z = 184$.

Having shown the instability of the uncharged vacuum in supercritical fields, let us return to the charged vacuum. This does not contain a 1s-eigenstate, but a collective 1s-state can be extracted from the continuum. The total charge distribution of the charged vacuum is given by: [2,4,5]

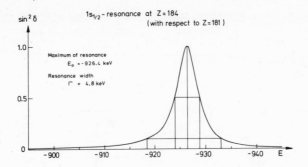

Fig.7: Spreading of the "K-shell" in the overcritical, charged vacuum of the element Z=181.

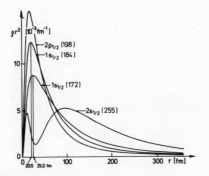

Fig.8: Charge distribution of the vacuum in overcritical fields (Z=184,198,155) and of the K-shell in under-critical fields (Z=172).

$$\rho_{ind}(\vec{x}) = - \frac{1}{2} \int_m^\infty \{\varphi_E^+(x)\varphi_E(x) - \varphi_{-E}^+(x)\varphi_{-E}(x)\}dE - \frac{1}{2}\rho_{BS}(\vec{x})$$

(45)

where the φ are the continuum solution to $Z_{cr}+Z'$ and ρ_{BS} is the charge distribution of the higher bound states. When the integral is restricted to the region of the re-sonance (33) we find the charge distribution of the old K-shell:

$$\rho_{1s}(x) = \int_{-E_{1s}-\Delta E}^{-E_{1s}+\Delta E} dE\{\varphi_E^+(x)\varphi_E(x) - \varphi_E^+(x)\varphi_{-E}(x)\}, \quad (46)$$

and the remaining part of (45) yields (approx.) the stan-dard vacuum polarization. Thus the K-shell has become part of the vacuum polarization,[2,5] so to say real part since $\int d^3x\, \rho_{1s}(x) = 2$. In contrast the familiar virtual part of vacuum polarization has a zero integral.

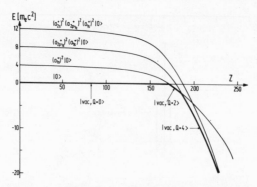

Fig.9:   The change of the vacuum to higher charge (iso-
spin) as the overcritical field increases.

Fig. 8 shows the charge distribution for several (collective) supercritical bound state resonances in comparison
to the 1s-eigenstate at Z_{cr}=172. Also for the higher bound
states there exist critical nuclear charges, e.g.
$Z_{cr}(2p_{1/2}) \approx 185$, $Z_{cr}(2s) \approx 245$. With the exception of
the $2p_{1/2}$ - criticality, however, all higher critical
points seem to be an academical problem, but can be treated in the very same way. Fig. 9 demonstrates the transition between the vacua of different charge (isospin) at
the critical points.

      4.   The Effects of Vacuum Polarization and Non-
             linearities in the Electromagnetic and Spinor
                               Fields

We are asking now for possible mechanisms which can prevent the diving of bound state levels into the negative
energy continuum, i.e. which stabilize the neutral vacuum also in overcritical fields. The first such mechanism to think of is the field effects themselves; in
particular the vacuum polarization, which is expected
to be of greatest significance for high z atoms. For
that it is necessary to investigate the vacuum polarization charge of eq.(45) and to renormalize it. This is
a quite mathematical game and has recently been carried
out by M. Guylassi[7]. His results indicate, that the
Uehling potential, which has already been calculated by
Pieper et al.,[1] shifts the diving point by $\Delta Z \approx -1$ and
the higher order vacuum polarization by $\Delta Z \approx + 1.5$, so
that a net shift of $\Delta Z \approx 0.5$ results. His main result
thus is, that vacuum polarization cannot prevent diving,
and practically does not alter the critical charge Z_{cr}.

This result has already been anticipated on quite physical grounds in refs.[2,8] by studying the effects of nonlinear electrodynamics. More precisely, consider a limiting field electrodynamics in the sense of Born and Infeld,[9] which is based on a Lagrangian of the type

$$\mathscr{L} = E_o^2 \left(\sqrt{1 - \frac{E^2 - B^2}{E_o^2}} - 1 \right) \tag{49}$$

which is constructed in analogy to the relativistic Lagrangian

$$\mathscr{L} = mc^2 \left(\sqrt{1 - \frac{v^2}{c^2}} - 1 \right) \tag{50}$$

which is a "limiting velocity Lagrangian". The Lagrangian (49) reduces for weak fields ($\frac{E^2 - B^2}{E_o^2} \ll 1$) to the Maxwell Lagrangian for free fields, i.e. $L_{max} = \frac{1}{2}(E^2 - B^2)$. Clearly, the analogy of (49) and (50) shows that E_o is a limiting field as c is the limiting velocity in special relativity. It is important to notice now, that a limiting field electrodynamics is a model electrodynamics for maximal polarization. If we plot, for example, the applied field E against the effective electric field E_{eff} (Fig.10), vacuum polarization can have at most the effect of saturating the applied field; i.e. leading to a limiting field of E_o.

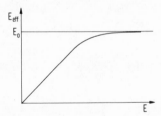

Fig.10: The effective field E_{eff} as a function of the applied field E in a model of maximum vacuum polarization.

The Born-Infeld electrodynamics as a prototype of a limiting field electrodynamics has been investigated by Rafelski et al.[10]. In such limiting field electrodynamics the energy, i.e. the T_{44}-component of the energy momentum tensor is integrable. Born requested

$$\int_{space} T_{44} d^3 x \overset{!}{=} m_o c^2 \tag{51}$$

where m_o is the electron mass. This leads to a limiting field $E_o \approx 10^{18} V/cm$. After recalculating with such a limiting field the modified electrostatic electron-nucleus interaction and coupling this into the Dirac equation, one still obtains diving, but Z_{cr} is shifted up to $Z_{cr} \approx 250$ (see Fig. 11).

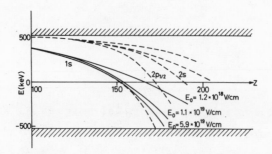

Fig.11:   The diving of the electronic K-shells in limiting field electrodynamics. The limiting field E_o is indicated. These theories serve as models for maximal vacuum polarization.

These relatively low limiting fields are, however, unrealistic, because such theories contradict high precision experiments. Soff et al.[8] have investigated the Lambshift experiments, the precision measurements of the electronic K-binding energies in Fm and Pb, and the spectra of muonic atoms of heavy nuclei, and found that the limiting field E_o has to be at least $E_o \geq 10^{20} V/cm$ in order to not disturb the obtained agreement between theory and experiments in these cases. With that value E_o the diving is practically occurring at the same critical charge as in the Maxwell theory, i.e. $\Delta Z_{cr} \approx 1$ (see Fig.11). Therefore we can conclude that 1) diving is not prevented by vacuum polarization, because it is even not prevented in the model of maximum vacuum polarization, i.e. limiting field electrodynamics and 2) that vacuum polarization can only shift the critical charge by not more than one charge unit.
Another mechanism to prevent diving (stabilize the vacuum) would occur if the Dirac field equations are changed by nonlinear term, e.g.

$$\gamma_\mu \left(\frac{\partial}{\partial x_\mu} - A_\mu \right) \psi + m \, \psi + \lambda_1 (\bar{\psi}\psi)\psi + \lambda_2 (\gamma_\mu \gamma_5 \bar{\psi}\gamma_\mu \gamma_5 \psi)\psi = 0 \tag{52}$$

Such nonlinear terms can, if the coupling constants λ_1 or λ_2 are strong enough, indeed stabilze the vacuum. This is quite plausible. Let us take for example, the

term $(\bar{\psi}\psi)\psi$. It acts like a repulsive potential when the
electron densities $\bar{\psi}\psi$ become high. This "counter potential" can cancel or weaken the Coulomb-potential so much that
a further increase in Z does not lead to more binding.
Waldeck[11] has shown, however, that such strong coupling
constants λ_1 and λ_2 again would contradict the above mentioned high precision experiments. In fact, one can determine upper limits for λ_1 and/or λ_2 from these experiments, and finds, that again diving is not prevented and
occurs at practically the same critical charge as in the
linear spinor theory. We can therefore conclude, that the
electron positron vacuum cannot be stable in overcritical
fields, and has to decay into a charged vacuum. We further comment, that we could exclude in this case of the
electron-positron fields and its interaction with the
electromagnetic fields violent nonlinear interactions,
because high precision measurements were available. This
is not the case for the strong interactions, i.e. the
pionic field and its coupling to the nuclear field. Therefore, in this latter case, our knowledge is much more restricted, predictions for the overcritical strong field
case much less reliable as in the present case. We will
come back to these questions in our other talk on shock
waves and pionization of matter.[12]

5.   Experimental Possibilities

It is clear that superheavy elements around Z=170 or 180
cannot be formed in the near future even though we predicted some years ago the existence of quasistable superheavy island around Z=164 (see ref.[13]). There is, however,
a promising way out of this difficulty. Underlying all
the various possible tests of QED of strong fields is the
fact that in heavy ion collisions at energies around the
Coulomb barrier the velocity of the ions, $v_{ion} \approx \frac{c}{20}$, is
relatively slow compared with the highly rela-
tivistic electrons ($v_e \approx c$) bound with equal or more than
twice their rest mass. Therefore there exists a kind of
adiabaticity

$$\frac{v_{ion}}{v_e} \sim \frac{1}{20} \tag{53}$$

which is not terribly good, but sufficient. In fact, we
will see that the deviations from adiabaticity bring advantages for the possible experiments which are absolutely necessary. This adiabaticity led us[14] to the prediction of intermediate superheavy molecules in the heavy
ion collision, which exist only during the time of collisions. The study of the intermediate molecules requires
the solution of the Two-Center-Dirac-Equation, to which
we will address ourselves now.

The motion of an electron under the influence of two
charged nuclei is one of the fundamental problems in
quantum mechanical theory of chemical binding.[15] Today
the solution of the one-electron spinless Schrödinger
equation for two point charges are known analytically[15]
and extensive numerical calculations have been done to
obtain binding energies in many-electron problems.[16,17]
All these treatments, however, are non-relativistic,
since the binding energies encountered in chemistry are
of the order of some eV and relativistic effects are negligible. This comes from the fact that in diatomic molecules formed by heavy atoms, e.g. Ni-Ni, AuI, the internuclear distance is of the range of a few Å and only
the outermost electronic orbitals overlap.
In heavy ion collisions with several MeV/nucleon scattering energy, the nuclei are for a short period of time
so close that even the K-electrons feel the attraction
of both centers. The typical collision time of heavy ions
inside the K-shell is $T_c \sim 10^{-19}$sec whereas the "orbiting"
time associated with a 1s-electron in very heavy atoms
is $T_e \sim 10^{-20}$sec. The same holds true if the ratio of
the corresponding velocities is examined. Therefore the
molecular electronic states have time to adjust to the
varying distance $\vec{R}$ between the nuclei. This process can
therefore be treated adiabatically in first approximation. The treatment, however, must be relativistic if
the total nuclear charge involved, i.e. $Z = Z_1 + Z_2$, is comparable to $\alpha^{-1} = 137$. Only one attempt has been made to
account for relativistic corrections by Luke et al.[18]
in ordinary molecular physics using first order perturbation theory in Z . This method, clearly, is inappli-
cable for systems with Z>90.

The relativistic wave equation for an electron (with mass
M) in the field of two electrical charges (located at
R and -R resp.) is given by[19]

$$[c\vec{\alpha}\cdot\vec{p}+\beta Mc^2 -E + V_1(\vec{r}-\vec{R})+V_2(\vec{r}+\vec{R})]\Psi = 0 \qquad (54)$$

The two charges are considered as fixed in space. It is
convenient to introduce prolate spheroidal coordinates
ξ,η,φ by defining

$$x = R[(\xi^2-1)(1-\eta^2)]^{1/2} \cos \varphi$$
$$y = R[(\xi^2-1)(1-\eta^2)]^{1/2} \sin \varphi \qquad (55)$$
$$z = R\xi\eta$$

where the z-axis cuts through the two charges, which are
foci of the ellipsoids ξ= const.. The φ-coordinate can
be easily separated by setting ($m+1/2$ is the quantum number

of angular momentum around the x-axis):

$$\Psi(r) = \begin{pmatrix} e^{im\varphi} & & & \\ & e^{i(m+1)\varphi} & & \\ & & ie^{im\varphi} & \\ & & & ie^{i(m+1)\varphi} \end{pmatrix} \Psi'\cdot(\xi,\eta) \qquad (56)$$

and one is left with the equation

$$\left[\frac{\hbar c}{R(\xi^2-\eta^2)} \begin{pmatrix} 0 & 0 & \pi_Z & \pi^- \\ 0 & 0 & \pi^+ & -\pi_Z \\ -\pi_Z & -\pi^- & 0 & 0 \\ -\pi^+ & \pi_Z & 0 & 0 \end{pmatrix} + \beta Mc^2 - E + V_1(\xi+\eta) + V_2(\xi-\eta) \right] \Psi'(\xi,\eta) = 0$$

$$(57)$$

with the operators

$$\pi^+ = W(\xi \frac{\partial}{\partial\xi} - \eta \frac{\partial}{\partial\eta}) - \frac{m}{W}(\xi^2-\eta^2)$$

$$\pi^- = W(\frac{\partial}{\partial\xi} - \eta \frac{\partial}{\partial\eta}) + \frac{m+1}{W}(\xi^2-\eta^2) \qquad (58)$$

$$\pi_Z = \eta(\xi^2-1)\frac{\partial}{\partial\xi} + \xi(1-\eta^2)\frac{\partial}{\partial\eta}$$

This equation is known not to be separable in the ξ and η coordinates and has to be solved numerically. We have chosen a diagonalization procedure with an appropriate set of basis functions. In analogy to non-relativistic two-center functions we have used the set

$$\psi^m_{n\ell s}(\xi,\eta) =$$

$$= (\xi^2-1)^{\frac{m+\varepsilon_s}{2}} e^{-\frac{\xi-1}{2a}} L_n^{m+\varepsilon_s}(\frac{\xi-1}{a}) P_e^{m+\varepsilon_s}(\eta) \chi_s \qquad (59)$$

where χ_s are the unit spin vectors, $\varepsilon_s = 0$ for s odd and $\varepsilon_s = 1$ for s even. L_n^α and p_e^α are the associated Laguerre and Legendre polynomials, respectively. a is a scaling factor to describe the asymptotic behaviour in the region $\xi \to \infty$. It is easy to see that eq.(59) form a complete set of basis function from the completeness of the L_n^α and p_e^α. Due to the common factor $(\xi^2-1)^{(m+\varepsilon_s)/2}$

which has been added to account for the structure of the
differential equation (57) in the vicinity of the z-axis
($\xi=1$), and the volume element $R^3(\xi^2-\eta^2)d\xi d\eta$ in spheroidal coordinates, however, the $\psi^m_{n\ell s}$ are not orthogonal:

$$\langle \psi^m_{n'\ell's'} | \psi^m_{n\ell} \rangle = \delta_{ss'} B^m_s(n'\ell',n\ell) \tag{60}$$

Taking matrix elements of Eq.(57) we have

$$\langle \psi^m_{n'\ell's'} | \frac{\hbar c}{R(\xi^2-\eta^2)} \begin{pmatrix} 0 & \pi \\ -\pi & 0 \end{pmatrix} + \beta Mc^2 + V_1(\xi+\eta) + V_2(\xi-\eta) | \psi^m_{n\ell s} \rangle =$$
$$\tag{61}$$

$$= A^m(n'\ell's',n\ell s) = E \, \delta_{ss'} \, B^m_s(n'\ell',n\ell)$$

as eigenvalue problem for E. It can be solved by standard procedures, i.e. first diagonalizing $B^m(n'\ell',n\ell)$
forming $[B^m]^{-1} A^m = C^m$ and subsequently diagonalizing
C^m. This method circumvents Schmidt's orthogonalization
of the basis functions $\psi^m_{n\ell s}$. The eigenvalues of C^m
give the eigenenergies of our problem, depending on
R and the explicit form of the potentials V_1 and V_2.
For point charge

$$V_1^P(\xi+\eta) = \frac{Z_1 e^2}{\xi+\eta} \qquad V_2^P(\xi-\eta) = \frac{Z_2 e^2}{\xi-\eta} \tag{62}$$

all matrix elements in eq.(61) can be evaluated analytically by use of recurrence formulae for the L^α and P^α_1
(see ref.[19]). This convenience is due to the special
choice of the basis functions eq.(59) and is lost by
even slight modifications, especially in the spin-orbit
coupling terms. For $Z_1 + Z_2 > \alpha^{-1}$ and close distances R
the extended structure of the charges becomes important and the additional matrix elements

$$\langle \psi^m_{n'\ell's} | V_i(\xi,\eta) - V_i^P(\xi,\eta) | \psi^m_{n\ell s} \rangle \quad , \quad i = 1,2$$

are computed by numerical integration over the nuclear
volumes. We have chosen homogeneously charged spheres
with a radius given by $r_N = 1.2 (0.00733Z^2 + 1.3Z + 63.6)^{1/3}$ fm (see ref.[13]).
For numerical calculations we have used a set of 100 basis
functions ($n=0,\ldots,4$; $\ell=m,\ldots,m+4$). The diagonalization
then gives about 15 correct eigenstates and several states
in the positive and negative energy continua. Figs.12-16
show the level schemes for the system $_{35}Br - _{35}Br$,

examined by Meyerhof et.al., Ni-Ni investigated by
Greenberg et.al.[21], $_{53}$I-$_{79}$Au and $_{92}$U-$_{92}$U investigated
by Mokler et.al.[22], Cl-Pb investigated by Burch et.al.[23].

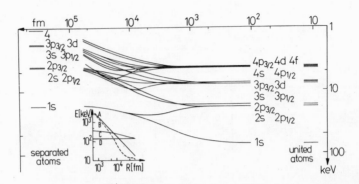

Fig. 12: The level scheme for the symmetrical system
$_{35}$Br-$_{35}$Br. Since parity is a good quantum number, adia-
           batic states of opposite parity may cross.
The small figure shows the molecular potential as function of the internuclear distance: A) the Coulomb repulsion between the nuclei; D) the electronic binding
energy in separate atoms; C) electronic energy in the
molecule; B) the total potential. The Br-Br-system has
been investigated by Meyerhof et al.[20]

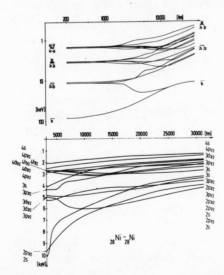

Fig.13: The level scheme for the symmetrical Ni-Ni-
system,[38] which has been investigated by Greenberg et.al.[21]

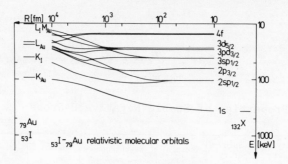

Fig.14: The level scheme for the asymmetrical system
$_{53}I$-$_{79}Au$.[38]. Parity is not a good quantum number. This
system has been investigated by P. Mokler
et.al.[22] and more recently by H. Gove et.al.[45]

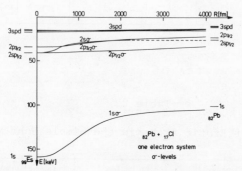

Fig.15: The level scheme for the very asymmetrical Cl-Pb-
system, which has been calculated by D.Burch et.al.[23]

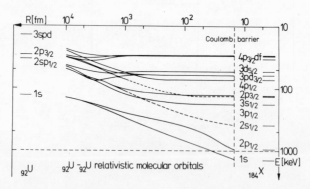

Fig.16: The $_{92}U$-$_{92}U$-system with extended nuclei. The
lowest molecular state reaches the negative energy
continuum at $R_{cr} \sim 34$ fm. It can be traced even further
down due to the discretization of the continuum states
by the mathematical method of diagonalization.

The I-Au and Cl-Pb systems are interesting because the
electronic states have no good parity and no level crossings are allowed. In the U-U system the nuclear charge
distribution plays an important role for R<60 fm. The
critical distance R_{cr}, i.e. the internuclear separation
where the deepest bound state acquires $1Mc^2$ binding energy and joins the negative energy continuum - is shifted
from 36 fm for point-like to 34 fm for extended nuclei.
The comparison of these results shows that for the
lighter molecular systems like Br-Br the asymptotic spectrum is already reached at distances of the order 10^3-10^4
fm, while for heavy systems like U-U the asymptotic spectrum is only reached when R=0. Thus, considering the dynamics, there exists a kind of a "run way" for the final
molecular spectra in light systems with a distance of the
order 10^3 fm along which the same K_α, K_β etc. lines can
be emitted, independent of R. This is probably the rea-
son for observing enhanced molecular X-rays, while the
great tail of molecular X-rays stems from the transition
region between 10^5 and 5×10^3 fm. The level schemes thus
obtained are exact for one electron in the electric
field of two motionless nuclei. In heavy ion scattering,
one will be confronted with many-electron problems in the
field of two (rapidly) moving nuclei. Then several corrections will arise: 1) Screening of outer electrons
by inner ones (Thomas-Fermi of Hartree-Fock methods). It
can be estimated from calculations in superheavy atoms[24]
to decrease the binding energy of 1s electron at R_{cr}
by 50 - 100 keV for the U-U molecular system.
2) Corrections due to non-adiabaticity, i.e. transitions
between different molecular levels that lead to ionization and unsharpening of the level energy. These effects
are treated by time-dependent perturbation theory on the
adiabatic basis[25,26].
These new two center energy diagrams can be applied to
the following problems: a) The exact shape of molecular
X-ray spectra obtained in high energetic collisions of
heavy ions[20,23,27,28]; b) the shape of positron autoionization spectra in the collisions of overcritical systems of heavy ions ($Z_1+Z_2>172$) (see ref.[26]); c) the
ionization cross section for inner molecular shells due
to Coulomb excitation[26], which is of eminent importance
for the positron autoionization experiments; d) deviations from the Rutherford scattering cross section
(especially angular distribution) due to effects of electronic molecular binding[30].
The latter effect is shown in Fig.12b for the Br-Br case
and shows that indeed strong deviations from the $\frac{Z^2\alpha}{R}$
potential occur.

    5.2 <u>X-rays from Intermediate Molecules</u>. - It is
quite clear from our discussion so far that the intermediate molecules must be observable through the emission of
their X-rays. Indeed, the first MO-X-rays (L-shell-X-rays)
have been observed by F. Saris and his coworkers[27]. The
first MO-X-rays for the M-shell, but for the superheavy
I-Au-system (see Fig.14) have been observed by Mokler,
Stein and Armbruster[22]. The MO-K-X-rays have been pio-
neered by W. Meyerhof and his cowerkers[20]. In these cases
one observed spectra as shown in Fig.17 for the monoatomic Ni-Ni-system which has been investigated by J.
Greenberg et.al. from the Yale University[21]. The long tail
of spectrum above the atomic K_α-,K_β-lines is interpreted
as being due to molecular X-rays. Clearly, as Greenberg
has shown, their exists no "united atom limit" for these
X-rays, if the background is suppressed sufficiently. This
is particularly so, since in this monoatomic system the
dipole-bremsstrahlung-background (nucleus-nucleus bremsstrahlung) vanishes. These and many other arguments made
it <u>necessary to look for a unique signature for the iden-</u>

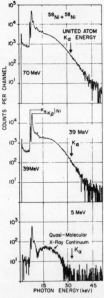

Fig.17: MO-K-X-ray spectrum observed for various ion energies
(70 MeV), 39 MeV, and 5 MeV) by J. Greenberg et.al. at Yale[33].
The united atom K_α-energy is indicated by an arrow. Clearly, the
MO-X-rays are observed far beyond this limit. This is an effect of
the collision broadening[32]. At 5 MeV ion energy the MO-X-rays join
the background below the K_α-united atom limit; as it must be, because
of the Two-Center-Level-Diagram and the small collision broading.

tification of molecular X-rays. This has recently been achieved by recognizing that their exist induced transitions in the intermediate molecules.[30-33] The experimentum crucis has been performed by Greenberg and collaborators[33] at Yale. We will describe this theoretical and experimental work here shortly.

Reports on the detection of molecular X-rays have been given by various experimental groups [20-23,27,28]. However, as mentioned above, background effects such as electronic and nuclear bremsstrahlung occur as competing effects in these measurements. Only recently Greenberg and Davis[21] have carried out experiments with monoisotopic Ni beam and target where nuclear dipole bremsstrahlung is forbidden.   The unique signature of molecular X-rays comes from the fact, that - contrary to atomic X-rays - in addition to the spontaneous X-rays also induced radiative transitions occur, which can be recognized by their asymmetric angular distribution relative to the beam axis. It is the purpose of this note to discuss the X-ray spectrum due to both spontaneous and induced transitions including the most important dynamical effects.

The adiabatic molecular states have been given by the Two-Center-Dirac-Equation discussed in the foregoing paragraph. Non-adiabatic effects and radiative coupling can be described by writing the generally covariant form of the Dirac equation[31,34] in a rotating, instantaneous coordinate system whose z-axis is along the line connecting the two scattering nuclei[19b]

$$\gamma^k \left(\frac{\partial}{\partial x^k} - \frac{ie}{\hbar c} A_k + \Gamma_k \right)\psi + \frac{m_e c}{\hbar} \psi = 0 \tag{63}$$

The $\Gamma_k = \frac{1}{4}\gamma^i(\gamma_{i/k} - \{^{\ell}_{ik}\}\gamma_{\ell}) - \frac{1}{32}$ trace $(\gamma\gamma^i\gamma_{i/k})\gamma$ are the generalized connections for spinor differentiation and the Dirac-matrices γ^k are determined via the metric g^{ik} of the coordinates in the rotating frame[31,19b]. Explicit evaluation for a system rotating with angular velocity $\vec{\omega}_{rot}(t)$ gives

$$i\hbar \frac{\partial \psi}{\partial t} = [c\vec{\alpha}(\vec{p} - \frac{e}{c}\vec{A}) + V + \beta m_e c^2]\psi - \vec{\omega}_{rot}\cdot[\vec{r}\times(\vec{p} - \frac{e}{c}\vec{A} + \frac{\hbar}{2}\sigma)]\psi - \frac{i\hbar}{4c}(\vec{\omega}_{rot}\times\vec{r})\cdot\vec{\alpha}\psi \tag{64}$$

There is a spontaneous photon emission via the operator $H_{rad\ spon} = -c\vec{\alpha}\ \vec{A} = -i\frac{e}{\hbar}[H,\vec{r}]\cdot\vec{A}$ and an induced emission caused by $H_{rad\ ind} = \frac{e}{c}\vec{\omega}_{rot}\cdot(\vec{r}\times\vec{A})$, where the former current is imaginary while the latter is real. Therefore no interference occurs and the cross section for the two processes is given in the approximation of Fermi's Golden Rule by[25,31,32]

$$\frac{d\sigma}{d\omega d\Omega_k d\Omega_{ion}} = \frac{\omega^3}{2\pi\hbar c^3}|\vec{d}_{f1}|^2 \sin^2\Theta_{kd}\left(\frac{d\sigma}{d\Omega}\right)_{ion\,Ruth},$$

$$\frac{d\sigma_{ind}}{d\omega d\Omega_k d\Omega_{ion}} = \frac{\omega\,\omega_{rot}^2}{2\pi\hbar c^3}|\frac{\vec{I}}{|I|}\times\vec{d}_{fi}|^2 \sin^2\Theta\left(\frac{d\sigma}{d\Omega}\right)_{ion\,Ruth}$$

(65)

Averaging over all possible orientations of the molecular dipole d_{fi} (closed shells!) gives a uniform distribution for the spontaneous radiation and for the induced radiation a distribution $\dfrac{d\sigma_{ind}}{d\omega\,d\Omega_k\,d\Omega_{ion}} \propto 1 - \dfrac{1}{2}\sin^2\Theta_{KI}$

where Θ_{KI} is the angle between the photon and the rotation axis $\dfrac{\vec{I}}{I}$, or if we integrate also over the azimuthal ion scattering angle

$$\frac{d\sigma_{ind}}{d\omega d\Omega_k \sin\Theta_{ion}\,d\Theta_{ion}} \propto \frac{1}{2} + \frac{1}{4}\sin^2\Theta_k$$

where Θ_k is the angle between photon and beam axis. Only the $2p_{3/2}\sigma$ and π and $2p_{1/2}\sigma$ to $1s_{1/2}\sigma$ transitions are included in our calculations. They are so far assumed to be equally occupied with electrons. Unequal population (alignment) of the $2p_{3/2}\pi$, $2p_{3/2}\sigma$ and $2p_{1/2}\sigma$ MO's with probabilities A^2, B^2 and C^2 respectively, leads to the following asymmetry as a function of photon energy $\hbar\omega$, photon scattering angle Θ_K and ion energy E

$$\eta(\omega,\theta_k,E)=\frac{d\sigma(\theta_k)}{d\omega d\theta_k}\bigg/\frac{d\sigma(\theta_k=0)}{d\omega d\theta_k} \quad -1=\frac{3}{2}\sin^2\theta_k\frac{(A^2-B^2)-(\gamma A^2-7/3B^2-c^2)\alpha_3^2}{3A^2+5\,B^2+6\beta c^2+(3A^2\mp B^2-3\gamma c^2)\alpha_3^2}$$

(66)

where $\beta=\dfrac{D_1^2}{D_3^2}$ is the ratio of the squared reduced dipole matrix elements of the $P_{1/2}\sigma$ and $2p_{3/2}\pi$ to the $1s_{1/2}\sigma$ states,

$$= \frac{D_1^2}{D_2^2}\frac{\alpha_1^2}{\alpha_2^2}, \text{ and } \alpha_\nu = \frac{v}{R_\nu(\omega)\omega} = \frac{\sqrt{2ME}}{R_\nu(\omega)\omega}, \quad R_\nu(\omega)=R_0\cdot\frac{\sqrt{\omega_s\omega_c}}{\omega-\omega_s}$$

is the (approximate) two center distance as a function of the photon energy. For the $2p_{1/2}$ MO one has $\hbar\omega_s=0$, $\hbar\omega_c=31.8$ keV and for the $2p_{3/2}$ MO, $\hbar\omega_s=7.5$ keV and $\hbar\omega_c = 32.2$ keV. $R_0=2100$ fm. Obviously η is largest perpendicular to the beam axis ($\theta_K=90°$). The comparison of (66) with experiments can serve to determine the alignments A^2/B^2, A^2/C^2. Due to the non-adiabatic effects the molecular states obtain a width $\Gamma(t)$ which is composed of three different parts[25]

$$\Gamma_{rad\ spon} = \oint_{nk\varepsilon} \left| e\sqrt{\frac{2\pi}{\hbar\omega}} \int_o^t dt'\ e^{i\int_o^{t'} d\tau'\omega_{mn}(\tau')+i\omega t'} \omega_{mn} <n(R)|\vec{r}\cdot\vec{A}|m(R)>\right|^2$$

$$\Gamma_{rad\ ind} = \oint_{nk\varepsilon} \left| e\sqrt{\frac{2\pi}{\hbar\omega}} \int_o^t dt'\ e^{i\int_o^{t'} d\tau'\omega_{mn}(\tau')+i\omega t'} \vec{\omega}_{rot}\times<n(R)|\vec{r}\cdot\vec{A}m(R)>\right|^2$$

$$\Gamma_{non\ rad} = \oint_{n=m} \left| \int_o^t dt'\ e^{i\int_o^{t'} d\tau\ \omega_{mn}(\tau)} <n(R)|\dot{R}\frac{\partial}{\partial R} + \frac{\vec{\omega}_{rot}\cdot\vec{j}}{\hbar}|m(R)>\right|^2 \qquad (67)$$

In our present calculations an effective parameter Γ has been chosen for the sum of these widths[25], which has been varied between 1 and 5 keV to establish the dependence of the spectra on the width which is a function of scattering energy ($\Gamma\sim\sqrt{E}$). It leads to a smooth transition from the molecular X-ray spectrum into the quadrupole bremsstrahlung background,[35] which can serve for its determination (see Fig.18b). Moreover, we have neglected interference effects of the radiation for high energetic X-ray radiation.[36] The cross sections are calculated here by integrating (65) along the trajectory and, for the single spectrum, by further integration over all ion scattering angles. The transition energies and dipole matrix elements are taken from the exact relativistic one-electron states for the Ni-Ni-system shown in Fig.13. For the symmetric Ni-Ni collision, the symmetrized Rutherford cross section is used. The results for a coincidence experiment (equal population assumed) at various scattering angles and energies are shown in Fig. 18a. The spectra at moderate forward angles exhibit a pronounced peak at the high energy end which is essentially due to the additional induced radiation that is strongly peaked at high photon energies. The induced radiation also grows with increasing scattering energy, whereas the spontaneous yield per K-vacancy drops. Of course, the K-vacancy production will increase with rising energy. For low scattering energy (e.g. 5 MeV) the molecule no longer reaches the asymptotic "runway" region and no peak structure occurs. For very forward or more backward angles ($\Theta<10°,\Theta>45°$) the contribution from induced radiation decreased with respect to the spontaneous X-ray yield. Especially for very forward angles ($\Theta<3°$) the runway region is forbidden and the peak vanishes altogether (Fig.18a). The width of the peak at the end of the

spectra at forward angles is almost completely determined
by the collision broadening $\Gamma(t)$ and not so much by the
shape of the level diagram. For the singles experiment
the spectra shapes look essentially exponential (Fig.18b).
The calculation assumes that a K-vacancy is brought into
the collision (two-step process). For a one-step process
(ionization and X-ray in the same collision) the impact
parameter dependence of K-vacancy production has to be
folded into the integration over Rutherford cross-section.
The end point of the spectrum is a function of scattering
energy, because Γ itself increases with scattering energies, e.g. 5 MeV in Fig.18b. The experimental endpoint of
the spectrum may even lie before the united atom limit as
the asymptotic region of the level diagram (Fig.13) is
not reached in the collision. For comparing theory with
experiment we note that the background of the tails of the
atomic K_α $_\beta$-lines, the electron bremsstrahlung, transitions
from higher MO's, interference effects between atomic and
MO-X-rays,[37] and also the contribution of the slower recoil ions to the MO-X-ray production[33] are expected to contribute mostly in the low energy part of the X-ray continuum. This leads to a greater "filling up" of the spectrum
shown in Fig.18b at lower energies. It will be discussed
in a forthcoming paper. The asymmetries (66) are shown in
Fig. 19 for various alignments.

For not too small scattering energies they show a practically linear dependence on energy,which is to a large extent compensated by the inverse energy dependence of the
width $\Gamma(t)$. They exhibit a strong dependence on photon energy $\hbar\omega$, which is entirely due to the induced transitions.
The observation of this effect by J.S. Greenberg and C.K.
Davis[32,33] uniquely proves the existence of molecular X-
rays and of induced transitions at the same time. We mention that recently P. Mokler and his collaborators[41] have
also observed the asymmetry for the molecular M-X-rays.
The most important $\hbar\omega$ -dependence of this asymmetry has
also been established, which independently supports the
observations of Greenberg et.al. and indicates further
the Coriolis-induced radiations. Also Groeneveld et.al.
find indications for the asymmetry in Ne-Si collisions,
as does Meyerhof et.al.[43] for the Kr-Zr-system at high
energies. These latter observations are less conclusive,
however, because of the thick targets used in these experiments and the thus resulting mixture of asymmetries
of two different molecular systems and the unknown Dopp-
ler- und recoil effects in the secondary and multiple
collisions.
The shape (peaking) of the coincidence spectra gives
great hope for future spectroscopy of two center orbitals:

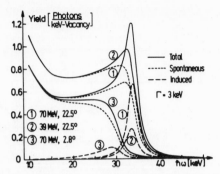

Fig.18: a) The MO-X-ray yield for coincidence experiments
with 70 MeV Ni-Ni at (1) 11.2° and (3) 1.4° scattering
angle, and (2) 39 MeV Ni-Ni at 11.2° scattering angle
(all quantities in the Lab system). The solid lines show
the total yield; the dotted and dashed lines the spontaneous and induced contributions, resp. Observe that
the peaks are strongly enhanced by the induced emission
mechanism.

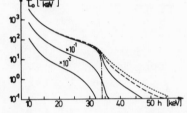

Fig.18: b) MO-X-ray cross sections for the singles experiment Ni-Ni. (1) yield for 70 MeV collision. The solid line
corresponds to $\Gamma=1$ keV, the dashed and dotted lines to
3 and 5 keV, resp. The dashed-dotted lines give the spectrum without collision broadening which sharply cuts off
at the limiting atom K_{α}-energy. (2) spectrum for 19 MeV
collision, (3) for 5 MeV, both with $\Gamma=1$ keV. All quantities in the Lab system.

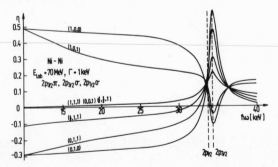

Fig.19: The asymmetry as a function of photon energy for
70 MeV Ni-Ni collisions. The alingnment (A^2,B^2,C^2) is indicated.

Shifting of the peak with heavy ion energy, which in turn
is related to the internuclear distance. In Ni-Ni collisions, e.g. it should ultimately be possible to determine
the center of the peak to within a few hundred eV. This
general feature should also hold for much heavier systems
because the induced radiation always comes from the
"compound atom"-region of the two center level diagram.
Experimentally this opens up the completely new field of
X-ray spectroscopy of superheavy molecules which is of
fundamental interest for quantum electrodynamics of strong
fields. For example, in a system with $Z_1+Z_2 \sim 150$ the contribution of vacuum polarization from the Uehling potential is expected to be of the order of 5 keV for the
$2p_{3/2}-1s_{1/2}$ transition and exceeds the selfenergy contribution. Thus it would be for the first time possible to
measure vacuum polarization as the dominant quantum
electrodynamical effect.

   5.3 Induced Decay of the Neutral Vacuum in Overcritical Fields.   The intermediate superheavy molecules
can also be used to test the predicted decay of the vacuum in overcritical fields. The idea is pictorially shown
in Fig.20
At large distances of the approaching nuclei the Fourier
frequencies descended from the changing Coulomb field
should give rise to ionization of the K-shells.Such K-
vacancies can also be produced by the Fano-Lichten[39] double collision process. At present it is unclear theoretically and experimentally which of the two processes is
contributing dominantly and how much. We assume here, because we do not know better at this time, that 1 K-hole
will be created in hundred collisions. After the field
has become overcritical at distances R<36 fm in e.g. an
U-U-collision, it will be populated by e^+e-creations.
The escaping positrons can then be observed experimentally as the indication for the decay of the vacuum.We
shall study this process now more quantitatively.

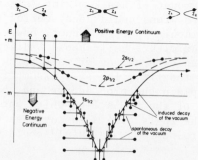

Fig.20: The dynamical processes occurring in connection
with the decay of the vacuum in overcritical fields.

Up to now calculations were done for the spontaneous positron creation (see sects. 2 and 3). The cross sections will be increased, however, by two effects due to the nonadiabaticity of the heavy ion collision: First, during the "diving"-process e^+e^--pairs are created in addition to the spontaneous ones by induced autoionization. Second, before and after the diving a large number of positrons will be created by induced transitions from the negative energy continuum to the $1s_{1/2}$-level. The latter transitions will occur even if there is no level-diving during the collision. To calculate the total transition amplitude, all the transitions may be, approximately grouped into, first, pre- and after-diving amplitudes, $C_{pA,E}$ and, secondly, the during diving amplitudes, C_D. Then we have

$$
C_{PA,E} = \int_{-\infty}^{-t_{CR}} dt' \, \exp \frac{i}{\hbar} \int_{-\infty}^{t'} dt''(E-E_{1s_{1/2}}(t'')) - \frac{1}{2\hbar}\int_{-\infty}^{t'} \Gamma(t'')dt'' M_E(t')
$$

$$
+ \int_{t_{CR}}^{\infty} dt' \, \exp \frac{i}{\hbar} \int_{-\infty}^{t'} dt''(E-E_{1s_{1/2}}(t'')) - \frac{1}{2\hbar} \int_{-\infty}^{t'}\Gamma(t'')dt'' M_E(t') \qquad (68)
$$

with

$$
M_E(t') = -\left\langle \psi_E(t') \left| \frac{\partial}{\partial t'} \right| \varphi(t') \right\rangle \qquad (69)
$$

and $\Gamma(t'')$ represents the decay of the $1s_{1/2}$-level due to the interaction with the continuum (to be determined later). $|\psi_E\rangle$ and $|\varphi\rangle$ are the wave functions for positron (negative energy continuum) and the $1s_{1/2}$-vacancy resp. The time integration can be replaced by an integration over dR/V_R along the ion hyperbola, where R is the distance of the two ions and V_R the radial velocity. The matrix element $M_E(t')$, can be computed by expanding the bound state, $|\varphi\rangle(t') = |\varphi(R)\rangle$, and the continuum states, $|\psi_E(t')\rangle = |\psi_E(r)\rangle$, which are eigenstates of the Hamiltonian, $H(R)$, about the $1s_{1/2}$ diving radius, $H(R)_{CR}$, i.e.

$$
|\varphi(R)\rangle = f(R) \left\{ |\varphi_{CR}\rangle + \int dE' g_{E'}(R) |\psi_{E',CR}\rangle \right\}
$$

$$
|\psi(R)\rangle = a(E,R)\left\{ |\varphi_{CR}\rangle + \int dE' b_{E'}(E,R)| \psi_{E',CR}\rangle \right\} \qquad (70)
$$

with the following normalizations

$$
\langle \varphi(R)|\varphi(R)\rangle = 1
$$
$$
\langle \psi_E(R)|\psi_{E'}(R)\rangle = \delta(E-E') \qquad (71)
$$
$$
\langle \psi_E(R)|\varphi(R)\rangle = 0
$$

In this expression (70), the higher bound states, i.e.
$2p_{1/2}$, $2s_{1/2}$,etc. have been neglected. Such effects are
expected to be most important at large ion separations
where the matrix element, $M_E(t)$ is small. The matrix
element (69) can now be written in terms of the expansion coefficients and reduces by use of eq.(71) to

$$M = [a (E,R)f(R) \int dE' b_E(E',R) [\tfrac{\partial}{\partial R} g_{E'}(R)]] v_R \tag{72}$$

To solve for the expansion coefficients, the following
matrix elements are needed

$$\varepsilon(R) = <\varphi_{CR} |V(R)| \varphi_{CR} >$$

$$V_E(R) = <\psi_{E,CR}| V(R)|\varphi_{CR}> = (\gamma(E)/_{2\pi})^{1/2} \varepsilon(R) \tag{73}$$

$$V_{EF'}(R) = <\psi_{E,CR}| V(R)|\psi_{E',CR}>$$

where $V(R) \equiv H(R) - H(R_{CR})$.

When the small contribution from the continuum-continuum
matrix elements, $V_{EE'}(R)$, are neglected, the set of coupled channel equations may be easily solved[25,26] for the expansion coefficients. Upon substituting these coefficients
into eq.(72), the matrix element $M_E(t)$ may be readily
computed. Once this matrix element is specified, the contribution from induced transitions to the total transition probability may be approximated by(eq.(68))

$$W_{PA} (E_p,E_I,\Theta) = | C_{PA}|^2 dE_p \tag{74}$$

where E_p is the positron energy. The dependence of this
probability on the ion energy E_I, and the scattering angle Θ enters through the time integration over the Rutherford trajectory.
The during diving amplitude C_D is given by[25,26]

$$C_D = \frac{i}{\hbar} \int_{-t_{CR}}^{t_{CR}} dt\, V_E(t) e^{\frac{i}{\hbar}\left[\int_{-\infty}^{t} dt'(E-E_{1s_{1/2}}(t'))\right] - \frac{1}{2\hbar} \int_{-\infty}^{t} \Gamma(t')dt'} \tag{75}$$

with the corresponding diving probability

$$W_D(E_p,E_I,\Theta) = |C_D|^2 dE_p \tag{76}$$

The total probability for producing a positron with an
energy between E_p and $E_p + dE_p$ is then

$$W_T (E_p,E_I,\Theta) = | C_{pA} + C_D|^2 dE_p \tag{77}$$

To compute these probabilities, the functions $\gamma(E)$ and
$\varepsilon(R)$ must be specified. The $\gamma(E)$ dependence was extracted
from the resonance in the one center continuum wave func-
tions. $2\pi| V_E(R)|^2$ depend on both the charge Z and the

separation R, as seen in Fig.21, the ration $\gamma(E) = M_e c^2 \Gamma_E(R)/\epsilon^2(R)$ depends only on energy to a good approximation. We used the approximation $\gamma(E)=\gamma_o(E-E_o)^2 e^{-\alpha(E-E_o)}$ where γ_o, E_o and α are parameters in the present calculations. While this energy dependence is based on one center calculation, we assume $\gamma(E)$ does not change appreciable when two center wave functions are used. The large width of $\gamma(E)$ enables one to use $\Gamma(R) = \bar{\gamma}_o \epsilon^2(R)$ in the calculation of the line broadening (eq.68). The remaining function to be determined,$\epsilon(R)$, was extracted from the Two-Center-Dirac-Model eigenvalues.

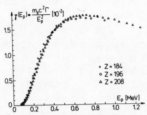

Fig. 21:   The $\gamma(E_p)$-dependence plotted for $180<Z<210$, for distances greater than 15 fm (i.e. below the Coulomb-barrier). Obviously, $\gamma(E_o)$ does not depend on Z and R.

The quantities $W_{PA}(E_p,\theta)$, $W_D(E_p,\theta)$ are shown for the U-U-system in Fig. 22a). They are related to the cross section for positron production by $\frac{d\sigma}{dE_p d\Omega_{ion}} = \frac{d\sigma_R}{d\Omega_{ion}} L_o W(E_p,\theta)$ where $d\sigma_R$ is the differential Rutherford cross section and L_o the initial K-hole probability, which has been taken in all our calculations to be $L_o = 10^{-2}$ (see Refs. 25,26). This number has no justification. It is a simple guess. The present results of Meyerhof[43] contradict this guess; he estimates from his data $L_o \approx 10^{-5}$. This in turn is in contradiction to results of Burch, Vandenbosch et. al.[44] who estimate from their findings in Cl-U-U-collisions $L_o \approx \frac{1}{2}$. Thus the situation is at present not quite clear, but encouraging and exctiting.
Fig. 22b) shows the total ionization probability $W_T(E_p,\theta)$. The full curves demonstrate that with decreasing ion energy $\bar{E}_I$, the energies E_p for the maximal positron cross sections are shifted to smaller values. This possibly allows to some extent a spectroscopy of the diving mechanism. For fixed ion energy and varying scattering angle θ, the energy maximum is only slightly shifted, as can be seen from the dashed curves.
The positron production cross section in the energy interval between E_p and $E_p + dE_p$ is given by integration of (3) over the angles

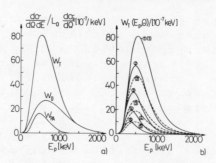

Fig.22: a) The probabilities $W_{PA}(E_p,\theta)$, $W_D(E_p,\theta)$ and
$W_T(E_p,\theta)$ for the system U→U in the case of central collision with E_I = 812.5 MeV.   b) The full lines show
$W_T(E_p,\theta)$ at θ= 180° for the system U→U and its dependence on R_{Min} (distance of closest approach in fm),which
corresponds to different ion energies (denoted in the
brackets, units are MeV).   (1) 15(815.5) (2) 20(609.4)
(3) 25(478.5) (4) 30(406.3) (5) 35(348.2).   In the
last case $W_T(E_p,\theta)\theta=W_{PA}(E_p,\theta)$ (no diving). The dashed
curves show for fixed ion energy E_I=815.5 MeV the θ-
dependence (1) θ=180° (2) θ=75°   (3) θ=50°
(4) θ=50° (5) θ=30°

$$\frac{d\sigma}{dE_p} = L_o \int_0^\pi W_T(E_p,\theta)d_R(\theta) \ . \tag{78}$$

For different ion energies, $d\sigma/dE_p$ is shown for U→U in
Fig. 23a). Compared with the purely spontaneous positron autoionization,[40] the dynamical, nondiabatic effects lead to a "smearing out" of the sharply peaked
excitation functions and to a considerable increase of
the cross section by nearly two orders of magnitude.
To get the total positron cross section σ one has to integrate (78) over the positron energy E_p. As a function of the ion energy E_I,σ is shown in Fig. 23 b).[x]

It is considerably larger than the corresponding one
for purely spontaneous positron autoionization. [40]
This is due to the induced transitions (nonadiabatic
effects). The vacuum thus decays - due to the nonadiabaticity introduced by the dynamics - as well spontaneously as by induction. In systems with higher total
charge Z_1+Z_2, the transitions to higher levels ($2p_{1/2}$,
$2s_{1/2}$) which come close to diving or are just diving
($2p_{1/2}$) must be taken into account and will further increase the cross sections. A spectroscopy of the diving
orbitals, i.e. of the structure of the charged vacuum,
should be possible by studying the positron

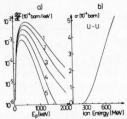

Fig.23: a) The positron production cross section $d\sigma/dE_p$ for the system U→U for various ion energies. The energies are the same as in Fig.22b).[x] b) The total prositron cross section in dependence of the ion energy.[x]

spectra in their dependence on heavy ion energy and their change with the total charge Z_1+Z_2 in the superheavy system. This, if done for lighter systems in a systematic fashion, should also yield information on the background effects. The vacuum structure can also be observed perhaps even more accurately, by spectroscopy of the molecular X-rays of the superheavy, overcritical system.

6.   Broken Symmetry and Summary

Let us come back now to the question we started with in the introduction, which is the breakdown of the Sommerfeld-finestructure formula for $\frac{Ze^2}{r}$ -potentials at $Z=137$. We can understand the physics at this "catastrophy" now as follows: If the finite radius of our supercritical system $(Z>137)$ is made smaller and smaller, i.e. if the finite nuclei are made more and more pointlike (highly compressed), the critical Z_{cr} (occurrence of diving) will shift towards smaller values and approach $Z=137$ in the limit R→0. All $s_{1/2}$-and $p_{1/2}$-configurations dive in the limit at $Z = \frac{1}{\alpha} = 137$ perpendicular, i.e. the normal vacuum becomes tremendously unstable and decays into an infinitely charged vacuum.[xx]   The change of the vacuum in overcritical fields is a most fundamental effect. It is a zero'th order effect unknown to the "weak-field-QED" commonly investigated. Its observation may shed now light to our understanding of field the-

---

[x] The abscissa of the analogous figure in our original publication[26] was erroneous by a factor 10^4. We thank W. E. Meyerhof for drawing our attention to this scale error. [xx] The positron production due to the decay of the vacuum in highly relativistic heavy ion collisions should, because of the point-like structure of the intermediate system due to shock waves[48], tremendously increase.

ories,and, in fact, may open a new and exciting field for
experimental investigation. The concept of overcritical
fields and the change of the vacuum state is not restricted to the electron-positron fields and electromagnetic
interactions. It may also occur for other spinor fields
(nucleons, quarks) and be caused by strong interactions.
Indeed, J.Wolf,[46] has constructed along these ideas a
quark-model of elementary particles, where nucleons,pions
etc. are considered as <u>charged quark vacua.</u> The overcritical phenomena happen, of course, also for boson fields.
Rafelski and Klein[47,48] have investigated and summarized
various attempts for the understanding of overcritical
pion fields. What is still lacking, and, in fact, quite
unknown, is the question to what extent such bosons are
real bosons. If a pion is, for example, built out of the
quark-antiquark Fermi-particles and if - as it seems -the
pionic orbitals near the overcritical phonemena are rather
small ($r_{pion} \approx 2$-3 fm) it could very well happen that the
complex structure of the bosons plays an important role.
This has to be seen in analogy with the α-particles, which
are the best massive bosons we know (superfluidity of
Helium etc.) and still are no bosons if put together in
e.g. an O^{16}-nucleus. In other words, there should occur
again a stabilization of the overcritical vacuum for complex bosons because of the Pauli-principle to which the
constituent particles are subject to.
The study of overcritical systems in superdense nuclear
matter occurring in shock waves[48] of relativistic heavy
ion collisions should thus shed light on very fundamental
questions.
We finally remark that we are dealing here with a master
example of a <u>broken symmetry</u>, which was invented mathematically by Goldstone[49], investigated and understood further by Higgs[50] and recently speculated about by Weinberg[51]
in connection with a unification of the theory of weak
electromagnetic interactions. To understand this comment,
let us make plausible the concept of broken symmetries with
the following remarks:
The theory of electromagnetism and the electron-positron
fields is generally gauge invariant. As a consequence of
that we are let to the continuity equation for the electric current, which in turn leads to charge conservation.
Similarly both the electromagnetic theory and its coupling with the electron-positron field are generally invariant under charge conjugation, which leads directly to
the symmetry between particles and antiparticles. Consider now our atoms built solely out of either protons (positive Z) with surrounding electrons and antiprotons (negative Z) with surrounding positrons. Then, for a given
undercritical charge Z we obtain the energy of the to-

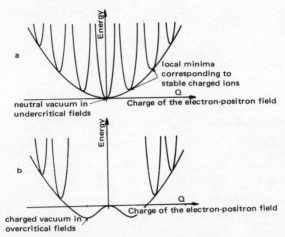

Fig.24:   Schematic drawing of the energy of the system
against charge, which is thought to be continous. The
quantization of charge at integer values can be imagined
to occur because of stabilization potentials. In case
a) the vacuum (groundstate) is symmetric under charge
conjugation. This symmetry is broken in overcritical
fields (case b) where the vacuum is charged.

tal system as a function of charge as shown in Fig.24a),
The groundstate of the total electron-positron field is
obviously belonging to charge O; i.e. the vacuum has zero charge. This state has still the charge conjugation
symmetry, as has the total system. The latter manifests
itself in the left-right symmetry of Fig.24a), while the
former is indicated by the left-right symmetry of the minimum at Q=0. The local minima at higher energy correspond to stable charged atoms. Because the charge is quantized, they are separated along the abscissa in integers
of e. These higher local minima as a function of e do no
more show charge conjugation symmetry - though the overall world, i.e. the total Fig. 24a) shows it. We say,
in these stable states the C-symmetry is spontaneously
broken. In overcritical fields now we have a situation
as shown in Fig. 24b). The groundstate has now become
charged, i.e. the vacuum is charged and does no more
show C-symmetry even though it exists globally. A tiny
physicist in an overcritical atom would perhaps not even
notice that overall C-symmetry exists, because his vacuum
(groundstate) does not show it. This situation is analogous to the magnetization of a ferromagnet at temperature T (undercritical case) and near zero temperature
(overcritical case).

References

1 W. Pieper and W. Greiner, Z.Physik 218 (1969)327

2 B. Müller, J. Rafelski, and W.Greiner, Z.Physik 257(1972)82 and 163

3 U. Fano, Phys.Rev.124 (1961)1866

4 P.G.Reinhard,W.Greiner, and A. Arenhövel,Nucl.Phys.A166(1971)173

5 J. Rafelski, B. Müller, and W.Greiner, Nucl.Phys.B68 (1974)585

6 B.Müller and J.Rafelski, On the Hartree-Fock Approximation in Quantum Electrodynamics, to be published

7 M. Guylassi, Thesis, Univ. of California at Berkeley and Lawrence Berkeley Lab.,1974

8 G. Soff, J. Rafelski, and W.Greiner, Phys.Rev. A7(1973)903

9 M. Born, Am.Inst. Henri Poincaré 7 (1937) 155; L. Infeld and B. Hofmann, Phys.Rev.51 (1937)765

10 J. Rafelski, B. Fulcher, and W. Greiner, Il Nuovo Cimento,Vol. 13B (1973)135

11 J. Waldeck, diploma thesis, Inst.f.Theoretische Physik der Univ. Frankfurt/Main, 1973

12 W. Scheid, J.Hofmann, and W. Greiner, Nuclear Shock Waves and Pionization of Matter, Berkeley Symposium on Physics with Relativistic Ions, July 1974

13 U. Mosel and W. Greiner, Z.Physik 217 (1968)256; G. Grumann, U. Mosel, B. Fink, and W. Greiner, Z.Physik 228(1969) 371; J. Grumann, T. Morovič, and W. Greiner, Z.Naturforschung 26a (1971) 643 ; A. Sobitcevsky, Proceedings of the Ronneby Conf., June 1974, see also contribution to the Int.Conf.on Nucl.Physics, Munich (1973)

14 The prediction of intermediate superheavy molecules in heavy ion collisions was made by the authors during the GSI-seminars 1969-1971 and first published in the article J.Rafelski, L. Fulcher, and W. Greiner, Phys.Rev.Lett. 27 (1971)958

15 W. Heitler, F.London, Z.Physik 44 (1927)455; E. Hylleraas, Z. Physik 71(1931)739; E. Teller, Z.Physik 61 (1930)458

16 K. Helfrich and H. Hartmann, Theoret.Chim. Acta 3 (1965)21

17 see e.g. F.T. Smith, Phys.Rev. 179 (1969) 111

18 S.K. Luke et.al., J.Chem. Phys. 50 (1969)1644

19 a) B. Müller, J. Rafelski, and W. Greiner, Phys.Lett.47B, No.1 (1973) 5 ; b) B. Müller and W. Greiner, The Two Center Dirac Equation,preprint, Inst.f.Theor.Physik der Univ.Frankfurt/ Main, to be published

[20] W.E. Meyerhof, F.K.Saylor, S.M. Lazarus, W.A.Little, and B.B. Triplett, Phys.Rev.Lett. 30(1973)1279; Phys.Rev.Lett.32(1974)502

[21] J.S. Greenberg, C.K.Davies, and P. Vincent, Phys.Rev.Lett.32 (1974)1215

[22] P. Mokler, H.K. Stein, and P. Armbruster, Phys.Rev.Lett.29(1972) 827

[23] D. Burch,W.B.Ingalls, H.Wieman, R. Vanderbosch, preprints,Univ. of Washington, Seattle (1973/74)

[24] B. Fricke, W. Greiner, and J.T. Weber, Thor.Chimica Acta 21(1971) 235 ; G.Soff and B. Fricke, GSI-report T 1-74 (1974)

[25] K. Smith, B.Müller, and W. Greiner, J.Phys.B. in print

[26] K. Smith, H. Peitz, B.Müller, and W. Greiner, Phys.Rev.Lett. 32(1974)554

[27] F.W. Saris, W.F.van der Weg, H.Tawara, and W.A.Laubert, Phys. Rev.Lett. 28(1972)717

[28] G.Presser, K.O.Groeneveld, and B. Knaf, Contribut.to the 4th Int. Conf.Atomic Physics, Heidelberg, July 1974, Abstracts, p. 621

[29] J.Rafelski, B.Müller, and W.Greiner, Lett.il Nuovo Cim.11(1972)469

[30] B. Müller, K. Smith, and W. Greiner, Phys.Lett.49B(1974)219

[31] B. Müller and W. Greiner, Phys.Rev.Lett. 33(1974)469

[32] J. Greenberg, C.K. Davies, and P. Vincent, Phys.Rev.Lett.33(1974) 473

[33] J. Greenberg, C.K. Davies,B.Müller, and W.Greiner, Proc.of the Int.Conf. on Interactions between Nuclei, Nashville,Tenn.June 1974

[34] H. Tetrode, Ann. Physik 50 (1928)336

[35] J. Reinhardt et.al., Multipole bremsstrahlung in heavy ion colli- sions, Inst.f.Theoretische Physik der Univ.Frankfurt/M.,March 1974

[36] Note that the expression for the time-dependence width and similar ones for the cross sections show rainbow effects due to the time dependence of the matrix elements, of $\omega_{mn}(t)$ and $\Gamma(t)$ itself. Since these interference effects show up where the phase differ- ence between contributions from approaching and separating ions along the scattering path are largest, they should occur most pronounced for low energetic transitions.

[37] K.Smith, B.Müller, and W.Greiner, Coherent effects in X-rays from intermediate molecular systems, preprint, Inst.f.Theor.Physik der Univ. Frankfurt/Main, July 1974

[38] We acknowledge the help of W. Betz in the calculations of the level diagrams.

[39] U.Fano and W. Lichten, Phys.Rev.Lett. 14(1965)627

[40] H.Peitz, B.Müller,J.Rafelski, and W.Greiner, Lett.Nuov.Cim.8 (1973)37

[41] P.Mokler et.al.,Inv.Paper at the 4th Int.Conf.on Atomic Physics, Heidelberg,July 1974, and Phys.Rev.Lett.33(1974)476

[42] K.O.Groeneveld and B.Knaf,Inst.f.Kernphysik der Univ.Frankfurt/M., private communication

[43] W.F.Meyerhof et.al.,Contr. to the 4th Int.Atomic Physics Conf., Heidelberg, July 1974, Abstracts, p. 625, 629,and 633

[44] D.Burch, V.Vandenbosch et.al.,preprint Univ.of Wash,Seattle(1974)

[45] H.E. Gove, F.C.Jundt, and H.Kubo, preprint Univ.Rochester,Spring 1974

[46] J.Wolf, diploma thesis, Inst.f.Theor.Physik der Univ.Frankfurt/M., June 1974

[47] J.Rafelski and A.Klein, Instabilities of Matter in Strong External Fields and at High Density, Center for Theoretical Studies, Miami, Fla.,Jan 1974

[48] See also W.Scheid, J.Hofmann, and W.Greiner, Nuclear Shock Waves and Pionization of Matter, Symp. on Physics with Relativistic Heavy Ions, Berkeley, July 1974

[49] J. Goldstone, Nuov.Cim. 19(1961)154

[50] P.W. Higgs, Phys.Rev. 145(1966) 2156

[51] S. Weinberg, Phys.Rev.Lett. 29(1972)388; Phys.Rev. D8(1973)4482

THE QUASI-MOLECULAR MODEL OF ATOMIC COLLISIONS*

William Lichten

Physics Department, Yale University

New Haven, CT   06520, U.S.A.

My mentor, Robert S. Mulliken, once said to me, "Molecules are just like a puzzle. They may seem hard to understand, but if you just think about them, you can always solve the problem." Can we be as confident about atomic collisions? Considering the progress we have made in the past dozen years, the answer seems to be yes.

What is the solution of a problem in atomic collisions? The elements are reliable experimental data, an ab initio calculation of the interactions in the colliding system and a solution of the coupled differential equations between the states that enter into the collision.

WHY A MOLECULAR MODEL?

Consider the example of K-shell excitation. The problem of collisions by fast, structureless projectiles, such as electrons, protons, or sometimes heavier nuclei, is solved. Born approximation, binary encounter and impact parameter models agree among themselves and with experiment. Figure 1 shows a fit of experimental data to a universal function derived from a modified Born approximation by Brandt and Laubert.[1] At high velocities (to the right of the diagram), or for light, structureless particles, theory fits a wide range of data. At the left hand side of the graph, where the particle velocity is low, or the bombarding nucleus is heavy, there is a discrepancy as large as a factor of 10^{14} between theory and experiment. Here the structure of the compound system of atom + projectile is all important. In recent years, good progress has been made in understanding this problem, which will occupy the remainder of this article.

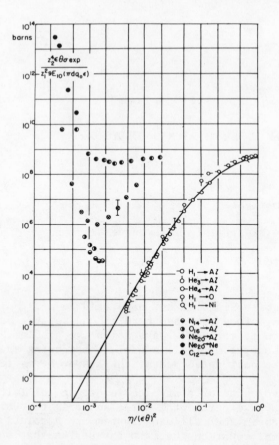

Fig. 1 K-shell ionization cross sections of various targets excited by the ions, as shown on the figure. The ordinate is a reduced
cross-section and the abscissa is a reduced velocity. The solid
line is the theoretical cross section in the Born approximation.
This figure is reproduced from ref. 1.

        Atomic systems have a shell structure in the electronic clouds
that surround the nuclei. The interactions in the <u>outer</u>, or valence,
shells are extremely delicate and complex. These make up the world
of chemistry and optical spectroscopy. Exciting discoveries have
been made in collisions involving outer shell electrons; great progress has been made in our understanding of these collisions. However each system requires a separate analysis and general principles
are hard to come by. For lack of space, we rule out a discussion of
this important field.
        Collisions involving <u>inner</u> shell electrons, on the other hand,
are universally understood by means of molecular orbital theory,
since the electrons act independently. Thus, a common thread ties
together atomic collisions over an enormous range between ultimates:

from proton–hydrogen encounters at thermal energies up to Uranium–
Uranium collisions at the Coulomb barrier of 1.6 GeV.

MOLECULAR STATES

We consider the molecular states which diagonalize the electronic Hamiltonian H_{el}, in the Born–Oppenheimer approximation of
slow nuclear velocities. If we neglect translational factors and
assume a classical path $\vec{R}(t)$ for the coordinate of the projectile
nucleus relative to the target nucleus. The time dependent Schrödinger equation becomes a set of coupled differential equations for
the state amplitudes

$$a_k: \quad i\dot{a}_k = H_{kk}\, a_k - i \sum_j V_{kj}\, a_j, \quad \text{where } H_{kk}(R)$$

is the diagonal electronic energy of the system in the <u>k</u>th state
and the coupling elements are the off-diagonal elements of the
Hamiltonian.

A major part of the problem is to find a suitable basis set
of approximate wave functions ψ_k, such that the off-diagonal matrix
elements ψ_k are small. As a nice way of defining the problem, we
consider the Landau–Zener model of level crossing (Fig. 2).

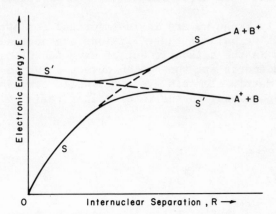

Fig. 2 Crossing of potential curves. Two potential curves for the
states S' and S may cross in a certain approximation (such as in a
single configuration molecular orbital theory). In a higher approximation, the curves repel each other. If the atoms approach each
other slowly in state S, an adiabatic transition from S to S' will
occur. If they approach each other rapidly, a diabatic transition
from S to S will occur.

As the internuclear distance R varies during a collision, the system goes through a pseudo-crossing of the two energy levels S' and S. Two extreme possibilities exist:

1- diabatic transition. If the transition is rapid, the systems cross without appreciable transition probability (solid curves on Fig. 2).

2- adiabatic transition. If the transition is slow, the states fail to cross (broken curves on Fig. 2).

The Landau-Zener formula gives a quantitative expression for the diabatic transition probability

$$P = e^{-v_0/v_R}, \text{ where } v_R = dR/dt$$

is the relative velocity of the two nuclei, and

$$v_0 = \frac{(2\pi H_{SS'})^2}{h \dfrac{d(H_{SS} - H_{SS'})}{dR}} \quad .$$

Note that for large v, the transition probability is unity; the transition is diabatic and the states cross. For v small, the transition probability is zero; the crossing is avoided.

Which velocity range we are in is important for our choice of our basis set. If we believe the collisions are near adiabatic, the correct basis is the adiabatic set of wave functions. The diagonal matrix elements are the broken curves shown in Fig. 2. The off-diagonal matrix elements are the dynamic coupling elements V_{jk} which are proportional to the internuclear radial velocity:

$$V_{kj}(\vec{R}) = \langle \psi_k | (\partial/\partial t) | \psi_j \rangle_\theta =$$

$$= v_R \langle \psi_k | (\frac{\partial}{\partial R}) | \psi_j \rangle$$

Note that the adiabatic states j,k change from one diabatic state S to the other S' or _vice versa_ during the transition, and at the crossing point the adiabatic states j,k are a 50-50 mixture of the diabatic states S,S'.

If we believe the collision is fast, the correct basis is the diabatic set of states S,S'. The diagonal matrix elements are the solid curves in Fig. 2 and the off-diagonal matrix elements is the static perturbation $H_{SS'} = \Delta E/2$ (See Fig. 2), where ΔE is the minimum splitting of the two adiabatic energy levels.

Likewise, the rotational motion of the nuclei mixes the adiabatic states via the dynamic coupling term:

$$V_{kj}(\vec{R}) = \langle\psi_k|\frac{\partial}{\partial t}|\psi_j\rangle_R = \omega\langle\psi_k|\frac{\partial}{\partial\theta}|\psi_j\rangle$$

$$= (\frac{v_{in}b}{R^2})\langle\psi_k|iL_y|\psi_j\rangle,$$

where ω is the instantaneous angular velocity, b the impact parameter, v_{in} the initial relative velocity and L_y the electronic angular momentum operator.[2]

THE BASIC SYSTEM: ONE-ELECTRON MOLECULES

Just as the hydrogen atom is basic to the study of atoms, the fundamental molecular systems are the one-electron, molecular ions H_2^+, HeH^{++}, He_2^{+++}, etc. Figures 1 and 2 of the article by J.E. Bayfield[3] (which will be referred to as JEB) show the adiabatic states, or molecular orbitals (MO) of an electron in the field of two protons, as calculated by Bates and co-workers.[4] The states at infinite internuclear separation ($R=\infty$) are those of the separated atoms (SA) $H+H^+$. The states at zero internuclear separation ($R=0$) are those of the united atom (UA) He^+. During the proton-hydrogen collision this molecule is formed. We now ask just how well we understand the dynamic behavior of this system.

The high velocity regime, in which the incident proton velocity is fast compared to the Bohr electron velocity (E>>25 keV) can be handled through perturbation calculations. The critical range of impact energies (E $\leq$ 25 keV) generally must be handled by means of calculations using a molecular basis set. A detailed description of the progress in this area is given in JEB.[3] This article will only touch on a few cases.

Everhart and coworkers[5] discovered oscillatory resonant charge exchange in proton-hydrogen collisions:

$$H^+ + H \rightarrow H^+ + H.$$

The charge state probability for fixed scattering angle was found to have a sinusoidal oscillation with reciprocal velocity. [Similar results were found for collisions of He^+ or He. (See Fig. 3)[6]] This oscillation is well understood as an interference between the $2p\sigma$ and $1s\sigma$ states, which are equal components of the initial wave function of $H+H^+$ (see Fig. 1 of JEB).

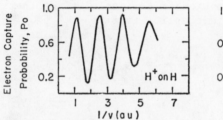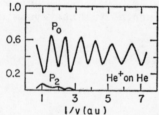

Fig. 3 Experimental results for charge exchange in wide angle scattering; P_0 is probability of the incident ion being scattered as a
neutral partice; P_2 is probability of being scattered as a doubly
ionized atom.

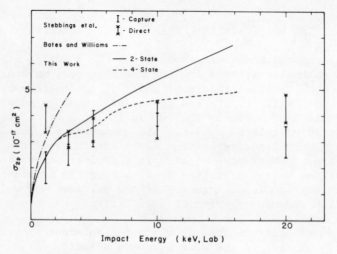

Fig. 4 Cross sections of H(2p) excitation following proton impact
on hydrogen atoms. Experimental observations of direct and capture cross sections are those of Ref. 10. Theory: see Ref. 9.

The excitation of the 2p state of H by very slow protons was
predicted correctly by Bates and Williams.[7] As the two nuclei
approach each other, the $2p\sigma$ molecular orbital is "promoted"[8] from
1s in the SA to 2p in the UA. The importance of this promotion is
that the $2p\sigma$ and $2p\pi$ MO's come very close to each other in energy
(see JEB, Fig. 2) thereby enhancing the rotational coupling between
the two states. Since the $2p\pi$ MO dissociates to $H^+ + H(2p)$, this
process results in excitation at very low impact energies.

Fig. 4 compares a calculation by Rosenthal[9] with experimental
results for the total cross section for excitation and capture into
the H(2p) state. The importance of the promotion effect is shown
by the extremely rapid rise at low impact energies. This is

emphasized by comparing Figure 4 with Figure 5, which shows the
results of Bayfield[11] for capture to the 2s state in proton-H
collisions. Note that the energy for maximum 2s capture occurs at
much larger energies than for 2p (Fig. 4) excitation and capture.
The reason for this is that there is no simple promotion mechanism
for 2s excitation or capture. The 2s state must be reached through
a chain of transitions, all of relatively high energy (see JEB,
Fig. 2), such as $2p\sigma_u \rightarrow 3p\sigma_u$, $2p\sigma_u \rightarrow 3p\pi_u \rightarrow 3p\sigma_u$, $2p\sigma_u \rightarrow 2p\pi_u \rightarrow$
$3p\sigma_u$, $2p\sigma_u \rightarrow \ldots \rightarrow 4f\sigma_u$. Thus, this process involves a rather com-
plex mixture of closely coupled states.

These conclusions are supported by the comparison of theory and
experiment. In the 2p excitation (Fig. 4), a simple, two-state
calculation agrees rather well with experiment up to about 6 keV
(half the velocity of the 1s electron). Mixing in two more states
has little effect below this energy, which verifies Bates and
Williams original conjecture[7] that this is basically a two state
process. Note that use of four states gives rather good agreement
with experiment almost up to the velocity of the 1s electron (Fig.
4).
Recently, other four state calculations of differential cross-
sections for H(2p) excitation by Chidichimo-Frank and Piacentini[12]
are in excellent agreement with the experimental work of Houver
et al.[13] However, the agreement for 2s excitation[12] is not as good,
again reflecting the multistate nature of the excitation.

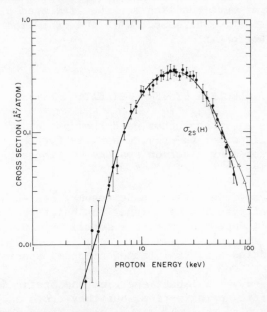

Fig. 5 Cross section σ_{2s}(H), by Bayfield (ref. 11). Note how this
differs in its energy dependence from σ_{2p}(H) (see Fig. 4).

Note that the theoretical calculations neglect the effect of the $1s\sigma_g$ MO, even though it makes 50% of the initial wave function. The reason for its small contribution is that it is energetically too deeply buried to allow an appreciable transition probability to higher electronic states.

Thorson et al.[14] have calculated the cross section for direct ionization of the H atom by incident protons at energies up to 500 eV (one seventh of the 1s electron velocity). In this energy range, the ionization proceeds almost entirely by a promotion mechanism $1s \to 2p\sigma$, $2p\pi \to$ continuum. The $1s\sigma_g \to$ continuum cross-section is 500 times smaller. Thus the major interaction leading to ionization in the p^+-H(1s) system is electron promotion and rotational coupling, the same as in the excitation of the 2p state.

As Thorson et al. point out,[14] quasi-bound molecular orbitals are a large component in the final state of the ionized electron. This was demonstrated experimentally by Rudd et al.[15] and Stolterfoht,[16] who found that the probability for an ejected electron in the forward direction is an order of magnitude greater than the predictions of the Born or binary encounter approximations. From the work of Oldham and successors,[17] we know this is "charge transfer into the continuum", where the electron is first transferred from the hydrogen atom to the incident proton, is carried along with the proton for a short time and then finally becomes a free electron moving slowly away from the proton. Recently Band[18] has used a molecular wave function for the scattered electrons and has obtained quantitative agreement with the experimental of Rudd et al.[15] and Stolterfoht.[16]

MANY ELECTRON SYSTEMS. DIABATIC STATES

A dozen years ago, the author used the concept of diabatic molecular states, in order to provide a basis for the theory of atomic collisions of many electron systems.[19] These have the following properties:[19]

1- The basis set consists of a Heisenberg representation of single configuration wave functions built up from molecular orbitals (independent-particle model);

2- the set includes virtual (autoionizable) states;

3- except at very low impact energies the states do not obey the adiabatic approximation, but they cross each other, in accordance with the well known Landau-Zener theory.

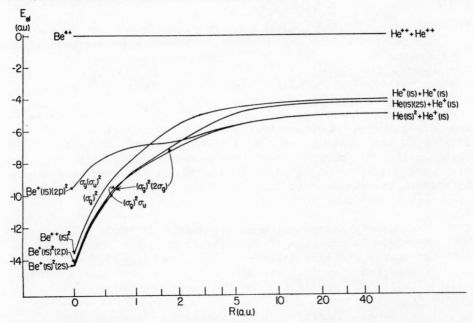

Fig. 6 Diabatic curves suitable for discussion of fast collisions
involving He ions. Crossing of states of like symmetry is permitted, in contrast with the noncrossing rule for adiabatic curves.
Thus, there is not the usual adiabatic correlation between states
of the united atom and separated atoms. Note in particular how the
$\sigma_g(\sigma_u)^2$ curve crosses the $(\sigma_g)^2(2\sigma_g)$ curve at R~2 a.u. An infinite
number of such crossings must be made before the $(\sigma_g)^2$ curve is
crossed. See Ref. 19 and Fig. 7.

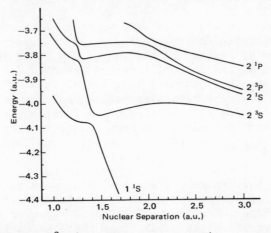

Fig. 7 Adiabatic $^2\Sigma_u^+$ curves of the He_2^+ molecule, as calculated
in Refs. 22 and 23. The legends on the right show the atomic
state of the SA He. This can be considered to be magnification
with greater detail of Fig. 6 for 1 a.u. $\leq$ R $\leq$ 3 a.u.

4- The energy of these states can be found from those of the neutral system, by calculating the MO energies (Koopman's theorem.).

A partial list of the application of the concepts follows:

A. The molecular states of $(He)_2^+$ and $(He)_2^{++}$. These simple systems furnished an important test case for the theory of atomic collisions in that they are simple, yet show most of the features of more complicated, many-electron collision partners. Diabatic states[19] are now believed to underlie almost all observed phenomena in both inelastic and elastic collisions involving helium atoms and ions. A few examples are:

1. Oscillatory charge exchange or elastic differential scattering. It was shown[19] that the frequency of oscillation of charge exchange in He^+-He collisions (Fig. 3) is accounted for by two, a priori calculable diabatic potential curves, $\sigma_g(\sigma_u)^2$, $^2\Sigma_g^+$ and $\sigma_g^2\sigma_u$, $^2\Sigma_u^+$, which interfere.

In an exhaustive study of elastic differential scattering in this system, Marchi and Smith[20] extracted these potentials from experimental data and concluded,

"A two-state theory, using the diabatic potential curves ... appears adequate to describe quantitatively all the main features seen experimentally in the elastic scattering of He^+ on He in the energy range up to 300 eV, including the non-zero minima of the interference pattern and the secondary oscillations."

2. Perturbations induced in elastic scattering by crossing of molecular states. Felix T. Smith et al.[21] interpreted a perturbation of elastic scattering of He^+ on He in terms of crossings of the potential curves of molecular states. In particular, the location, relative slope of the curves and magnitude of the repulsion between curves was found to agree well with the theory. Since then, the existence of these pseudo crossings has been quantitatively calculated by a priori methods by Rosenthal[22] and Bardsley.[23] (See Fig. 7)

3. Double charge exchange – Diabatic states were used to make simple predictions of the combined double and single charge exchange in He^{++}-He collisions.[19] Keever and Everhart[24] verified the predictions that the oscillatory frequency of double charge exchange should be half of that of single charge exchange. Also, the absolute frequencies of oscillation were in reasonable agreement with energies

derived from diabatic molecular states.[19,24]

4. Rosenthal oscillations – In many cases of excitation of electrons in outer shells of atoms, the total cross section oscillates with energy of the projectile.[25] Many of these results can be understood by the Rosenthal model.[22] In the case of He^+ on He, Rosenthal found that the primary excitation mechanism was via the diabatic $^2\Sigma_g^+$ state postulated by the author.[19] He then postulated other diabatic molecular states to provide parallel, interfering exit channels. A priori calculations by Rosenthal verified both the primary excitation mechanism and the interfering exit channels.[22,23]

5. Higher energy experiments: inelastic excitation – Barat and coworkers have extended the range of projectile energies for inelastic He^+ and He collisions.[26] McCarroll and coworkers[27] have performed quantitative calculations to explain these experimental results in terms of rotational coupling at small internuclear distances out of the diabatic $^2\Sigma_u^+$ and $^2\Sigma_g^+$ states.[19]

IMPORTANCE OF DIABATIC STATES

The pseudo-crossings shown in Fig. 6 and 7 are only the lowest few in a series. Close examination of Fig. 6 shows that there must be an infinite number of such crossings made by diabatic states during a collision.[19] A calculation based on adiabatic states is not only a poor approximation for these collisions; it is, in principle, an impossible way of solving the problem. Thus, it must be emphasized that the concept of diabatic states, based on the MO model,[19] is a complete departure from the adiabatic approximation.

The physical basis for diabatic states is the molecular orbital approximation, that treats the electrons as independent particles. The adiabatic interactions arise from smaller electronic interactions that can be neglected in collisions of energies above a few hundred electron volts. The properties (1), (2), (3), (4) (p. 8-9) of diabatic states are all necessarily present in inner shell excitation processes, including topics of current interest such as MO X-rays and positron formation.

Alternative attempts to define diabatic states or less physical and more purely mathematical grounds have been made by Smith[28] and others[29] but a recent paper[30] by Gabriel and Taulbjerg disputes the validity these definitions.[28] Thus, we are left with the definitions (1), (2), (3), (4) (p. 257) of diabatic states, which are based on the independent particle model for the electrons.

QUASI-RESONANT PROCESSES.   DEMKOV FORMULA

Another important form of excitation occurs when there are near degenerate states at large internuclear distances. In 1963, Demkov[31] made a model of this excitation in which two energy level curves are parallel, with a separation $E_1 = E_2 - E_1$, at large internuclear distance R. At small internuclear distances a perturbation of the form H_{12} = const. $e^{-R/\lambda}$ splits the levels apart. Over the transition region ($R \approx \lambda$) the eigenfunctions change into a pair of gerade and ungerade linear combinations of the separated atom wave functions (phase shift of   in the wave function). It is this change in the wave function which induces transitions between the adiabatic eigenstates. In contrast with the Landau-Zener case, there is no such thing as a diabatic state. Rapid collisions lead to an equal mixture of the wave functions of the two states.[31,32,33] The Demkov formula for K shell excitation probability w becomes, as modified empirically by Olson[32] (in atomic units)

$$w = \text{sech}^2(x) \sin^2 (\int_{-\infty}^{+\infty} H_{12}\ dt)$$

where $2x = v_0/v$

where $v_0 = \dfrac{\pi(I_1 - I_2)}{\left[(I_1 + I_2)\right]^{1/2}},$

$$H_{12} = (I_1 I_2)^{1/2} (\kappa_1 + \kappa_2)R\ e[-\tfrac{1}{2}(\kappa_1 + \kappa_2)R],$$

and $\begin{cases} \kappa_1 = \sqrt{2I_1} \\ \kappa_2 = \sqrt{2I_2} \end{cases}.$

where I_1, I_2 are the ionization energies and v_1, v_2 the velocities of the projectile and target, respectively. This process is important for understanding near resonant charge exchange or excitation transfer.

It is important to note that this process, which is quite different from the Landau-Zener model, has the same asymptotic formula in the low velocity limit:

$$P \sim e^{-v_0/v}.$$

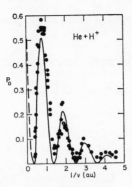

Fig. 8 Fit of experimental He+H$^+$ data to the empirical formula in the text (see p. 12).

The fact that experimental data fit this formula is no evidence that pseudocrossings exist. In fact, this asymptotic relation may be model-independent entirely.

Figure 8 shows a empirical fit made by the author[33] to data for proton-helium collisions, measured by Helbig and Everhart.[34] The empirical values are (in atomic units)

$$\int_{-\infty}^{+\infty} E_{12}\, dR = 5.71(1.15) \text{ and } v_0 = 0.85(1.06)$$

The predicted values from the modified Demkov formula are given in parentheses. The good agreement between experiment and the model for v_0 is related to the simplicity of calculating energies, overlap integrals, etc. at asymptotically large values of R. The poor agreement for the phase integral comes from the inability of simple models to get the details of the molecular interactions when the wavefunctions of the two atoms strongly overlap each other.

INNER SHELL EXCITATION

The concepts of diabatic molecular states, developed in ref. 19, were extended to collisions of complex atoms with a shell structure by means of the electron promotion model (also called the "Fano-Lichten" or MO model).[35-38] In principal, this model should hold for inner shell excitation at ion velocities small compared to those of the inner-shell electrons.[37]

The electron promotion model contains the following elements:[35-38]

1- The quasi-molecular states are diabatic during the collision. As in the case of the He_2^+ quasi-molecule, consideration of the adiabatic <u>states</u> of the system would be unthinkable.

2- In addition, the molecular orbitals themselves are diabatic in the sense that they are one-electron-like and can cross other MO's of the same symmetry ($\sigma_g-\sigma_g, \pi_u-\pi_u$, etc.). The correlation diagrams between UA and SA follows from the nodal properties of the one-electron MO's which allow crossings.

3- The adiabatic interactions which tend to avoid crossings are now core penetration effects, which result from departures of the coulomb fields of the two bare nuclei. The relative importance of these effects is proportional to the ratio of

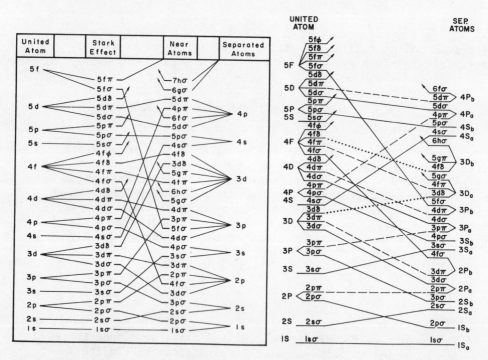

Fig. 9 Correlation diagram for a symmetric, diatomic system with many electrons.

Correlation diagram for many electrons in the field of two slightly differently charged nuclei. Source: Ref. 38.

sub-shell splitting (i.e. 2s→2p, 3s→3p→3d, etc.) to separation of states of different principle quantum numbers (i.e. 2s→3s→4s, etc).

4- Transitions at sufficiently low velocities tend to occur via Bates-type coupling (see p. 005) between near by states of the UA, by Demkov-type coupling (see page 012) at the SA, or near adiabatic tendencies of the Landau-Zener type (see page 004) in the transition region. Thus major excitations (from one principle quantum number to another) occur only upon promotion of the diabatic, one-electron-like MO's. Transitions that occur between un-promoted MO's of differing principle quantum number occur at higher velocities and represent a breakdown of the promotion model.

This model, from its very inception, was surrounded with controversy and criticism. Diametrically opposite interpretations in terms of statistical or collective effects were made.[39,40] In other cases, apparent contradictions were found between the predictions of the model and experiment. These controversies now seem to be settled. It is now agreed that the model is correct within the range originally stated[37]. In some cases where it should not hold (e.g., outer shell excitation or very high nuclear velocities) it has still been of limited value. Documentation of these statements is contained in numerous review articles.[41-46a] A few examples will be given here.

1. <u>Discrete energy losses</u> - Shortly after the appearance of the now historic observation of discrete energy losses of the order of 200 eV in Ar^+-Ar collisions by Afrosimov, Gordeev, Panov and Federenko[40] and Kessel and Everhart,[47] the electron promotion model, we have seen, asserted these discrete energy losses resulted from excitation of inner shell electrons via promotion of diabatic molecular orbitals.[35-37] Auger electrons and X-rays are well known products of inner shell vacancies. Auger electrons were predicted[35] and such electrons were observed by Rudd,[43] Everhart and coworkers,[45] Ogurtsov[44] and others in Ar^+-Ar collisions and in many other systems. Saris and many others have observed the corresponding X-rays.[42,46] Many of these X-rays and auger electron experiments followed and confirmed the principles of the electron promotion model.[35-37]

2. <u>Exit channels</u> - Empty atomic orbitals act as exit channels for promoted electrons.[37] In particular, K-shell excitation in Ne-Ne, Ne^+-Ne, Ne^{++}-Ne collisions was predicted[37] to be in the ratio of 0:1:2. The observed ratios of 0.6:1.0:2.0 agreed in part[48] with the theoretical predictions. The deviations in the case of Ne could arise from metastable atoms formed in the charge

exchange process at high energy.[99] Since then, more definitive
tests by Fastrup et al.[49,50] showed less than 1% excitation in the
K-shell of Na^+ projectiles, where the possibility of formation of
metastable projectiles was not present.

Recently, this group[50] has made a careful study of the limits
of applicability of the Electron Promotion Model in K-shell excitation. The breakdown in the predicted ratios, (see previous
paragraph and Fig. 10) occurs at velocities v ~ 0.4 a.u. (~4 keV/
nucleon). These are characteristic of velocities of electrons in
the outer shells and arise from Demkov-type couplings which break
down the MO correlations in the exit channels, even though MO
correlations in the inner shells presumably remain good up to the
MeV energy range.

The work of Fastrup et al.[49,50] is an example of a general and
very important principle. <u>The probability of inner shell excitation depends critically on vacancies present in the exit channel
MO's.</u> These vacancies are present before the collision, as in the
case of stripped ions, or are induced in the early stages of the
collision. This is a serious limitation on our ability to predict
the outcome of collisions, even though the basic inner-shell processes are well understood.

3. <u>K-Shell Excitation</u> - In the case of K-shell excitation,
the conditions are optimal for application of the electron promotion model, as the orbitals are very close to pure hydrogenic; also
there is no ambiguity whatsoever about the construction of correlation diagrams. Nevertheless, the extension of the electron promotion model to asymmetric collisions by Barat and the author[38] came
surrounded by a cloud of attacks by critics, as there seemed to be
almost universal disagreement between experimental results and the
predictions of the model.[38]

Kavanagh et al.[51] found very large carbon K-shell excitation
by Xe^+ and Kr^+ projectiles, whereas the model unambiguously predicted none, since all exit channels were closed.[37-38] The contradiction was eliminated by Taulbjerg and Sigmund, who showed that
results actually were caused by collisions of recoil carbon atoms
within the solid target.[52]

Similarly, Larkins,[53] Brandt and Laubert[54] pointed to discrepancies between experimental results[54] and the electron promotion
model for K-shell excitation in asymmetric collisions. These
results[54] also are spurious and are caused by recoil effects in
solid targets.[55] Tawara and Kistemaker[55] also have shown by a
careful study that spurious K-shell X-rays arose in Ne^+-Ar collisions due to Ne beam inpurities in the Ar gas target.

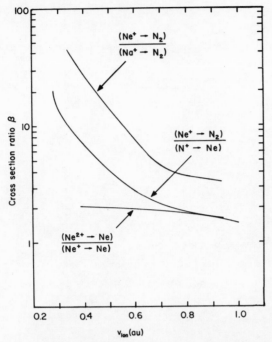

Fig. 10 Test of the exit channels mechanism of the electron promotion model. The theoretical cross section ratio is infinite for
the upper two curves; it is 2 for the lower curve. The results
agree with theory at low energies. At higher energies, the model
breaks down. Source: Ref. 50.

Finally, Meyerhof has shown that, consistent with earlier
discussions[31,33,38] that the "spurious" K-shell excitation in the
heavier atom of the collision pair occurs through charge exchange
after the electron promotion.[56] He has shown that the modified
Demkov equations (see p. 012)[31,32] give a quantitative prediction
of a wide range of experimental data in which both the asymmetry
Z_2-Z_1 and projectile velocity are changed. Fig. 11 shows the
remarkable agreement between data of Fastrup et al.[57] and Meyerhof's
application[56] of Demkov's formula,[31] as modified by Olson.[32]

Thus, apparent inconsistencies between K-shell excitation
experiments and the electron promotion model have been removed[100].

4- Asymmetric collisions.

Swapping - The extension of the electron promotion model

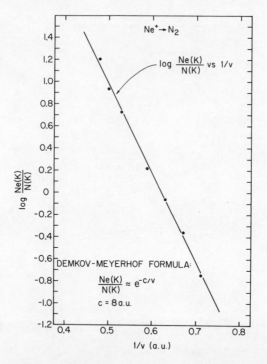

Fig. 11 Test by Fastrup and coworkers of the Demkov formula,
modified by Olson, and applied by Meyerhof. Source of experimental
data: Ref 57; Theory: Refs. 31, 32, 56.

to asymmetric atomic collisions[38] answered certain general
questions:

A) Why do light particles, such as protons and α particles
   act only through coulomb excitation and not through
   electron promotion?

B) Why are there sharp thresholds for excitation of electrons by heavy ion collisions?

C) Why do cross sections for excitation reach a maximum for
   atomic numbers where energy levels of collision partners
   match each other?

The promotion model[38] gave specific answers to these questions.
"Swapping" of correlations (see Fig. 9 - right side) occurs when
the energy of a level of atom b relative to atom a is changed.

This occurs, for example, as the atomic number of one collision partner is changed. In particular, the energy levels of helium and hydrogen atoms are far above the inner shells of all other atoms. Thus no promotion effects produce inner-shell excitation in proton or α-particle collisions. By considering the nature of the promotion process and the coulomb repulsion between nuclei, the electron promotion model also answers questions (B) and (C).[38]

5. <u>Quantitative Calculations</u> – An outgrowth of the electron promotion model has been the stimulus to calculation of molecular orbital energy levels at small internuclear distances, an area that has been lain untouched every since the beginning of quantum mechanics. Such calculations have been performed by Thulstrup and Johansen,[58] Mulliken,[59] Briggs and Hayns,[60] Barat and Sidis,[61] Larkins,[53] and Eichler and Wille[62] (see Figs. 12 and 13). In general, their results confirm the earlier work[35-38], but furnish valuable details, which only can be given accurately by computer (compare Figs. 12 and 13 with Fig. 9). There is no question that

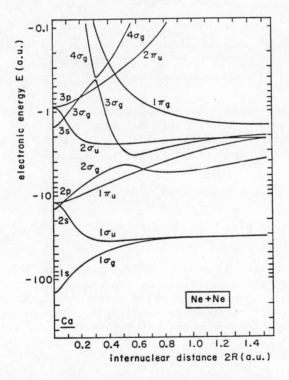

Fig. 12   Adiabatic molecular orbitals of a symmetric system, compare this with the left hand side of Fig. 9. Note that the diabatic MO 3dσ changes from $3\sigma_g$ at large internuclear distances to $4\sigma_g$ at smaller distances. Source: Ref. 62.

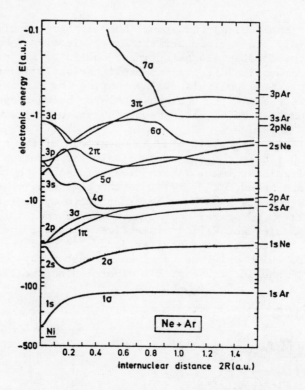

Fig. 13 Adiabatic molecular orbitals of an asymmetric system.
Compare with right hand side of Figure 9. Note again how the 3dσ MO
changes character, in running from 4σ→5σ→6σ. Likewise, note how
the 2σ and 3σ MO's repel each other for very small internuclear
distance.

future progress in this area will depend on quantitative, computer-
based calculations.[63]

An important advance was made by Briggs and Macek, who made
a quantitative, a priori calculation for the K-shell excitation
cross section for Ne$^+$-Ne collisions above threshold.[64] This
calculation showed that the promotion model could be made quantitative with no adjustable parameters. The computation was extremely
simple, basically a simple scaling of results for the one electron
2pσ→2pπ transition in protons on hydrogen.[7,9,66]

Figure 14 compares the total cross-section, as calculated by
Briggs and Macek[64] with experimental results by Cacek et al.[65]
Figure 15 compares the theoretical differential cross-section for

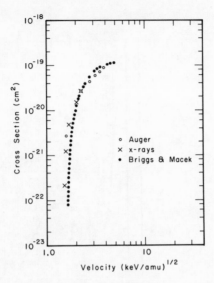

Fig. 14 Comparison of <u>ab initio</u> calculation of total cross-section for Ne$^+$ + Ne K-shell excitation. Source: theory-ref. 64; experiment: ref. 65.

1s→2p excitation in the proton hydrogen case by Knudson and Thorson [66] with a further comparison of theory[64] with experimental results for Ne$^+$-Ne collisions by Sackmann, Lutz and Briggs.[67] This approach furnishes a promising avenue for pursuit of the electron promotion model. One important factor is the simplicity that results from the fact that the many electron system is free from the degeneracies of final states which plague one-electron calculations.

MOLECULAR ORBITAL X-RAYS

Two recent developments in atomic collisions have excited the imagination of scientists: molecular orbital X-rays and positron formation. These new fields of interest are closely related to the quasi-molecular system formed during the collision.

"Molecular-orbital X-rays" (MO X-rays) consist of emitted photons caused by a transition between two MO's of the molecular-collision complex.[68] Saris postulated a two-step excitation mechanism.[69] An inner-shell vacancy is produced in the projectile in a first collision with an atom. In a solid target, it is possible for a second collision to occur before the inner-shell vacancy has

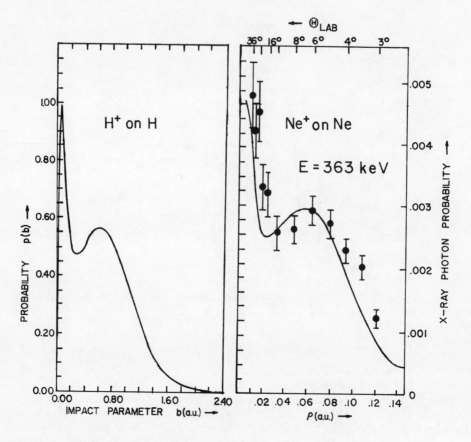

Fig. 15 Comparison of theory and experiment for K-shell excitation.
Left: One electron system-probability for 2p excitation for protons on H at 250 eV. Source: Ref. 66. Right: Many electron
system (Ne$^+$ or Ne). Source: Ref. 67.

decayed. Because of electron promotion, the MO X-ray can have a
larger energy than the X-ray of the SA. There are many experimental
reports of such MO X-rays.[70-79] (See Fig. 16)[101]

   One of the major problems is identification of the MO rays in
a unique manner. The lack of structure makes this very difficult.
In lighter atoms, the end point of the continuum radiation should
be at the united atom limit of the splitting between the two energy
levels. Figure 17 shows an analysis of the data of McDonald et
al.,[73] which shows that the measured end point is plausibly related
to estimated energies. Recently, (1974), Bissinger and Feldman[76]

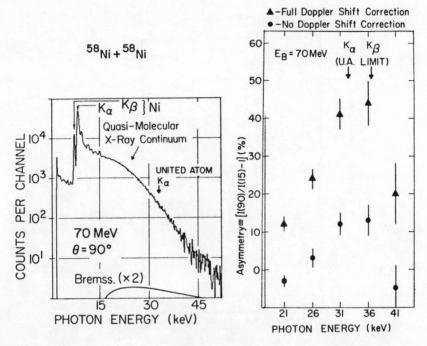

Fig. 16 Left: MO X-rays. An isotopically pure ^{58}Ni target is bombarded with a ^{58}Ni beam to eliminate nuclear dipole bremsstrahlung. Right: Angular asymmetry as a function of photon energy. Source: Ref. 82.

have performed a similar experiment with MO X-radiation from L shell vacancies in heavier elements. They varied the atomic number of the projectile and also achieved plausible end points for their spectra.

Meyerhof et al.[74,75] have studied MO X-rays ($2p\sigma, 2p\pi$, etc. → $1s\sigma$) in heavier systems including 30-60 MeV Br projectiles on Br targets. They have calculated the spectral distribution of their radiation and claim good agreement with experiment. Davis and Greenberg[77] have repeated their experiments and dispute the validity of some of their results and their interpretation.[74,75] Among other things the effects of Heisenberg broadening, which becomes severe in high energy collisions, were neglected.[74,75]

A correct calculation of Heisenberg broadening, which affects the MO X-ray spectrum most near its end point, requires a full Fourier integration of the emitted spectrum, averaged over all impact parameters. Briggs and Macek[83] have performed this calcu-

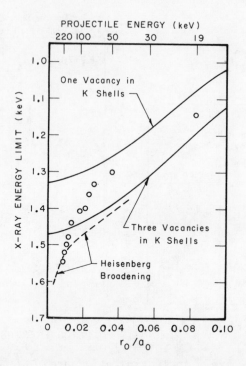

Fig. 17 Comparison of experimental X-ray cutoff energies (from
Ref. 73) with estimated energies [from scaled energy levels (Ref.
64)] in the $C-C^+$ system. The Heisenberg broadening is a rough
estimate based on the uncertainty principle and should not be
taken as quantitatively correct.

lation and find that the broadening is proportional to the square
root of the projectile velocity, contrary to expectation based on
simple uncertainty-principle arguments.

Greenberg et al.[82] (see Fig. 16) have predicted, and confirmed
by experimental observation, that an induced MO X-ray will result
from collisions. This induced X-ray has a signature of a directional asymmetry which is strongest at the united atom limit. The
importance of this result is that it furnishes a means of identi-
fying the MO X-ray uniquely.

Other possible means of identification would be interference
fringes[80] in the MO X-rays (see Fig. 18). Also, the end point of
the spectrum can be enhanced[80] by selection of widely scattered
atoms in concidence with the emitted photons (Fig. 18).

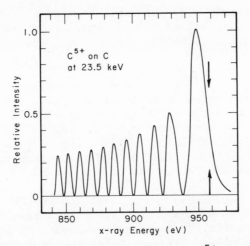

Fig. 18 Spectral energy distribution for C^{5+} on C at 23.5 keV
(6^3 a.u. in the center-of-mass frame of reference). Note that the
energy scale is greatly expanded. The classical turning point is
shown by arrows. Source: Ref. 80.

Molecular emission during collisions would be almost universal,[80] with a cross section with an order of magnitude ~10^{-21} cm^2,
for all collision partners. Thus for ions and atoms for Z~1, the
MO X-ray cross section is many orders of magnitude smaller than
cross sections for excitation of X-rays in the separated atoms.
Autoionization processes[81] during the collision are much more
intense than MO X-rays for Z~1.

For atoms with large Z, MO X-rays would compete with separated
atom X-rays, if vacancies could be produced. However, the two
step process postulated by Saris[69] no longer operates for very
heavy atoms, as there are few vacant MO's available to act as exit
channels for promoted electrons. Thus the intensity of MO X-rays
remains very small at all atomic numbers, and the cross-section for
MO X-rays decreases far below the figure of 10^{-21} cm^2 (= 1000 barns),
given above.[86]

MECHANISM OF VACANCY PRODUCTION IN COLLISIONS BETWEEN
HEAVY ATOMS

Kessel[45] has observed large energy losses (~2 keV) in collisions
of 1.5   MeV Argon ions with Ar atoms. Garcia et al.[41] have
interpreted these results in terms of a multiple promotion of L-
shell electrons. However the impact parameter dependence[45] makes
K-shell excitation much more likely.[84]   Because exit channels are

closed in this collision, a direct excitation of a K-shell electron
is likely to be via transitions already known to exist in one
electron systems, such as $2p\sigma \to 3p\pi$ or $2p\sigma \to$continuum, as Saris et al.
have pointed out.[85]

Meyerhof et al.[74,75,86] have postulated similar processes for
the vacancy production which is necessary to produce MO X-rays and
positrons in atomic collisions. Here the requirement is even more
stringent, in that direct excitation out of the $1s\sigma$ MO is required.
Because there is only one, limited calculation of this process,[14]
Meyerhof has used a semi-empirical method for estimating the
probability of direct excitation out of the $1s\sigma$ MO.[86]

Burch[87] et al. have taken another approach to the $1s\sigma$ excitation, by treating it as a binary encounter. The results[87] disagree
violently with Meyerhof's.[86] An example relevant to positron formation[88] is the case of 1.6 GeV U on U, at the Coulomb barrier.
Here Meyerhof estimates a vacancy probability of 5×10^{-5}; Burch[87] gives
a figure of 0.2; Greiner and co-workers[88,89] assume a figure of
0.01.

RELATIVISTIC EFFECTS. SUPERHEAVY ATOMS. POSITRON FORMATION IN ATOMIC COLLISIONS

Datz et al. have found sub-shell differentiation of excitation cross sections within the fine-structure levels of a given
subshell in collisions of I ions with solid targets.[90] Barat and
the author[38] constructed relativistic MO correlation diagrams which
show a differentiation among sub-shells. However, a detailed
explanation of the experimental results[90] is yet to be made.

The correlation diagrams of Barat and the author[38] are correct
up to the united atomic number $Z \sim 1/\alpha = 137$. Above this value,
nuclear size effects become important and change the level
order.[91] Müller et al.[92] have solved the two center Dirac equation with nuclear size effects and have constructed correlation
diagrams up through united atomic numbers of Z=184 (U on U).

Mokler et al.[71,72] and Jundt et al.[79] have seen MO X-rays from
united atoms with superheavy atomic numbers (Z>137) [101].

The problem of superheavy atoms (Z>137) has long fascinated
physicists because of the lack of a solution to the Dirac equation.[93-96] For sufficiently large Z, the 1s level "dives" into
the positron continuum.

In 1969, Gershtein and Zeldovitch[97] considered a slow collision between two bare uranium nuclei. They pointed out that, as the

lowest bound level dove into the positron continuum, the system could pick up one or more electron and emit a corresponding number of positrons. Shortly thereafter, Zeldovich and Popov[98] and Greiner and coworkers[88] gave more realistic and detailed discussions of the conditions under which positrons are formed.

The effects of the quasi-molecular are of very high importance in this process. It should be clear from the previous discussion in this article, that there can be no better case for applying one electron MO's, as Müller et al. have done.[92] Greiner et al.[89] found that nuclear motion induced transitions from the positron continuum into 1s enhance positron formation by two orders of magnitude, a good illustration of the essential nature of a molecular analysis.

However our understanding of the molecular picture needs to be improved to understand positron formation.

Greiner et al. have ignored the effect of induced transitions into empty MO's or the electron continuum. These transitions should have overall effects which are of the same order of magnitude as those considered previously.[89]

There are serious disagreements as to the order of magnitude of the expected positron yield. Greiner et al.[89] estimate a cross section for $U+U \rightarrow e^+ + e^-$ to be 5 barns at 1600 MeV laboratory kinetic energy, at the Coulomb barrier. Meyerhof[86] claims the actual cross section is only 1.3 µb. The difference of more than six orders of magnitude is due in part to the differing estimates of vacancy formation in the $1s\sigma$ MO, which were discussed in the previous section. Clearly, until this problem is solved, we shall not have reliable estimates for the possibility of a very important experiment [102].

Furthermore, the shapes of MO and positron spectra depend critically on the impact parameter dependence of the probability of vacancy formation in the $1s\sigma$ MO. Until we have better information about this process, published curves for these processes, both as to shape and order of magnitude, must be treated with extreme caution.

CONCLUSIONS

We have seen how the electron promotion model of atomic collisions started out in an atmosphere of contention and confusion. Now it has been thoroughly tested and accepted. However its limits, and how it connects with other approximations (Born, binary encounter etc.) have not been explored in depth.

Today, atomic physics has entered into a new, exciting phase, again full of conflicting claims and statements. History will no doubt repeat itself and we shall soon know the truth. But to find that truth, there is a great need for understanding in greater detail the mechanism of the direct, molecular excitation mechanisms, which lie outside the promotion model. There is no "royal road to learning" to find the answer to these questions. A great many detailed theoretical computations are needed to solve these problems. Then shall we see the light [103]

Acknowledgment

The author is indepted to Jack Greenberg for many helpful discussions.

*Research supported in part by the National Science Foundation under Grant No. NSF GP27714 and Grant No. NSF GP43979.

REFERENCES

1. W. Brandt and R. Laubert, Phys. Rev. Letters $\underline{24}$, 1037 (1970).

2. It is questionable whether the concept of diabatic states can be applied to level crossings in which there is a change in angular momentum ($\sigma-\pi$, $\pi-\delta$, etc.) The author is not aware of any such set of diabatic states.

3. J.E. Bayfield, accompanying article, p. 397.

4. D.R. Bates and R.H.G. Reid, in Advances in Atomic and Molecular Physics, $\underline{4}$, 13 (1968); D.R. Bates, K. Ledsham and A.L. Stewart, Phil. Trans. Roy. Soc. (Lond) $\underline{246}$, 215 (1953).

5. G.J. Lockwood and E. Everhart, Phys. Rev. $\underline{125}$, 567 (1962).

6. F.P. Ziemba and E. Everhart, Phys. Rev. Letters $\underline{2}$, 299 (1959).

7. D.R. Bates and D.A. Williams, Proc. Phys. Soc. (London) $\underline{83}$, 425 (1964).

8. F. Hund, Z. Physik $\underline{40}$, 742 (1927); R.S. Mulliken, Phys. Rev. $\underline{32}$, 186 (1928).

9. H. Rosenthal, Phys. Rev. Letters $\underline{27}$, 635 (1971).

10. R.F. Stebbings, R.A. Young, C.L. Oxley, and H. Ehrhardt, Phys. Rev. $\underline{138}$, A1312 (1965).

11. J.E. Bayfield, Phys. Rev. $\underline{185}$, 105 (1969).

12. M.C. Chidichimo-Frank and R.D. Piacentini, J. Phys. B: Atom. Molec. Phys. $\underline{7}$, 548 (1974).

13. J.C. Houver, J. Fayeton, M. Abignoli, and M. Barat, Phys. Rev. Letters $\underline{28}$, 1433 (1972); also J.C. Houver, J. Fayeton and M. Barat, Proc. VIII International Conference on the Phys. of Electronic and Atomic Collisions, Belgrade (1973), p. 759; also M. Barat (private communication).

14. V. SethuRaman, W.R. Thorson and C.F. Lebeda, Phys. Rev. A $\underline{8}$, 1316 (1973). References to several earlier papers by Thorson et al. are given here.

15. M.E. Rudd, C.A. Sautter and C.L. Bailey, Phys. Rev. $\underline{151}$, 20 (1966); G.B. Crooks and M.E. Rudd, Phys. Rev. A $\underline{3}$, 1628 (1971).

16. N. Stolterfoht, Z. Phys. $\underline{248}$, 81 (1971).

17. W.J.B. Oldham, Jr., Phys. Rev. 140, A1477 (1965); 161, 1
     (1967). See also A. Salin, J. Phys. B, Atom. Molec. Phys. 2,
     631 (1969); J. Macek, Phys. Rev. A 1, 235 (1970) and T.F.M.
     Bonsen and D. Banks, J. Phys. B, Atom. Molec. Phys. 4, 706
     (1971).

18. Y.B. Band (preprint).

19. W. Lichten, Phys. Rev. 131, 229-238 (1963).

20. R.P. Marchi and Felix Smith, Phys. Rev. 139, A1025 (1965).

21. F.T. Smith, D.C. Lorents, W. Aberth and R.P. Marchi, Phys.
     Rev. Letters 15, 742 (1965).

22. H. Rosenthal and H. Foley, Phys. Rev. Letters 23, 1480 (1969).
     H. Rosenthal, Phys. Rev. A 4, 1030 (1971).

23. J.N. Bardsley, Phys. Rev. A 3, 1317 (1971).

24. W.C. Keever and E. Everhart, Phys. Rev. 150, 43 (1966).

25. M. Lipeles, R. Novick, and N. Tolk, Phys. Rev. Letters 15,
     815 (1965); S. Dworetsky, R. Novick, W.W. Smith, and N. Tolk,
     Phys. Rev. Letters 18, 939 (1967). See also references 22
     and 38 for listing of references.

26. M. Barat, D. Dhuicq. R. Francois, R. McCarroll, R.D.
     Piacentini, A. Salin, J. Phys. B 5, 1343 (1972).

27. R. McCarroll and R.D. Piacentini, J. Phys. B 4, 1026 (1971).

28. F.T. Smith, Phys. Rev. 179, 111 (1969).

29. B.A. Lippman and T.F. O'Malley, Phys. Rev. A 2, 2115 (1970);
     T.F. O'Malley, Phys. Rev. 150, 14 (1966); 162, 98 (1967);
     T.F. O'Malley and H.S. Taylor, ibid, 176, 207 (1968); V. Sidis
     and H. LeFebvre-Brion, J. Phys. B, Atom. Molec. Phys. 4, 1040
     (1971); B. Andresen and S.E. Nielsen, Mol. Phys. 21, 523
     (1971).

30. H. Gabriel and K. Taulbjerg, Phys. Rev. (July, 1974 issue,
     in press).

31. Yu. N. Demkov, J. Exptl. Theoret. Phys. (U.S.S.R.) 45, 195
     (1963) (English Translation: Soviet Physics, JETP 18, 138
     (1964)).

32. R.E. Olson, Phys. Rev. A 6, 1822 (1972).

33. W. Lichten, Phys. Rev. 139, A27 (1965).

34. H.F. Helbig and E. Everhart, Phys. Rev. 136, A674 (1964).

35. U. Fano and W. Lichten, Phys. Rev. Letters 14, 627 (1965).

36. W. Lichten, Advances in Chemical Physics 13, 41 (1967).

37. W. Lichten, Phys. Rev. 164, 131 (1967).

38. M. Barat and W. Lichten, Phys. Rev. A 6, 211 (1972).

39. W. Brandt and S. Lundgvist, Physics Letters 4, 47 (1963), Arkiv
     För Fysik 28, 399 (1964); J. Quant. Spectroscopy Radiative
     Transfer 4, 679 (1964); A. Russek and M. Tom Thomas, Phys.
     Rev. 109, 2015 (1958); 114, 1538 (1959); A. Russek and J.B.
     Bulman, Phys. Rev. 122, 506 (1961); A. Russek 132, 246 (1963);
     M. Ya. Amusia, Phys. Lett. 14, 36 (1965), Zh. Tekh. Fiz. 36,
     1409 (1966) [English Transl: Soviet Physics,-Tech. Phys. 11,
     1053 (1967)].

40. A.V. Afrosimov, Yu. S. Gordeev, M.N. Panov, and N.V. Federenko,
     Zh. Tekh. Fiz. 34, 1613 [1964-English: Soviet Phys-Technical
     Physics 9, 1248 (1965)].

41. J.D. Garcia, R.J. Fortner, and T.M. Kavanagh, Revs. Mod.
     Physics 45, 111 (1973).

42. F.W. Saris, "Characteristic X-Ray Production in Heavy-Ion-
     Atom Collisions." in VII ICPEAC, Invited Talks and Progress
     Reports, T.R. Govers and F.J. deHeer, eds., (North-Holland,
     Amsterdam, 1971).

43. M.E. Rudd, "Introduction to Inner-Shell Excitation and De-
     excitation Processes, ref. 42, see also, "Mechanisms of Inner
     Shell Excitation and De-excitation in Multiply Ionized Beams."
     p. 1485, Proceedings of the International Conference on Inner
     Shell Ionization Phenomena and Future Applications, CONF-
     720404, U.S. Atomic Energy Commission, Technical Information
     Center, Oak Ridge, Tenn. (1973).

44. G.N. Ogurtsov, Reviews of Modern Physics 44, 1 (1972).

45. Quentin C. Kessel, "Coincidence Measurements," in Case Studies
     in Atomic Physics, Vol. I, 401 (1969), edited by M.R.C.
     McDowell and E.W. McDaniel; Q.C. Kessel and B. Fastrup, "The
     Production of Inner-Shell Vacancies in Heavy Ion-Atom
     Collisions." Ibid, 3, 139 (1973).

46. For several other review articles and discussions of current
     research see the Proceedings listed in footnote 43, Part B-
     Heavy Ion-Atom and Atom-Atom Collisions, articles by F.T.
     Smith, D.G. Lorents and R.E. Olson, p. 1175; B. Fastrup,
     p. 1188; J.S. Briggs, p. 1209; J.A. Cairns, p. 1223; F.W.
     Saris, I.V. Mitchell, D.C. Santry, J.A. Davies and R. Laubert,
     p. 1255; P.H. Mokler, H.J. Stein, and P. Armbruster, p. 1283;
     V.S. Afrosimov, p. 1297; F.W. Bingham, p. 1320; T.M. Kavanagh,
     R.J. Fortner and R.C. Der, p. 1332; F.C. Jundt, H. Kubo and
     K.H. Purser, p. 1450. All these articles discuss application
     of the electron promotion model to collision experiments.

46a. See "Theory of Charged-Particle Excitation," D.H. Madison and
     E. Merzbacher in Atomic Inner-Shell Process,B. Crasemann, ed.
     (Academic Press, to be published).

47. Q.C. Kessel and E. Everhart, Phys. Rev. $\underline{146}$, 16 (1966).

48. M.P. McCaughey, E.J. Knystautas, H.C. Hayden, and E. Everhart,
     Phys. Rev. Letters $\underline{21}$, 65 (1968).

49. B. Fastrup. G. Hermann, and Q.C. Kessel, Phys. Rev. Letters
     $\underline{27}$, 771 (1971).

50. B. Fastrup, E. Bøving, G.A. Larsen and P. Dahl, J. Phys. B:
     Atom. Molec. Phys. $\underline{7}$, L206 (1974).

51. T.M. Kavanagh, M.E. Cunningham, R.C. Der, R.J. Fortner, J.M.
     Khan, J. Zaharis, and J.D. Garcia, Phys. Rev. Letters $\underline{25}$,
     1473 (1970).

52. K. Taulbjerg and P. Sigmund, Phys. Rev. A $\underline{5}$, 1285 (1972);
     see also K. Taulbjerg, B. Fastrup and E. Laegsgaard, Phys.
     Rev. A $\underline{8}$, 1814 (1973); also J. Macek, J.A. Cairns and J.S.
     Briggs, Phys. Rev. Letters $\underline{28}$, 1298 (1972); G. Bissinger and
     L.C. Feldman, Phys. Rev. A $\underline{8}$, 1624 (1973).

53. F.P. Larkins, J. Phys. B $\underline{5}$, 571 (1972). FICAP abstracts,p.613.

54. W. Brandt and R. Laubert, Phys. Lett. A $\underline{43}$, 53 (1973).

55. H. Tawara and J. Kistemaker, Phys. Lett. A $\underline{41}$, 287 (1972).

56. W.E. Meyerhof, Phys. Rev. Letters $\underline{31}$, 1341 (1973). For
     experiments on the K-vacancy sharing mechanism, see H. Kubo,
     F.C. Jundt and K.H. Purser, Phys. Rev. Letters $\underline{31}$, 674 (1973);
     H. Ziem, N. Stolterfocht, and D. Ridder, Paper to be presented
     at Helsinki Conference on X-ray Processes in Matter (July 1974).

57. B. Fastrup (private communication).

58. E.W. Thulstrup and H. Johansen, Phys. Rev. A 6, 206 (1972).

59. R.S. Mulliken, Chem. Phys. Letters 14, 137 (1972).

60. J.S. Briggs and M.R. Hayns, J. Phys. B 6, 514 (1973).

61. V. Sidis and M. Barat, VIII ICPEAC, p. 200 (1973), J. Phys. B
    (in press)

62a. J. Eichler and U. Wille, Phys. Rev. Letters 33, 56 (1974);
  b. J. Eichler and U. Wille, FICAP - Abstracts, p. 613.

63. Larkins, Ref. 53, has claimed that his diabatic correlations
    do not agree with that of the electron promotion model. How-
    ever, inspection of his calculations shows that his steps in
    internuclear distance are too coarse to support his conclusions.
    His conclusions, in part, have been based on an apparent
    Landau-Zener crossing of the $2\sigma_g$ and $3\sigma_g$ curves (see Fig. 12
    at 2R = 0.6 a.u.). B. Müller (Dissertation, Frankfurt
    University, 1973) has shown that this type of pseudo-crossing
    does not obey the Landau-Zener theory, but follows the Demkov
    theory instead. That is, the eigenfunctions have a phase
    shift of $\pi/4$ rather than $\pi/2$ in going through the pseudo-
    crossing. Thus there is no diabatic correlation possible at
    this pseudocrossing at very high velocities, where the promo-
    tion model breaks down.

64. J.S. Briggs and J.H. Macek, J. Phys. B 5, 579 (1972); 6, 982
    (1973).

65. R.K. Cacek, Q.C. Kessel and M.E. Rudd, Phys. Rev. A 2, 1327
    (1970).

66. S.K. Knudson and W.R. Thorson, Canad. J. of Physics 48, 313
    (1970).

67. S. Sackmann, H.O. Lutz, and J. Briggs, Phys. Rev. Letters 32,
    805 (1974).

68. F.W. Saris, W.F. van der Weg, H. Tawara, and R. Laubert, Phys.
    Rev. Letters 28, 717 (1972).

69. F.W. Saris, in Seventh ICPEAC, Invited Talks and Progress
    Reports, edited by T.R. Govers and F.J. deHeer (North-Holland,
    Amsterdam, 1971.

70. F.W. Saris, I.V. Mitchell, D.C. Santry, J.A. Davies, and R.
    Laubert, in Proceedings of the International Conference on
    Inner Shell Ionization Phenomena, edited by R.W. Fink, S.T.
    Manson, J.M. Palms, and P.V. Rao, CONF-720404 (U.S. Atomic
    Energy Commision, Oak Ridge, Tenn., 1973), p. 1255.

71. P.H. Mokler, H.J. Stein, and P. Armbruster, Phys. Rev. Letters
     29, 827 (1972).

72. P.H. Mokler, H.J. Stein, and P. Armbruster, in Ref. 70,
     p. 1283.

73. J.R. MacDonald, M.D. Brown, and T. Chiao, Phys. Rev. Letters
     30, 471 (1973).

74. W.E. Meyerhof, T.K. Saylor, S.M. Lazerus, W.A. Little, B.B.
     Triplett, and L.F. Chase, Jr., Phys. Rev. Letters 30, 1279
     (1973); erratum, 30, 1279 (1973).

75. W.E. Meyerhof, T.K. Saylor, S.M. Lazerus, W.A. Little, B.B.
     Triplett, L.F. Chase, Jr., and R. Anholt, Phys. Rev. Letters
     32, 1279 (1974).

76. G. Bissinger and L.C. Feldman, Phys. Rev. Letters 33, 1 (1974);
     Phys. Rev. A 8, 1624 (1973).

77. C.K. Davis and J.S. Greenberg, Phys. Rev. Letters 32, 1215
     (1974).

78. J.A. Cairns, A.D. Marwick, J. Macek and J.S. Briggs, Phys.
     Rev. Letters 32, 509 (1974).

79. F.C. Jundt, H. Kubo, H.E. Gove, University of Rochester
     Nuclear Structure Laboratory report UR-NSRL-81 (1974).

80. W. Lichten, Phys. Rev. A 9, 1458 (1974).

81. G. Gerber and A. Niehaus, Phys. Rev. Letters 31, 1231 (1973).

82. J.S. Greenberg (private communication); J.S. Greenberg, C.K.
     Davis, B. Müller and W. Greiner (to be published in the Proceed-
     ings of the Int. Conf. on Reactions between Complex Nuclei,
     Nashville, 1974); J.S. Greenberg, C.K. Davis, P. Vincent, FICAP
     Abstracts, p. 617.

83. J.S. Briggs and J.H. Macek, (private communication). Similar
     conclusions have been reached by K. Smith, B. Müller, and W.
     Greiner (unpublished).

84. An X-ray-atom-coincidence experiment would be decisive (Q.C.
     Kessel - private communication).

85. F.W. Saris, C. Foster, A. Langenberg, and J.V. Eck (preprint).

86. W.E. Meyerhof, Bull. Am. Phys. Soc. 19, 663 (1974); Physical
     Review (in press - September, 1974 issue).

87. D. Burch, W.B. Ingalls, H. Wieman, and R. Vandenbosch (preprint).

88. See accompanying article by W. Greiner for a summary and references.

89. K. Smith, H. Peitz, B. Müller and W. Greiner, Phys. Rev. Letters 32, 554 (1974).

90. S. Datz, C.D. Moak, B.R. Appleton, and T.A. Carlson, Phys. Rev. Letters 27, 363 (1971).

91. W. Pieper and W. Greiner, Z. Physik 218, 327 (1969); L.P. Fulcher and W. Greiner, Lettere al Nuovo Cimento 2, 279 (1971); V.S. Popov, Sov. Jour. of Nucl. Phys. 12, 235 (1971) [Yad. Fiz. 12, 429 (1970)].

92. B. Müller, J. Rafelski and W. Greiner, Physics Letters 47B, 5 (1973); B. Müller, Dissertation, U. of Frankfurt (1973).

93. For a historical review, see S.J. Brodsky, SLAC-PUB-1337 (Nov. 1973-Stanford Linear Accelerator Center).

94. L. Schiff, H. Snyder, and J. Weinberg, Phys. Rev. 57, 315 (1940).

95. I. Ya. Pomeranchuk, Ya. A. Smorodinsky, Jour. Phys. USSR 9, 97 (1945).

96. E. Fermi, Nuclear Physics, University of Chicago Press, (1950).

97. S.S. Gershtein and Ya. B. Zeldovich, Sov. Phys. JETP 30, 358 (1970) [Zh. Eksp. Teor. Fiz. 57, 654 (1969)].

98. Y.B. Zeldovich and V.S. Popov, Sov. Phys. Usp. 14, 673 (1972).

99. Recently, Mssrs. Aagaard, Bøving and Fastrup (private communication) have measured the Ne-Ne K-shell excitation cross-section as a function of energy. They confirm the earlier Ne/Ne^+ ratio of 0.6 at a velocity of 0.78 a.u. (300 KeV). But at lower impact energies, the ratio falls off. At 0.35 a.u. it is down to 0.1. These results are in excellent agreement with the $(Ne^+ - N_2)/(N^+ - Ne)$ curve on fig. 10, and show that the apparent discrepancy with the promotion model was due to Demkov-type excitation of the exit channels.

100. Further confirmation has been made by ab initio calculations of one-electron wavefunctions for the asymmetric case by Mssrs. Taulbjerg, Vaaben and Fastrup (private communication). They have solved the coupled differential equations for the 1s - 2p excitation and agree with the Demkov-Olson-Meyerhof formula to within about 20 %.
     In addition, N. Stolterfoth, P. Ziem and D. Ridder (private

communication) further confirm the DOM formula in several combinations of asymmetric ion-atom collisions with gas targets.

101. For detailed discussions of MO x-rays, see the papers in this
      volume by F.J. de Heer (p. 287) and P. Mokler (p. 301).

102. Prof. W. Greiner has kindly informed me that a numerical
      error in the calculated cross sections of ref. 89 has been
      made. The corrected cross sections should be 500 barns at
      1600 MeV.

103. At FICAP, several groups presented results which are not dis-
      cussed in this paper, but have important bearing on it. The
      page number refers to the FICAP book of <u>Abstracts of</u> Contri-
      buted Papers (Heidelberg, July 22-26, 1974).
           G. Presser, K.O. Gröneveld and B. Knaf have seen MO x-rays
      arising from 1sσ vacancies arising in collisions with UA
      Z = 30. The agreement of measured energy endpoints with MO
      energies is similar to that of Fig. 17. (p. 621)
           Meyerhof and coworkers (p. 633) and Mokler and coworkers[101]
      also report observation of MO x-ray anisotropy, as predicted
      by Greiner and coworkers.[88,89]
           Meyerhof has raised the following questions about the
      postulate of induced MO x-rays[82]:
           1. One should look for the expected velocity dependence of
              the asymmetry observed by Greenberg et al.[82] (fig. 16),
              and also that
           2. the effect of Heisenberg broadening should reduce the
              theoretical asymmetry below the observed value.
      Also, the actual observed asymmetry is small. Only a large
      Doppler correction makes it comparable to the predicted va-
      lues (**Fig.**16). Finally, the appearance of a rotational
      effect near the UA seems unphysical. It is well known that
      the electronic wave function cannot follow the sudden rota-
      tion of the internuclear axis at small distances, which ex-
      plains the unit transition probability at small impact para-
      meters shown in Fig. 15. The predicted sign and magnitude of
      the asymmetry[82] depends on the number of electrons in each
      MO[82], which is not known.
           Thus, at present, many questions are unanswered. Neverthe-
      less, as Greenberg pointed out in the discussion, the pre-
      sence of the asymmetry near the united atom limit is good
      evidence that the "non-characteristic x-ray" is really a mo-
      lecular effect.
           H.L. Betz (pp. 670,674) has studied a process which he
      calls "radiative electron capture" in solid targets. ⸀t con-
      sists of filling of vacancies in projectile ions with elec-
      trons from the target. These processes are similar to MO x-
      rays, except that capture is from continuum orbitals. Betz

also observed an anisotropy in the angular distribution of
these x-rays. The results do not agree with theory.

M. Barat and coworkers (p. 637) report elastic and inelastic excitation of outer shells of rare gas atoms in neutral-
neutral collisions. The authors found it necessary to extend
the electron promotion model to understand their data.

B. Fricke (p. 641) has calculated ab initio correlation
diagrams for heavy ion scattering experiments. He includes
the effect of electron interactions and corrects for relativistic effects.

P. Dahl, G. Hermann and M. Rødro have found alignment in
inner shell excitation in Z^+-Ar collisions $(Z \sim 18)$, in fairly
close agreement with the assumption of excitation via the
promoted $4f\sigma$ MO. The existence of this alignment may have implications for the anisotropy of MO x-ray radiation.

RADIATIVE TRANSITIONS IN QUASI-MOLECULES

F.W. Saris and F.J. de Heer

FOM-Institute for Atomic and Molecular Physics

Amsterdam, The Netherlands

1. INTRODUCTION

In 1972 a broad X-ray band was reported to be observed during Ar bombardment of C, Al, Si and Fe [1]. These new X-rays were speculatively interpreted as arising from the radiative decay of a projectile 2p vacancy in the quasi-molecule transiently formed in the projectile-target atom encounter. Since then experimental studies of ion-induced X-ray spectra were no longer confined to the prominent characteristic lines. Measurements of cross sections for characteristic X-ray production have certainly stimulated the development of the molecular model for ion-atom collision processes. Yet the study of the continuum part of the X-ray spectrum may give more detailed information on the electronic states of the transient molecule. In the past two years the X-ray bands have been subject to intense investigations by various research groups [2,3,4,5,6,7]. It is the aim of this paper to review these experimental studies and to briefly discuss some prospects of a direct experimental verification of calculated diabatic molecular orbitals. We shall confine ourselves to data obtained at beam energies $\leq$ 1 MeV, for high energy data will be discussed in the following paper by Mokler. For a theoretical review of the theory of the Quasi-Molecular Model of Atomic Collisions (and of Molecular Orbital X-rays) one is referred to the preceeding paper by Lichten.

2. EXPERIMENTAL

Let us first sketch a typical experimental set-up for X-ray measurements, see fig. 1. A mass and energy analyzed ion beam in the energy range of the order of 100 keV impinges on a solid target

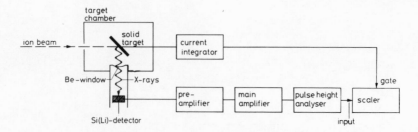

Fig. 1. Experimental set-up for X-ray measurements.

of high purity. The beam current is measured on the target, which
is biased such as to prevent secondary electrons from escaping.
It appears that it is important to record the X-ray spectrum as a
function of beam dose. The emitted X-rays are registered and
energy analyzed by a thin-window solid state Si(Li) X-ray detector
with a typical resolution at FWHM of about 0.2 keV depending on
the X-ray energy. In general the pulse-handling system is on line
with the computer. It will be illustrated further on that it is
important to measure the detection efficiency of the counter as
function of the X-ray energy. One cannot simply rely on transmission
data for the Be counter window and the gold contact layer. Transmission and detection efficiency in the low energy X-ray region may
be affected by dead surface layers of the counter as well as by
surface layers on the counter window formed by sputtering during
the course of these and previous solid target bombardments [2].
Application of apertures in front of the X-ray detector may suppress
charge-loss effects in the counter edges and therefore improve
signal to background ratios [8].

3. RESULTS AND DISCUSSION

3.1. Ar-Ar Collisions in Solids

Fig. 2 shows the X-ray spectra of ref. 1 recorded when argon
ions are incident on silicon and carbon targets. We observe that,
for example, ~ 100 keV argon ions produce a broad X-ray band near
a channel position which corresponds to an X-ray energy in the
range of 1 keV. The X-ray band, in particular its high energy side,
shifts to higher X-ray energies, when the beam energy is increased.
As we shall see later, the low energy side is entirely determined
by the strong X-ray absorption in the counter window. So the bandwidth is very much larger than the FWHM for characteristic X-ray
lines, and cannot be due to target impurities either. Also from the
other targets, Al and Fe, broad X-ray bands, similar in width and

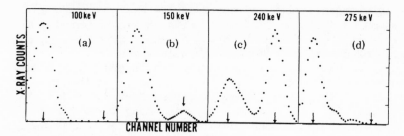

Fig. 2. Typical recorded spectra when argon ions are incident on silicon, (a)-(c), and carbon, (d), targets. The first arrow marks the channel number corresponding to the centroid of a 1.0-keV X-ray. The second arrow marks the Si(K) X-ray (1.74 keV) in (a)-(c) and the Ar(K) X-ray (2.96 keV) in (d) (ref. 1).

intensity, were observed. Therefore one cannot seek the origin of this X-ray band in ion-target interactions alone. The X-rays must then be due to Ar-Ar interactions (Ar being implanted into the solid), which means that the intensity must increase with ion dose. Indeed this was observed experimentally, as will also be discussed further on.

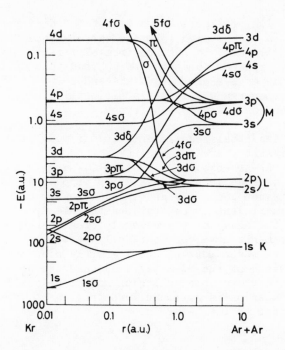

Fig. 3. Energy levels of molecular orbitals for the Ar-Ar system (see preceding article of Lichten).

In explaining the origin of these new X-rays we view the colliding atoms as a diatomic molecule with varying internuclear distance. For the Ar-Ar system Lichten has constructed the energy level diagram of diabatic molecular orbitals shown in fig. 3. At large internuclear distance the energy levels are those of Ar, while at zero internuclear distance they are those of the united atom Kr. At intermediate distances, electrons form molecular orbitals (MO). In this scheme the X-ray band near 1 keV X-ray energy can be attributed to the radiative filling of a previously formed vacancy carried into the $2p\pi$ MO during an Ar-Ar encounter. For this process to occur we require that the Ar projectile has a 2p vacancy prior to such an Ar-Ar collision. During the collision this Ar 2p vacancy may follow the $2p\pi$ MO (see fig. 3) as the inter-nuclear distance shrinks it ends up as a 2p vacancy in the elec-tronic configuration of the united atom Kr. It is interesting to note that the lifetime of a 2p vacancy in Kr is one order of magnitude shorter than in Ar, viz. $\sim 4 \times 10^{-16}$ sec versus 4×10^{-15} sec [1]). Consequently in close Ar-Ar encounters the life-time of the vacancy in the $2p\pi$ MO could be comparable to the collision time, which is $\sim 10^{-16}$ sec for argon ions of keV energies. Moreover the observation of the radiative transition should be aided by the fact that the fluorescence yield of the Kr L-shell is some hundred times larger than that of the Ar L-shell [1]).

From inspection of fig. 3 we note that the difference between the 3d and 2p MO's varies with internuclear distance from a few hundred eV up to 1.6 keV. So the spectrum of X-rays emitted during the collision is expected to appear as a broad band on the high energy side of the Ar L-shell X-ray. The end-point of the band will shift with impact energy until, at sufficiently violent collisions, the quasi-krypton configuration is formed.

One of the essential requirements in our model for the emission of these MO-X-rays is that a 2p vacancy in the Ar projectile be brought into the Ar-Ar encounter. Most solid targets can create such conditions because in the penetration process Ar-L-shell vacancies are created copiously and in addition Ar atoms become implanted with sufficient concentrations to make Ar-Ar encounters possible before the decay of a previously created L-shell vacancy. The above requirements have been carefully tested by various authors. In the next sections we shall shortly discuss experimental studies of the width of the MO-X-ray band, and the dose dependence in connection with the double collision model. We shall also see that MO-X-ray emission is not merely restricted to symmetric collisions, neither to the L-shells.

3.2. MO-X-Ray Band Width

In an attempt to better understand the MO-X-ray spectral shape
Bissinger and Feldman [4]) examined the band using a high-transmission-
window proportional counter, since it is clear that the Si(Li)
detector Be window is distorting the observed spectra on the low-
energy side [2]). The proportional-counter spectra obtained for 400
and 1000 keV Ar beams on Si are shown in fig. 4. These spectra do
not show the fall off of the band at the low energy side. Instead
they show a broad distribution, monotonically decreasing with increasing X-ray energy. The enhanced radiation transition rate near
the united atom limit, which we discussed in section 3.1., is not
reflected in an increased X-ray intensity near the united atom X-
ray energy. This is probably due to the very small cross section
for reaching this united atom limit in collisions which have
relatively small impact parameters compared with collisions
leading to X-rays of lower energy. In addition, since one has been
using thick solid targets in these experiments, the X-ray spectra
are integrated over the projectile energy until very low energies
due to the stopping by the target atoms.

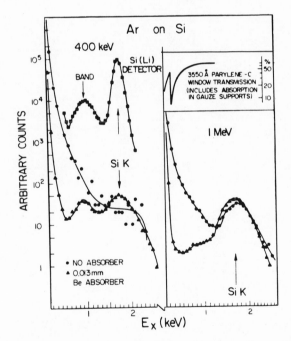

Fig. 4. Proportional counter spectra obtained with 400-keV and
1.0-MeV Ar on Si. Insert shows window transmission over the 100-eV
to 1.2-keV photon energy range. Also shown is a Si(Li) detector
spectrum for 400-keV Ar (ref. 4).

3.3. Symmetric Systems

From the dose dependence and strong similarity of spectra from different targets, the authors of ref. 1 concluded that the MO-X-rays they observed originated from collisions of Ar ions with previously implanted Ar atoms. This was immediately tested at Kansas State University [3]), where X-ray spectra were recorded during Ar impact on a solid Ar target, and at the Chalk River Nuclear Labs [2]), where a solid argon target was simulated using KCl (K is one atomic number higher, Cl one lower than Ar). The X-ray intensities in these spectra were very much higher and no longer ion-dose dependent, but the band seemed to extend to slightly higher X-ray energies than in the Ar implanted case. This can probably be explained by a density effect. In the solid argon targets the Ar projectile which has acquired an L-shell vacancy will already produce Ar-Ar MO-X-rays in a collision with the nearest neighbour. It does not have to go a long way to find another argon atom as in the case of implanted targets. Therefore the average energy in Ar-Ar collisions will be

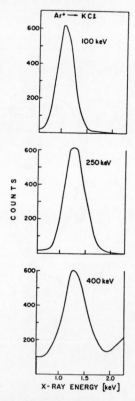

Fig. 5.   X-ray spectra for 100, 250 and 400 keV argon ions incident on KCl. The intensities are normalized in order to compare the shapes of the X-ray band. Data points omitted for clarity (ref. 2).

higher in the case of a solid argon target. So the mean internuclear
distance will become smaller resulting in higher energy X-rays as
we show now.

Fig. 5 illustrates that an increase of the Ar impact energy
from 100 keV to 250 keV clearly shifts the high energy endpoint of
the MO-X-ray band by ~ 250 eV. A further increase to 400 keV impact
energy does not affect the band very much but does produce K-shell
X-rays as well as a weak continuum ranging from 1.6 keV (united
atom L-X-ray energy) to 3.0 keV (separated atom K-shell X-ray
energy) [2]. This latter effect is probably due to $2p\pi$ vacancies
that are brought into the quasi-krypton 2p shell and then transferred into the $2p\sigma$ MO via rotational coupling, see fig. 3. Therefore one expects K-shell X-rays as well as a weak signal from the
radiative filling of the vacancies in this $2p\sigma$ MO as soon as the
threshold energy for this rotational coupling is surpassed
(> 250 keV).

3.4. Asymmetric Systems

The requirements for observing X-rays from transient molecules
can be met in many systems, and detection should be feasible in
symmetric as well as asymmetric projectile-target combinations.
Indeed MO-X-ray spectra from asymmetric systems were readily
observed by the above mentioned groups for low dose Ar-Al [3],
Ar-Si [2] and Ar-Ti [2] collisions. A comparison of the spectra from
Ar-Si, fig. 6, and Ar-Ti, fig. 7, clearly shows a dependence of
the position of the X-ray band on the atomic number of the united
atom transiently formed in these collisions.

Cairns et al. [6] performed measurements on the role of implanted argon in the formation of Ar-L, Si-K and MO-X-rays when Si is
bombarded by 270 keV Ar ions. They reported that the intensities
of the characteristic X-ray lines as well as that of the MO-X-ray
continuum were linearly proportional with ion dose. In contrast
with the authors of refs. 1 and 2 they concluded that even at high
doses an important part of the MO X-rays comes from Ar-Si collisions.
However there is no garantee in ref. 6 that in measuring the total
X-ray intensity of the continuum X-rays one has taken into account
the positive shift of the high energy side with ion dose. This shift
in the X-ray spectrum has been reported by Bissinger and Feldman [4],
see fig. 8. From the data of ref. 4 and ref. 2 we infer that at a
low dose < 10^{15} ions/cm^2 Ar-L vacancies are produced in Ar-Si
collisions and also MO-X-rays originate from Ar-Si encounters. As
the Ar dose increases L-vacancies will be produced relatively more
in Ar-Ar collisions due to the larger cross sections [9], thus giving
rise to an increase in the MO-X-ray signal from Ar-Si collisions.
In addition MO-X-rays from Ar-Ar collisions will appear causing the

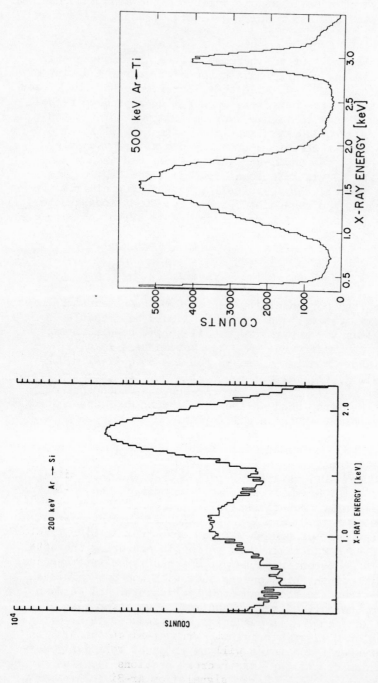

Fig. 7. X-ray spectrum recorded when 500 keV Ar ions are incident on Ti. A broad X-ray distribution is observed along with the Ar-K_α line (2.96 keV) (ref. 2).

Fig. 6. X-ray spectrum recorded when 200 keV argon ions are incident on silicon. The argon dose was kept below 10^{15} cm^{-2} in order to avoid Ar-Ar collisions inside the silicon target (ref. 2).

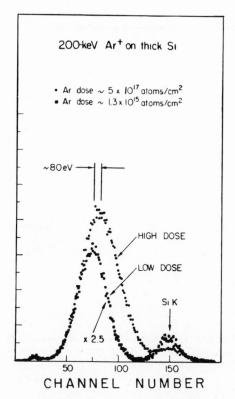

Fig. 8. Spectra obtained with Si(Li) detector for Ar⁺ on thick Si;
Ar dosage effect on the band shape.

high energy side of the MO-X-ray band to shift to higher X-ray
energies and the <u>total</u> intensity in the band will increase further.
One will be able to detect the proper quadratic dependence of the
MO-X-ray intensity on Ar ion dose <u>only</u> if the <u>full</u> spectrum of
Ar-Ar MO-X-rays is recorded and corrected for changes in detection
efficiency and then integrated.

In the molecular model one does not expect to ionize the Ar
L-shell in Ar-Be collisions, because the 1s state of Be has a
smaller binding energy than the 2p state of Ar. Hence, the absence
of MO-X-rays in this collision system, as observed by Bissinger
and Feldman [4]) for low argon doses, leads strong support for this
model as well as our model for the production of MO-X-rays.

Recently Bissinger and Feldman [7]) published a systematic study
of end-point energy variations in the MO-X-ray bands produced by
S, Cl, Ar, K, Ca and Ti bombardment of Si. MO-X-ray spectra were
recorded for beam doses $< 10^{16}$ atoms/cm². The end-point energy was

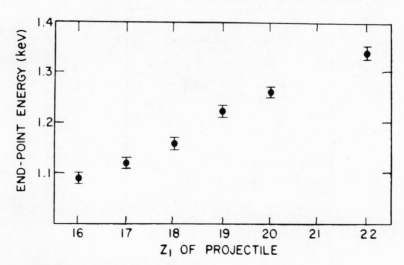

Fig. 9. End-point energies for S, Cl, Ar, K, Ca, and Ti on thick Si (ref. 7).

defined as the intersection of the line drawn tangent to the high-energy side of the MO band at the half-maximum and the abcissa on a linear scale. The results are shown in Fig. 9. The observed increase of the experimental end-point energies with the projectile atomic number is in qualitative agreement with the increase in L-shell X-ray energy of the different united atoms formed in these collisions with Si. The authors discuss the effects of various systematic errors in their data analysis and propose an interesting way of data reduction for which we refer to their publication [7]).

3.5. MO-X-rays from Transitions Involving the K-shell

K-shell X-ray bands have been observed by McDonald et al. [5]) for 19-240 keV C^+ ion impact on graphite. In fig. 10 the level diagram is given for the C-C system. In a C-C collision a K-electron can be promoted from the $2p\sigma$ MO to the $2p\pi$ MO via rotational coupling. If a second collision occurs before the K-hole is decayed, it can follow the $1s\sigma$ MO and a radiative transition may occur between the $2p\sigma$ or $2p\pi$ orbital and the $1s\sigma$ orbital. McDonald et al. have determined the end-point energy in the observed X-ray band in the same way as described above and the result is plotted in fig. 11 as a function of the impact energy as well as the distance of closest approach in a head-on collision. The latter is calculated for the impact energies used by applying a screened Coulomb potential. It is clear that the maximum X-ray energy in the band should correspond to transitions that occur near the distance of closest

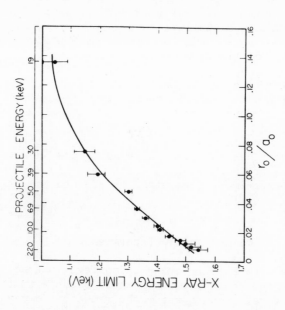
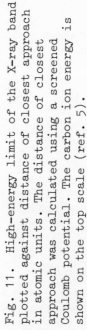

Fig. 11. High-energy limit of the X-ray band plotted against distance of closest approach in atomic units. The distance of closest approach was calculated using a screened Coulomb potential. The carbon ion energy is shown on the top scale (ref. 5).

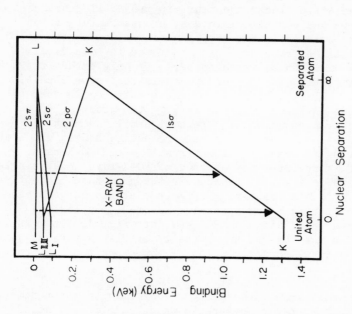

Fig. 10. Electronic energy level diagram for the C-C collision system with the transitions of the observed X-ray band illustrated (ref. 5).

approach in the collision. Hence the $1s\sigma$ energy level for the
transient C-C system is mapped in fig. 11. However it is observed
that the united atom limit of Mg K (1.3 keV) is surpassed. This
could be due to multiple L-shell vacancies. Moreover, collision-
time broadening effects may cause an appreciable photon intensity
beyond this united atom limit. For instance: the united atom (Mg)
K-shell radius is ~ 0.04 Å and the velocity of a 200 keV C^+ ion is
~ 2×10^{16} Å/s. Then, due to the Heisenberg uncertainty relation,
the width of the united atom X-ray energy is ~ 300 eV (see also
the preceeding paper by Lichten).

Also MO-X-rays from K-shells as well as from M-levels of much
heavier systems have been observed in recent years. However, these
heavier systems require beam energies well into the MeV range. For
a review of these experimental results we refer to the following
paper by Mokler.

3.6. Gas Target Experiments

Solid targets were used in order to achieve in a double
collision process the necessary inner-shell vacancy in the projectile after the first encounter and the radiative decay in the
second encounter. However, the use of solid targets opens the
possibility to produce continuum X-rays due to MO-X-rays from
recoil effects, radiative electron capture in projectile vacancies
that do not exist in gas target collisions, bremsstrahlung from
secondary electrons created in the stopping of the ion in the solid,
etc. Since these solid effects have been reviewed in some detail
elsewhere [10]) this will not be repeated here. We only emphasize
the possibility of meeting the necessary double collision condition
by the use of a molecular gas target. Double scattering in one
molecule can produce an inner-shell vacancy in the projectile
during the first collision which will persist until the next
collision with the second atom of the same molecule. Radiative decay
of the vacancy during the second encounter will yield MO-X-rays.
The authors of ref. 1 reported the absence of MO-X-rays under single
collision conditions in Ar-Ar collisions in the gas phase. Recently
strong evidence for double scattering in molecular gas targets has
been obtained by comparing X-ray spectra from $Ar-Cl_2$ collisions
with Ar-HCl and Ar-Ar collisions at 300 keV, see fig. 12. Only in
the $Ar-Cl_2$ case does one observe an intensity of Cl-K X-rays as well
as MO-X-rays. For a detailed description of the experiment see ref.
11.

Clearly this opens some interesting ways for further experimental investigations of MO-X-ray spectra in which complicated solid
effects are eliminated. Briggs [12]) and Macek and Briggs [13]), using
the theory of line broadening (see invited talks of Lichten and
Gallagher), have calculated the spectral shape and intensity of

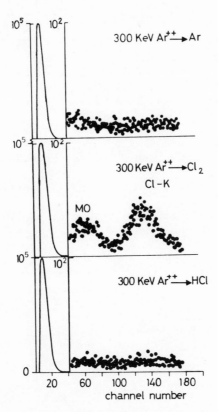

Fig. 12. The X-ray spectra for 300 keV Ar$^+$ on Cl$_2$, HCl and Ar.
L-X-ray yields, on left of spectra, are identical. Ar$^+$ on Cl$_2$
spectra shows Cl-K X-ray and additional feature at X-ray energies
below 1.5 keV (ref. 11).

MO-X-rays per N-K-shell vacancy in N-N collisions, using ab initio
calculations of the N-N molecular energy levels. Measurements of
the MO-X-ray spectra emitted in collisions of N$^+$ with N$_2$ gas should
yield a direct test of these calculated MO energy levels and
transition rates. Such measurements are in progress in our
institute.

        This work is part of the research program of the Stichting
voor Fundamenteel Onderzoek der Materie (Foundation for Fundamental
Research on Matter) and was made possible by financial support from
the Nederlandse Organisatie voor Zuiver-Wetenschappelijk Onderzoek
(Netherlands Organization for the Advancement of Pure Research).

REFERENCES

1) F.W. Saris, W.F. van der Weg, H. Tawara, and R. Laubert, Phys.Rev.Letters 28 (1972) 717.

2) F.W. Saris, I.V. Mitchell, D.C. Santry, J.A. Davies and R. Laubert, Proceedings of the Int.Conf. on Inner-Shell Ionization Phenomena and Future Applications, Atlanta, Georgia 1972, USAEC-Conf.720404, p.1255.

3) J.R. MacDonald and M.D. Brown, Phys.Rev.Lett. 29 (1972) 4.

4) G. Bissinger and L.C. Feldman, Phys.Rev. A8 (1973) 1624.

5) J.R. MacDonald, M.D. Brown and J. Chiao, Phys.Rev.Letters 30 (1973) 471.

6) J.A. Cairns, A.D. Marwick, J. Macek and J.S. Briggs, Phys.Rev. Letters 32 (1974) 509.

7) G. Bissinger, L.C. Feldman, Phys.Rev.Lett. 33 (1974) 1.

8) J.F. Chemin, I.V. Mitchell and F.W. Saris, J.Appl.Phys. 45 (1974) 532.

9) F.W. Saris, Physica 52 (1971) 290.

10) F.W. Saris, Proceedings of the Int.Conf.on Atomic Collisions in Solids, Gatlinburg 1973, to be published.

11) F.W. Saris, C. Foster, A. Langenberg and J. v. Eck, J.Phys. B., to be published.

12) J. Briggs, J.Phys. B. 7 (1974) 47.

13) J. Macek and J. Briggs, J.Phys. B., to be published.

SUPERHEAVY QUASIMOLECULES

P.H. Mokler and S. Hagmann
GSI, Darmstadt and Universität zu Köln

P. Armbruster and G. Kraft
GSI, Darmstadt

H.J. Stein and K. Rashid
Kernforschungsanlage Jülich

and

B. Fricke
Gesamthochschule Kassel, Germany

In November 1971 F. Saris, from the FOM Institute in Amsterdam, reported in a seminar talk at Kernforschungsanlage Jülich on his new experimental finding of noncharacteristic x rays from Ar→Ar collisions.[1] These noncharacteristic x rays are interpreted in the framework of the electron promotion model[2] as radiative transitions between molecular orbitals (MO) in the transiently formed collision molecule. A review of this work is given in this volume by F.J. de Heer, Amsterdam.[3] The first experimental results of Saris and coworkers inspired us to look for such noncharacteristic, i. e., quasimolecular x rays for the case of really heavy collision system.[4] Additionally, there were suggestions from the theory group of W. Greiner, Frankfurt, to investigate extremely heavy collision systems.[5] For such heavy ion-atom collisions - as, e. g., U→U - new features are expected. A review on these new phenomena is given in this volume by W. Greiner.[6] As stated there, it is of most interest to look for transitions to the quasimolecular K level in the heaviest collision systems - either radiative transitions or even autoionizing transitions. But until now only radiative transitions in quasimolecules are observed.

This report deals with the noncharacteristic radiation from extremely heavy collision systems, as, e. g., I→Au. Main emphasis is given to the quasimolecular M radiation. In this case, experiments and theory indicate even quasiatomic radiation. Quasimolecular L and K radiations are only summarized very briefly.

SPECTRAL EVIDENCE ON QUASIATOMIC M RADIATION

Experimentally, the detection of M-x rays from superheavy quasimolecules seemed to be most favourable.[4,7] Hence, we started to search for this radiation using the I beam from the Tandem accelerator of the University of Cologne (= Universität zu Köln). We used a simple experimental set-up, shown in Fig. 1. Thin targets of heavy elements, as e. g., Yb, Au, Pb, Th, and U were bombarded with monoenergetic I beams between 5 and 60 MeV. The emitted x rays were detected by 30-mm^2 large Si(Li) detectors placed perpendicular to beam axis. The resolution of the best detector used was 170 eV FWHM at 5.9 keV. In order to reduce the high rate of low-energy x rays Al foils could be inserted between target and detector. For cross section determination a special designed heavy-ion beam monitor was added just in front of the target chamber.[8] Here the primary ion intensity is measured via x rays produced in an off-beam-axis rotating thick target intersecting the ion beam

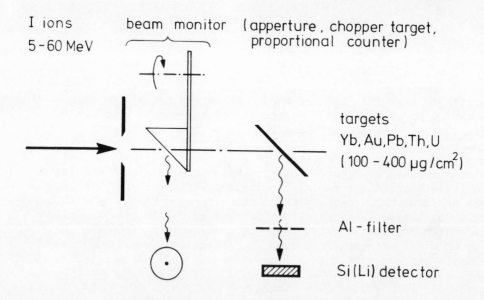

Fig. 1. Sketch of experimental set-up

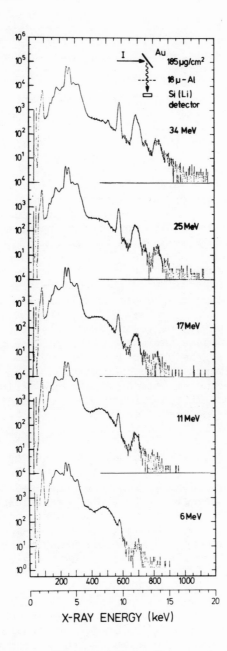

Fig. 2. X-ray spectra from I → Au collisions for
        bombarding energies between 6 and 34 MeV.
        An 18-μm Al filter was inserted between
        target and detector.

periodically (beam chopper). The x rays are detected by a flow-mode
proportional counter. The whole arrangement is calibrated for each
beam energy with a well-shielded and electron-suppressed Faraday
cup instead of the target.

In Fig. 2 x-ray spectra from I bombardment on a 200-μg/cm^2
thick, self-supporting Au target are shown. The spectra were taken
at impact energies of 34, 25, 17,11, and 6 MeV. In all cases an
18-μm Al foil was used to filter out low-energy M-x rays of both
the collision partners. (The low-energy peak at about 1.5 keV is
due to the K fluorescence radiation from the Al filter). Besides
the characteristic L radiation of target atoms and incident ions
a noncharacteristic x-ray band is seen in the 8-keV region. This
noncharacteristic x-ray band displays a pronounced peak structure
at low impact energies. With increasing impact energy the peak
broadens and merges finally with the underlying continuum.

It is this broad peak structure appearing at low impact energies which has been attributed to radiative transitions to the M
shell in the corresponding superheavy quasiatom with an effective
atomic number of 132; $Z_{eff}=Z_1+Z_2$.[4] The expression quasiatom stands
here for a transiently formed quasimolecule at very small internuclear distances where the molecular level energies approach the
binding energies of the united-atom system. Due to the different
electron-shell dimensions the quasiatomic region may be reached
at different internuclear distances for the various electron
shells. That means, if the quasiatomic approximation is already
fulfilled for the "united" M shell it must not be valid for the
united L and K shell.

For a Z = 132 quasiatom we expect Mα and Mβ transition energies of 7.6 and 8.5 keV, respectively.[9] The broad peak seen at
low impact energies, Fig. 2, is indeed centered nearby the expected transition energies giving first experimental evidence on
superheavy quasiatomic M radiation. In this connection it is also
of interest to note that the peak position does not significantly
vary with impact energy. (If there is a dependence the peak position might slightly shift to lower x-ray energies with increasing
impact energy.)

Another feature seen in the spectra is: The ratio between
band intensity and I-L x-ray intensity remains about constant for
the impact energies shown. In contrast, the ratio of Au-L and
I-L intensity decreases appreciably with decreasing impact energy.
This can also be seen from the cross sections plotted in Fig. 3.[8]
At an impact energy of 11 MeV the cross sections for I-L, MO, and
Au-L x-ray production are about 4 kbarn, 40 barn, 6 barn, respectively, having an energy dependence E^α with α = 1.6, 1.6, and 3,
respectively.

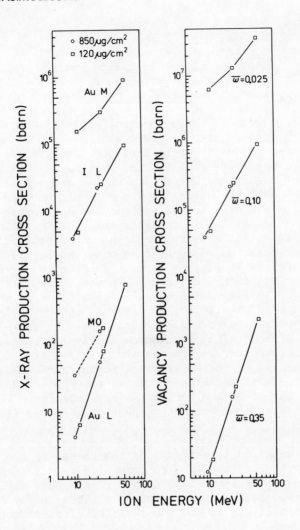

Fig. 3. X-ray production and ionization cross sections
for I → Au collisions, cf. Ref. 8.

    There have been many objections to our original measurements
and their interpretation. These objections were all experimentally
checked by independent measurements done by H. E. Gove, F. C. Jundt,
and H. Kubo from Rochester.[10] This group confirmed our results
and agrees fully with our interpretation. They considered the following effects as possibly competing with quasimolecular x-ray
emission:[11]

a) Electronic pulse pile-up effects,
b) Target impurities,
c) Radiation from recoiling Au - Au collisions,
d) Addition of low and high-energy line tails,
e) Different bremsstrahlungs effects, and
f) Radiative electron capture.

The Rochester group measured spectra, excluding pile-up effects. The cross sections they measured agreed with our measurements within a factor of 2. They found the bremsstrahlungs cross sections to be about 3 orders of magnitude smaller than the observed x-ray emission in the 8-keV region.[11] Target impurities and radiative electron capture were also excluded: The Rochester group investigated inverted collision systems, i. e., systems where the role of the target atom and incident ion is interchanged. In Fig. 4a the spectra for I→Au and Au→Sn collisions are compared for corresponding impact energies. In both cases an x-ray band is observed at slightly different x-ray energies. This gives also evidence of a negligible contribution from recoiling collisions. Moreover, they tested this point once more by investigating Au→Au collisions, Fig. 4b. Here, no x-ray band is seen at all in the concerned energy region. This fact may also demonstrate negligible contributions from electron bremsstrahlungs effects. They studied carefully the influence of different absorbers on the spectra and obtained a more or less "true spectral shape" of the x-ray band. Furthermore, they tested the dependence of x-ray emission on target thickness and found a non-proportionality of MO and target L x-ray emission. (The non-proportionality extends up to unexplained large target thicknesses).

With these detailed investigations the Rochester group stated definitively: up to now there are no competing processes known which contribute to the 8-keV band in any significant way. The only remaining explanation for the origin of these x-ray bands are radiative transitions in the quasimolecule formed during collision.

We have continued the investigation of this superheavy quasimolecular or quasiatomic M radiation at the Cologne Tandem accelerator. We were interested in the true spectral shape of the x-ray band at extremely low impact energies, and we also looked for the x-ray band in similar collision systems.

The true spectral shape at 6-MeV I impact on a thin gold target can be extracted from Fig. 5a and 5b. The spectrum in Fig. 5a was taken without any Al filter. Absorption due to detector and target-chamber windows (0.3-mil Be, 5-mm air, 6-μm HOSTAPHAN) and target self-absorption are only significant for lower-energy x rays. In the energy region between 7 and 9 keV the total transmission (including target about 95 %) varies by less than 2 % per keV.

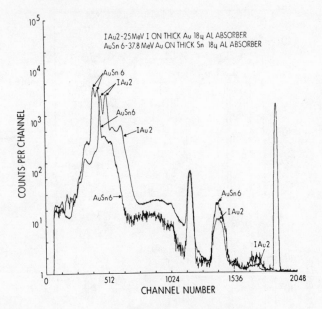

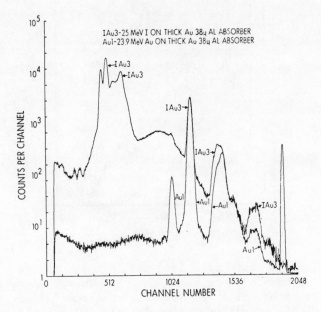

Fig. 4. X-ray spectra(a) from I→Au and Au→Sn collisions, (b) from I→Au and Au→Au collisions. The spectra are taken from Ref. 10 (Rochester).

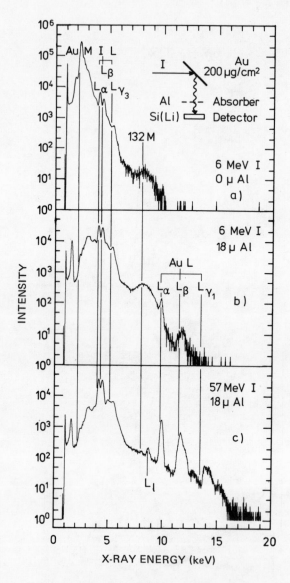

Fig. 5. X-ray spectra from I→Au collisions. Spectra
(a) and (b) were taken at 6-MeV impact energy;
spectrum (c) at 57 MeV. For spectrum (a) no
Al filter was used, for spectra (b) and (c) an
18-μm Al filter was used.

The count rates applied were low enough to ensure negligible pile-
up contributions. Spectrum (a) definitively proves the existence
of a real peak structure for the 8-keV band at low impact energies.
Spectrum (b) once more demonstrates this structure with better

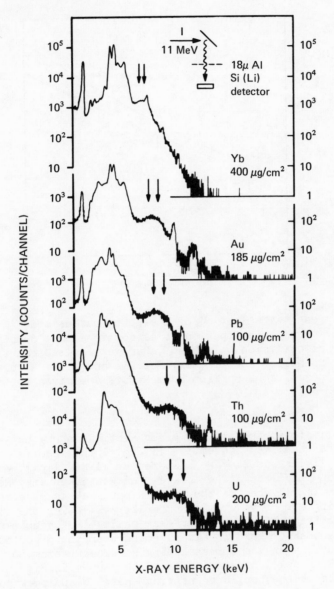

Fig. 6. X-ray spectra for 11 MeV I impact on Yb, Au, Pb,
Th, and U. A 18-μm Al filter was used.

statistics using an 18-μm Al filter. At high impact energies the
peak structure disappears, cf. spectrum (c).

Similar x-ray bands with an equivalent behaviour on impact
energies are found for neighbouring collision systems, from I→Yb

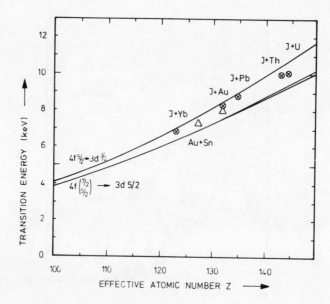

Fig. 7. Peak position(keV) of the noncharacteristic
       x rays as a function of effective atomic number
       of collision molecules (Δ are extracted from the
       Rochester spectra, cf. Ref. 10 and 11). Calculated
       transition energies from Ref. 9 are given for com-
       parison

up to I→U collisions. Spectra and cross sections show the same
features as for the I→Au system.[12] In Fig. 6 the spectra for the
different collision systems are compared at 11-MeV impact energy
using an 18-μm Al filter. (The Yb spectrum was taken at 6-MeV impact energy). In all spectra the broad x-ray peak is observed.
With increasing effective atomic number $Z_{eff}=Z_1+Z_2$ of the collision
system the position of the band or peak shifts to higher x-ray
energies. The peak or band is always found near the expected Mα
and Mβ transitions in the corresponding superheavy atom.[9] These
quasiatomic M transitions are indicated by arrows in Fig. 6. The
coincidence between expected transition energies and peak positions
is shown once more in Fig. 7. The full lines show the calculated
Mα and Mβ transition energies as a function of the atomic number;[9]
the data points show the peak positions. In order to obtain the
peak positions, the spectra in Fig. 6 were roughly decomposed by
hand in a descending continuum and a peak. This procedure shifts
the peak position slightly to higher x-ray energies. Anyhow, the
peak positions are correct within at least 1/2 keV. The inverted
system Au→Sn measured by the Rochester group fits also well in
this picture. This coincidence between peak position and ex-

pected transition energies is - beside the spectral shape of the
x-ray band - the most striking feature extracted from simple x-ray
spectra. Both facts may give strong experimental evidence of the
existence of superheavy quasiatomic M radiation.

ELECTRON PROMOTION IN THE I → Au QUASIMOLECULE

At the applied impact energies the orbital velocities of the
inner-shell electrons are large compared with the collision velocity, e. g., at 11-MeV impact energy the orbital velocity of the
I-L electrons is about 10 times larger than the collision velocity.
Hence, for the inner shell ionization the electron promotion model
is applicable and the spectra can be discussed in the framework of
diabatic correlation diagrams.[2] Originally we used a diabatic
correlation diagram including spin-orbit coupling, Fig. 8, in order
to interprete the I-Au spectra.[4] A simplified level diagram,
shown at the top of Fig. 9,[13] may clarify the things a little.
Due to the large L ionization cross sections of the flying I ions,
see Fig. 3, and due to the lifetime of an I-L vacancy ($\sim 2 \cdot 10^{-16}$sec)
[14] a certain fraction of the I beam in a dense target material
always carries a I-L vacancy. For 11-MeV I impact on a solid Au target this fraction is about $3 \cdot 10^{-4}$.[8] Considering this fraction of
L excited I ions in the beam alone we find that the I-L vacancy can
be transferred in a collision via the united M levels to the L shell
of the separated Au atom, finally causing the characterstic Au-L
radiation. On its way to the L shell of the target atom the vacancy
will also have a chance to decay during the collision itself. Hence,
quasimolecular x rays, and in particular, MO transitions to the
"united 3d levels" may be seen.

The spectral shape of this quasimolecular x-ray emission is
determined by the actual level-energy dependence on internuclear
distance. But looking on the rough handdrawn level diagram in Fig. 9,
we would not expect a priori to find a peak-structure in the molecular x-ray spectrum. This statement may also be true taking into
account a small enhancement of transition probability and fluorescence yield due to increasing transition energies at small internuclear distances.

There seems to be a second discrepancy, the cross sections.
The quasiatomic M cross section can be estimated by comparing the
collisional lifetime of the quasiatom (M-shell approximation) and
radiative lifetime of the vacancy in the quasiatom.[4,8] The vacancy lifetime for spontaneous radiative decay is expected to be
about 10^{-16} sec.[8] Assuming a straight-line path for the I ion
and a quasiatomic M-shell radius of about $3.5 \cdot 10^{-10}$cm the collisional lifetime of the quasiatom for the M shell is expected to be
$\approx 10^{-18}$ sec at 11 MeV. With these values the quasiatomic M x-ray

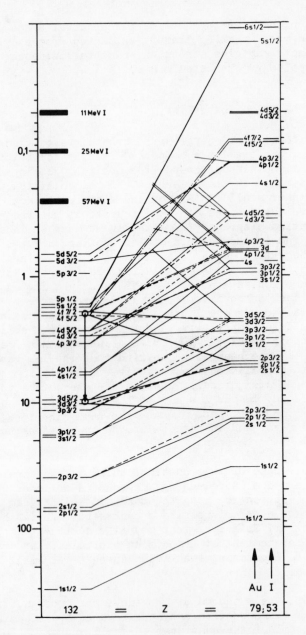

Fig. 8. Diabatic correlation diagramm (cf. Ref. 4)

cross section is estimated to be around 1 barn. The measured cross
section for the x-ray band is about 40 times larger at 11-MeV I
impact on Au.

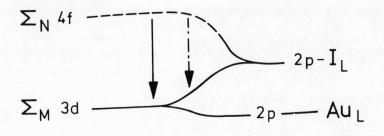

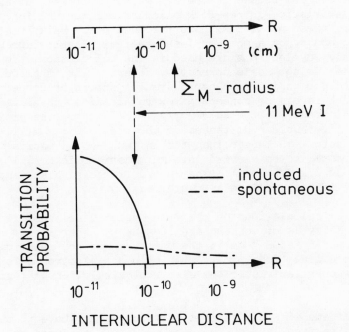

Fig. 9. Sketch of a simplified level diagram for
11 MeV I → Au collisions

For higher impact energies we find the same discrepancy. In
this connection it is interesting to point out that in the straight-
line approximation the ratio of I-L/MO cross sections does not
depend on impact energy as it is experimentally observed for all
our investigated collision systems. The cross section discrepancy
might slightly be decreased at low impact energies due to large
angle scattering giving possibly an enhancement for quasiatomic
radiation.[15]

The main discrepancy in cross sections, we think, may be caused
by two different effects: (1) The lifetime of the vacancy in the

quasimolecule may at small internuclear distances be drastically
"reduced by collisionally induced deexcitation". This possibility
was postulated in Ref. 4; the sketch at the bottom of Fig. 9 indicates this idea. Such an induced emission may explain both spectral
shape and cross section. (2) The quasiatomic approximation may
hold true for larger internuclear distances enhancing quasiatomic
x-ray emission. We believe that both effects do contribute to the
pronounced x-ray band.

In order to study the second point more reliably, molecular-
level calculations are needed. Recently, we have carried out ab
initio non-relativistic Hartree-Fock calculations for the I-Au
molecule. Using these results we have introduced the relativistic
corrections to construct a more or less realistic level diagram,
Fig. 10.[16] Only 20 electrons are used in the calculation, thus
binding energies do not scale correctly. Nevertheless, the trend
of level dependence on internuclear distances should be correct.
For the united M levels of concern an extremely wide and flat
minimum is indicated. For the 3d 3/2 3/2 level, e. g., the minimum
extends up to internuclear distances of about $6 \cdot 10^{-10}$ cm, i. e.,
roughly by a factor 1.5 larger than the united M-shell radius. The
vacancy is expected to be carried into the quasiatomic M shell
mainly by this 3d 3/2 3/2 level. (The 3d 5/2 1/2 level also displays
a minimum, but not as wide and as flat).

Other level calculations on the I-Au system display similar
features. The M levels of concern are flat up to about $8 \cdot 10^{-10}$ cm,
$6 \cdot 10^{-10}$ cm, and $5 \cdot 10^{-10}$ cm in the framework of non-relativistic
multi-electron calculations,[16] relativistic one-electron calculations,[17] and non-relativistic one-electron calculations,[18] respectively.

These flat and wide minima most certainly contribute to the
large MO cross sections and to the peak structure observed at low
impact energies. Moreover, at low impact energies the dynamics of
crossings with deeper levels may yield a cut-off on the high-energy edge of the noncharacteristic x-ray spectrum. At higher impact
energies the vacancy may have a higher probability to jump to deeper levels finally causing the Au-L radiation. Hence, a high energy
cut-off will no more be found for the noncharacteristic radiation.
Due to this effect and due to collision broadening[15,19,20] the observed peak structure merges into the broad descending continuum
with increasing impact energies. The dynamics of the crossings
involved may be deduced by the measured cross section ratios of
I-L, MO, and Au-L x rays.

Our experimental findings concerning peak structure and large
cross sections may be qualitatively explained by the electron
promotion model together with calculated multi-electron level

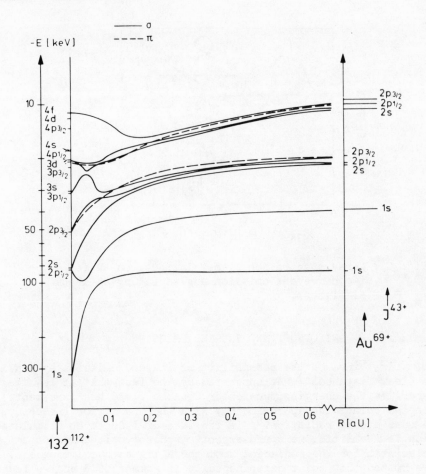

Fig. 10. Relativistic multi-electron level calculation
for the I-Au molecule, cf. Ref. 16.

diagrams. Quantitative agreement, or disagreement, can only be
yielded by comparing the measured spectral shape with a calculated spectrum as it is already done for light collision systems.[15,19]
Furthermore, the MO x-ray cross section calculations need better
approximation. Nevertheless, we think that there will still remain
a small discrepancy, which may be explained by x-ray contributions
due to induced or stimulated emission, see point (1).

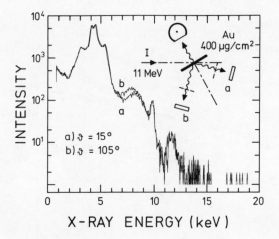

Fig. 11. X-ray spectra for 11-MeV I impact on Au for
        two different emission angles (using an 18-μm
        Al filter).

ANISOTROPY AND INDUCED EMISSION

Our first idea on the possibility of induced emission at small
internuclear distances[4] was supported by the fact that at low impact energies the changing two-center Coulomb field has fundamental
frequencies coinciding approximately with the transition frequency
of the quasiatomic radiation.[8] If the idea of induced or stimulated
emission is realistic, we would expect a nonisotropic emission
characteristic for the radiation from the alligned quasimolecules.
We have tested this point experimentally in summer 1973 and we found
a nonisotropic emission for the x-ray band. [21,22] In Fig. 11,
two spectra taken at different emission angles ϑ are shown for
11-MeV I impact using an 18-μm Al filter. Target angle and emission
angles are chosen in such a way that the target selfabsorption is
equal in both cases. The spectra are normalized to equal beam
intensity by means of x rays registered in a proportional counter
at a fixed position. A significant difference in both spectra is
seen for the 8-keV band only.

For 11-MeV I impact we have measured this anisotropy of the
x-ray band at emission angles from 15° to 105°, Fig. 12. X rays
between 7.1 and 9.5 keV are preferentially emitted perpendicular
to the beam axis. The measured emission characteristic of the
x-ray band can be approximated by a $\sin^2$ function yielding a mean
anisotropy of 16 ± 4 % ($2\{I(90^\circ)-I(0^\circ)\}/\{I(90^\circ)+I(0^\circ)\}$).[22]

At the time these measurements were carried out we had no
well-founded theoretical model to explain these data. Meanwhile,

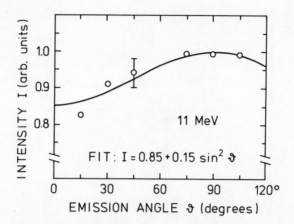

Fig. 12. Integral anisotropy of the noncharacteristic
         x-ray emission between 7.1 and 9.5 keV at
         11-MeV I impact on Au.

B. Müller, R. Kent-Smith, and W. Greiner worked out a model for the
occurence of induced transitions in heavy-ion-atom collisions.[20,23,6]
In this model, the Coriolis forces due to the rotation of the internuclear axis are responsible for induced transitions. The induced
transition probability increases with the square of the angular
velocity of the internuclear axis, i. e., with its rotation frequency. Neglecting back-scattering the angular velocity is largest
at distances of closest approach. Thus the induced emission will
enhance transitions at small internuclear distances - in agreement
with our preliminary ideas, cf. Fig. 9. Hence, at not too high
impact energies quasiatomic radiation should be observed. Moreover,
the measured emission characteristic for the x-ray band fits well
in the model of induced emission.[20,23]

     Looking once more at the spectra in Fig. 12 we find that the
spectral shape of the x-ray band varies with the emission angle.
The anisotropy is largest in the middle of the x-ray band and
vanishes on both sides. In Fig. 13 the differential anisotropy is
given (cf. Ref. 23). The various curves show the intensity ratio
of MO x-ray emission at different emission angles depending on
x-ray energy. All curves are normalized to an emission angle of 15°.
All curves display a maximum at the same x-ray energy, increasing
in amplitude with increasing emission angle (roughly according the
$\sin^2 \vartheta$ - law). The maximum anisotropy is about 35 % at 7.5 keV x-
ray energy. This anisotropy maximum is situated almost exactly at
the calculated M-transition energy. As postulated by the theory
of B. Müller et. al. this differential anisotropy or asymmetry is
the most striking feature giving evidence of true quasiatomic radiation.[20,23]

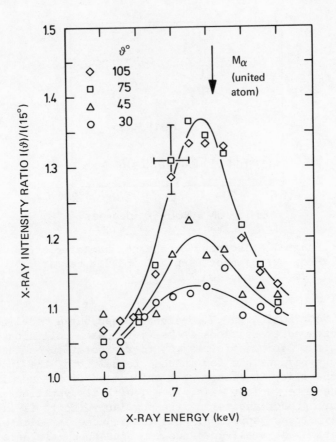

Fig. 13. Differential anisotropy of the noncharacteristic
x-ray emission at 11-MeV I impact on Au.

Neglecting a possible nonisotropic emission for the spontaneous
decay of the vacancy in the I-Au quasimolecule the noncharacteristic
x-ray emission in the 8-keV region may be understood as follows:
The difference in energy position of spectral peak and anisotropy
peak may indicate two different components contributing to the
x-ray band: (1) Spontaneous quasimolecular and quasiatomic emission.
               (2) Induced quasiatomic emission.
The spontaneous emission linked with the special energy dependence
of the levels on internuclear axis, see Fig. 9, may cause the broad
peak in the x-ray spectrum; whereas the induced emission may cause
the anisotropy peak.

     Further measurements and calculations are needed to ascertain
this interpretation. We hope to be able to produce more experimental
data in the near future. Nevertheless, we think that the available

measurements are very encouraging for a possible spectroscopy in
the superheavy region without the need to produce stable superheavy elements.

QUASIMOLECULAR L AND K RADIATION

The shown measurements are related only to the united M shell.
However, theoretically, quasiatomic L and particularly K radiation
are of most interest. A brief review on quasimolecular K and L
radiation is given in this section starting with investigations
concerning the quasimolecular L radiation. In contrast to the well-

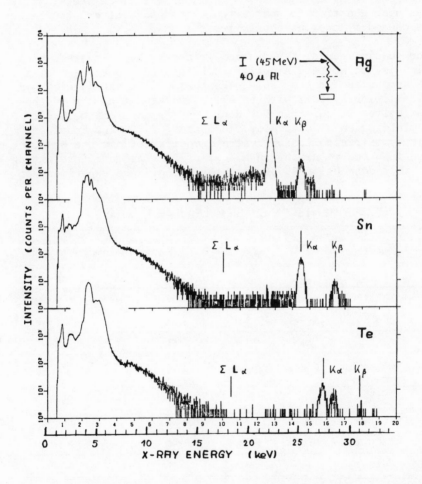

Fig. 14. X-ray spectra for 45-MeV I impact on Ag, Sn,
and Te targets (using a 40-μm Al filter).

studied light collision systems, as e. g. Ar→Ar, very little is
known on quasimolecular L radiation from real heavy collision systems. This fact is mainly due to the small cross sections. If in
this case the double-collision mechanism is still in act it is
necessary for the bombarding particle to be the heavier collision
partner and to carry an L vacancy into the second collision. So far,
the heaviest system investigated on quasimolecular L radiation is
the Au→Sn system measured by the Rochester group.[11] We tried similar measurements on lighter systems, Fig. 14. Here spectra for
45-MeV I bombardment on Ag, Sn, and Te are shown, using a 40-μm Al
filter. In all spectra a descending continuum of noncharacteristic
x rays is present. Equivalent spectra were found by Presser et. al.
for recoiling Ag→Ag collisions.[24] But, in all these cases, it is
still an open question to what extent the noncharacteristic radiation is caused by quasimolecular transitions.

For quasimolecular K radiation from very heavy systems the
situation is not so bad. A lot of work is being done. Part of the
work is summarized in the table. The table deals mainly with symmetric collision systems with an effective atomic number of the
combined system larger than 50. Most work is done on the Br→Br
region yielding an effective atomic number of 70. The heaviest
reliable symmetric collision system investigated, is the Nb→Nb
measured in Dubna resulting in a Z_{eff} of 82. The most interesting
work in this connection is perhaps the coincidence experiment on

EXPERIMENTS ON QUASI-MOLECULAR K X RAYS

projectile Z_1	energy (MeV/amu)	target Z_2	K-binding energy (keV)[a]	eff. atomic number Z_1+Z_2	rel. yield Y(MO)/Y(K)	reference (et. al.)	
$_6C$	$(1,5....20)\cdot10^{-3}$	$_6C$	0.28	12	$(5....50)\cdot10^{-4}$	Macdonald	25)
$_{28}Ni$	0.7..... 1.2	$_{28}Ni$	8.33	56	—	Greenberg	26)
$_{10}Ne$	$1.5\cdot10^{-2}...1$	$_{29}Cu \rightarrow _{29}Cu$ [b]	8.98	58	—	Groeneveld	24)
$_{32}Ge$	1.1	$_{32}Ge$	11.10	64	$4.4\cdot10^{-5}$	Gippner	27)
$_{35}Br$	0.4.....0.75 / 0.4..... 1.3	$K_{35}Br$	13.47	70	$(0.6....1.4)\cdot10^{-5}$ / —	Meyerhof / Davis	28) / 29)
$_{36}Kr$	2.4	$_{40}Zr$	18.00	76	—	Meyerhof	30)
$_{41}Nb$	4.3..... 6.5	$_{41}Nb$	18.99	82	$(1.....2)\,10^{-5}$	Gippner	31)
$_{53}I$	0.7	$Na_{55}I$	33.17	106	—	Meyerhof [c]	32)
$_{17}Cl$	1.7	$_{82}Pb$	88.01	99	—	Burch [d]	33)

a) target atoms   b) recoils   c) uncertain due to γ-ray background
d) possibly recoil contributions   (Pb-Pb united L x rays)

the Cl→Pb carried out by Burch et. al.. They tried to supress
the characteristic lines by selecting only backward scattering
collisions.[33] And indeed they get a noncharacteristic x-ray
enhancement ranging up to the right energy position. These measurements partly avoid one difficulty which all heavy quasimolecular
K x-ray investigations have, namely, to determine an end-point of
an endless desceding x-ray continuum. This fact would make impossible
a quasiatomic K spectroscopy from simple x-ray spectra. More elaborated work is needed, e. g., coincidence measurements as first
applied by D. Burch et. al., (cf. also the paper of W. Lichten[15])
or anisotropy measurements as shown in this paper for the united
M-shell. Meanwhile, such anisotropy measurements can be compared
with the model on rationally-induced emission developed by the theory
group in Frankfurt.[20,23] Recently, two anisotropy measurements
were made on quasiatomic K radiation, one from the Stanford group
on the Kr→Zr collision system[30] and one from the Yale group on
the Ni→Ni system.[34] Both measurements show - similarly to our results - an anisotropy maximum around the expected quasiatomic transition energies.

References:

1) F.W. Saris, W.F. von der Weg, H. Tawara, and R. Laubert,
    Phys. Rev. Lett. 28, 717 (1972); F.W. Saris, I. V. Mitchell,
    D.C. Santry, J.A. Davies, and R. Laubert, in Proceedings of the
    International Conference on Inner-Shell Ionization Phenomena,
    Atlanta, Georgia, 1972, CONF-720404 (Vol. 2) 1255 (1973).

2) U. Fano and W. Lichten, Phys. Rev. Lett. 14, 627 (1965);
    W. Lichten, Phys. Rev. 164, 131 (1967);
    M. Barat and W. Lichten, Phys. Rev. A 6, 211 (1972);
    W. Lichten, this issue.

3) F.J. de Heer, this issue

4) P.H. Mokler, H.J. Stein, and P. Armbruster, Phys. Rev. Lett. 29,
    827 (1972); P.H. Mokler, H.J. Stein, and P. Armbruster, in Pro-
    ceedings of the International Conference on Inner-Shell Ioniza-
    tion Phenomena, Atlanta, Georgia, 1972, CONF-720404 (Vol. 2)
    1283 (1973).

5) J. Rafelski, L.P. Fulcher, and W. Greiner, Phys. Rev. Lett. 27,
    958 (1971).

6) W. Greiner, this issue.

7) P. Armbruster, G. Kraft, P.H. Mokler, B. Fricke, and H.J. Stein, in Proceedings of the Nobel Symposium on Superheavy Elements, to be published; the paper was presented by P. Armbruster in Ronneby, Sweden, June 1974.

8) H.J. Stein, in report of Gesellschaft für Schwerionenforschung Darmstadt, GSI 73-11, 106 (1973).

9) B. Fricke, and G. Soff, report of Gesellschaft für Schwerionen-forschung Darmstadt, GSI-T1-74 (1974).

10) H.E. Gove, F.C. Jundt, and H. Kubo, Bull. Am. Phys. Soc. Vol. 18, No. 4, 559 (1973);F.C. Jundt and H.E. Gove, private communication.

11) H.E. Gove, F.C. Jundt, and H. Kubo, report of the University of Rochester, UR-NSRL-81 (1974).

12) S. Hagmann, P. Armbruster, G. Kraft, P.H. Mokler, and H.J. Stein, to be published.

13) P.H. Mokler, H.J. Stein, P. Armbruster, Verhandlungen DPG (VI) 8, 51 (1973); Fig. 9 was shown at spring meeting of the German Physical Society, Heidelberg, Febr. 1973.

14) E.J. McGuire, Phys. Rev. A $\underline{3}$, 587 (1971); E.J. McGuire, Phys. Rev. A $\underline{5}$, 1043 (1972).

15) W. Lichten, Phys. Rev. A $\underline{9}$, 1458 (1974).

16) B. Fricke, in Abstracts of Contributed Papers to F.I.C.A.P. Heidelberg, 641 (1974);
    B. Fricke and K. Rashid, to be published.

17) B. Müller, J. Rafelski, and W. Greiner, in report of Gesell-schaft für Schwerionenforschung Darmstadt, GSI-73-11, 95 (1973).

18) H. Hartmann and K. Helfrich, Theor. Chim. Acta $\underline{10}$, 406 (1968); K. Helfrich and H. Hartmann, Theor. Chim. Acta $\overline{16}$, 263 (1970); cf. B. Fricke, in report of Gesellschaft für Schwerionenfor-schung Darmstadt, GSI 73-11, 88 (1973).

19) J.S. Briggs, J. Phys. B $\underline{7}$, 47 (1974).

20) B. Müller, R. Kent-Smith, and W. Greiner, Phys. Lett. $\underline{49B}$, 219 (1974).

21) G. Kraft, H.J. Stein, P.H. Mokler, and S. Hagmann, Verhand-lungen DPG (VI) 9, 77 (1974).

22) P.H. Mokler, G. Kraft, and H.J. Stein, in Contributed Papers
     to International Conference on Reactions Between Complex Nuclei,
     Nashville, 134 (1974).

23) B. Müller, and W. Greiner, in print, Phys. Rev. Lett. (1974);
     B. Müller, R. Kent-Smith, and W. Greiner, in Abstracts of
     Contributed Papers to F.I.C.A.P., Heidelberg, 608 (1974);
     R. Kent-Smith, B. Müller, and W. Greiner, in print, Journal of
     Phys. B (1974). Fruitful discussions with Prof. Greiner and
     coworkers are gratefully acknowledged.

24) G. Presser, K.O. Groeneveld, and B. Knaf, in Abstracts of
     Contributed Papers to F.I.C.A.P., Heidelberg, 621 (1974);
     K.O. Groeneveld, B. Knaf, and G. Presser, in annual report 1973
     of Institut für Kernphysik, Frankfurt, IKF-32, 77 (1974).

25) J.R. Macdonald, M.D. Brown, and T. Chiao, Phys. Rev. Lett. 30,
     471 (1973).

26) J.S. Greenberg, C.K. Davis, and P. Vincent, in Contributed
     Papers to International Conference on Reactions Between Com-
     plex Nuclei, Nashville, 135 (1974).

27) P. Gippner, K.H. Kaun, F. Stary, W. Schulze, and Yu. P. Tretyakov,
     JINR preprint E7-7636, Dubna (1973).

28) W.E. Meyerhof, T.K. Saylor, S.M. Lazarus, W.A. Little,
     B.B. Triplett, and L.F. Chase, Phys. Rev. Lett. 30, 1279 (1973);
     W.E. Meyerhof et. al., Phys. Rev. Lett. 32, 502 (1974).

29) C.K. Davis, and J.S. Greenberg, Phys. Rev. Lett. 32, 1215 (1974).

30) W.E. Meyerhof, R. Anholt, F.S. Stephens, and R. Diamond, in
     Abstract of Contributed Papers to F.I.C.A.P., Heidelberg, 633
     (1974).

31) P. Gippner, K.H. Kaun, H. Sodan, F. Stary, W. Schulze, and
     Yu. Tretyakov, in Contributed Papers to International Confe-
     rence on Reactions Between Complex Nuclei, Nashville, 20 (1974).

32) W.E. Meyerhof, T.K. Saylor, S.M. Lazarus, A. Little,
     B.B. Triplett, L.F. Chase, and R. Anholt, private communication
     by W.E. Meyerhof.

33) D. Burch, W.B. Ingalls, R. Vandenbosch, and H. Wieman, in
     Annual Report 1973, University of Washington, Nuclear Physics
     Lab., 161 (1974).

34) J.S. Greenberg, C.K. Davis, and P. Vincent, in Abstracts of
    Contributed Papers to F.I.C.A.P., Heidelberg, 617 (1974);
    and to be published.

MODEL AND PSEUDOPOTENTIAL CALCULATIONS

A. Dalgarno

Center for Astrophysics

60 Garden Street, Cambridge, Massachusetts 02138

1. INTRODUCTION

The structure of the Periodic Table and the concept of valence demonstrate the qualitative value of methods which avoid a detailed description of the inner shell core electrons and the introduction of model and pseudopotential methods stemmed from the recognition that many atomic and molecular properties are determined quantitatively by the interactions of only the outermost electrons of the atomic or molecular system. Detailed reviews of the applications of model and pseudopotential methods have been presented by Weeks, Hazi and Rice (1969) and by Bardsley (1974).

Although in use earlier, the first formal theoretical developments appear to be those of Gombas (1935) and Hellmann (1935). The basic theory was greatly clarified in more recent papers by Phillips and Kleinman (1959), by Cohen and Heine (1961) and by Austin, Heine and Shaw (1962).

There are various ways of approaching the problem which can be classified either as a pseudopotential or as a model potential procedure. The pseudopotential procedure begins with the recognition that the exclusion of the outer shell valence electrons from the core is produced mathematically by the orthogonality requirement that the orbitals of the valence electrons ϕ_v be orthogonal to the ϕ_α of the core electrons. Thus $\langle \phi_v | \phi_\alpha \rangle = 0$. The same effect can be achieved physically by a short-range repulsive potential.

The equation for the orbital of the valence electron

$$H\phi_v = (T+v)\phi_v = E_v\phi_v$$

where

$$\langle \phi_v | \phi_\alpha \rangle = 0$$

may be rewritten in the form (cf. Phillips and Kleinman 1959)

$$(H + V_R) \chi_v = E_v \chi_v$$

where V_R is a pseudopotential defined by

$$V_R = \sum_{\text{core}} (E_v - E_\alpha) \langle \phi_\alpha | \chi_v \rangle \phi_\alpha \quad ,$$

E_α is the eigenvalue of the α core electron, and χ_v is a pseudo-orbital

$$\chi_v = \phi_v + \sum_{\text{core}} \theta_\alpha \phi_\alpha$$

where θ_α are arbitrary coefficients which express the departure from orthogonality of χ_v. The pseudopotential V_R can also be written more generally in the form

$$V_R = \sum_\alpha | \phi_\alpha \rangle \langle F_\alpha |$$

where F_α are arbitrary functions.

We have presented the analysis within the independent particle approximation. The generalisation in which the pseudopotential projects onto any chosen subspace is straightforward (Weeks et al. 1959, Tully 1969, Kleiner and McWeeny 1973).

The equation containing the pseudopotential is merely a restatement of the original equation in which the orthogonality constraint has been replaced by a modification of the Hamiltonian. The valence eigenvalues are the same and the valence orbitals are obtained directly from the pseudo-orbitals by projecting out the core orbitals.

The pseudopotential is non-local, energy-dependent, angular-momentum dependent and it contains a large degree of arbitrariness. The strength of the pseudopotential method lies in that arbitrariness. It is usually convenient to choose χ_v so that it is a smooth function in the core region. The kinetic energy is decreased and the corresponding pseudopotential is a repulsive short-range potential that prevents penetration of the core region.

Various procedures have been suggested for the theoretical calculation of χ_v and V_R (cf. Bardsley 1974). Recent studies are those of Melius and Goddard (1973) and Kleiner and McWeeny (1973).

However most applications of pseudopotential methods avoid the
calculation of V_R by appealing to experimental data for its construction. Customarily V_R is written as some simple function of
position containing several parameters that are determined by requiring that the eigenvalues E_v equal the measured binding energies.
V_R depends in general on symmetries such as orbital and spin
angular momenta and is defined as a sum involving projection
operators.

With this last empirical step, the pseudopotential method is
similar in practice to the model potential method. The model
potential approach requires that the spectrum of the model
Hamiltonian coincide with the exact spectrum. The analysis can be
developed by writing the eigenfunctions of the systems as sums of
antisymmetrised products of the inner core eigenfunctions Φ_α and
the valence eigenfunctions Φ_v. In order to obtain formally an
energy-independent local potential it appears necessary to assume
that the excitation energies of the valence electrons are small
compared to those of the core and to assume that exchange interactions between the core and the valence electrons can be neglected
(Bottcher and Dalgarno 1974). The first assumption is usually a
satisfactory approximation. The second may not be but it is the
expectation of the model potential method that the effects of the
exchange can be reproduced by an empirical local interaction.

A formal description of the model potential can be obtained
by expanding as before and projecting on to the subspace of the
valence electrons. This procedure is commonly used in scattering
theory where one projects on to the subspace of the scattered
electron and it leads to the introduction of the optical potential.
At large distances from the nucleus, the leading term of the optical
potential becomes real, local and energy-independent and indeed is
the polarization interaction.

The explicit construction of a model potential can also be
carried out by holding the valence electrons at fixed distances
from the nucleus and calculating the eigenvalues, which will
depend parametrically on the positions of the valence electrons.
The eigenvalues define a local potential in which the valence
electrons move (cf. Bersuker 1958, Caves and Dalgarno 1973).   In
molecular physics, this procedure is the Born-Oppenheimer
approximation.

All these various approaches are valuable in suggesting the
form that an empirical representation of a model potential should
take. In particular, they show that asymptotically the model
potential tends to the polarization potential, decreasing as the
inverse fourth power of the distance from the nucleus.

2. APPLICATIONS

In most applications, very simple forms have been adopted to
represent the pseudopotentials or the model potentials. The
simplest example is the widely used Coulomb approximation of Bates
and Damgaard (1949) and Burgess and Seaton (1960) in which
$V(r) = -Zeff/r$, where Zeff is an effective nuclear charge chosen
to reproduce the binding energy. The Coulomb approximation is only
valid asymptotically. An extension in which a term $\lambda(\lambda+)/r^2$ is
added to V(r) overcomes this limitation and it has been used
recently to predict a large number of oscillator strengths of
atomic systems (Simons 1974). A second example is the more complicated problem of classifying the Rydberg levels of formaldehyde and
carbon dioxide which has been considered by Betts and McKoy (1974)
using the Abarenkov-Heine model potential for each atom

$$V_m(r) = A_m \quad r < r_o$$
$$= \frac{\delta z}{r} \quad r > r_o$$

and assuming that the molecular model potential is the sum of the
atomic potentials. The Abarenkov-Heine potential has been criti-
cised by Giulano and Ruggeri (1969) who suggest the alternative form

$$V(r) = \frac{-Z}{r+A/r^2} - \frac{Nexp(-\alpha r)}{r} \quad .$$

Other forms have been used, some of which are designed to
simplify the subsequent computations (cf. Schwartz and Switalski
197 , Bonifacic and Huzinaga 1974). It is also of considerable
interest to use pseudopotentials and model potentials of greater
sophistication in order that the limits of accuracy can be assessed
quantitatively. In the work of Weisheit (1972) and Norcross (1972)
on the alkali metals, potentials are used which reproduce a wide
range of binding energy data and which include the long range
dipole and quadrupole interactions that should be present.

The calculations of Weisheit (1972) of the photoionization
cross section of sodium are reproduced in figure 1. There occurs a
discrepancy at shorter wavelengths which is due presumably to
dynamic interactions with the core electrons and cannot be removed
by a local potential method but for low energies of ejection and
for the discrete oscillator strengths, the results are good. It is
not easy to predict the correct photoionization cross sections of
the alkali metal atoms by conventional methods (Labahn and Garbaty
1974) and the success of the model potential method is encouraging.

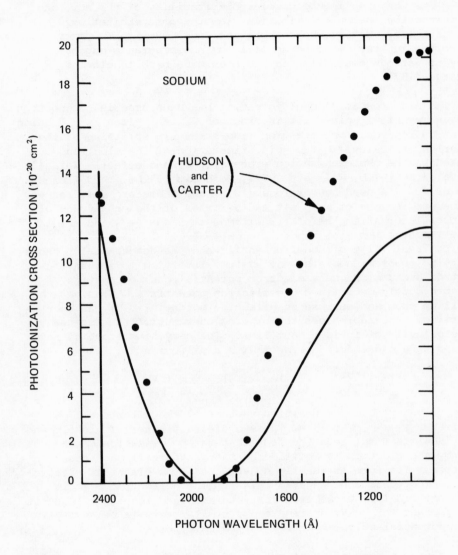

Figure 1. The photoionization cross section of sodium. The filled circles are the measurements of Hudson and Carter (from Weisheit 1972).

The theoretical development of the model potential method shows that for consistency in calculating dipole matrix elements the dipole operator $\underset{\sim}{r}$ should be replaced asymptotically by $\underset{\sim}{r}(1-\alpha_d/r^3)$ where α_d is the dipole polarizability of the core and the correction is the contribution from the core excitations (Bersuker 1957, Hameed, Herzenberg and James 1968). In practice, the correction term must be modified by a short-range cut-off. A study of various possibilities has been made by Weisheit and Dalgarno (1971a).

The electrostatic model potential has been used in conjunction with an empirical spin-orbit interaction $V_{so} = \underset{\sim}{\ell} \cdot \underset{\sim}{s}(z-\sigma)/r^3$, σ being chosen to reproduce the measured fine-structure splitting of the ^2P series, to calculate the spin orientation of the electrons ejected in the photoionization process (the Fano effect) (Weisheit and Dalgarno 1971b, Weisheit 1972a, Weisheit 1972b). A different form of V_{so} was used by Norcross (1973) and there are differences in the predictions of Weisheit and Norcross of the magnitudes of the oscillator strengths of transitions to high $^2P_{1/2}$ and $^2P_{3/2}$ Rydberg states.

The real value of model potential and pseudopotential methods in atomic problems emerges for systems involving more than one electron because the many electron potential can be constructed from previously determined one electron potentials. Consider the beryllium atom and suppose an effective potential V(r) has been determined that reproduces the binding energies of the excited Rydberg levels of ionized beryllium. The beryllium atom is then reduced to a two-electron system with Hamiltonian

$$H = -\tfrac{1}{2}\nabla_1^2 - \tfrac{1}{2}\nabla_2^2 + V(r_1) + V(r_2) + \frac{1}{r_{12}} + V_c$$

where $\underset{\sim}{r}_1$ and $\underset{\sim}{r}_2$ are the position vectors of the outer shell electrons, $r_{12} = |\underset{\sim}{r}_1 - \underset{\sim}{r}_2|$ and V_c is a dielectric term (Chisholm and Öpik 1964) that reflects the fact that the polarization of the core by one electron is modified by the pressure of the second electron. Its form is readily constructed from classical electrostatics (Bottcher and Dalgarno 1974a). The dielectric term is often omitted (cf. Cavaliere, Ferrante and Geracitano 1972) but it is important for heavy systems and a similar term is necessary in model potential calculations of interatomic forces.

A large number of model and pseudopotential calculations of two electron systems have been carried out. Some of the more recent papers are those of Schwartz (1971), Simons (1971), McGinn (1972), Szasz and McGinn (1972), Cavaliere et al. (1972), Fung and Matese (1972), Victor and Laughlin (1972, 1973, 1974), Bardsley and Junker (1974), Norcross (1974) and Stewart, Laughlin and Victor (1974). Many of the papers discuss the electron affinities of the alkali metal atoms and table 1 lists some values for Li and Na. The table includes the experimental values of Patterson et al. (1974).

Table 1

Electron affinities of Li and Na

(eV)

Reference	Li	Na
Patterson et al. (1974)	0.620 ± 0.007	0.548 ± 0.004
Schwarz (1971)	0.62	0.54
Victor and Laughlin (1972a)	0.591	--
Cavaliere et al. (1972)	--	0.535
Fung and Matese (1972)	--	0.567
Norcross (1974)	0.614	0.538
Bardsley and Junker (1974)	0.60	0.53
Stewart et al. (1974)	0.601	0.566

The agreement is good although neither Schwarz (1971) nor
Cavaliere et al. (1972) included the dielectric contribution. In
the calculations of Victor and Laughlin (1972) and Stewart et al.
(1972), the dielectric contribution is included but it appears that
the ultimate convergence limit has not been reached. In order to
avoid variational collapse, they restrict the trial functions to
products of orbitals of the single particle model potential.
Norcross (1974) proceeds alternatively by solving a set of coupled
differential equations similar to those of scattering theory
(Seaton and Wilson 1972), a procedure which may have approached
close to the limiting values of the model.

The model potentials of Norcross (1974) and Stewart et al.
(1974) also demonstrate the existences of bound ^3P states of the
negative ions   except for Li⁻ and Fung and Matese (1952) and
Stewart et al. (1974) have used the model potential method to
identify doubly-excited onto detaching resonances.

Spin-orbit interactions in two-electron systems have been
studied by Victor and Laughlin (1974) who assumed that the spin-
orbit interaction is given simply by the sum of the single electron
interactions determined empirically (cf. Weisheit 1972). Victor
and Laughlin were able to calculate the effects of spin-orbit
interactions on the decay of the metastable ^3P states of beryllium
and magnesium. For beryllium they calculate a lifetime of 3.72 sec
but no experimental value exists. For magnesium, they calculate a
lifetime of 4.6m sec somewhat smaller than an unpublished measurement
of 5.4 msec by Mitchell (1973) but larger than the measurement,
1.8 msec, of Wright, Dawson and Balling (1974). The lifetimes are
sensitive measures of the degree of singlet triplet mixing and the
comparison shows that empirical model potential methods are capable
of describing the electron density distribution near the nucleus
with useful accuracy.

Model and pseudopotential methods have been widely used in the calculation of interatomic potentials (cf. Bardsley 1974) for which it is assumed that the molecular Hamiltonian is the sum of the atomic model Hamiltonians and a direct core-core interaction term, which is often approximated as R^{-1} where R is the core-core distance. There is an additional term analogous to the dielectric term of a two-electron system (Bottcher and Dalgarno 1974) whose inclusion is essential if the correct long-range interaction is to be obtained (Dalgarno, Bottcher and Victor 1970). It is of particular importance for the interactions between neutral species and it is contained in the model calculations for alkali metal atoms and inert gas atoms of Baylis (1969), Bottcher, Dalgarno and Wright (1973) and Pascale and Vandenplanque (1974).

Considerable attention has been given to the alkali molecular ions for which a description as a single electron moving in the field of two polarizable cores is particularly appropriate. The magnitude of the contribution of the dielectric term to the dissociation energy of Li_2^+ is only 0.02 eV (Dalgarno et al. 1970). Differences between the various model potential values are much larger. Table 2 is a summary of the results. The most refined model potential calculations are in harmony with each other but about 0.28 eV below the experimental value listed by Gaydon (1968). However Bardsley (1974) has argued persuasively that the correct experimental value is 1.29 eV.

Table 2

Dissociation energy of Li_2^+ in eV

Reference	Dissociation energy
Struve (1970)	0.98 or 1.44
Dalgarno et al. (1970)	1.28
Kahn and Goddard (1972)	1.22
Schwartz and Switalski (1972)	1.23
Bardsley and Junker (1974)	1.31
Bottcher and Dalgarno (1974b)	1.29

In the cases of heavier ionic systems and of neutral systems, further study is needed to establish the limits of accuracy that can be attained. It is clear however that interatomic potentials that are of useful accuracy at intermediate and large distances should be derivable by a careful choice of model Hamiltonian with relatively little labour. Figure 2 illustrates some unpublished model potential calculations of the ground and several excited states of MgHe by Bottcher, Docken and Dalgarno (1974).

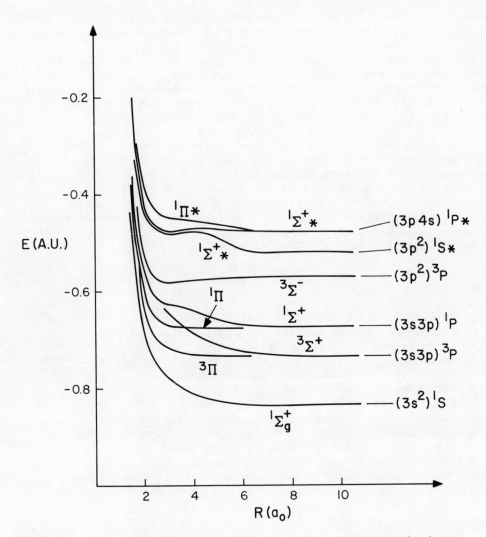

Figure 2. Potential energy curves for MgHe. The dissociating states of Mg are labelled. The estimates specify autoionizing states.

3. CONCLUSION

This account has been brief and much recent work has been omitted. The examples were chosen selectively to demonstrate the range of applicability of pseudopotential and model potentials. A significant role for such methods may be in their association with other methods such as multichannel quantum defect theory, Z-expansion theories and various extrapolation methods. This association is demonstrated by the extension of the close-coupling methods of scattering theory to bound state problems (Seaton and Wilson 1972, Norcross 1974). Based as they are on a frozen core approximation, they can be regarded as a pseudopotential method.

The fact that the pseudopotential is repulsive has yet to be fully exploited in multicenter molecular problems. The pseudo-potential method suggests that most of the multicenter nature arises from the requirements of orthogonality to the multicenter core orbitals. The pseudoorbitals may be nearly spherical and substantial simplification should be possible, as was emphasized by Schneider and Berry (1969) in a simple pseudopotential calculation of the photoionization of molecular nitrogen.

REFERENCES

B. J. Austin, V. Heine and L. J. Shain 1962 Phys. Rev. 127, 276.
J. N. Bardsley 1974 Case Studies in Atomic Physics in press.
J. N. Bardsley and B. R. Junker 1974 to be published.
D. R. Bates and A. Damgaard 1949 Phil. Trans. Roy. Soc. A242, 101.
W. E. Baylis 1969 J. Chem. Phys. 51, 2665.
I. B. Bersuker 1957 Opt. Spectr. 3, 97.
T. C. Betts and V. McKoy 1974 J. Chem. Phys. 60, 2947.
V. Bonifacic and S. Huzinaga 1974 J. Chem. Phys. 60, ´´79.
C. Bottcher and A. Dalgarno 1974a Proc. Roy. Soc. A in press.
C. Bottcher and A. Dalgarno 1974b in preparation.
C. Bottcher, A. Dalgarno and E. L. Wright 1973 Phys. Rev. A7, 1606.
C. Bottcher, K. K. Docken and A. Dalgarno 1974 to be published.
A. Burgess and M. J. Seaton 1960 Mon. Not. Roy. Astron. Soc. 120, 121.
P. Cavaliere, G. Ferrante and R. Geracitano 1972 Nuovo Cimento 9B, 96.
T. Caves and A. Dalgarno 1972 J. Quant. Spectr. Rad. Trans. 12, 1539.
C. D. H. Chisholm and U. Öpik 1964 Proc. Phys. Soc. 83, 541.
M. H. Cohen and V. Heine 1961 Phys. Rev. 122, 1821.
A. C. Fung and J. J. Matese 1972 Phys. Rev. A5, 22.
P. Gombas 1935 Z. Physik 94, 473.
S. Hameed, A. Herzenberg and M. G. Janes 1968 J. Phys. B. (Atom. Mol. Phys.) 1, 822.

H. Hellmann 1935 Acta Physicochim. URSS $\underline{1}$, 913.

L. R. Kahn and W. A. Goddard 1972 J. Chem. Phys. $\underline{56}$, 2685.

M. Kleiner and R. McWeeny 1973 Chem. Phys. Lett. $\underline{19}$, 476.

R. W. Labahn and E. A. Garbaty 1974 Phys. Rev. $\underline{A9}$, 2255.

W. C. Lineberger 1974 Phys. Rev. Lett. $\underline{32}$, 189.

C. Melius and W. A. Goddard 1973 preprint.

C. J. Mitchell 1973 to be published.

D. W. Norcross 1973 Phys. Rev. $\underline{A7}$, 606.

D. W. Norcross 1974 Phys. Rev. Lett. $\underline{32}$, 192.

J. Pascale and J. Vandenplanque 1974 J. Chem. Phys. $\underline{60}$, 2278.

J. C. Phillips and L. Kleinman 1959 Phys. Rev. $\underline{116}$, 207.

B. Schneider and R. S. Berry 1969 Phys. Rev. $\underline{182}$, 141.

M. E. Schwartz and J. D. Switalski 1972 J. Chem. Phys. $\underline{57}$, 4125.

W. H. E. Schwarz 1971 Chem. Phys. Lett. $\underline{10}$, 478.

M. J. Seaton and P. M. H. Wilson 1972 J. Phys. B. (Atom. Mol. Phys.)
      $\underline{5}$, L175.

G. Simons 1971 J. Chem. Phys. $\underline{55}$, 756.

G. Simons 1974 J. Chem. Phys. $\underline{60}$, 645.

R. F. Stewart, C. Laughlin and G. A. Victor 1974 Chem. Phys. Lett.
      in press.

W. S. Struve 1970 Chem. Phys. Lett. $\underline{7}$, 382.

J. Tully 1969 Phys. Rev. $\underline{181}$, 7.

G. A. Victor and C. Laughlin 1972a Chem. Phys. Lett. $\underline{14}$, 74.

G. A. Victor and C. Laughlin 1973 Nucl. Inst. Methods $\underline{110}$, 189.

G. A. Victor and C. Laughlin 1974 Ap. J. in press.

J. D. Weeks, A. Hazi and S. A. Rice 1969 Adv. Chem. Phys. $\underline{16}$, 283.

J. C. Weisheit 1972a Phys. Rev. $\underline{A5}$, 1961.

J. C. Weisheit 1972b J. Quant. Spectr. Rad. Trans. $\underline{12}$, 1241.

J. C. Weisheit and A. Dalgarno 1971a Chem. Phys. Lett. $\underline{9}$, 517.

J. C. Weisheit and A. Dalgarno 1971b Phys. Rev. Lett. $\underline{11}$, 701.

J. J. Wright, J. F. Dawson and L. C. Balling (1974) Phys. Rev. $\underline{A9}$,
      83.

EIKONAL-BORN SERIES METHODS IN ELECTRON-ATOM SCATTERING

F.W. Byron, Jr.

Department of Physics

University of Massachusetts, Amherst, Mass. 01002

There have been two developments in recent atomic collision theory which are particularly important from a fundamental point of view. The first of these was the use of the closure approximation to study the second term in the many-particle Born series, primarily by Moiseiwitsch and his collaborators[1]. This enabled one to envision the possibility of systematically correcting the well-known and often used first Born approximation. The second development was the introduction of the Glauber approximation[2] into atomic physics. By making certain key simplifications of the closure method through the use of eikonal concepts, this approximation gave a particularly attractive way of summing parts of perturbation theory to <u>all</u> orders.

The first published work on this method as applied to atomic physics goes back to the study of electron-hydrogen elastic scattering by Franco in 1968[3]. The method was extended to direct inelastic scattering by Ghosh and Sill[4] by Tai, Bassel, Gerjuoy and Franco[5] and by Byron[6] who looked at some basic excitation processes in hydrogen and helium and, in a more speculative manner, to rearrangement collisions by Byron and Joachain[7]. A recent review article by Gerjuoy and Thomas[8] discusses in considerable detail the application of the Glauber approximation to atomic physics as it has developed in the past 5 years. The review by Joachain and Quigg[9] places the Glauber approximation in the general context of multiple scattering collision theory.

In this paper, I want to discuss some work which has been done recently to bring out the relationship between these two developments. The basic idea is to study the Glauber approximation as a multiple scattering series and to relate this "Glauber series" to the Born series in hopes of elucidating the range of validity of both methods.

To begin with let me give a very brief resumé of some of the key concepts mentioned above. First of all let us consider potential theory, or equivalently, electron-atom scattering with all channels neglected except the elastic channel. In this case the Glauber approximation is nothing more than the familiar eikonal approximation discussed in many textbooks. This approximation gives a remarkably compact, unitary representation of the scattering amplitude in the large wave number as follows:

$$f_E = \frac{p}{2\pi i} \int e^{i\vec{K}\cdot\vec{b}} [e^{i\chi} - 1] \, d\vec{b} \quad . \tag{1a}$$

Here p is the wave number of the incident electron, and

$$\chi(\vec{b}) = -\frac{1}{p} \int_{-\infty}^{\infty} V(\vec{b}, z) \, dz \quad . \tag{1b}$$

The z-axis is chosen perpendicular to the momentum transfer, following the remarkably fruitful suggestion of Glauber.

The closure approximation, mentioned above in connection with the second Born approximation, is simply an attempt to render tractable the many-body Green's function

$$G^{MB}(\vec{r}, \vec{r}', \xi, \xi') = -\frac{1}{4\pi} \frac{e^{ip_n|\vec{r}-\vec{r}'|}}{|\vec{r}-\vec{r}'|} \phi_n(\xi) \phi_n^*(\xi') \quad , \tag{2}$$

where $p_n = [p^2 - (\varepsilon_n-\varepsilon_i)]^{1/2}$. In Eq. (2) ε_i is the internal energy of the target in its initial state, and ε_n is its energy in an arbitrary state, n. All energies are in Rydbergs. The ensemble of target coordinates is denoted by ξ, and ϕ_n is a typical internal target state. The closure approximation replaces the infinitude of energy differences, $(\varepsilon_n-\varepsilon_i)$, by a single average difference Δ_i. Then one writes for all n

$$p_n = \bar{p} = (p^2 - \Delta_i)^{1/2} \quad ,$$

and the target state sum in G^{MB} reduces to the closure relation, giving

$$G^{MB} \cong - \frac{1}{4\pi} \frac{e^{i\bar{p}|\vec{r}-\vec{r}'|}}{|\vec{r}-\vec{r}'|} \delta(\xi-\xi') \quad . \tag{3}$$

With G^{MB} in this closure form the second Born term for direct processes,

$$\bar{f}_{B2} = - (2\pi)^2 <f|V\ G^{MB}\ V|i> , \tag{4}$$

can be evaluated analytically for many cases of interest.

Finally, the Glauber approximation is just a simple combination of the eikonal and closure approximations in which one conjectures that for p large one may plausibly take $\bar{p} \simeq p$ in Eq. (3), so that

$$G^{MB} = G^{free}\ \delta(\xi-\xi') \tag{5}$$

Because of the δ- function each term in perturbation theory reduces to potential scattering by a "frozen target" potential in which the target coordinates are just parameters. If we treat the scattering by this special potential in the above-mentioned eikonal approximation then we find for the many-body scattering amplitude

$$f_G = \frac{p}{2\pi i} \int e^{i\vec{K}\cdot\vec{b}} <f_B|e^{i\chi} - 1|i_B)\ d\vec{b} \tag{6a}$$

where χ is given according to Eq. (1b) by

$$\chi = - \frac{1}{p} \int_{-\infty}^{\infty} V(\vec{b},\ z,\ \xi)\ dz , \tag{6b}$$

the z-axis being taken perpendicular to the momentum transfer. The subscript B in Eq. (6a) is a reminder that here the states i and f refer only to the bound part of the asymptotic initial and final states. Since χ can be evaluated analytically for electron scattering by any atom this gives a very attractive, elegant closed form expression for the multi-particle scattering amplitude. The

properties of f_G, given by Eq. (6a), have been studied in great
detail by Gerjuoy and his collaborators[8].

Let us now consider the Glauber approximation as a multiple
scattering series[10-14]. If we expand the exponential in Eq. (6a)
we find that

$$f_G = \sum_n \bar{f}_{Gn} \tag{7a}$$

with

$$\bar{f}_{Gn} = \frac{1}{2\pi i} \left(\frac{i}{p}\right)^{n-1} \int e^{i\vec{K}\cdot\vec{b}} <f_B|\chi^n|i_B> d\vec{b} \quad . \tag{7b}$$

Since χ is proportional to V we see that the terms of this
"Glauber series" bear a one-to-one relationship with the terms of
the Born series. However, we note that the terms above are alternatively purely real and purely imaginary, unlike the Born series
whose terms have both real and imaginary parts after first order.
The details of this statement depend on the symmetry of the functions ϕ_i and ϕ_f. For example, if ϕ_i and ϕ_f are both s-states,
then Eq. (7b) reduces to

$$\bar{f}_{Gn} = \left(\frac{i}{p}\right)^{n-1} \int J_o(Kb) <f_B|\chi^n|i_B> b\,db \quad . \tag{7c}$$

For non-spherically symmetric states one will have higher order
Bessel functions in the expression for $\bar{f}_{Gn}$.

Tables I and II summarize the situation for elastic electron-
atom scattering. There are three points to be made in reference
to these tables. First, note that for large momentum transfer
(Table II), apart from slowly varying factors like $\ln K$, the
momentum transfer dependence of all terms is essentially the same,
K^{-2}, so that in discussing the importance of various terms only
the p-dependence is of much importance. Second, the real part of
the second Born term, which has no analogue in the Glauber series
will in general (apart from questions concerning powers of $\ln K$
at large angles) be just as important as Im $\bar{f}_{B2}$ and Re $\bar{f}_{B3}$ in
giving the leading, systematic correction to the first Born approximation in the large p limit. Third, from the Glauber point of view
we see that the third Born term is just as important as Re $\bar{f}_{B2}$ and
Im $\bar{f}_{B2}$ in correcting the first Born term, so that if one wants the
leading correction to the first Born term one needs not just the

Table I. A comparison of the lowest order terms of the Born and Glauber series for momentum transfers less than or of the order of magnitude of unity.

Term	Born Series	Glauber Series
$\text{Re } \bar{f}_1$	~ 1	~ 1
$\text{Im } \bar{f}_1$	0	0
$\text{Re } \bar{f}_2$	$\sim \begin{array}{l} p^{-1}(K \lesssim p^{-1}) \\ \\ p^{-2}(K \gtrsim p^{-1}) \end{array}$	0
$\text{Im } \bar{f}_2$	$\sim \begin{array}{l} p^{-1}\ \ell n\ (K < p^{-1}) \\ \\ p^{-1}\quad (K > p^{-1}) \end{array}$	$\sim p^{-1}\ \ell n\ K$
$\text{Re } \bar{f}_3$	$\sim p^{-2}$	$\sim p^{-2}$
$\text{Im } \bar{f}_3$	$\sim p^{-3}$	0

second Born term but also the real part of the third Born term which is terribly difficult to evaluate. Given that the third (or n^{th}) Glauber term is rather easy to evaluate this obviously suggests that we should replace $\text{Re } \bar{f}_{B3}$ by $\text{Re } \bar{f}_{G3}$ and write the direct scattering amplitude as

$$f_{EBS} = \bar{f}_{B1} + \text{Re } \bar{f}_{B2} + \text{Im } \bar{f}_{B2} + \text{Re } \bar{f}_{G3} \qquad (8)$$

Table II. A comparison of the lowest orders of the Born and
Glauber series for large momentum transfers.

Term	Born Series	Glauber Series
Re $\bar{f}_1$	$\sim K^{-2}$	$\sim K^{-2}$
Im $\bar{f}_1$	0	0
Re f_2	$\sim p^{-2} K^{-2}$	0
Im f_2	$\sim p^{-1} K^{-2} \ell n\ K$	$\sim p^{-1} K^{-2} \ell n\ K$
Re f_3	$\sim p^{-2} K^{-2} \ell n^2\ K$	$\sim p^{-2} K^{-2} \ell n^2\ K$
Im f_3	$\sim p^{-3} K^{-2} \ell n\ K$	0

A major question can immediately be raised concerning the
validity of Eq. (8). The eikonal approximation from which the
Glauber approximation derives is usually considered, following
the work of Glauber, to be a small angle approximation. Is then
Eq. (8) restricted to small angles? Probably not. In the past
few years much study has been expended on the eikonal approximation. For the present discussion of particular relevance is the
work of Byron, Joachain and Mund[10,11] who showed that for a superposition of Yukawa potentials the second term in the Born series
and the corresponding term in the eikonal series (i.e., the
expansion of Eq. (1) as a multiple scattering series) are equal
in the large p limit for all angles, and similarly for the third
and fourth order terms. This relationship has been conjectured
to hold for all orders.

However, it is also known from the work of Byron, Joachain
and Mund that this all-angles property certainly does not hold
for long range potentials of the polarization or van der Waals
type. Fortunately, when such potentials occur in atomic physics

(e.g. in electron–atom scattering) their effect is primarily at small angles whereas large angle scattering is generally controlled by static potentials which are of the Yukawa type mentioned above.

Of course, the proof of the pudding is in the eating, so let us look at the second Born term compared with the second Glauber term for two cases of interest. Figure 1 shows this comparison for electron–hydrogen elastic scattering[12]. Notice how well the two terms in question agree. We also show the result obtained for Im $\bar{f}_{B2}$ by using only the ground state in the many-body Green's function. Obviously this is a very poor approximation at small angles and shows that an optical model calculation using only the static optical potential would give poor results at small angles.

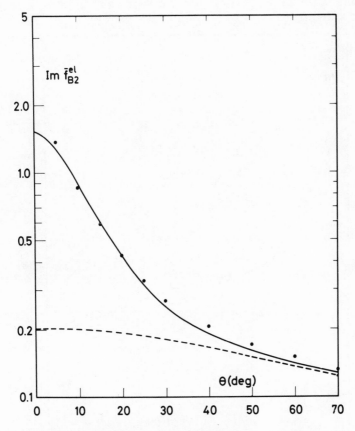

Fig. 1 The imaginary part of the second Born term for e^- + H elastic scattering. The solid curve is the closure approximation with the elastic intermediate state included exactly; the dots are the Glauber approximation; and the dashed curve illustrates what happens when only the elastic intermediate state is included in evaluating $\bar{f}_{B2}$. The incident electron energy is 100 eV.

I make this last remark because I think it worth noting that
an important use of ab initio multiple scattering techniques
is precisely to shed light on questions concerning the validity of
simple models. If such a model fails badly in some region or another
in giving the lowest orders of perturbation theory, then at least
at intermediate and high energies its results should be viewed with
some caution. Perhaps a more interesting case from this point of
view is shown in Fig. 2 which illustrates the excitation of atomic
hydrogen to the 2s state[14]. Once again we note the extraordinary
agreement between the Born and Glauber second order terms. The
dashed curve shows the approximation to the second order term obtained by keeping just the 1s and 2s states in the many-particle
Green's function. This suggests strongly that for this process the
distorted wave Born approximation will not be very reliable at
small angles.

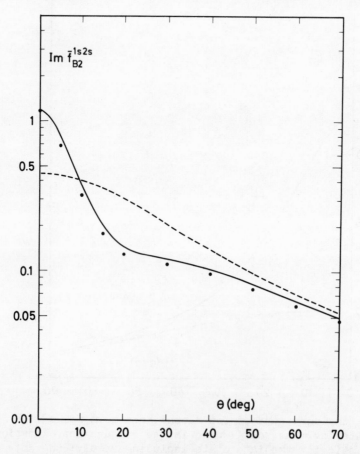

Fig. 2 The same as Fig. 1 but for the excitation of atomic hydrogen
to the 2s state. Thus there are two elastic intermediate states.

Returning to our main point, we see that the imaginary part of the second Born term is given well at all angles by the corresponding Glauber term. Indeed, the same is found for elastic scattering in helium[12] and for the excitation of helium to the 2^1S state[15]. It is also, of course, well known from the work of Glauber that the first terms in the Born and Glauber series are in fact identically equal because of the choice of z-axis made by Glauber. Thus encouraged, let us conjecture that just as in potential theory there is a similar correspondence between Re $\bar{f}_{B3}$ and Re $\bar{f}_{G3}$ and use the easily calculated Glauber term to write the eikonal-Born series (EBS) scattering amplitude as in Eq. (8).

What are the consequences of Eq. (8)? First, let us consider large angles when exchange effects can be neglected to the order in which we are interested. Here results can be displayed analytically. One finds for electron-hydrogen elastic scattering in three cases of interest,

$$f_{ST} = \frac{2}{K^2} - \frac{1}{p^2 K^2} [4 \ln^2 \frac{K}{2} + 4 \ln \frac{K}{2} - 1] + \frac{i}{p K^2} [4 \ln \frac{K}{2} + 2] \quad (9a)$$

$$f_G = \frac{2}{K^2} - \frac{1}{p^2 K^2} [4 \ln^2 \frac{K}{2} + 4 \ln \frac{K}{2} + \frac{\pi^2}{3}] + \frac{i}{p K^2} [4 \ln \frac{K}{2} + 2] \quad (9b)$$

$$f_{EBS} = \frac{2}{K^2} - \frac{1}{p^2 K^2} [4 \ln^2 \frac{K}{2} + 4 \ln \frac{K}{2} + \frac{\pi^2}{3} - 2] + \frac{i}{p K^2} [4 \ln \frac{K}{2} + 2]$$

$$(9c)$$

The quantity f_{ST} given in Eq. (9a) is what would be obtained for the elastic scattering amplitude if we used only the elastic state as an intermediate state, i.e., if we regarded the electron as being simply scattered by the static potential of the hydrogen atom. Equation (9b) gives the Glauber amplitude, and Eq. (9c) shows the EBS result, obtained from Eq. (8). We have kept terms in the amplitudes only through order p^{-2}. These three expressions are nearly identical, but if we square them to obtain the differential cross-section we find that all the terms involving $\ln K$ cancel to order p^{-2} and one has simply

$$\frac{d\sigma_{ST}}{d\Omega} = \frac{4}{K^4} [1 + 2/p^2 + \ldots] \quad (10a)$$

$$\frac{d\sigma_G}{d\Omega} = \frac{4}{K^4} \left[1 - \left(\frac{\pi^2}{3} - 1\right)/p^2 + \ldots \right] \tag{10b}$$

$$\frac{d\sigma_{EBS}}{d\Omega} = \frac{4}{K^4} \left[1 - \left(\frac{\pi^2}{3} - 3\right)/p^2 + \ldots \right] \tag{10c}$$

We see from these equations the great importance of obtaining all terms which contribute to order p^{-2}. By a numerical coincidence, we find in this case that since $(\pi^2/3 - 3) \simeq 0$, the full (EBS) result through order p^{-2} gives almost no correction to the first Born approximation while the other two results give significant corrections, nearly equal in magnitude and opposite in sign. As an extreme example of the need for consistency in correcting the first Born approximation, we note that if one imagined that the second Born term alone (real plus imaginary parts) gave the leading correction then one would obtain

$$\frac{d\sigma_{B2}}{d\Omega} = \frac{4}{K^4} \left[1 + \left(4 \ln^2 \frac{K}{2} + 4 \ln \frac{K}{2} + 3\right) + \ldots \right] \tag{10d}$$

which differs in a very important way from the EBS result of Eq. (10c).

Finally, we remark that from Eqs. (10a), (10b) and (10c) the critical importance of absolute measurements in assessing theoretical results is clear. Each of these three expressions has the form

$$\frac{d\sigma}{d\Omega} = C(p) \frac{d\sigma_{Born}}{d\Omega} \quad , \tag{11}$$

so measurements which must be normalized at each energy will not distinguish between the three results in question.

To illustrate that some complications arise as one goes to inelastic problems we consider briefly results which have been obtained recently for the excitation of atomic hydrogen to the 2s state[14]. Again looking at the large momentum transfer case, we show in Table III the leading terms of the EBS and Glauber series. One is struck immediately in Table III by a major change from the elastic scattering case shown in Table II, namely, the first Born term no longer has the same momentum transfer dependence as the other terms, but falls off very rapidly, becoming quickly insigni-

ficant as K becomes much greater than unity. Thus, in this region
the first Born term no longer gives the asymptotic form of the
scattering amplitude as it does in elastic scattering and potential
scattering. From Table III it is clear that the imaginary part of
the second Born term dominates the amplitude in the large momentum
transfer limit. A quick check of the order of magnitude of the

Table III. A comparison of the leading terms in the Born and
Glauber series for the excitation of atomic hydrogen to the 2s
state in the large momentum transfer limit. We make use of the
definitions $C = 2^{13/2}\ 3^{-4}\ K^{-2}$ and $L = \ln (2K/3)$.

Term	Glauber Series	Born Series
$\mathrm{Re}\ \bar{f}_1$	$-2^{7/2}/K^6$	$-2^{7/2}/K^6$
$\mathrm{Im}\ \bar{f}_1$	0	0
$\mathrm{Re}\ \bar{f}_2$	0	$3C/8p^2$
$\mathrm{Im}\ \bar{f}_2$	$-C/p$	$-C/p$
$\mathrm{Re}\ \bar{f}_3$	$C(2L + 1)/p^2$	$C(2L + 1)/p^2$
$\mathrm{Im}\ \bar{f}_3$	0	$O(p^{-3}\ K^{-2})$
$\mathrm{Re}\ \bar{f}_4$	0	$O(p^{-4}\ K^{-2})$
$\mathrm{Im}\ \bar{f}_4$	$C(2L^2 + 2L + \pi^2/6)/p^3$	$C(2L^2 + 2L + \pi^2/6)/p^3$

other terms in Table III also reveals that keeping terms through Re f_3, i.e., following Eq. (8), no longer gives the leading term plus the first correction. It gives only the leading term and part of the first correction.

If we put the results of Table III together, we find the following results for the differential cross-sections:

$$\frac{d\sigma_G}{d\Omega} = (2^{13}/3^8) \frac{1}{p^2 K^4} [1 - (\frac{\pi^2}{3} - 1)/p^2 + \ldots] \qquad (12a)$$

$$\frac{d\sigma_{EBS}}{d\Omega} = (2^{13}/3^8) \frac{1}{p^2 K^4} [1 + (\frac{3}{2} \ell n \frac{2K}{3} + A)/p^2 + \ldots] \quad (12b)$$

In order to have a precise value for the constant A in Eq. (12b) we have to know the value of the term Im f_{B3} in Table III. In addition, the exchange amplitude makes its appearance in this order, also giving a contribution to A. Thus, we have been able to obtain only the weakly dominant part of the leading correction to the asymptotic form of the differential cross-section. Further work will be necessary to have an expression which can be used with confidence at intermediate energies to correct the leading term in the cross-section at large angles. At small angles, where the first Born term is of order unity, these difficulties do not arise, and Eq. (8) gives a correct description of the leading (first Born) term and the first correction to it, if account is taken of exchange.

Let us now consider the situation at small and intermediate angles where, as just mentioned, the effects of exchange are not negligible. In the first Born approximation to the exchange amplitude, which we shall denote by g_{B1}, the contribution from the part of the potential describing the interaction between the incident and exchanged electrons gives a term of order p^{-2} in the large p limit. The other parts of the potential give much smaller contributions which we neglect. Since this term of order p^{-2} falls off like K^{-4} for K large, it is negligible in this region compared with the corrections we have been discussing before. In fact, just as for 2s excitation, the elastic and 1s - 2s exchange amplitudes will actually be dominated for large K by second order exchange terms which vary like $p^{-3} K^{-2}$; these terms are also negligible to the order to which we are presently working. Thus, although the term we keep would not, for example, give a correct spin-flip differential cross-section in hydrogen at all angles, it is sufficient for our purposes. As soon as we go beyond the order discussed here these questions become important: the constant A in Eq. (12b) and the term of order p^{-4} in the elastic amplitude will both depend on the second-order part of the appropriate exchange amplitude.

With the dominant term isolated, we have for the exchange amplitude for electron-hydrogen scattering in the large-p limit,

$$\bar{g}_{B1} = - (2/p^2) \int e^{i\vec{K}\cdot\vec{r}} \phi_i(\vec{r}) \phi_f^*(\vec{r}) d^3r \tag{13}$$

with a very similar expression for electron-helium scattering. This

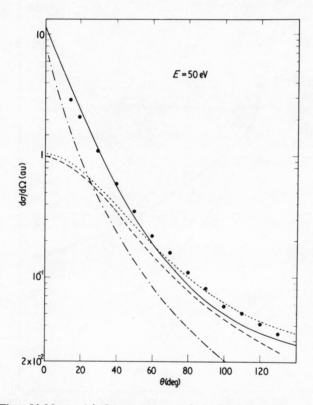

Fig. 3 The differential cross-section for the elastic scattering of electrons by atomic hyrogen. The filled circles are the experimental points of reference 16. The solid curve is the EBS result; the dash-dotted curve is the Glauber approximation; the dashed curve is the first Born approximation; and the dotted curve gives the result of an optical model calculation using just the lowest order, static optical potential.

is generally called the Ochkur approximation. Using Eq. (13) one
forms the singlet and triplet amplitudes for hydrogen or the single
full amplitude (f - g) for helium from which the differential cross-
sections are obtained. Figure 3 shows some recent results on the
elastic scattering of electrons by atomic hydrogen at 50 ev. The
experimental points are those of Teubner et al[16]; the agreement
with the EBS result is seen to be good. Figure 4 shows at 100 ev
the angular distribution of electrons exciting atomic hydrogen to
the 2s and 2p states (which are indistinguishable in energy loss).
The experimental results are those of Williams and Willis[17]. Finally,

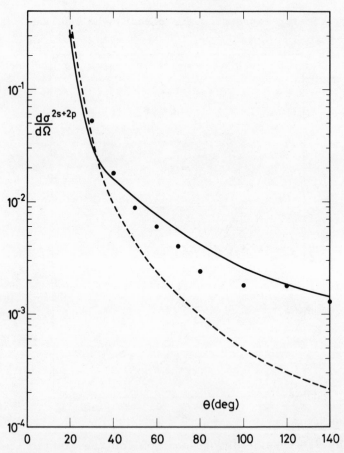

Fig. 4 The differential cross-section for exciting the degenerate
2s and 2p states of atomic hydrogen by 100 ev electrons. The filled
circles are the experimental points of reference 17; the solid
curve is the EBS result; and the dashed curve is the Glauber result.
Over most of the angular range shown the first Born approximation
bears no relation to experiment and hence is not shown.

Figs. 5 and 6 show the situation for the elastic scattering of electrons by helium at 400 ev. The EBS results are compared with the recent experimental points of Bromberg[18], Jansen et al[19] and Crooks and Rudd[20]. The agreement between theory and experiment is encouraging, being good over the entire angular range, and gives strong support to the picture of elastic scattering given in this report.

Much of the work discussed above was done while the author was

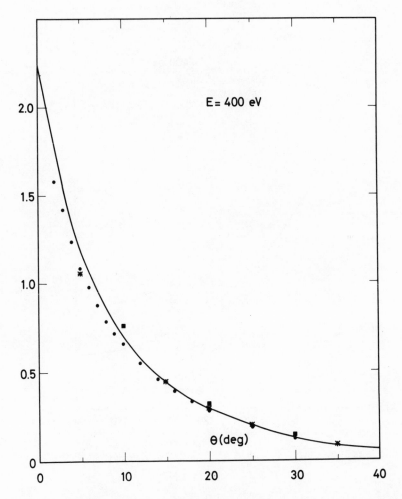

Fig. 5 The differential cross-section for the scattering of 400 ev electrons by helium for angles less than 40°. The solid curve is the EBS result. The dots are the experimental results of reference 18, the asterisks are those of reference 19, and the squares are those of reference 20.

a Fulbright Research Scholar at the Université Libre de Bruxelles.
He is grateful to the Fulbright-Hays Commission for its support.
It is a particular pleasure to be able to thank Professor C. J.
Joachain for his hospitality in Brussels and for many interesting
discussions on electron-atom scattering.

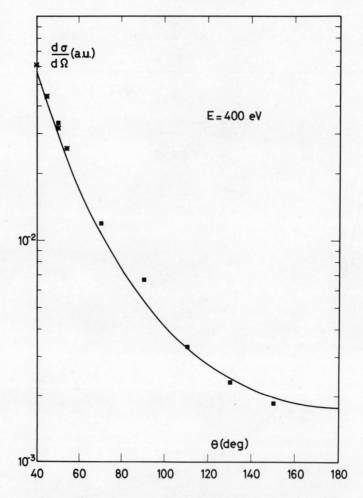

Fig. 6 The same as Fig. 5 but for scattering angles greater than 40.⁰

REFERENCES

1. A. R. Holt and B. L. Moiseiwitsch, Proc. Phys. Soc. (London) B1, 36 (1968).
2. R. J. Glauber in Lectures in Theoretical Physics (New York, Interscience, 1959) vol. 1, p.315.
3. Victor Franco, Phys. Rev. Lett. 20, 709 (1968).
4. A. S. Ghosh and N. C. Sil, Indian J. Phys. 43, 490 (1969); ibid. 44, 153 (1970).
5. H. Tai, R. H. Bassel, E. Gerjuoy and V. Franco, Phys. Rev. A1, 1819 (1970).
6. F. W. Byron, Jr., Phys. Rev. A4, 1907 (1971).
7. F. W. Byron, Jr. and C. J. Joachain, Phys. Letters 38A, 185 (1972).
8. E. Gerjuoy and B. K. Thomas, SRCC Report Nr. 194, Univ. of Pittsburgh (1974).
9. C. J. Joachain and C. Quigg, Rev. Mod. Phys. 46, 279 (1974).
10. F. W. Byron, Jr. and C. J. Joachain, Physica 66, 33 (1973).
11. F. W. Byron, Jr., C. J. Joachain and E. H. Mund, Phys. Rev. D8, 2622 (1973).
12. F. W. Byron, Jr. and C. J. Joachain, Phys. Rev. A8, 1267 (1973).
13. F. W. Byron, Jr. and C. J. Joachain, J. Phys. B7, L212 (1974).
14. F. W. Byron, Jr. and L. J. Latour (unpublished).
15. F. W. Byron, Jr. and C. J. Joachain (unpublished).
16. P. J. O. Teubner, C. R. Lloyd and E. Weigold, J. Phys. B6, L134 (1973).
17. J. F. Williams, Proceedings of the International Symposium on Electron and Photon Interactions with Atoms (to be published).
18. J. P. Bromberg, J. Chem. Phys. (to be published).
19. R. H. J. Jansen, F. J. de Heer, H. J. Luyken, B van Wingerden and H. J. Blaauw, FOM-Instituut voor Atoom-en Molecuulfysika Report Nr. 35693 (1974).
20. G. B. Crooks and M. E. Rudd, Bull. Amer. Phys. Soc. 17, 131 (1972); G. B. Crooks, thesis, University of Nebraska.

INTERACTION OF LOW ENERGY POSITRONS WITH GASEOUS ATOMS AND MOLECULES[*]

P.G.Coleman,T.C.Griffith, G.R.Heyland,and T.L.Killeen

Department of Physics & Astronomy, University College

London, Gower Street, London W C 1

1. INTRODUCTION

There has been substantial progress in the study of positron interactions with atoms and molecules in the last two years. Reliable measurements of the total cross-sections[1-8] for positrons of controllable energy in the range 2-400 eV in various gases are now available and these can be compared with some recent theoretical calculations especially in the case of helium. Below 2 eV improvements in the technique of measuring positron lifetime spectra[9,10] are yielding new information on the behaviour of thermalized positrons in various gases and gaseous mixtures at different pressures.

Following from the initial work of Costello et al[11], where positrons of well-defined energy near 1 eV were seen to be emitted after the moderation of the energy of fast positrons in gold foils, substantial beams of positrons are now available for scattering experiments. These beams have recently been obtained either with strong radioactive sources and energy-selecting spectrometers (Jaduszliwer et al[5] and Pendyala et al.[12]) or in a time of flight method with a weak source as described by Coleman et al.[13]. Although the latter method is limited by its source strength of only 100μCi or so to a beam intensity of 3-4 positrons/sec., the absence of serious background problems and the timing feature more than offsets any disadvantages it might appear to have in relation to the spectrometer methods.

---

[*] Invited paper presented by T.C.Griffith.

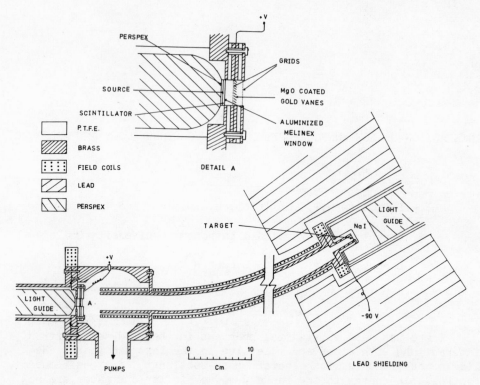

Figure 1.   Schematic diagram of the flight tube and counters.

2.   BEAM EXPERIMENTS

The principles of the spectrometer system for low energy positron
scattering measurements were outlined at the Boulder Conference [14]
two years ago. It is therefore appropriate to concentrate in this
report on the time of flight technique of Coleman et al. [13] which is
illustrated in Fig.1.   Positrons from a Na[22] source are detected in
a thin plastic scintillator giving a start pulse to the timing
sequence as they enter the flight tube. At the entrance to the
tube energy moderation occurs and the yield of monoenergetic low
energy positrons has been greatly enhanced by the backscattering of
fast positrons from gold vanes coated with a thin layer of
magnesium oxide granules. The 1.0 ± 0.5 eV positrons emitted from
the magnesium oxide can be accelerated to any energy by the suitable
application of a D.C. potential. The positrons are confined to
the axis by a longitudinal magnetic field and annihilate at the end
of the 1 meter flight path on an aluminium target which in turn lies
within the well of an NaI counter which detects the annihilation
γ -rays. This provides the stop pulse for the time to amplitude
converter of the timing system whose output is fed to a multichannel

analyser where the time of flight spectrum of the positrons showing
the low energy peak is accumulated and displayed. At low energies
the width of this peak reflects the energy spread and angular
distribution of the positrons emitted from the MgO. At higher
energies, where all the positrons have been accelerated and move
essentially along the axis, the peak width due to these effects is
negligible and the observed width here reflects a time spread of
some 8 nsec in the re-emission of the fast positrons at 1 eV from
the MgO.

The mechanism behind the emission of the positrons from MgO
is not understood but it may have a close connection with the large
amount of orthopositronium also observed under similar conditions
from MgO (Paulin and Ambrosino [15], Curry and Schawlow [16]) and more
recently from metal surfaces at high temperatures (Canter et al. [17]).
As a possible explanation for the emission of slow positrons from
metal surfaces Tong[19] has suggested that the emission is due to a
negative work function but as yet there is not conclusive support
for this theory. With MgO two groups of positrons of low energy
are observed - the prompt group which are timed as the peak and
another group of roughly equal intensity emitted from the source
but having no definite time relation (within the time range of the
t.a.c.) to the fast positron pulse and therefore indistinguishable
from background. The yield of low energy positrons in the peak is
1 in 10^5 of those emitted from the source.

The time spectrum of positrons in the peak and elsewhere has
to be deduced from the spectrum accumulated in the multichannel
analyser. This observed spectrum consists of signal events, random
coincidence events and events due to signal starts converted by
random stops. An accurate method for deducing the true signal
distribution has been developed by Coleman et al.[18]. The method is
based on counting the total number of start and stop pulses coupled
with accurate calibration of the timing system and is applicable to
both the time of flight spectra and the lifetime spectra to be
discussed later on.

Total cross-sections as a function of energy in the range
2 - 400 eV are obtained by measurement of the attenuation of the
positrons in the peak when the flight tube is alternately evacuated
and filled with high purity gas to known pressures. The gases under
investigation were leaked through the system at a controlled rate
such that the pressure differential between the two ends of the
tube, measured with both a Baratron and a McLeod gauge, was no
greater than 10%. The mean gas density is obtained to an accuracy
of about 2% for most gases. At each energy the attenuation can be
measured as a function of gas pressure and the cross-sections obtained from the gradient of the logarithmic plot of the number of
counts within the defined limits of the peak versus pressure. At

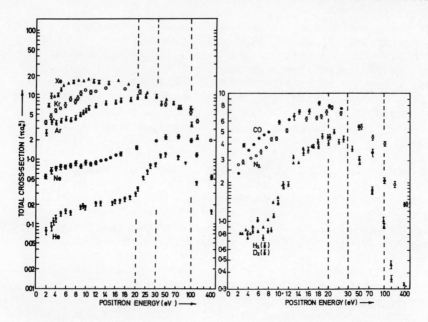

Figure 2. Measured total cross-sections for the rare gases[3] and some molecular gases[6].

attenuations of up to 10 : 1 no departure from exponentiality has been noticed and much of the recent data has been taken at a constant attenuation of approximately 3 : 1 as this gives minimum statistical error in the cross-section. The cross-sections measured in this way by Canter et al.[4] and Coleman et al.[6] are reproduced in Fig.2. This data was not corrected for errors arising from forward scattering within the defined limits of the peaks used to evaluate the cross-sections but arguments were given that in the rare gases at energies below 20 eV the corrections are not large and recent work has vindicated this viewpoint.

     In the new results presented in this report an allowance has been made for small angle scattering and for inelastic scattering involving small energy losses. A closer look at the spectra helps to clarify the nature of the corrections that have to be applied. The magnitude of the longitudinal magnetic field is such that all the forward moving positrons from the moderator in both the vacuum and gas runs will reach the target at the end of the flight path. Only those positrons scattered through angles greater than 90° in the gas fail to do so but all the other scattered positrons will be displaced on the spectrum towards longer time intervals. If the displacement in time exceeds the t.a.c. time range, as may be the

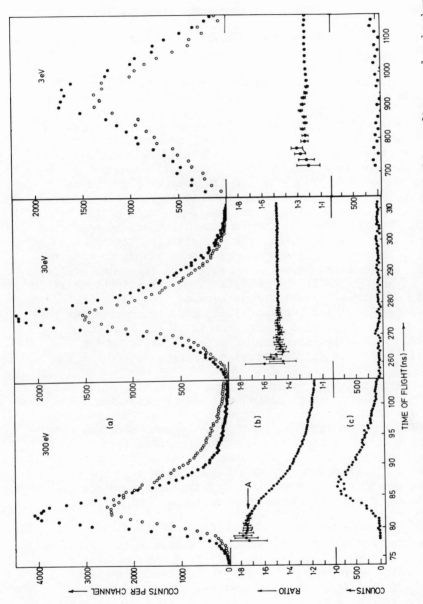

Figure 3. (a) The time spectrum for 3,30,and 300eV positrons in helium after random background correction. Experimental points: ●,for vacuum runs, o, for gas runs. (b) Attenuation (ratio) obtained as discussed in the text.(c) Time distribution of the scattered positrons derived from the difference between the gas and scaled down vacuum runs.

case for scattering angles close to 90°, the event will be recorded
as a random event. From a comparison of the spectra for the gas and
vacuum runs the true cross-sections can be deduced without defining
specific limits on the peaks. The procedure is illustrated in
Fig.3 where it is assumed that as one moves across the peak from
right to left (towards shorter time intervals) progressively fewer
positrons can be scattered into each channel. Thus the attenuation
should approach the true value near the left-hand edge of the peak.
Starting from channel 1 on the left of the peak the attenuation
for channel n , deduced from the integral counts for gas and vacuum
in channels 1 to n , is plotted for each channel. The change in
attenuation on traversing the peak is illustrated in Fig. 3b. At
energies below 30 eV there is no measurable change in the attenuation and the cross-section cited in the earlier reports are not
seriously in error below this energy. At higher energies the
attenuation changes significantly and the effect increases in
prominence as the energy increases. In this region the intercept
from the best fit to the curve at channel 1 is taken as the true
attenuation. Fig.3c shows the time distribution of the scattered
positrons. At low energies there are no scattered positrons to be
seen near the peak but at higher energies they assume prominence
as a tail to the peak amounting to about 66% of the total scattering
at 400eV. At present it is not profitable to attempt any detailed
interpretation of these distributions because the scattering occurs
along the entire length of the gas column and moreover is due to a
mixture of elastic and inelastic processes. A new system is
currently being assembled with the object of localising the
scattering close to the source end of the flight tube.

    The most interesting feature of these observations is that in
helium and neon and probably most other gases at energies below 20
to 30 eV there is no measurable change in the shape of the peaks
between vacuum and gas runs and that the attenuation between any
defined limits on the peaks gives the true cross-sections. In all
cases, therefore, very strong large angle or backward scattering is
suggested at energies between 4 and 20 eV. Below 4 eV the angular
distribution and energy spread of the primary positrons may be
smearing out any change in shape and some caution is necessary in
the interpretation of data in this region.

    From a theoretical point of view the most important results
are undoubtedly those for helium and in Fig. 4 the cross-sections
at energies below 20 eV are compared with the measurements of
Jaduszliwer and Paul [7] and some theoretical calculations by
Humberston [20] and Bransden et al.[21]. Using Kohn's variational
method Humberston has calculated more accurate S-wave phase shifts
than those of Houston and Drachman [22]. When used with the lowest
set of P and D-wave phase shifts of Drachman [23] these new S-wave
phases give very good agreement with the experimental data at low

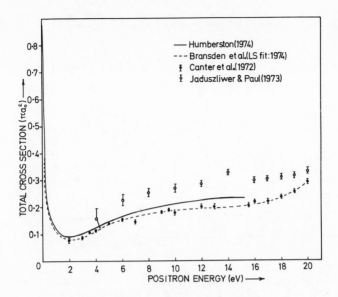

Figure 4. Comparison of the measured positron-helium total cross-sections with theory at energies up to 20 eV.

energies. The agreement is further enhanced by Bransden's least squares fit to the data using the low energy effective range expansion.

$$Q_{tot}(k^2) = a_o + a_1 k + a_2 k^2 \ell n\ k + a_3 k^2 + a_4 k^4$$

The analytic fit to the data, as recently shown by Humberston[24], has the same energy dependence below 2eV as the theoretical curve obtained with Humberston's phase shifts[20] and this extrapolates, within the errors, to the zero energy scattering length A = -0.472 a.u. The spectra at 3eV in Fig.3 as well as others at energies up to 20 eV show very little evidence for forward scattering in helium. This observation may well favour P and D wave phase shifts of smaller magnitude than those used by Humberston[24] and improve the agreement with the data of Canter et al.[4]. The cross-section for positronium formation, σ_{Ps},   can be estimated from the data between 18 and 20 eV in Fig. 4. Assuming a constant value for the elastic cross-section between 16 and 20 eV values of 0.02 ± 0.01 πa_o^2 and 0.07 ± 0.03 πa_o^2 can be assigned to σ_{Ps} at 18 eV and 20eV respectively for the data of Canter et al.[4].

In Fig. 5 the corrected positron-helium cross-sections over the energy range 2 - 400 eV are compared with the Born approximation calculations in the region 300-400 eV. Bransden et al.[21] had suggested that the earlier cross-sections of Canter et al.[3] were increasingly in error at energies above 50 eV. The corrected data

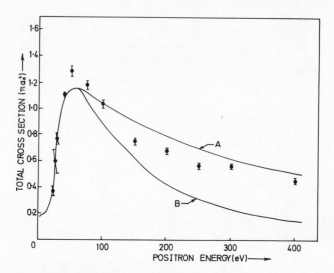

Figure 5. Corrected positron-helium total cross-sections at
energies up to 400 eV. Curve B - earlier measurements of Canter et
al.[3]. Curve A - suggested theoretical curve of Bransden et al.[21].
●, corrected cross-sections.

show that this is so and the true cross-sections are close to
Bransden et al.'s extrapolated values. Bransden et al.[21] use the sum
rule based on a forward dispersion relation

$$-(A + f_B) = (2\pi)^{-1} \int_0^\infty Q_{tot}(k)\, dk$$

to arrive at these conclusions. f_B is the Born approximation for
the elastic scattering amplitude in the forward direction taken as
-0.791 from Bransden and McDowell's work[25] and A is the scattering
length given the theoretical value of -0.472. $-(A + f_B)$ is therefore 1.26 ± 0.06 and a recent evaluation of the integral by Bransden
(private communication) using our corrected data gives

$$(2\pi)^{-1} \int_0^\infty Q_{tot}(k)dk = 1.21 \pm 0.05.$$

Hence, within the errors, there is now complete agreement with the
sum rule. Some recent data by Jaduszliwer and Paul (private
communication) covering an energy range up to 270 eV are appreciably
higher than our cross-sections the lowest value for their integral
being 1.34. There has been a consistent disagreement between the
data of Canter et al.[3,4] and of Jaduszliwer and Paul [7,8], the latter
showing higher cross-sections for helium and neon but only slightly
higher for argon. The results (representative sample) in Fig.3

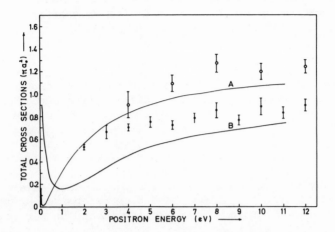

Figure 6.   Comparison of the total cross-sections for neon with the calculations of Montgomery and Labahn[26].   A - (2p-d); B - (2p-d) and (2s-p)"norm".   Experimental points: o, for Jaduszliwer and Paul[7];   •, Canter et al.[4].

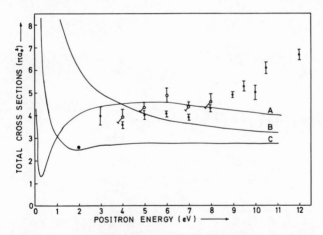

Figure 7.   Comparison of the total cross-sections for argon with the calculations of Montgomery and Labahn[26].   A - (3p-d) and (3p-s) and (3s-p);   B - (3p-d);   C - (3p-d) "norm".   Experimental points: o, for Jaduszliwer and Paul[7];   •, Canter et al.[4].

show that there is no convincing evidence for any strong forward
peaking of the angular distributions in helium or neon in our
data, in contrast to the observations reported by Jaduszliwer and
Paul [8].

Turning to neon and argon the most recent calculations are
those of Montgomery and Labahn [26] using the polarized orbital
method.   In Figs. 6 and 7 their phase shifts (private communication)
have been used to evaluate the total cross-sections that are compared with the measurements.   Although not checked in detail yet
it is reasonable to assume that the data of Canter et al. [4] used in
these comparisons do not need correction for forward scattering.

In Fig. 8 some new cross-sections for positrons scattered
in O_2 and CO_2 are presented. These have been determined using the
method described for the recent helium data. Both the corrected
and uncorrected values are shown giving some indication of the

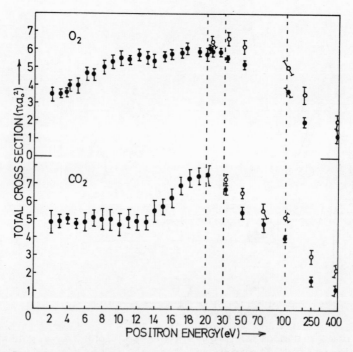

Figure 8.   Total cross-sections for positrons in O_2 and CO_2.
•, uncorrected values;   o, corrected values.

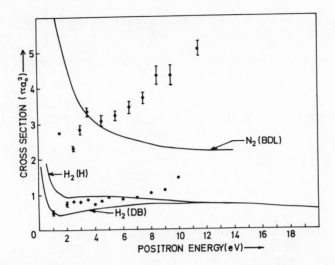

Figure 9. Comparison of the total cross-sections of Coleman et al.[6] for H_2 and N_2 with the calculations of H - Hara[27]; DB - Darewych and Baille [28] and BDL - Baille et al.[29].

correction needed for the data of Coleman et al.[6] for N_2, CO, H_2 and D_2. The sharp increase in the cross-section for CO_2 at 14 eV is a distinctive feature and probably relates to the ionization potential of 14.4 eV.

Figure 9 compares theory and experiment for some of the molecular gases. The data for N_2 and H_2 shown in Figure 2 is compared with some calculations by Hara[27], Darewych and Baille [28] and Baille et al.[29]. The theory only applies to elastic scattering so that the comparison is only meaningful at low energies. All that can be said is that at the lower energies there is tolerable agreement between theory and experiment as far as they go.

3. POSITRON LIFETIME SPECTRA

The measurement of lifetime spectra for positrons in gases is comparable with swarm type experiments for electrons. The energy of fast positrons from a radioactive source (usually Na^{22}) is moderated in a vessel filled with gas under pressure and the life history of each positron from its emission to annihilation is examined. The result is to build up a lifetime spectrum with its characteristic prompt peak and shoulder followed by the equilibrium region consisting of the free positron, orthopositronium (O-Ps), and background components. Careful analysis of such spectra for the rare gases using a characteristic transport equation to describe the positron "slowing down" process at different pressures, temperature (see for example Canter and Heyland [30], Roellig and Kelly [31]), and in the presence of applied electric fields can yield information about cross-sections in an energy region which is not yet accessible to beam experiments.

Hara and Fraser have recently attempted to bridge the gap between the beam cross-section measurements and the lifetime data for argon. The effective number of annihilation electrons, Z_{eff}, is determined as a function of velocity from the work of Lee and Jones [33]. The diffusion equation can then be used to estimate the scattering length and S-wave phase shift which can in turn be extrapolated to the region where it can be matched to direct cross-section measurements.

Some measure of success has been achieved by our group in improving the conventional methods of measuring lifetime spectra. Two factors have contributed to this improvement: (a) An increase in the rate of accumulation of data so that the same statistical accuracy can now be achieved in a much shorter time. Typically about $5 . 10^7$ positron annihilations are distributed through the spectrum in about 10 hours. (b) Exact evaluation and correction for the background to the data due to random coincidence events and to conversion of signal starts by random stops as described by Coleman et al. [18]. In all previous work these backgrounds have been kept to a low level using weak sources with consequent loss of statistical accuracy.

The rate of accumulation of data has been increased using two methods: (1) the start pulse for the timing sequence was derived from a thin scintillator in a manner similar to that illustrated in Fig.1. but with the flight tube replaced by the pressure vessel in which those positrons traversing the counter are brought to rest their annihilation γ-rays being detected in the usual way as described by Coleman et al. [9]. Such a system was used by Coleman and Griffith [34] for an accurate determination of the vacuum lifetime of O - Ps. (2) a much simpler and equally effective system based on confining the gas volume to a chamber of small dimensions thus enabling the use of large scintillators subtending large solid angles at the effective volume of the gas. Start and stop pulses are derived in the conventional manner, viz., detecting the prompt 1.28 MeV γ-ray from Na^{22} as a start with one counter and one of the annihilation γ-rays as a stop with the other counter. The inside of the chamber is gold plated thus utilizing the back scattering of positrons from the walls and source holder to increase their effective path length in the gas. Details of this system and some preliminary results have been reported by Coleman et al. [10].

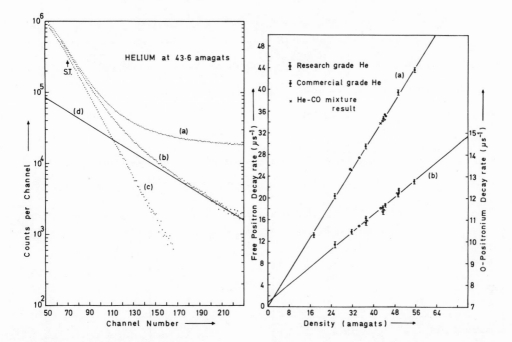

Figure 10. Lifetime spectrum
for helium at 43.6 amagats.
(a) raw data (b) data after full
random background correction
(c) data after subtraction of
background plus linear fit to
O-Ps component (d) linear fit
to O-Ps component.

Figure 11. (a) Equilibrium free
positron decay rate in helium as
a function of gas density (b) O-Ps
decay rate in helium as a function
of gas density.

All the recent data has been taken with the second system for
which gas purity is more easily maintained. The data with most
topical interest has been that for helium and particularly an
accurate determination of Z_{eff} for thermalized free positrons at
room temperature. This number provides an independent check on
the helium wave function used to describe the total cross-sections
determined in the beam experiments. In order to obtain accurate
phase shifts and annihilation rates the wave function has to be a
good approximation when the positron and electron are close together
(annihilation) as well as at the larger positron atom separations.
Humberston's wave function [20] gives S-wave phases in good agreement
with the cross-section data (Fig.4) and it has recently been shown
that the same wave function predicts a value for Z_{eff} of 4.0 at
the zero momentum limit. The recently measured value at room

TABLE 1

Gas	Density Range (Amagats)	Z_{eff}	0-Ps quenching ($\mu s^{-1} Amagat^{-1}$)	Shoulder width[40] (ns Amagats)
Helium	18 - 55	3.94 ±0.04	0.1010 ±0.0003	1700 ± 50
Neon	7 - 39	5.99 ±0.08	0.189 ±0.006	~1700 (Provisional)
Argon	3 - 28	26.77 ±0.09	0.276 ±0.003	362 ± 5
Krypton	1 - 7	64.6 ±0.8	0.387 ± 0.009	296 ± 6
(Hydrogen (Deuterium	12 - 39	13.64 ±0.09	0.150 ±0.001	-
Carbon Monoxide	1.5- 12	69.44 ±0.05	0.239 ±0.001	-
Nitrogen	7 - 28	28.89 ±0.11	0.2097 ±0.0007	14 ± 2

temperature is 3.94 ± 0.02 in good agreement with theory. Other recent measurements range from 3.96 ± 0.04 at 77°K by Roellig and Kelly[35], 3.84 by Tao and Kelly[36], 3.68 ± 0.03 at 77°K by Leung and Paul[37] to 3.63 ± 0.04 by Lee et al.[38] at room temperature. Some of the present helium data from which Z_{eff} has been deduced is shown in Figs. 10 and 11. Fig.10 shows various stages in the processing of the raw lifetime data to give the equilibrium annihilation rate at 43.6 amagats and Fig.11 demonstrates the dependence of the decay rates on density. There is no evidence here for any departure from linearity of the 0-Ps decay rate as was reported by Leung and Paul[37] and hence no support for the existence of any longer lived component in the data. The extrapolation of line (a) in Fig.11 to liquid helium density gives a lifetime of 1.9 nsec in good agreement with the directly measured lifetime in liquid helium[39]. A similar extrapolation of line

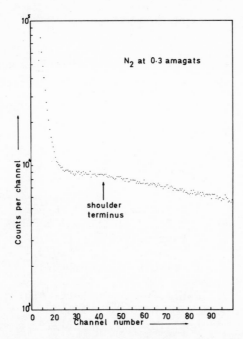

Figure 12. Lifetime spectrum for positrons in nitrogen showing
the shoulder. Full background corrections have been applied.

(b) gives 13.4 nsec at liquid helium density in sharp contrast to
the directly measured value of 88 nsec. This provides further
evidence for the existence of Ps induced zeropoint energy bubbles
in liquid Helium as reported by Roellig[39].

        Along with the helium data values for Z_{eff}, shoulder widths
and 0-Ps quenching coefficients have been determined for other
rare gases with improved accuracy. Table 1 gives a summary of
these parameters together with similar data for the molecular
gases that have also been investigated. In this table is included
the first measurement of the shoulder width in neon. Fig.12 shows
the shoulder observed in nitrogen confirming the observation of
the shoulder width in this gas by Tao[41]. Several other distinctive
features in N_2 and CO_2 are being investigated at present but the
data is not suitable for presentation at this stage.

ACKNOWLEDGEMENTS

The authors gratefully acknowledge Sir Harrie Massey's continued interest and encouragement; Dr. J. W. Humberston for many invaluable discussions; Messrs A. Agathangelou, F.J.Atkinson and A.G. Walker for excellent technical assistance and finally the Science Research Council for the financial support for this work.

REFERENCES

1. D.G.Costello, D.E.Groce, D.F.Herring and J.Wm.McGowan
                     Can.J.Phys. $\underline{50}$, 23, (1972)
2. K.F.Canter, P.G.Coleman, T.C.Griffith and G.R.Heyland
                     J.Phys.B.$\underline{5}$, L167 (1972)
3.   ditto   J.Phys.B.$\underline{6}$, L201, (1973)
4.   ditto   Appl.Phys. $\underline{3}$, 249,(1974)
5. B.Jaduszliwer, W.C.Keever and D.A.L. Paul
                     Can.J.Phys. $\underline{50}$, 1414, (1972)
6. P.G.Coleman, T.C.Griffith and G.R.Heyland
                     Appl.Phys. $\underline{4}$, 89, (1974)
7. B. Jaduszliwer and D.A.L.Paul
                     Can.J.Phys. $\underline{51}$, 1565,(1973)
8.   ditto   Can.J.Phys. $\underline{52}$, 272, (1974)
9. P.G.Coleman, T.C.Griffith and G.R.Heyland
                     J.Phys.E. $\underline{5}$, 376, (1972)
10. P.G.Coleman, T.C.Griffith, G.R.Heyland and T.L.Killeen
                     Appl.Phys.$\underline{3}$, 271, (1974)
11. D.G.Costello, D.E.Groce, D.F.Herring and J.Wm.McGowan
                     Phys.Rev. B $\underline{5}$, 1433 (1972)
12. S.Pendyala, P.W.Zitzewitz, J.Wm.McGowan and P.H.R.Orth
                     Phys.Letters $\underline{43A}$, 298,(1973)
13. P.G.Coleman, T.C.Griffith and G.R.Heyland
                     Proc.Roy.Soc., Ser A $\underline{331}$, 561, (1973)
14. W.C.Keever, B.Jaduszliwer and D.A.L.Paul
            Atomic Physics 3.Edited by S.J.Smith and G.K.Walters
            (Plenum, New York, 1973) p.561
15. R.Paulin and G.Ambrosino   J.Phys.(Paris) $\underline{29}$, 263, (1968)
16. S.M.Curry and A.L.Schawlow   Phys.Lett. $\underline{37A}$, 5,(1971)
17. K.F.Canter, A.P.Mills and S.Berko   Phys.Rev.Lett. $\underline{33}$, 7,(1974)
18. P.G.Coleman, T.C. Griffith and G.R.Heyland
                     Appl.Phys. $\underline{4}$ Oct. (1974)
19. B.Y.Tong   Phys.Rev. B$\underline{5}$, 1436,(1972)
20. J.W. Humberston   J.Phys.B.$\underline{6}$, L305 (1973)
21. B.H. Bransden, P.K.Hutt and K. H. Winters.
                     J.Phys.B. $\underline{7}$, L129, (1973)
22. S.K.Houston and R.J.Drachman   Phys.Rev. $\underline{A3}$, 1335,(1971)

23. R.J. Drachman   Phys.Rev. $\underline{144}$, 25, (1966)
24. J.W. Humberston   J.Phys.B. $\underline{7}$, L286, (1974)
25. B.H. Bransden and M.R.C. McDowell
                            J.Phys.B. $\underline{2}$, L1187, (1969)
26. R.E.Montgomery and R.W. Labahn
                            Can.J.Phys. $\underline{48}$, 1288, (1970)
27. S. Hara   J.Phys.B., $\underline{7}$, 607, (1974)
28. J.W. Darewych and P. Baille
                            Can.J.Phys. $\underline{52}$, 667, (1974)
29. P.Baille, J.W.Darewych and J.G.Lodge
                            J.Phys.B. $\underline{7}$, L1, (1974)
30. K.F. Canter and G. R. Heyland
                            Appl.Phys. $\underline{4}$, Oct. , (1974)
31. L.O.Roellig and T.M. Kelly
                            Phys.Rev.Lett., $\underline{15}$, 746 (1965)
32. S. Hara and P. A. Fraser   J.Phys.B. (1974) in press
33. G.F.Lee and G. Jones   Can.J.Phys. $\underline{52}$, 17, (1974)
34. P.G. Coleman and T.C. Griffith
                            J.Phys.B., $\underline{6}$, 2155, (1973)
35. L.O.Roellig and T. M. Kelly (1967). Private communication
             quoted by P.A.Fraser in "Advances in Atomic and
             Molecular Physics" $\underline{4}$, 63,(New York,Academic Press)
                            (1968)
36. S.J.Tao and T.M. Kelly   Phys.Rev., $\underline{185}$, 135, (1969)
37. C.Y.Leung and D.A.L. Paul J.Phys.B., $\underline{2}$, 1278, (1969)
38. G.F. Lee, P.H.R. Orth and G. Jones
                            Phys.Letters, $\underline{A28}$, 674, (1969)
39. L.O. Roellig, "Positron Annihilation" (New York, Academic Press)
                            p. 127 (1967)
40. D.A.L. Paul and C.Y Leung Can.J.Phys., $\underline{46}$, 2779, (1968)
41. S.J.Tao   Phys.Rev. A., $\underline{2}$, 1669, (1970)

THRESHOLD PROCESSES IN ELECTRON-HELIUM SCATTERING

F. H. Read

Laboratoire de Physique et Optique Corpusculaires
Université de Paris VI

Permanent address: Physics Department, Schuster
Laboratory, University of Manchester, England

Quite recently new threshold phenomena have been observed in electron-atom scattering, and previously known threshold effects have been seen with greater clarity. In this talk four separate threshold processes will be discussed, all of them occurring in electron-helium scattering.

It has long been suspected (see for example Wannier[1], Fano[2]) that long range inter-electron correlations may play a decisive role in determining the form of the electron-atom ionization cross-section near to threshold, and more recently (Fano[3]) it has been realized that these same correlations may also influence the form of the threshold excitation cross sections of high quantum number Rydberg states. These correlations and the recent experimental evidence for them (Cvejanović and Read[4]) will be discussed in the first part of the talk. It has also been found recently (Hicks et al[5], Smith et al[6]) that long range post-collision interactions between electrons can influence the excitation cross-sections of Rydberg states when the incident electron energy is near the threshold energy of autoionizing states. This phenomenon, and also that of cusps in elastic scattering cross-sections, will be dealt with in the latter part of the talk.

THRESHOLD IONIZATION

Consider the process of ionization of a neutral atom by an incident electron which has only a small energy excess E above the ionization energy I of the atom. The two outgoing electrons from the ionization event will then each have only a small kinetic energy. It has been pointed out by Wannier[1] and Rau[7] and others that in this ionization process a dynamic screening exists which results in a reduction in the ionization cross section. This occurs because although two electrons may start from the collision region with comparable speeds, the slower one will become increasingly less screened, and the faster one increasingly more screened, from the residual ion, and hence their energy difference will tend to increase with time: this means that the probability that both will escape is reduced, while the probability that one will be captured by the residual ion is increased. The result of this, according to the classical theory of Wannier[1], and its quantum mechanical and semi-classical analogues[7,8], is that the total ionization cross-section should be proportional to E^n, where n is 1.127 for the ionization of a neutral atom.

The inter-electron interactions which give this effect extend over distances of the order of E^{-1} (in atomic units) and they give rise also to strong correlations between the directions of the outgoing electrons[1,7,9], causing the most probable angle θ_{12} between them to be π, with a mean spread $\Delta\theta_{12}$ about this position of the order of $E^{1/4}$. This also implies that the maximum angular momentum of each electron is of the order of $E^{-1/4}$.

Experimental studies involving scattered or ionization electrons of very low energy are particularly difficult to carry out, but advantage may be taken of the low velocities of the electrons by using time-of-flight coincidence methods. An experiment of this type, for electron-helium ionization, has recently been completed by Cvejanović and Read[4], and although the results are not definitive they are certainly consistent with an angular correlation having the energy dependence $E^{1/4}$ (in the range of E from 0.2 to 3.0 eV), and they also show that the distribution of the available energy E between the two electrons is approximately uniform (as predicted by all the various theories) in the range of E from 0.2 to 0.8 eV.

The counting rates in these coincidence experiments are rather low, but it has also been possible to carry out a non-coincidence experiment with much higher counting rates and hence a higher accuracy. This has been done by arranging to detect only those electrons in a narrow and constant band of energies near zero energy[4]. In these circumstances the yield of detected electrons is proportional to $\sigma \times E^{-1}$, if the distribution of individual energies is indeed uniform, and one would then expect this yield to have the energy dependence $E^{0.127}$ according to the Wannier-Rau-Peterkop theory[1,7,8].

Figure 1 shows a spectrum obtained in this way. The curve drawn on the figure has the energy dependence $E^{0.127}$ and can be seen to fit the data points quite well. A more detailed analysis

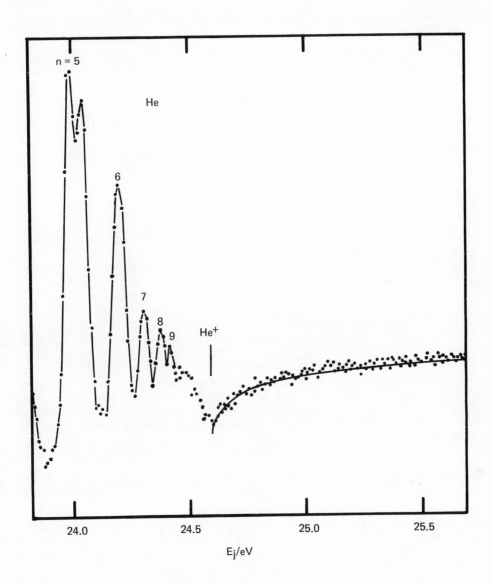

Figure 1. A threshold excitation spectrum showing the yield of electrons of very low energy (≤ 50 meV) from electron–helium scattering, as a function of the incident electron energy in the neighbourhood of the ionization energy of helium.

gives an exponent 0.131 ± 0.019 over the range of E from 0.2 to
1.7 eV. This result is again consistent with the Wannier-Rau-
Peterkop model. Although there are considerable experimental
difficulties in extending the energy ranges of these experiments,
and in maintaining high energy resolutions, it is clearly important to try to do so.

Previous experimental studies[10-14] have aimed at finding the
exponent n by measuring the energy dependence of the total yield
of ions. For example, Marchand et al.[14] have found that n is 1.16
$\pm$ 0.03 within 1 eV of threshold (in electron-helium ionization) and
that it decreases to 1.02 ± 0.02 at 12 eV above threshold.

THRESHOLD EXCITATION OF HIGH N STATES

The peaks below the ionization energy in Figure 1 correspond
to the threshold excitation of the high n Rydberg states of neutral
helium, and the average yield of these states appears to be approximately a mirror image of the yield of low energy electrons above
the ionization threshold. It has been pointed out by Fano[3] that
this depression of the threshold excitation cross-sections below
the ionization energy is caused by the same dynamic screening and
radial correlation instability which depresses the ionization cross-
section above the threshold. High n states of energy I-E, for
which the mean ion-electron separation is of the order of E^{-1}, can
be excited near their thresholds only if the two outgoing electrons
can both reach distances of this order, and as in the case of
threshold ionization the probability of this happening is reduced
by the dynamic screening between the electrons. In effect a cusp
is produced in the yield of low energy electrons. Other cusps of
this type, but less pronounced, have been observed[15] in Ar and N_2.

The energy density of states of high quantum number n is
proportional to n^3. As in the case of ionization above the
threshold, the angular correlations between the outgoing electrons
imply that these high n states can be excited with angular momenta
ℓ up to about $(-E)^{-1/4}$, which is approximately $n^{1/2}$. The effective
density of $(n\ell)$ states which can be excited is therefore proportional to $n^{3.5}$, and if one takes a somewhat different point of view
and interprets the yield of these states by ascribing a dependence
n^m to each, one finds by analyzing the spectrum of Figure 1 between
24.2 and 24.5 eV that m has the value -4.0 ± 0.25. In terms of the
energy -E the averaged yield has the dependence $(-E)^p$, where
$p = 0.25 \pm 0.13$.

CORRELATIONS NEAR AUTOIONIZATION THRESHOLDS

A different phenomenon, but one which also involves long
range inter-electron interactions near thresholds, has recently

been observed[5,6] in electron-helium scattering in the range of
incident energies from 57 to 60 eV. In this region of energies,
which includes the $(2s^2)^1S$ and $(2s2p)^3P$ autoionizing state thresholds, the possible reaction products are (i) elastically
scattered electrons, (ii) inelastically scattered electrons leading to excited bound states of helium, (iii) low energy inelastic
electrons leading to the autoionizing states, (iv) electrons
ejected from these autoionizing states, (v) ionization electrons,
and (vi) photons from the bound states. In fact all these
reaction products have been studied, using four different pieces
of apparatus. The three main results of these studies may be
summarized by saying (i) that when the autoionizing states are
excited near their thresholds the electrons ejected from them
are observed to have increased energies, (ii) that in threshold
excitation spectra the peaks corresponding to these autoionizing
states are displaced from their expected positions towards higher
energies, and (iii) that the excitation cross-sections for high n
states of neutral helium contain irregularities at energies just
above the autoionization energies, and the positions of these
irregularities change with the state being excited.

These phenomena may all be understood in terms of a post-
collision interaction model, as follows. When a short lived
autoionizing state of helium is excited near its threshold by
electron impact, the inelastic electron does not move far away
before the state decays by ejecting an electron. The coulomb
interaction between the slow moving inelastic electron and the
much faster ejected electron then causes an increase in the average
energy of the ejected electron and a corresponding decrease in
the energy of the inelastic electron. The change in energy, $f(E_1)$,
is a function of the energy E_1 which the inelastic electron would
have had in the absence of this interaction, and it is also a
function of the lifetime of the autoionizing state, as well perhaps as other parameters of the state such as its polarizability.
This post-collision interaction model was first proposed by Barker
and Berry[16] to explain the observation of energy changes of ejected
electrons resulting from positive ion impact on helium and neon.

The observed dependence of $f(E_1)$ on E_1 for the $(2s^2)^1S$ state
of helium is shown in Figure 2. The data for this figure have
been obtained from three types of scattering experiment, and three
separate regions of energy may be identified in this figure.
Firstly when $f(E_1)$ is less than E_1 the final energy of the inelastic
electron is positive and it continues to be free, and the ejected
electrons are observed to have an increased energy. Secondly when
$f(E_1)$ is approximately equal to E_1 the inelastic electrons have a
small energy and are observed in threshold excitation spectra at
an incident energy which is above the true energy of the autoionizing state, thus giving rise to a "displaced" threshold. The third

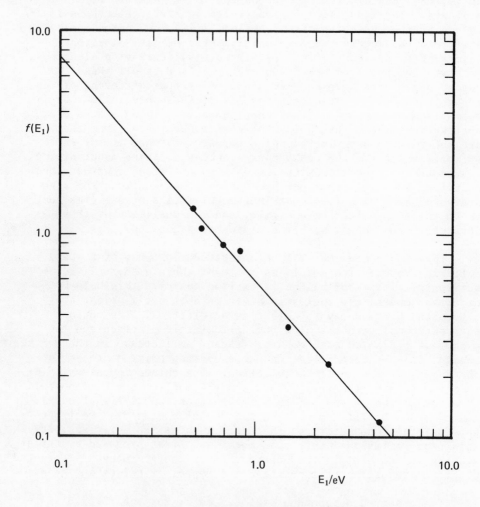

Figure 2. The observed dependence of the energy
lost by the inelastic electrons which have excited
the $(2s^2)^1S$ autoionizing state of helium. E_1 is the
energy which an inelastic electron would have had
in the absence of the post-collision interaction
between it and the ejected electron.

region is that in which the energy change $f(E_1)$ is greater than the available energy E_1. In this case the inelastic electron is effectively captured by the residual ion to form an excited bound state, usually having a high value of the quantum number n. This amounts to an extra mechanism for exciting these states, and it causes structure in the excitation cross-sections for them when the incident electron energy is just above the energy of a short lived autoionizing state. The energy at which the structure appears is a function of the quantum number n, and it decreases as the number n decreases. This effect has been observed for the states, with quantum numbers 4, 5 and 8.

It can be seen from Figure 2 that although the data have been obtained from three quite different scattering experiments, the value of $f(E_1)$ changes reasonably smoothly with E_1. The line drawn on the figure has a slope of −1.2, which is different from the value −0.5 predicted by the model of Barker and Berry[16], although the range of validity of this particular model is rather limited[5]. A more detailed theory of the mechanism of this post-collision interaction is obviously needed.

It should be noted that although the observed structures in the excitation cross-sections for the high n states look super-ficially like resonance structures, they are in fact quite differ-ent since they are not due to the formation of He⁻ during the collision but are caused instead by an inter-electron interaction after the collision.

CUSPS IN ELASTIC SCATTERING

The discontinuities (called cusps or rounded steps) which can occur in scattering cross-sections at energies above and below the opening of new exit channels are also caused in a sense by the long range effects of the real or virtual inelastic electron. With the advent of improved energy resolutions in electron-atom scattering experiments it is now becoming possible to resolve such features more clearly and hence to derive quantities such as phase shifts from analysis of their shapes.

The 2^3S cusp in helium has been studied by Cvejanović et al[17] with an energy resolution of 18 meV (FWHM), and an analysis of the measurements at a scattering angle of $90°$ (at which the effects of p-wave scattering are absent) has given the value $107 ± 11°$ for the s-wave phase shift at this energy. Cusp positions can also provide an accurate means of calibrating energy scales, and in this way it has been found[17] that the central energy of the 2^2S resonance in helium is $19.367 ± 0.009$ eV. Another recent study of this same cusp has been made by Golden et al[18] with a somewhat

poorer energy resolution. It is clear that there is much yet to
be learned from the study of electron-atom scattering cusps.

REFERENCES

1.   G.H. Wannier, Phys. Rev. 90 (1953) 817-25.

2.   U. Fano, Comm. At. Mol. Phys. 1 (1970) 159-65.

3.   U. Fano, J. Phys. B 7 (1974) (to be published).

4.   S. Cvejanović and F.H. Read, J. Phys. B 7 (1974) (to be
     published).

5.   P.J. Hicks, S. Cvejanović, J. Comer, F.H. Read, and J. Sharp,
     Vacuum, (to be published).

6.   A.J. Smith, P.J. Hicks, F.H. Read, S. Cvejanović, G.C.M.
     King, J. Comer and J. Sharp (to be published).

7.   A.R.P. Rau, Phys. Rev. A 4 (1971) 207-20.

8.   R. Peterkop, J. Phys. B 4 (1971) 513-21.

9.   I. Vinkalns and M. Gailitis, Abst. 5th Int. Conf. Phys.
     Electronic and Atomic Coll. (Leningrad: Nauka, 1967) 648-50.

10. J.W. McGowan and E. Clarke, Phys. Rev. 167 (1968) 43-51.

11. C.E. Brion and G.E. Thomas, Phys. Rev. Lett. 20 (1968) 241-2.

12. C.E. Brion and G.E. Thomas, Abst. 5th Int. Conf. Phys.
     Electronic and Atomic Coll. (Leningrad: Nauka, 1967) 53-5.

13. G.J. Krige, S.M. Gordon and P.C. Haarhoff, Z.f.Nat. 23a
     (1968) 1383-5.

14. P. Marchand, C. Paquet and P. Marmet, Phys. Rev. 180 (1969)
     123-32.

15. S. Cvejanović and F.H. Read, J. Phys. B 7 (1974) (to be
     published).

16. R.B. Barker and H.W. Berry, Phys. Rev. 151 (1966) 14.

17. S. Cvejanović, J. Comer and F.H. Read, J. Phys. B 7 (1974)
     468-77.

18. D.E. Golden, F.D. Schowengerdt and J. Macek, J. Phys. B 7
     (1974) 478-87.

ELECTRON SCATTERING BY LASER EXCITED ATOMS IN WELL DEFINED STATES[+]

I. V. Hertel

Fachbereich Physik der Universität Trier

Kaiserslautern, Kaiserslautern, Germany

I. INTRODUCTION

For many decades, spectroscopists have dreamed about a powerful, tunable monochromatic light source. During the last few years this dream became true due to the development of argon laser pumped CW dye lasers and we have already heard at the present conference what beautiful experiments are possible, when applying this tool to high resolution spectroscopy. However, also to atomic collision physics it seems to open up a whole new area of investigations.

It is now possible to prepare a high percentage of atoms in an atomic beam in a short living excited state, and to perform successful scattering experiments by irradiating the atoms with laser light of the resonance frequency in the collision region.

The studies now possible may be directed at two different aspects of collision physics:

1. Processes may be observed, hitherto unaccessible to the experiment. Outside the scope of this talk which is on electron scattering let me just mention one example from the heavy particle field. The quenching of resonance radiation has been subject of many studies during the last years without leading to a de-

---

+ A. Stammatovich and W. Stoll (1974) have participated essentially in the work reported here.

finite understanding of the basic mechanisms involved.
A crossed beam experiment with excited alkali atoms
and subsequent energy analysis of the scattered particles will yield information on vibrational excitation of the molecules and allow an understanding of
this process.

2. A more basic difference between conventional
scattering experiments and collision studies involving laser excited atoms is, however, the optical
pumping process when applying intense light for the
excitation. The atoms may thus be prepared in well
defined states and a whole set of details of the
scattering dynamics may be revealed which was up to
now hidden in the averaging process incorporated
when measuring ordinary differential cross sections.

I hope to be able to illustrate the latter aspect in
one particular example in the present paper.

II. OPTICAL PUMPING PROCESSES

First however, we have to discuss the optical pumping process which is the basis of these experiments.
Let us consider as an example the sodium atom whose
resonance 3^2P-state is easily excited with a rhodamin
6G-laser. Since we have to deal with an atomic beam
displaying a negligible Doppler width, hyperfine structure is to be taken into consideration. Figure 1 shows
the term scheme of the sodium atom which has a nuclear
spin I = 3/2.

The 3^2P 3/2 state has a total angular momentum
F = 0, 1, 2, 3 and the 3^2S 1/2 ground state may have
F = 1 or 2. Since the selection rule for optical
transitions is ΔF = 0, $\pm$ 1, for most of the transitions
depopulation of one of the ground states and population
of the other occurs due to optical pumping, leaving no
atoms in the excited state. Only the transition
3^2S. 1/2 F' = 2 $\longrightarrow$ 3^2P 3/2 F = 3 allows the excited
state to decay only to the original ground state and
leads to a non zero stationary population of the excited state; its maximum being 31% as shown in Figure 2.
Our experimental conditions correspond to the right
half of figure 2.

How does the magnetic sublevel population look as
a result of this pumping process. We have to distinguish two cases:

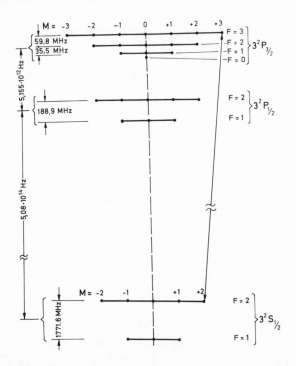

Figure 1:   Energy level diagram of the hyperfine
            splitting of the sodium 3^2S and 3^2P states.

1. Circularly polarized (σ^+) light implies a selection
   rule $\Delta M_F = 1$. Thus, in the course of the pumping,
   the positive magnetic quantum number ($M_F \geqq -1$) sublevels are populated preferentially. In the stationary state the probability (suitably normalized) to
   find excited atoms in a magnetic quantum number M_F
   state is

$$W \ (M_F) = \delta_{F, M_F}$$

   This refers of course to a photon frame which is defined by the incident light direction being the Z-
   axis. Due to the unequal population of positive and
   negative M_F levels, we have produced an oriented
   atom.

2. Linear polarized π-light imposes vertical transitions
   (Figure 1) with $\Delta M_F = 0$ to the atom. This leaves the
   $M_F = \pm 3$ excited state unpopulated and due to spontaneous transitions in the course of the optical
   pumping an overpopulation of states with small M_F is
   produced as shown in Figure 3. This corresponds

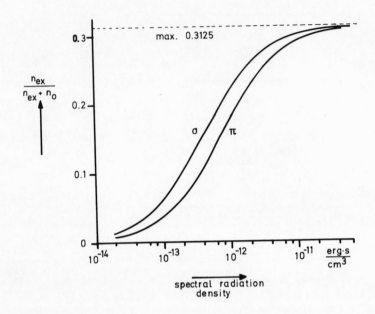

Figure 2:   Stationary probability of finding sodium
            atoms in the $3^2P3/2$ F=3 state as a function
            of the exciting spectral radiation density
            for σ and π light.

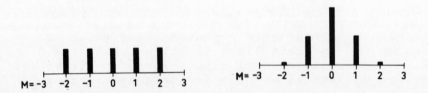

Figure 3: Magnetic sublevel population for pumping with
          π -light. Left: First moment of the pumping
          process, right: Stationary conditions.

to an alignment of the atom. The photon frame refers
here to a Z-axis parallel to the electric vector of
the laser beam. The so excited atom must be represented
by an incoherent superposition of pure states with a
probability to find the magnetic quantum number M_F

$$W(M_F) = (F-1 \; M_F \; F-M_F \; | \; 2F - 1 \; 0)^2$$

under stationary conditions.

We may describe the result of the pumping in terms
of multipole moments of the excited atom. In the photon frame only one component of each moment can be non
zero and may be expressed by

$$\mathcal{W}(k) = \sum (-)^{k-F-M_F} \; W(M_F) \; (F \; -M_F \; F \; M_F \; k \; 0 \;)$$

It can be seen that moments up to k = 2F can be non
vanishing. $\mathcal{W}(1)$ corresponds to the non zero component
of the familiar orientation vector

$$\mathcal{W}(1) \propto O_o^{ph},$$

while $\mathcal{W}(2)$ is related to the alignment tensor

$$\mathcal{W}(2) \propto A_o^{ph}.$$

Obviously for linear polarized light only the even moments are non vanishing while for circular light any of
the 2F + 1 = 7 terms are finite.

At a first view, stationary conditions seem to be
a good assumption. The atoms stay in the pumping and
scattering region for around 10^{-6} sec corresponding to
approx. 100 pumping cycles which are determined essentially by the spontaneous decay time. A solution of
the time dependent pumping equations clarifies the
situation: Figure 4 shows a result of such a calculation
for different pumping intensities. The multipole
moment $\mathcal{W}(2)$ (alignment) normalized to $\mathcal{W}(o)$ is plotted as a function of time in units of the natural life
time. The radiation densities corresponding to our experiment are between the dashed and the full curve.
The result is somewhat unexpected since stationarity
is reached only after more than 15 pumping cycles for
high intensities. For lower intensities as found in the
beam wings the dashed line shows much slower pumping.
Thus, averaged over the collision region $\mathcal{W}(2)$ may well

be 20% less than in the stationary case. This will
have to be considered when evaluating our experiments.

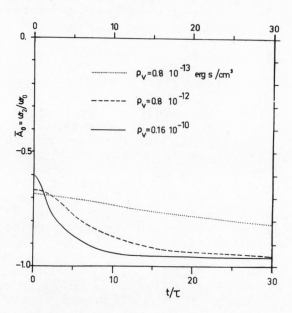

Figure 4: Multipole moments of laser excited sodium
          atoms as a function of time for various
          pumping intensities.

III. EXPERIMENT

Let me now turn to the experiment. Without going
into details (see Hertel and Stoll 1974, a,b) I just
show the scattering geometry in figure 5. The atom
beam is excited by linear polarized light coming from
the right along the scattering plane. The latter is
defined by the incoming and the scattered electron, the
"collision frame" is given by Z_{col} , Y_{col} , X_{col} . The
scattering angle ϑ_{col} and the angle $\Theta_{\hat{n}}$ between incident light direction and collision frame may be varied
independently as well as the angle ψ of the polarization vector, with respect to the collision plane.
So we have many variable parameters defining the
orientation of the photon frame with respect to the
collision frame. Different bits of collision dynamics

may be projected out of the scattering wave function
by imposing it on the multipole moments produced by
the optical pumping.

   Apart from the excitation devices a "standard"
crossed beam experiment is used. Energy loss spectra

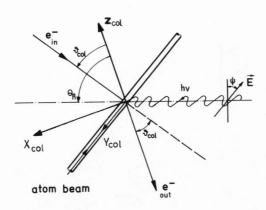

Figure 5:   Schematic diagram of the collision geometry

may be measured without and with laser excitation of
the atoms as shown in figure 6.   One can clearly distinguish several processes of significant cross sections for the inelastic electron scattering of the excited 3^2P state. This shows a comparison of scattering
without (upper fig.6) and with (lower fig.6) light.
In the following we will focus our attention on the
superelastic cross section $3^2P \longrightarrow 3^2S$ which is the
inverse process to excitation of the resonance line by
electron impact. Since we may vary, however, the
scattering geometry, more detailed information can be
obtained.

   It should be mentioned here that supplementary
information can be obtained from electron-photon coincidence experiments as, e.g., reported at this conference by Kleinpoppen (1974).

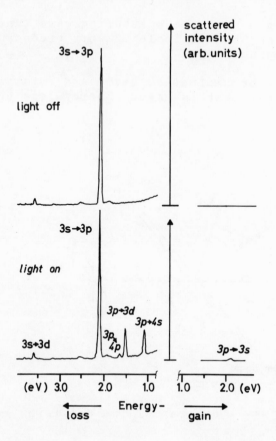

Figure 6: Typical energy loss spectra for the electron
scattering by sodium atoms with and without
laser excitation.

IV. RESULTS AND INTERPRETATION

The "Perfect Scattering Experiment" has often been
discussed (e.g.,by Bederson 1970) meaning an experimental observation of all amplitudes which theorists calculate to describe the collision processes. The principle
of detailed balance allows us to compare directly these
amplitudes for our deexcitation process and the inverse.
Total decoupling of nuclear spin, electron spin, and orbital angular momentum is usually assumed during the
collision time. Then there are three amplitudes

(see figure 7) for momentum transfer $\Delta M = 0, \pm 1$, two
of which are not independent. Direct (f) and exchange
(g) amplitudes govern the process (four independent complex amplitudes).

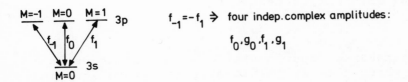

Figure 7:   Scattering amplitudes for an s → p transition.

     What is measured usually? In figure 8 we see a
comparison of different types of experiments, while the
differential excitation cross section gives only a
fixed combination of amplitude squares, experiments
with spin analysis reveal ratios and possibly phases
between combinations of direct to exchange amplitudes.
The phases and ratios of f_0, g_0 and f_1, g_1 however,
are still covered. Here the deexcitation of laser
excited atoms gives information about. The double sum
containing the scattering geometry in $K_{MM'}$, allows to
measure different combinations of f_1, f_0, g_0 ,$f_{-1}f_1$,etc.

inelast. diff. cross section	spin flip cross section	deexitation of aligned atoms
$\sigma \propto \sum\limits_{M=-1}^{1} \left(f_M f_M^* + \frac{f_M^* g_M + f_M g_M^*}{2} + g g^* \right)$	$\sigma \propto \sum\limits_{M=-1}^{1} g_M g_M^*$	$\propto \sum\limits_{M,M'=-1}^{1} K_{MM'} \cdot \left(f_M f_{M'}^* + \frac{f_{M'}^* g_M - f_M^* g_{M'}}{2} + g_M g_{M'}^* \right)$
⟹ magnitude of com- bined amplitudes	⟹ ratio: direct to exchange amplitude combin.	⟹ ratio and phase betw. $f_0 \cdot g_0$ and f_1, g_1–com- bination

Figure 8:   Amplitude combinations measured in different
experiments.

Experimentally, the question is: What is most easily varied? Let us look at the scattering geometry (fig.5) again. The polarization vector may be rotated without difficulties variing Ψ for fixed $\Theta_{\hat{n}}$. A theoretical analysis (Macek and Hertel (1974)) which makes use of the multipole expansion language developed recently by Fano and Macek (1973), shows that the scattering intensity depends on Ψ as

$$I(\Psi) = C_1 + C_2 \cos 2\Psi .$$

The corresponding experimental results are shown in figure 9 which clearly illustrates the above equation.

The physics is contained in the ratio for $I(\Psi = 0)/I(\Psi = 90°)$. How does it change with variing direction of light incidence? There is one trivial case, forward scattering with $\vartheta_{col} = 0$. Since no angular momentum can be transferred here f_1 and g_1 vanish.

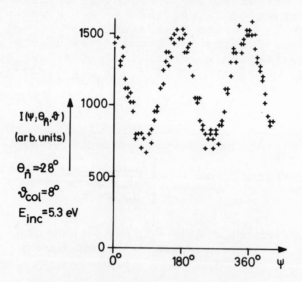

Figure 9: Electron scattering intensity as a function of the angle Ψ between polarization vector and collision plane.

Figure 10 shows the calculations. The dashed line corresponds to stationary pumping while the full curve
corresponds to a best fit to our experimental points
leading to an alignment $\mathcal{W}(2)$ which is 20% less than
stationary. This was to be expected from the previous
discussion. We may now turn to $\vartheta_{col} \neq 0$. A calculation
may use Moores' and Norcross' (1972) results, to whom
I am very much indebted for letting me have their complex amplitudes. For $\vartheta_{col} = 10°$ in fig. 11 the full
curve shows the result. In order to see how sensitive
our method is, we decrease with brute force the amplitudes f_0 and g_0 by a factor α : We see that the curve
rotates around zero, standing upright for $f_0 = g_0 = 0$
corresponding to maximum angular momentum transfer.

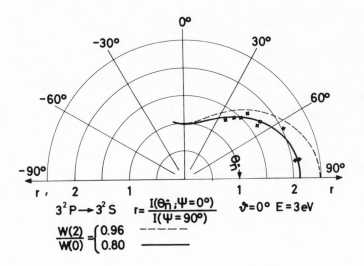

Figure 10: Scattering intensity calculated with diffe-
            rent sets of amplitudes.

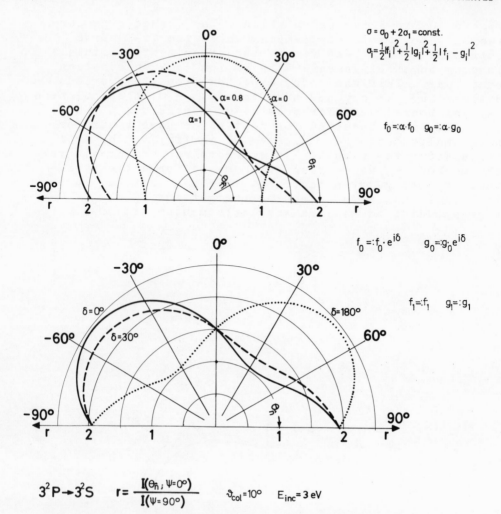

$$\sigma = \sigma_0 + 2\sigma_1 = \text{const.}$$
$$\sigma_i = \tfrac{1}{2}|f_i|^2 + \tfrac{1}{2}|g_i|^2 + \tfrac{1}{2}|f_i - g_i|^2$$

$$f_0 =: \alpha \cdot f_0 \qquad g_0 =: \alpha \cdot g_0$$

$$f_0 =: f_0 \cdot e^{i\delta} \qquad g_0 =: g_0 \, e^{i\delta}$$

$$f_1 =: f_1 \qquad g_1 =: g_1$$

$$3^2P \rightarrow 3^2S \qquad r = \frac{I(\theta_{\bar{n}};\, \psi = 0°)}{I(\psi = 90°)} \qquad \vartheta_{col} = 10° \qquad E_{inc} = 3\,\text{eV}$$

Figure 11:   Scattering intensity calculated with
different sets of amplitudes.

In figure 11 the result of changing the phases is
indicated to which our experiment is also sensitive.
We may now compare experiment and theory, using $\vartheta(2)$
as determined before: Figure 12 shows $\vartheta_{col} = 10°$
results for 3 eV incident energy. The full line is
given by Moores' and Norcross' amplitudes. They may
need a slight change to fit the experiments better
(dashed). The results are, however, not yet conclusive.

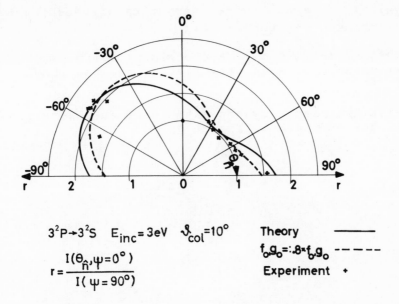

Figure 12: Comparison of measured and calculated results.
Full curve: Moores' and Norcross' amplitudes.

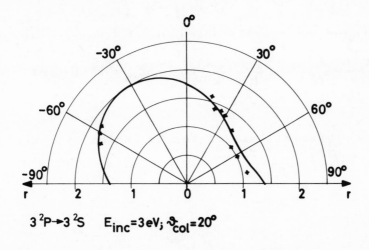

Figure: 13: Comparison of measured and calculated
scattering intensities.

For ϑ_{col} = 20° we get very nice agreement between experiment and theory (fig.13).

Further results of the experiment for 6 eV
scattering and the same deexcitation process shows
figure 14.

Finally I may show one example of an excitation
process $3^2P \longrightarrow 4^2S$ in figure 15.

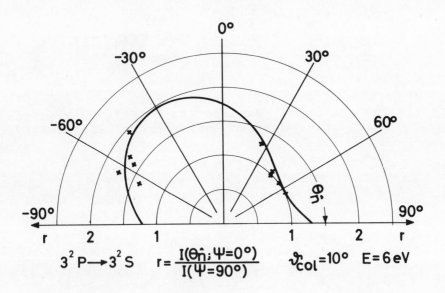

$$3^2P \longrightarrow 3^2S \qquad r = \frac{I(\theta_{\hat{n}}; \Psi = 0°)}{I(\Psi = 90°)} \qquad \vartheta_{col} = 10° \quad E = 6\,eV$$

Figure 14:   Experimental scattering intensities with
             arbitrary fit.

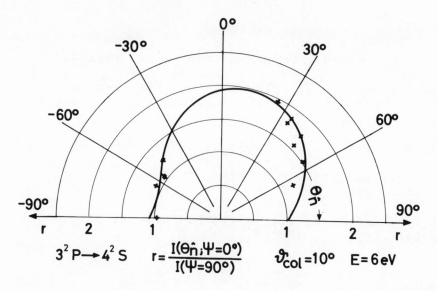

Figure 15:   Experimental scattering intensities for
             $3^2P \longrightarrow 4^2S$ excitation.

V.   CONCLUSION

    I have shown you some of the possibilities which
are offered by scattering of laser excited atoms. I
hope that this technique may be applied to many more
problems, perhaps also in heavy particle collision
physics where, e.g., the influence of spin-orbit interaction could be investigated as well as the applicability range of the Born or Glauber theory without
knowing detailed dynamical calculations.

REFERENCES

Bederson B., (1970)   Comments Atom. Molec.Phys.$\underline{2}$,160

Eminyan M., MacAdam K., Slevin J. and Kleinpoppen H.,
    1973 Abstr. VIII ICPEAC   ICPEAC (Beograd: Institute
    of Physics), 318

Fano U. and Macek J. H., 1973, Rev. Mod. Phys. $\underline{45}$, 553

Hertel I. V. and Stoll W., 1974a, J. Phys. B:Atom.
    molec. Phys. $\underline{7}$, 570

Hertel I.V. and Stoll W., 1974b, J. Phys. B: Atom.
    molec. Phys. $\underline{7}$, 583

Macek J. and Hertel I. V., 1974, J. Phys. B:Atom.
    molec. Phys. $\underline{7}$, in press

Moores D. and Norcross D., 1972, J. Phys. B:Atom.
    molec. Phys. $\underline{5}$, 1482

Stoll W., 1974, Dissertation, Kaiserslautern

ELECTRON TRANSFER IN SIMPLE ATOMIC COLLISIONS: RECENT THEORY,

EXPERIMENT, AND APPLICATIONS*

James E. Bayfield

J.W. Gibbs Laboratory, Yale University

New Haven, Connecticut, U.S.A.

I.   INTRODUCTION

The search for the mechanisms underlying the transfer of
particles and/or energy between interacting atoms or molecules continues to fascinate many investigators in physics, chemistry and
biology. Most interesting are those processes which are unusally probable and highly specific, requiring special conditions such as
internal energy resonance during some portion of the collisional interaction time. The general questions are:Under what conditions can
particles and/or energy be transferred? How can we control these
conditions and study the processes in the laboratory?;and Can we
make predictive calculations to verify our physical understanding
and eliminate the need to do uninteresting experiments? The
answers to these questions are not generally available except for
some simple atom-atom collisions and a few triatomic cases such as
$H^+ + H_2$. And we shall see that much remains to be understood even
for those simple systems extensively studied in the four years
following the closing dates for past monographs on electron
transfer.[1-5]

As interest in particle transfer processes continues to broaden
far beyond the small group of atomic collision physicists of two
decades ago, it seems appropriate to more carefully define our
terms so that a biophysicist, for example, would initially understand
them correctly. Thus this paper uses the term "electron transfer,"
rather than "electron capture" which is a different thing to nuclear
chemists, or "charge exchange" which does not specify the transfered
particle (proton transfer occurs in chemistry!)

*Research supported in part by the Air Force Office of Scientific
Research under Contract No.   F44620-71-C-0042.

Although the goal of a unified understanding of particle
rearrangement processes remains, today's experiments and calculations usually involve limited ranges of collision velocity. So we
shall continue tradition and divide this paper into parts concerned
with low, intermediate and high energy collisions respectively.
The center of the intermediate energy region we take to be at collision velocities equal to the atomic velocity v_o of the active
electron to be transferred. Low velocities and high velocities are,
respectively, considerably below and above v_o; what we mean by
"considerably" is one of the questions underlying today's research.
A prevailing feeling among investigators is that low velocities are
those for which the most accurate simple picture of the collision
is the transient formation of a composite quasimolecule. This
molecular picture we call "MS" (for molecular states) and note that
its usefulness now appears of broader range than once believed.
By high velocities one usually has in mind a different collision
picture with the separate colliding atoms somehow retaining more of
their identity during a rapid, perturbative collision too short for
sizeable transition amplitudes to develop. We call this atomic
picture "AS" (for atomic states).

The intermediate energy region $v \approx v_o$ for simple heavy particle atomic collisions is characterized by collision energies in
the keV region, two or three orders of magnitude larger than typical
atomic electron binding energies. Thus an infinity of inelastic
scattering channels are energetically quite open and possibly
important. This is a real challenge both for the experimentalist
who must strive to observe each channel separately and for the
theorist who must find a dynamical treatment of a generally highly
coupled multistate three-or-more body problem.

A major qualitative aid in determining favored scattering
channels is the idea that collisionally induced changes in atomic
electron binding energies require dynamics that become more difficult to achieve as the electronic energy change or defect ΔE
increases. Considerable experimental and theoretical verification
of this is discussed below. However, the definition of ΔE to be
used for instance with the adiabatic criterion[6] for the energy at
which an inelastic cross section is maximum is now recognized to
be generally dependent upon specifics of the process, with a value
to be taken at the largest separation R of the collision partners
where a large transition amplitude develops.

A well understood special electron transfer process is
that of resonance, where ΔE at $R=\infty$ is both relevant and zero. The
classical example is the process $H^+ + H(1s) \rightarrow H(1s) + H^+$, for which
the H_2^+ molecular states of the entire collision system are correlated in Fig. 1 and their Born-Oppenheimer electronic potential
energies drawn in Fig. 2. This resonance process dominates the

experimental total H^+ + H(1s) electron transfer cross section σ_{10}(H)
and the results of theory agree well with experiments as is seen in
Fig. 3.   The experiments include merged beams[7] and crossed beams[8]
studies at low energies, as well as beam-target work at the intermediate and higher energies.[9-11]   An accurate experimental test of
theory here requires knowledge of excited state contributions to
the data, and recent work has concentrated on these truly inelastic
processes.   A good comparison also exists for the measured cross
section σ_{20}(He), nominally for the resonant process He^{++} + $He(1s^2)$ →
$He(1s^2)$ + He^{++} (see Fig. 20).   However, much of this report relates
to present theoretical problems with some of the non-resonant or
truly inelastic processes.

II.   THEORETICAL APPROACHES AND PHYSICAL PICTURES

It is valuable for us to look at some of the problems theory
encounters when AS and MS physical intuition is put into good mathematical terms.   The time-dependent non-relativistic Schroedinger
equation is to be solved in the three-body case using the Hamiltonian

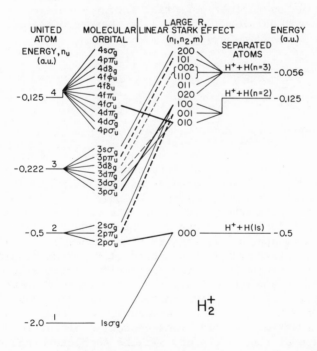

Fig. 1 Correlation diagram for the H^+ + H collision system.

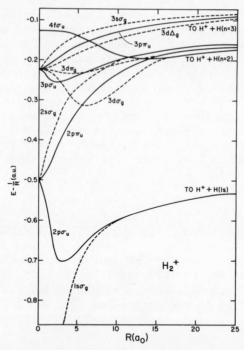

Fig. 2   The Born-Oppenheimer potential energy levels of H_2^+.   Note
the internuclear repulsion R^{-1} (a.u.) has been subtracted.

$$- \frac{1}{2} \hbar^2 \sum_{i=1}^{3} m_i^{-1} \nabla_i^2 + \sum_{i=1}^{3} \sum_{j>i} \frac{e_i e_j}{|\vec{R}_i - \vec{R}_j|} \qquad (1)$$

The center of mass motion is then separated off[17] and one of several
possible sets of relative coordinates chosen.[18]   Among these is the
set of Eyring-Polyani skewed coordinates that eliminates all kinetic-energy cross terms while making the Coulomb potentials into
functions of more than one coordinate.   The center-of-mass of the
nuclei (CMN) coordinates reverses the situation.   It is at this
point that serious approximations are first made; a common procedure
is to employ the impact parameter method (IPM) which assumes classical straight-line trajectories for the nuclei moving with constant
velocity, a reasonable approximation for small angle scattering at
energies above 100 eV.   A transformation to relative coordinates
rotating with the internuclear axis then is followed by an expansion in a basis set of wavefunctions chosen according to one's
physical picture of the collision.   In the AS picture and using the
IPM, the expansion is in traveling electronic atomic   orbitals

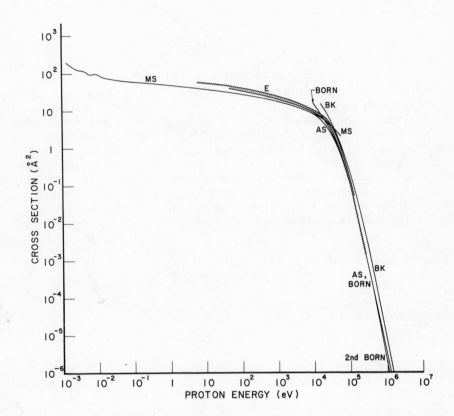

Figure 3.   The data "E" for total electron transfer in H^++H(1s)
collisions is compared with AS^{12}, $MS^{13,14}$ and high energy OBK^{15}
and First Born^{16} predictions.

ϕ_n, ψ_m which are exact solutions of the asymptotic electronic Hamiltonian for the electron about one nucleus A or the other B:[19]

$$\phi_n\ (\vec{r}_A,t)=\ U_n^A\ (\vec{r}_A)e^{ip\vec{v}\cdot\vec{r}}\ e^{-i(\epsilon_n^A\ +\ \frac{1}{2}\ p^2v^2)t} \tag{2}$$

$$\psi_m(r_B,t)=\ U_m^B(\vec{r}_B)e^{-iq\vec{v}\cdot\vec{r}}\ e^{-i(\epsilon_m^B\ +\ \frac{1}{2}\ q^2v^2)t} \tag{3}$$

where $p=M_B/(M_A+M_B)$, $p+q=1$ and the U_n are eigenfunctions with eigenvalues ϵ_n^B of the time-independent electronic Schroedinger equation for the isolated, stationary atom. In the MS picture one alternatively expands in traveling linear combinations of Born-Oppenheimer molecular states obtained by first solving the clamped-nuclei problem, $H\chi_K(\vec{r},R) = E_K(R)\psi_K(\vec{r},\vec{R})$. The linear combination is chosen so as to leave the electron asymptotically in a chosen state on a given atom. In either case the expansion combined with the original full Schroedinger equation leads to a set of coupled equations for the time-dependent expansion coefficients, with coupling matrix elements K_{nm} depending on the picture. For the case $M_A = M_B$, the AS picture has[3]

$$K_{nm}=\ \int\psi_m^*(\vec{r}_B)e^{i\vec{v}\cdot\vec{r}}\ (\frac{Z_B}{R}\ -\ \frac{1}{r_A}\ +\ \dot{\theta}L_{yB})\phi_n\ (\vec{r}_A)d\vec{r}_B\ e^{i(\epsilon_m^B-\epsilon_n^A)t} \tag{4}$$

whereas in MS

$$K_{nm}\ =\ [E_m^A(R)\ -\ \epsilon_m^A]\ \int\chi_n^{B*}\chi_m^A\ e^{i\vec{v}\cdot\vec{r}}\ d\vec{r}_B$$

$$+\int\chi_n^{B*}\ e^{i\vec{v}\cdot\vec{r}}\ [\dot{\theta}L_y\ -\ iv\ \frac{\partial}{\partial Z_R}]\ \chi_m^A\ d\vec{r}_B\ . \tag{5}$$

Here $\dot{\theta} = \frac{v}{R^2}$ is the rate of rotation of the internuclear axis, L_y is an electronic angular momentum operator, and $\frac{\partial}{\partial Z_R}$ is the radial component of nuclear relative momentum operator. Thus in MS we have two kinds of kinetic-energy induced transitions, due to rotational coupling ($\dot{\theta}L_y$) and radial coupling ($iv\ \frac{\partial}{\partial Z_R}$), but no transitions between (molecular) states arising from Coulomb forces. In AS, however, we have rotational coupling and Coulomb-force induced transitions between (atomic) states, but no radial motion coupling. Thus the choice of physical picture leads to specific

breakup of the collision problem, putting part of the physics into
the basis wavefunctions and the rest into the coupling operators.
At present the relationships between pictures are not well-developed,
and general criteria for the usefulness of a given approach do not
exist. Accurate numerical methods for evaluating the K_{mn} are being
found[20] and only a few close-coupling calculations have included the
$e^{i\vec{v}\cdot\vec{r}}$ electron momentum factors. At low velocities v these are not
expected to be important. Tests for the accuracy of numerical
close-coupling integrations are just being developed, with careful
detailed-balance checks being a necessary but not always sufficient
test for accuracy.[21]

Within AS close-coupling theory one can define a true high-
energy limit. Here a perturbation theory approach for the flow of
electron flux from the initial state into the various final states
permits neglect of back-coupling terms to the initial state as well
as phase shift effects arising from "distortion" of the instantaneous electronic energies by the collisional interactions.[22] This
leads to the familiar Brinkman-Kramers (OBK) and First Born approximation results,[23] for which various possible improvements have been
suggested, including various distorted wave approximations[24,25] and
orthogonalization corrections.[26] The general usefulness of such
ideas is far from established, as we shall see in Section V below;
but then that might be said for AS close-coupling theory also. Only
the low-energy MS theory seems at present to be established as useable
with some confidence.

A presently active theoretical field is the development of
Faddeev's three body equations[27] for the atomic particle rearrangement process. So far only a Born-like high-energy expansion has
been derived, with results also discussed in Section V. A problem
here has been a lack of experimental data for comparison above a
few hundred keV collision energy, but this situation can be expected to change as nuclear physics accelerators become more available
for atomic collision experiments.

III. PROGRESS AT LOW ENERGIES

Work with the molecular picture (MS) begins with molecular
state correlation diagrams[28] and accurate molecular energy levels;
considerable effort has gone into the latter for H_2^+,[29]
HeH^{++}[30] and H_2.[31] Aspects of the one-electron molecule ZeZ' with
$Z>2$ have also been studied with emphasis on the nature of true
potential-energy level crossings in such molecules.[32] A few of
the coupling matrix elements of the type of equation 5 have also
been accurately computed for the H_2^+[33] and HeH^{++}[34] systems. If
the translation factors $e^{-\vec{v}\cdot\vec{r}}$ in K_{mn} are ignored, spurious long

range (large R) behavior results, with rotational couplings sometimes varying asymptotically as R^{-1} and radial couplings becoming
a non-zero constant in some cases. The evidence is that the
spurious coupling effects cancel in a proper close-coupling calculation;[35] this is directly seen in a formulation of the problem not
using the usual rotating CMN coordinate system, but rather mixed
coordinates centered about one atomic center or the others.[36]
Figure 4 shows this for the famous $2p\sigma_u$-$2p\pi_u$ rotational coupling
which produces $2p_{\pm 1}$ atoms in low energy H^+-$\bar{H}(1s)$ collisions. The
work of Ref. 36 (CPW) also uncovered an atomic-type of coupling
("translational coupling") lost in earlier work that ignored the
translational factors; thus MS close-coupling results up to present
are all incorrect by an amount increasing as the velocity increases
towards the intermediate energy region.

Because of recent work on He^{++}-$H(1s)$ collisions, we show in
Figures 5, 6 and 7 the correlation diagram, potential energies and
couplings for HeH^{++}. Comparing these with Figures 1 and 2 for H_2^+
shows the changes arising first from parity no longer being a good
quantum number for asymmetric systems, and second the shifts in electronic binding energy in He^+ relative to H. In an He^{++}-$H(1s)$ electron
transfer collision, the initial $2p\sigma$ state is primarily coupled to
the nearby $2p\pi$, $2s\sigma$ and $3d\sigma$ states leading to $He^+(n=2)$.

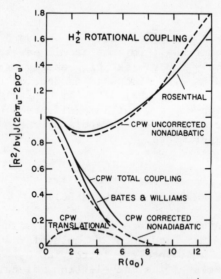

Fig. 4 Rotational $2p\pi_u$-$2p\sigma_u$ coupling in H_2^+. The spurious behavior
of the element computed by Rosenthal without $e^{i\vec{v}\cdot\vec{r}}$ is extracted by
Chen, Ponce and Watson (CPW uncorrected) and a second contribution
cancels this to give the CPW corrected curve. The total coupling
is close to an old calculation by Bates and Williams.

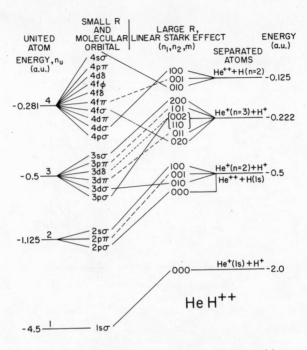

Fig. 5 Correlation diagram for HeH^{++}.

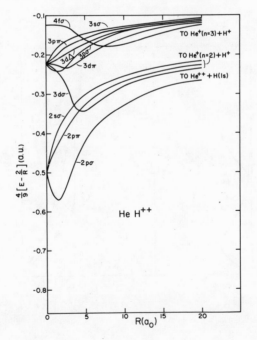

Fig. 6 Born–Oppenheimer potential energy curves for HeH^{++}.

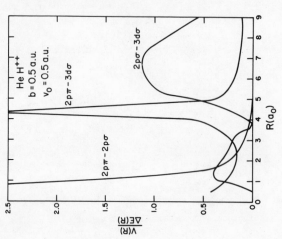
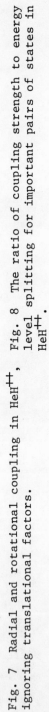

Fig. 8 The ratio of coupling strength to energy level splitting for important pairs of states in HeH^{++}.

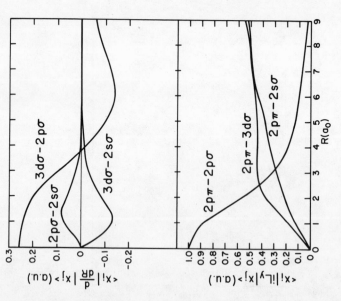

Fig. 7 Radial and rotational coupling in HeH^{++}, ignoring translational factors.

Recent H^++H(1s) MS close-coupling calculations are ($1s\sigma_g$, $2p\sigma_u$, $2p\pi_u$) calculations [33,37,38] and a ($1s\sigma_g$, $2p\sigma_u$, $2p\pi_u$, $3p\sigma_u$, $3p\pi_u$) calculation. [39] All these include rotational couplings only. In addition, some two-state resonant electron transfer cross sections have been obtained for H^++H(nℓ) collisions with n$\leq$5. [40] One very recent He^{++}-H(1s) $\rightarrow$ He^+(n=2)+H^+ calculation including $2p\sigma$, $2p\pi$ and $3d\sigma$ states[34] identifies three important coupling regions indicated in Figure 8; at R~$7a_o$ a long range $2p\sigma$-$3d\sigma$ radial coupling is active, at R~$4a_o$ rotational coupling near the $2p\pi$-$3d\sigma$ level crossing is important, and a number of couplings are simultaneously involved at smaller R.

New experiments have been performed on both the H^+-H and He^{++}-H systems. Figure 9 compares typical H^+-H(1s) differential scattering data [41] with 3-state MS[38] and 4-state AS IPM [42] calculated curves. The resonant transfer oscillations and rotationally induced n=2 atom production is very well described by the MS theory. Figure 10 compares data for fractional differential electron transfer into the 2s state [43] with the 5-state MS [39] and an 4-state AS[42,66] calculation. Again the MS result is not bad, and would perhaps be better if the $3d\sigma_g$-$2s\sigma_g$ and $2p\sigma_u$-$4f\sigma_u$ radial couplings were included.

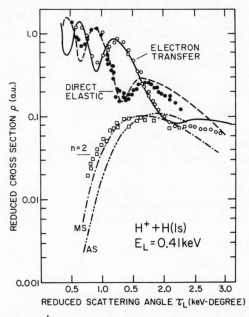

Fig. 9 Differential H^+-H scattering, theory and experiment. The reduced cross section is $\rho= \sigma \cdot \sin\theta_L \cdot \theta_L$ and reduced angle is $\tau_L=E_L\theta_L$, where E_L and θ_L are laboratory collision energy and scattering angle respectively. Transfer and direct excitation to n=2 were not resolved from one another.

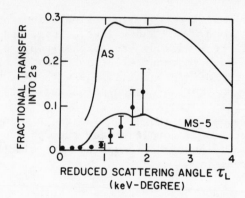

Fig. 10. Differential fractional electron transfer into H(2s), theory and experiment for E_L = 6 keV.

Two new sets of data[44,45] for He^{++}-H(1s) collisions now accompany the earlier work.[8] A comparison of the total electron transfer cross section σ_{21}(H) into all states is made in Figure 11 with theory predicting mainly n=2 state production. The present experimental situation is uncertain above 15 keV, with the Yale data agreeing with the MS results[34] and the Belfast data siding with a number of AS calculations. The MS calculations here are believed reliable without concern for translation factor problems, since in asymmetric systems the states around different nuclei are uncoupled asymptotically and the transfer probability is computed as one minus the elastic scattering probability. The $2p\sigma$-$3d\sigma$ radial coupling dominates the total cross section.

Electron-transfer collisions of protons with excited hydrogen atoms with principal quantum number n have recently become of interest because of merged-beams data for $n \approx 47$.[50] The mentioned $n \leq 5$ 2-state MS results[40] compare favorably at n=5 with calculations using Smirnov's approximation.[51] However the latter theory appears to break down at n=47. The $n \approx 47$ data does agree with a theoretical curve obtained from the n=1 data using classical scaling rules.[52]

Of great interest for some time has been an explanation for the observed energy dependence for the collision $H^+ + H^-(1s^2) \rightarrow 2H$ at low energies. Not all of the possibly relevant H_2 potential energy curves are yet known[32], although the major linear-combination-of-atomic-states (LCAS) components of the adiabatic states were discussed by Mulliken[52] as tabulated in Figure 12. The singlet states 3, 7, 10 and 14 mix at intermediate internuclear separations R with the ionic singlet states; there are also triplet ionic states of both parities to be considered. In face of this, recent theory has

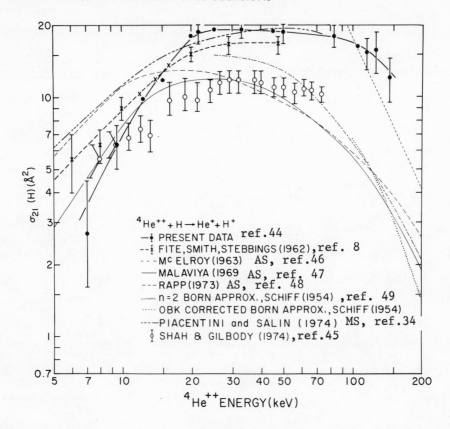

Fig. 11 Total electron transfer in He^{++}-H(1s) collisions. The data of Fite, Smith and Stebbings and of Shah and Gilbody are for incident $^3He^{++}$, plotted here for equivalent relative collision velocity.

concentrated on an LCAO (AS-like) picture of the problem, with recent computations[53] of the exchange splittings Δ at the ionic-covalent level crossings of Figure 13, and a calculation including $H^-(1s^2)$, $H(2s, 2p_0, 2p_{\pm 1})$ states.[54] The latter result finds $2p_{\pm 1}$ formation dominant with the 2s state next in importance. Coupling of all the excited states with the initial ionic state is found to be quite significant up to the large internuclear separation of 15 Å. Improvement over the old Bates and Lewis calculation[55] is found in the 0.2 - 6 keV energy range (see Figure 14). At lower energies the n=3 states are believed important.

PREDOMINANT LCAS TERMS FOR ADIABATIC H_2 ENERGY LEVELS

(In order of increasing asymptotic total energy)

Very large R	Rather large R	$R \approx R_e$
1. $(1s \cdot 1s)_g \, {}^1\Sigma_g^+$	.	$(1s \cdot 1s)_g$
2. $(1s \times 1s)_u \, {}^3\Sigma_u^+$	$(1x \times 1s)_u$	$(1s \times 2p\sigma)_u$
3. $(1s \cdot 2p\sigma)_u \, {}^1\Sigma_1^+$	$(1s \cdot 2di')_u \quad (1s \cdot 1s)_u \quad (H^+H^-) \quad (1s \cdot 2di)_u$	$(1s \cdot 2p\sigma)_u$
4. $(1s \times 2p\sigma)_g \, {}^3\Sigma_g^+$	$(1s \times 2di')_g$	$(1s \times 2s)_g$
5. $(1s \cdot 2p\pi)_u \, {}^3\pi_u$		$(1s \cdot 2p\pi)_u$
6. $(1s \times 2p\pi)_g \, {}^1\pi_g$	$(1s \times 2p\pi)_g$	$(1s \times 3d\pi)_u$
7. $(1s \cdot 2s)_g \, {}^1\Sigma_g^+$	$(1s \cdot 2di')_g \quad (1s \times 1s)_g \quad (H^+H^-) \quad (1s \cdot 2di)_g$	$(1s \cdot 2s)_g$
8. $(1s \times 2s)_u \, {}^3\Sigma_u^+$	$(1s \times 2di')_u$	$(1s \times 3p\sigma)_u$
9. $(1s \times 2s)_g \, {}^3\Sigma_g^+$	$(1s \times 2di)_g$	$(1s \times 3d\sigma)_g$
10. $(1s \times 2s)_u \, {}^1\Sigma_u^+$	$(1s \cdot 2di)_u \quad (1s \cdot 1s)_u \quad (H^+H^-) \quad (1s \cdot 2di')_u$	$(1s \cdot 3p\sigma)_u$
11. $(1s \times 2p\pi)_u \, {}^1\pi_u$		$(1s \times 2p\pi)_u$
12. $(1s \cdot 2p\pi)_g \, {}^3\pi_g$	$(1s \cdot 2p\pi)_g$	$(1s \cdot 3d\pi)_g$
13. $(1s \times 2p\sigma)_u \, {}^3\Sigma_u^+$	$(1s \times 2di)_u$	$(1s \times 4f\sigma)_u$
14. $(1s \cdot 2p\sigma)_g \, {}^1\Sigma_g^+$	$(1s \cdot 2di)_g \quad (1s \times 1s)_g \quad (H^+H^-) \quad (1s \cdot 2di')_g$	$(1s \cdot 3d\sigma)_g$
15. $(1s \cdot 1s)_u \, (H^+H^-) \, {}^1\Sigma_u^+$	$(1s \cdot 2di)_u \quad (1s \cdot 2p\sigma)_u$	$(1s \cdot 4f\sigma)_u$
16. $(1s \times 1s)_g \, (H^+H^-) \, {}^3\Sigma_g^+$	$(1s \times 2di)_g \quad (2p\sigma \times 2p\sigma)_g$	

Notes:

1) The last two states are asymptotically ionic only if crossings with higher n=3 covalent states are ignored.

2) States 7 and 14 possess double minima because of double crossing with $(1s \times 1s)_g \, (H^+H^-)$.

3) States 6 and 11 exhibit dispersion-force maxima at large r.

4) $(1s2di') \equiv \frac{1}{\sqrt{2}} \left[1s2s - 1s2p\sigma \right]$

Fig. 12.

IV.   RECENT RESULTS AT INTERMEDIATE ENERGIES.

Close-coupled AS calculations have been extensively developed for H^+-H(1s) collisions since the pioneering attempt of Wilets and Gallaher.[59] Unreal structure in and non-reproducibility of some early results have ushered in a new era of concern about numerical accuracy in the computer integrations that trace the coupled amplitudes together through the collision. The old test of

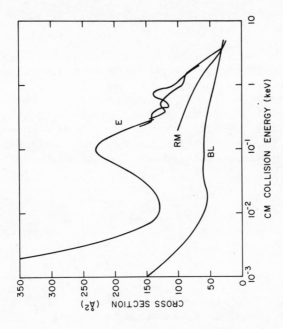

Fig. 14 Theory and experiment (E) for $H^+ + H^-$ → 2H. The data are from references 56–58, and the theoretical curves are BL: Ref. 55, and RM: Ref. 54.

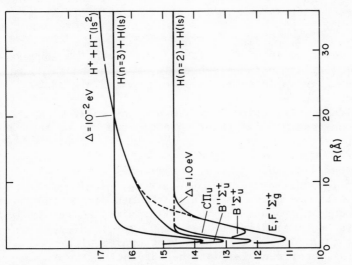

Fig. 13 Some potential energy curves for the $H^+ + H^-$ → 2H problem. Recent values for the exchange splittings Δ for diabatic states are given along with some adiabatic curves at smaller R.

conservation of total probability is inadequate, and recently detailed-balancing checks have been found better but still not always a guarantee of accuracy.[21]

It was early recognized that an expansion in a small number of separated-atom AS states would not lead to good wavefunctions at small R. Schemes to incorporate some of the MS small R capability into AS theory continue to be investigated. The idea is to introduce non-physical states into the expansion that are orthogonal to the physical (hydrogenic) ones, but yet have a large overlap with the united atom wavefunctions. A quite successful combination of 3s, $3p_0$ and $3p_{\pm1}$ pseudostates with radial parts

$$\overline{3s} = \sqrt{\frac{600}{17}}\, e^{-r}(1 - \frac{16}{15}\, r + \frac{1}{5}\, r^2)$$

$$\overline{3p} = \frac{1}{2}\sqrt{\frac{5}{6}}\, e^{-r/2}\, (1 - \frac{1}{5}\, r)$$

together with the hydrogenic 1s, 2s, $2p_0$ and $2p_{\pm1}$ states has lead to the best $H^+ +H(1s)$ AS close-coupling results yet.[61] Some care in the choice of pseudostates is needed if final physical probabilities are not asymptotically time-dependent.[62] Sturmian states also are still of interest,[62,] as well as polarized orbital basis states[64] similar to those used in electron scattering to account for static atomic polarization induced by the incoming charged particle. Another approach is to let the effective charge Z of the hydrogenic states vary towards the united atom limit as R decreases.[65]

A practical problem in AS calculations has been the long range character of the coupling between sublevels of an n-manifold, which forces careful small-step integration out to quite large $R\sim20a_0$ for $H^+ +H$ electron transfer into the n=2 states.[42] The possibility of using spheroidal wavefunctions[67] which are the correct AS limits of the MS wavefunctions remains to be investigated beyond the case of simple target excitation.[68]

Several recent (1s, 2s, $2p_0$, $2p_{\pm1}$) $H^+ +H$ AS-4 calculations[60,61,66,42] in addition to the pseudostate calculation(AS-7P)[61] can be compared with the data, especially the new direct and transfer data for 2s and 2p production.[63] Similarly new AS results for $He^{++}-H(1s)$ collisions including [$He^+(2s, 2p_0, 2p_{\pm1})$, $H(1s)$][47,19] and in addition H(2s), $2p_0$, $2p_{\pm1})$[48] can be compared with recent data[44,45] especially for transfer into $He^+(2s)$.

Figure 15 shows the good agreement for the H^+-H case, with even oscillations in the 2s transfer to direct cross section ratio

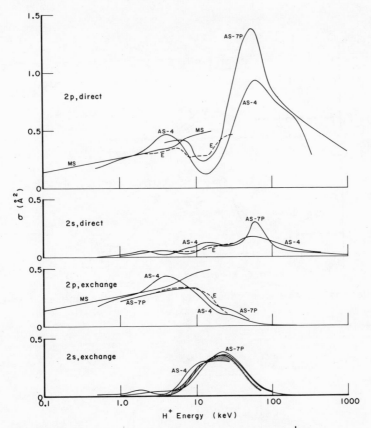

Fig. 15 Theory and experiment (E, ref. 63) for $H^+ + H \rightarrow H(n=2)$ transfer and direct excitation. Theoretical curves are AS-4: Ref. 42,
AS-7P: Ref. 61; MS: Ref. 33.

both predicted and observed (see Figure 16). A long-standing
problem with AS calculations has been a maximum in the 2s transfer
cross section below 6 keV which is sensitive to details at small
R. Pseudostates improve the situation, but MS results best agree
with experiment[69] (see Figure 17).

The $He^+(2s)$ production question in He^{++}-H(1s) collisions is in
a state of considerable disarray, as can be seen in Figure 18. The
two sets of experimental data disagree at their higher energies,
and no theoretical result seems correct, AS, MS or otherwise. The
differences in theoretical predictions become even greater in
the case of the highly non-resonant transfer into $He^+(1s)$, which is
sensitive to small amounts of coupling with the favored $He^+(n=2)$
and He^{++}-H(1s) channels (see Figure 5 and reference 71).

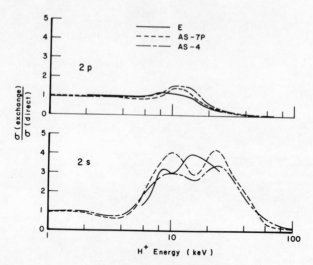

Fig. 16 Ratios of the corresponding direct and transfer curves of
Figure 15.

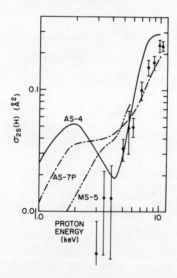

Fig. 17 Low Energy Cross Section for $H^+ + H(1s) \rightarrow H(2s) + H^+$. As in
Figure 16, except MS-5: Ref. 39 and the data points are from Ref.
69.

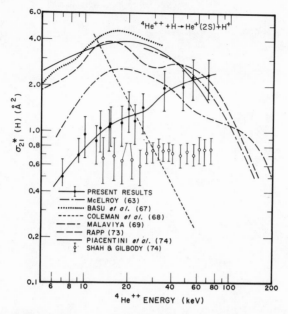

Fig. 18.   Theory versus experiment for He[+](2s) production in He[++]-
H(1s) collisions. The data are "present results", ref. 71
uncertain by ±40%; and "Shah and Gilbody," ref. 45, uncertain by
a factor of two. Using AS theory are "McElroy", ref. 46; "Malaviya",
ref. 47, and "Rapp", ref. 48. The MS calculation of Piacentini and
Salin (ref. 34) is not expected to properly divide up the He[+](2s)
and He[+](2p) fluxes because of difficulties with momentum transfer
factors. A molecular-like variational calculation of Basu et al.
also ignore these factors (ref. 73). The curve of Coleman et al.
(ref. 74) comes from a high-energy impulse approximation.

        A considerable amount of work also has been done on He[++]-He
collisions, where the experiments are easier but good two-electron
theory much harder. Figure 19 compares recent data for total one
and two electron transfer, while comparisons with theory are made
in Figures 20 and 21. The single transfer cross section is difficult to calculate because no scattering channel is close to resonant, although He[+](n=2) is somewhat favored at intermediate energies.

        Recently some data has been obtained for the He[+](2s) part of
the He[++]-He single electron transfer cross section.[84,85] The
factor of three difference in cross section scales between the Yale
and Belfast results is barely within the respective ±40% and ± a
factor of two quoted uncertainties dominated by He[+] Lyman-alpha
detection efficiency problems. Nevertheless published theoretical
values for this partial cross section are limited to a CDW result

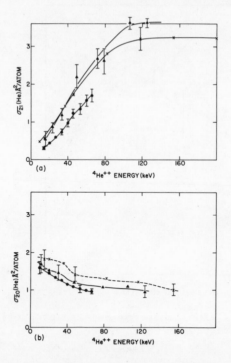

Fig. 19. Data for the single (σ_{21}) and double (σ_{20}) electron transfer cross sections for He^{++} + He: X, ref. 75; ▲ ref. 76, ● ref. 77.

of 81 $\overset{o}{A}{}^2$ at 25 keV, two orders of magnitude too large.[83]

Kinetic energy loss or gain techniques for resolving final
states of different principal quantum number in collisions such as
He^{++}+He → He^{+}(n)+He^{+} recently have been used not only for total
cross section measurements [85-86] but also in differential scattering studies. [87] An interesting variation of this is "double electron
transfer translational spectroscopy" where kinetic energy losses
in the formation of various neutral atoms are detected by first
attaching an electron via a second electron transfer collision and
then energy-analyzing the resultant negative ions. Thus the processes
H^{+}+He → H(1s,2p)+He^{+} can be separated from one another even if the
H(2p) atoms radiatively decay back to the 1s states. [88]

A great deal of experimental work has gone into the study of
single electron transfer in the H^{+}+He(1s^2) system. Available
theoretically are two-state [H^{+}+He(1s^2), H(1s)+He^{+}(1s)][89,90] and five
state [H^{+}+He(1s^2), H(1s,2s,2p$_o$,2p$_{\pm1}$)+He^{+}(1s)][91] AS results as well
as a two-state MS result. [92] A look at the HeH^{+} molecular potential

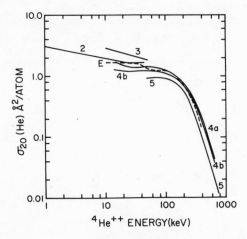
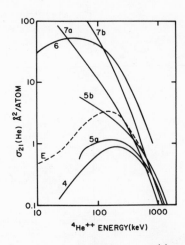

Figs. 20 and 21.   Double and single electron transfer in He^{++}+He, theory and experiment (E).   The theoretical curves are: 2, two-state MS of ref. 78; 3, two-state variational, ref. 79; 4, three-state [(He^{++}-He(1s^2), He$^+$(1s)+He$^+$(1s)] AS from ref. 80 with two different approximate matrix elements; 5, nine-state [He^{++}-He(1s^2), He$^+$(1s, 2s, 2p$_0$, 2p$_{\pm 1}$) + He$^+$(1s)] AS from ref. 81 with 5a being the He$^+$(1s) part of σ_{21} and 5b being the sum over n=1 and 2 final states; 6, high-energy OBK approximation of ref. 75; 7, high-energy CDW approximation (see section V) of ref. 83, with 7a again the He$^+$(1s) part of σ_{21} and 7b the total over n=1 and 2.

energy curves[93] shows this collision system to be an ideal model case where all electron transfer channels have large internal energy defects increasing with the principal quantum number of the final H-atom state.   The cross sections are of course all small, especially at lower energies.

New total cross section data for H$^+$+He → H(2s,2p)+He$^+$ have been reported,[94-96] and concern about corrections arising from the polarization of the detected Lyman-alpha radiation[97] is often expressed in discussing differences in 2s results.   However, such light detection problems should not affect the ratio of cross sections for two different target gases, shown for He and Ar in Figure 23.   Clearly other aspects of such experiments (such as collecting all of the H(2s) scattered atoms) are not always completely under control.   Thus critical data reviews such as that by Thomas[5] are of great value.

The total H$^+$+He cross sections for transfer into 3s, 3p, 3d and 4s have been measured by observing the spatial optical decay distributions downstream from the gas target.   It is interesting to

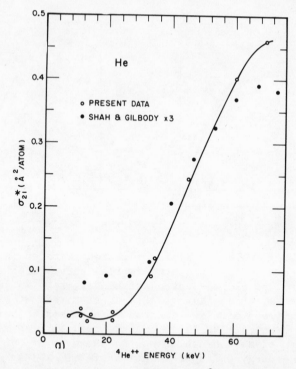

Fig. 22. The cross section for He^{++}+He(1s^2) → He$^+$(2s)+He$^+$. "Present
data" is from ref. 84; the "Shah and Gilbody" ^{3}He^{++} data reported
in ref. 77 is scaled for equivalent collision velocity.

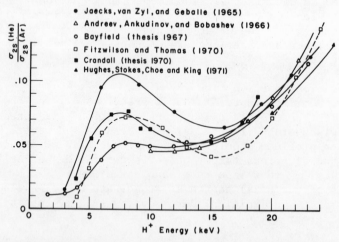

Fig. 23. The ratio of cross sections for electron transfer into
H(2s) for H$^+$ incident on He and Ar gases. The references in the
listed order are numbers 98, 99, 100, 94, 95 and 96.

compare the data with the n and ℓ dependence of the cross section for $H^+ + H(1s) \rightarrow H(n\ell) + H^+$ as predicted by the high-energy Brinkman-Kramers (OBK) approximation:[101-102]

$$\sigma(n\ell) = \frac{2^{18}}{5n^3} \frac{v^8}{[v^2+(\frac{n-1}{n})^2]^5 [v^2+(\frac{n+1}{n})^2]^5} K_{n\ell}(v)\ \pi a_o^2,$$

with $K_{n\ell}(v) = \frac{5}{5+\ell} \frac{2^{2\ell}}{(2\ell+1)!!(2\ell-1)!!} [\frac{4v^2}{v^4+2(\frac{n^2+1}{n^2})v^2+(\frac{n^2-1}{n^2})^2}]^\ell$

and the collision velocity v is in atomic units. These equations suggest the rule of thumb that $\sigma(n\ell)$ varies as n^{-3} and is dominated by $\ell=0$ and 1. Figures 24 and 25 confirm this for $H^+ + He$ collisions.

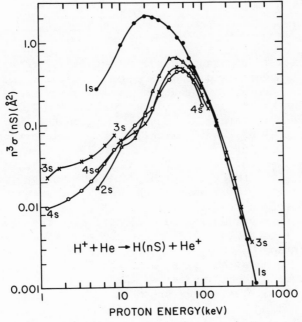

Fig. 24. Experimental cross sections for $H^+ + He \rightarrow H(nS) + He^+$ weighted by the factor n^3. The low energy data is from ref. 103, that at intermediate energies from refs. 104 and 105, and the high energy 3s data from ref. 106. The "Jackson-Schiff" OBK n^{-3} scaling rule is good for the excited S states at all energies, as well as for n=1 at high energies.

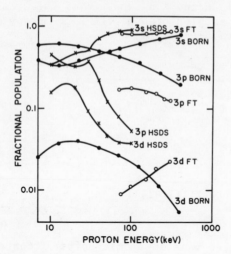

Fig. 25. Fractional population of the H(n=3) sublevels in H^++He collisions. The data of refs. 104 (HSDS) and 106 (FT) are compared with calculations in the First Born approximation, ref. 107.

Differential electron-transfer scattering experiments in H^++He collisions have been conducted at intermediate energies and scattering angles of a few degrees. The 2s and 2p components have been studied in addition to the total transfer probability P_o.[108-110] The earlier P_o data of Everhart is confirmed, with the incorrect oscillation amplitudes of the AS calculation[91] and incorrect phase of the MS work[92] remaining as problems within H^+-He theory. An interesting result is that the 1s, 2s and 2p electron transfer probabilities oscillate together as the collision energy is varied while the scattering angle is held fixed. (The quasiresonant nature of H^+-He electron transfer collisions is also discussed by W. Lichten within this volume.) The explanation of the low energy substructure in the total 2s and 2p cross sections (see Figure 24) lies in n=2 population occuring in a second transition after electron transfer to H(1s) has first occurred.[70] The same intermediate state argument should apply for the n=2 differential data, although the nature of the needed intermediate state (s) is not yet known.

Recent interest has developed in electron transfer into highly-excited states, both as production mechanisms for big atoms needed for other kinds of physics and for further information on cascade contributions to measured total electron-transfer cross sections. Figure 26 compares recent data with some high-energy theory results.

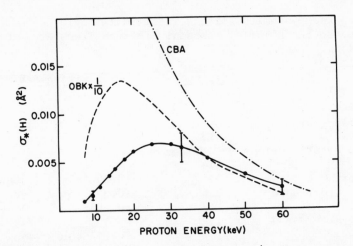

Fig. 26.   The sum of cross sections for H^++H(1s) electron transfer into the band of highly excited states H(n), $13\leq n \leq 28$.   The data points are from ref. 111, the very large Brinkman-Kramers (OBK) results from ref. 102, and the orthogonalized First Born results (CBA) from ref. 82.

V.   DEVELOPMENTS AT HIGH ENERGY

Great theoretical interest continues in the high energy behavior of the H^+-H(1s) electron transfer cross section (see Figure 3).[2,4] Unfortunately no data exists for the $E \gtrsim 500$ keV energy region of interest.   Table I lists predicted high energy limiting behaviors for the many non-relativistic high energy approximations.   A total cross section experiment to differentiate between these predictions would require state-of-the-art precision in addition to a capability to measure very small cross sections. Alternatively very high resolution differential scattering would distinguish between the various theories.   A cancellation of contributions to the transfer amplitude occurs in the First Born approximation which combines parts arising from the electron-proton and proton-proton Coulomb interactions.   The resultant zero in the differential cross section disappears to varying degrees when approximations of higher order in one interaction or the other are employed. (See Figure 27).

Table I. High Energy Limits for the Total Small Angle
   Contribution to the H⁺+H(1s) Resonant Electron Transfer
   Cross Section.

Approximation	Reference	Limiting Behavior
"Bates" Two-State AS	72	σ_{OBK}
Brinkman-Kramers	112	"
First Order Born	113	$0.661\ \sigma_{OBK}$
Orthogonalized First Order Born	114	"
Second Order Born	115	$(0.2946+5\pi v\,2^{-12})\sigma_{OBK}$
Estimated Third Order Born	116	$(0.319\ +5\pi v\,2^{-12})\sigma_{OBK}$
Impulse	117	$(0.2946+5\pi v\,2^{-11})\sigma_{OBK}$
Pradhan's Impulse	118	σ_{OBK}
First Order CDW	119	$(0.2946+5\pi v\,2^{-11})\sigma_{OBK}$
Second Order CDW	119	$(0.2946+5\pi v\,2^{-12})\sigma_{OBK}$
Geltman's Distorted Wave	120	$0.810\ \sigma_{OBK}$
First Order FWMS	121	$0.661\ \sigma_{OBK}$

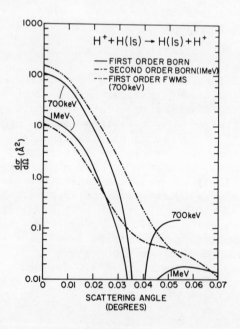

Fig. 27. Theoretical high-energy predictions for the small angle
differential scattering for resonant H⁺+H(1s) electron transfer.
The first order Born curves are from refs. 115 and 121, the second
order Born curve from ref. 115, and the first order Faddeev-
Watson multiple scattering (FWMS) approximation result from
ref. 121.

By CDW in the table we mean the continuum distorted wave approximation introduced by Cheshire.[119] This approach takes into account the proper boundary conditions for the high-energy transfer problem; these lead to asymptotic phases not included in the plane-wave First Born approximation. The CDW distorted waves are $\Psi_i = \phi_i L'_i$, where the ϕ_i are the usual AS traveling atomic orbitals of equations (2) and (3), and the L'_i are approximate solutions for that part of the Hamiltonian not diagonalized in obtaining the ϕ_i. The L'_i are exact solutions for the reduced problem except for the neglect of one gradient (∇_r) term. For the target atom state they look like

$$L'(s,R,t) = \exp(\pi/2v)\Gamma(1-i/v) \, {}_1F_1 \, [i/v,1,i(vs+v\cdot s)]\cdot$$

$$\cdot \exp[-(i/v)\ln(vR-v^2 t)]$$

where s is the electron position vector relative to the target atom nucleus and ${}_1F_1$ is a hypergeometric function. A good number of CDW calculations have been made in the last few years, with considerable success in an energy region between intermediate and high; examples are the cross sections for $H^+ + H(1s) \rightarrow H(2s) + H^+$[122] and $H(1s) + H(1s) \rightarrow H^- + H^+$.[123] Comparison with experiment for the H+H process is made in Figure 28.

VI. NEW APPLICATIONS OF SIMPLE ELECTRON TRANSFER
PROCESSES

Specific electron transfer processes are being used increasingly for the favored production of excited atomic states. Brute force methods such as electron impact are rarely as efficient. There are also special cases of interest in ground state as well as excited state processes, such as fast muonium atom production for possible future fine and hyperfine structure measurements, see Figure 29.

Two applications proposed recently involve a preferred formation of $He^+(n=2)$ ions. The first is an extension of known electron-transfer techniques using the chain $H^+ + Cs \rightarrow H(2s)$, $H(2s) + X \rightarrow H^\pm$ with the intermediate H(2s) being first electronic – and then nuclear-spin polarized using spin-selective, hyperfine transitions in externally applied fields.[129] For spin-polarized ^3He nuclei the chain is now $He^{++} + X_1 \rightarrow He^+(2s)$, $He^+(2s) + X_2 \rightarrow He^{++}$ where X_1 at present seems optimally chosen as H(1s) below 10keV

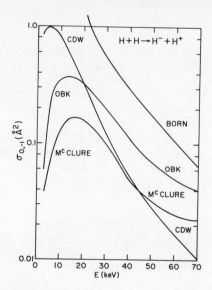

Figure 28. The cross section for H(1s)+H(1s) → H⁻+H⁺. The experimental curve is due to McClure, ref. 125; the OBK and CDW curves
are from ref. 123, and the First Born calculation is from ref. 124.
One unusual feature is the higher value of the Born result over
that of OBK.

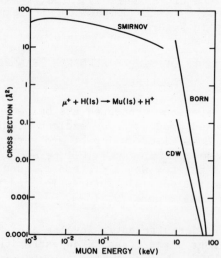

Figure 29. Calculated cross sections for ground state muonium Mu(1s)
production. Ref. 126 uses Smirnov's low-energy approximation for
nearly-resonant processes. The high energy First Born and CDW
calculations are from refs. 127 and 128. The Born and CDW workers
both report fractional Mu(2s) formation of 10% to 20% in the indicated energy range. On a equivalent velocity scale the predicted
cross sections for incident H⁺ and μ⁺ are similar, but with the
muonium formation up to a factor of two smaller.

and He around 60 keV (see Figure 31). A good stripper favoring
$He^+(2s)$ over $He^+(1s)$ is N_2 gas, although many targets may work
here.[130] Some of the data and theory of Figures 18 and 11 are
combined in Figure 30 to show curves for fractional electron
transfer into the 2s state in He^{++} + H(1s) collisions. Clearly
the theory does not yet appear reliable here, at least for
separating out the 2s production from the 2p part. We note that
specific electron-transfer collisions offer possibilities not only
for use in polarized ion sources, but also in the design of
polarimeters to measure beam polarization at keV energies.[131]

The second $He^+(n=2)$ application is in the pumping of a soft
X-ray laser. $He^+(2p)$ decays with the emission of Lyman alpha
radiation at 304 A. A proposed laser[132] using He^{++}+H(1s) → $He^+(2p)$
utilizes a scheme by McCorkle[133] to scan an incident He^{++} beam along a
target H(1s) thermal beam at the speed of light. Thus the
$He^+(2p)$ ions needed to build up a traveling uv laser pulse are
continually created where they are needed, at the pulse wavefront.
The theories of Figure 30 all predict population inversion of
$He^+(2p)$ over $He^+(1s)$ in the low to intermediate keV energy range,
but no direct experimental verification of this has been made. As
a totally stripped ion of charge Z incident on an H(1s) target

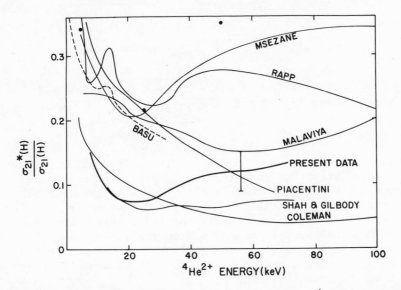

Figure 30. Fractional electron transfer into $He^+(2s)$ in He^{++}+H(1s)
collisions, theory and experiment. The data are: "Present data",
refs. 44 and 71; Shah & Gilbody, ref. 45. Theoretical curves are:
"Piacentini", an MS result of ref. 34; Basu, ref. 73, Malaviya,
ref. 47; Rapp, ref, 48; "Msezane", ref. 19, and in the impulse
approximation, "Coleman", ref. 74.

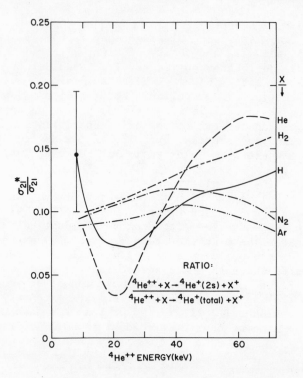

Figure 31.   Fractional electron transfer into $He^+(2s)$ for He^{++} incident on various gases, from refs. 44, 71 and 84.

undergoes accidentally resonant electron-transfer into a hydrogenic state with n=Z, extension of this X-ray laser idea to higher photon energies may be possible. No data or theory exists for Z > 2.

A final application to be mentioned is the formation of fast beams of excited atoms and ions for use in precision measurement of atomic fine structure and Lamb shifts[134], as well as for multiphoton ionization studies of loosely-bound atoms in what for them can be extremely high externally-applied electromagnetic fields.[135] For the bound-bound two photon transition case the Doppler effect can be exactly cancelled for fast atoms at microwave frequencies by using a standing wave field, as is done for thermal atoms at laser frequencies.[136,137]

Although atomic hydrogen targets[138] are less convenient to use than a usual gas collision target, the advantages of having only one electron available for electron transfer provides a higher level of final beam purity that may be helpful in a number of these applications.

REFERENCES

1. J.P. Coleman and M.R.C. McDowell, Introduction to the Theory of Ion-Atom Collisions, Amsterdam, North-Holland (1970).

2. R.A. Mapleton, Theory of Charge Exchange, New York, Wiley-Interscience (1972).

3. B.H. Bransden, Reports on Progress in Physics $\underline{35}$, 949 (1972).

4. J.C.Y. Chen, Case Studies in Atomic Physics $\underline{3}$, 305 (1973).

5. E.W. Thomas, Excitation in Heavy Particle Collisions, New York, Wiley-Interscience (1972).

6. J.B. Hasted, Physics of Atomic Collisions, New York, American Elsevier (1972), section 12.2.

7. V.A. Belyaev, B.G. Brezhnev and E.M. Erastov, Sov. Phys. JETP $\underline{25}$, 777 (1967).

8. W.L. Fite, A.C.H. Smith and R.F. Stebbings, Proc. Roy. Soc. London $\underline{A268}$, 527 (1962).

9. G.W. McClure, Phys. Rev. $\underline{148}$, 47 (1966).

10. A.B. Wittkower, G. Ryding and H.B. Gilbody, Proc. Phys. Soc. $\underline{89}$, 541 (1966).

11. J.E. Bayfield, Phys. Rev. $\underline{185}$, 105 (1969).

12. R. McCarroll, Proc. Roy. Soc. London $\underline{A264}$, 547 (1961).

13. A.F. Ferguson, Proc. Roy. Soc. London $\underline{A264}$, 540 (1961).

14. F.J. Smith, Proc. Phys. Soc. $\underline{84}$, 889 (1964); $\underline{92}$, 866 (1967).

15. J.R. Oppenheimer, Phys. Rev. $\underline{32}$, 361 (1928).

16. J.D. Jackson and H. Schiff, Phys. Rev. $\underline{89}$, 359 (1953); D.R. Bates and A. Dalgarno, Proc. Phys. Soc. $\underline{A65}$, 191 (1952).

17. G. Hunter, B.F. Gray and H.O. Pritchard, J. Chem. Phys. $\underline{45}$, 3806 (1966).

18. R.T. Pack and J.O. Hirschfelder, J. Chem. Phys. $\underline{49}$, 4009 (1968); $\underline{52}$, 521 (1970), $\underline{52}$, 4198 (1970); A.M. Halpern, Phys. Rev. $\underline{186}$, 14 (1969); H. Laue, J. Chem. Phys. $\underline{46}$, 3034 (1967).

19. A. Msezane and D.F. Gallaher, J. Phys. B6, 2334 (1973).

20. L.A. Parcell, J. Phys. B5, 1478 (1972); Computer Physics 5,283
     (1973);D.F.Gallaher and A. Msezane,J.Phys. B7, 127 (1974).

21. H. Tai and E. Gerjouy,J. Phys. B6, 1426 (1973); E. Gerjouy,
     Case Studies in Atomic Physics 3, 1 (1972).

22. D.R. Bates and R. McCarroll, Proc. Roy. Soc. London A245,
     125 (1958).

23. R.A. Mapleton, Proc. Phys. Soc. 83, 895 (1964).

24. I.M. Cheshire, Proc. Phys. Soc. 84, 89 (1964).

25. R. Shakeshaft, J. Phys. B6, 2315 (1973).

26. Y. Band, Phys. Rev. A8, 243 (1973).

27. L.D. Faddeev, Sov. Phys. JETP 12, 319 (1961); A.M. Brodskii,
     V.S. Potapov and V.V. Tolmachev, Sov. Phys. JETP 31, 144
     (1970).

28. M. Barat and W. Lichten, Phys. Rev. A6, 211 (1972).

29. D.R. Bates and R.H.G. Reid, Adv. Atomic Molec. Physics 4,
     13 (1968); M.M. Madsen and J.M. Peek, Atomic Data 2, 171
     (1971).

30. K. Helfrich and H. Hartman, Theoret. Chim. Acta (Berlin)
     16, 263 (1970); L.Y. Wilson and G.A. Gallup, J. Chem.
     Phys. 45, 586 (1966).

31. T.E. Sharp, Atomic Data 2, 119 (1971).

32. L.I. Ponomarev and T.P. Puzynina, Sov. Phys. JETP 25, 846
     (1967); I.V. Komarov and S. Yu. Slavyanov, J. Phys. B2,
     1066 (1968).

33. H. Rosenthal, Phys. Rev. Letters 27, 635 (1971); Proc. VIIth
     ICPEAC, Amsterdam, North-Holland (1971), page 105.

34. R.D. Piacentini and A. Salin, J. Phys. B (1974).

35. D.R. Bates and D. Sprevak, J. Phys. B4, L47 (1971).

36. J.C.Y. Chen, V.H. Ponce and K.M. Watson, J. Phys. B6, 965
     (1973).

37. S.K. Knudson and W.R. Thorson, Can. J. Phys. $\underline{48}$, 313 (1970).

38. R. McCarroll and R.D. Piacentini, J. Phys. $\underline{B3}$, 1336 (1970).

39. M.C. Chidichimo-Frank and R.D. Piacentini, J. Phys. $\underline{B7}$, 548 (1974).

40. D.R. Bates and R.H.G. Reid, J. Phys. $\underline{B2}$, 851 (1969).

41. J.C. Houver, J. Fayeton, M. Abignoli and M. Barat, Phys.Rev. Letters $\underline{28}$, 1433 (1972); J.C. Houver, J. Fayeton and M. Barat, J. Phys. $\underline{B6}$, 2118 (1973); also J. Phys. $\underline{8}$ (1974).

42. C. Gaussorges and A. Salin, J. Phys. $\underline{B4}$, 503 (1971).

43. J. Bayfield, Phys. Rev. Letters $\underline{25}$, 1 (1970).

44. J.E. Bayfield and G.A. Khayrallah, "Direct Experimental Test of the Close-Coupling Calculations for $^{4}He^{++}$ - H(1s) Collisions," Bull. Am. Phys. Soc. $\underline{18}$, 1516 (1973) and to be published.

45. M.B. Shah and H.B. Gilbody, J. Phys. $\underline{B7}$, 630 (1974).

46. M.B. McElroy, Proc. Roy. Soc. London $\underline{A272}$, 542 (1963).

47. V. Malaviya, J. Phys. $\underline{B2}$, 843 (1969).

48. D. Rapp, J. Chem. Phys. $\underline{58}$, 2043 (1973).

49. H. Schiff, Can. J. Physics $\underline{32}$, 393 (1954).

50. P.M. Koch and J.E. Bayfield, "Merged Beams Measurement of Electron Loss in eV Energy Collisions of Highly-Excited Atoms with Protons" Abs. IV ICAP, Heidelberg (1974), p. 495

51. B.M. Smirnov, Sov. Phys.-JETP $\underline{19}$, 692 (1964).

52. R.S. Mulliken, Jour. Am. Chem. Soc. $\underline{88}$, 1849 (1966); Phys. Rev. $\underline{120}$, 1674 (1960).

53. R.K. Janev and A. Salin, J. Phys. $\underline{B5}$, 177 (1972).

54. K. Roy and S.C. Mukherjee, Phys. Rev. $\underline{A7}$, 130 (1973).

55. D.R. Bates and J.T. Lewis, Proc. Phys. Soc. $\underline{A68}$, 173 (1955).

56. J. Moseley, W. Aberth and J.R. Peterson, Phys. Rev. Letters
     24, 435 (1970).

57. R.D. Rundel, K.L. Aitken and M.F.A. Harrison, J. Phys. B2,
     954 (1969).

58. T.D. Gaily and M.F.A. Harrison, J. Phys. B3, L25 (1970).

59. L. Wilets and D.F. Gallaher, Phys. Rev. 147, 13 (1966).

60. D. Rapp and D. Dinwiddie, J. Chem. Phys. 57, 4919 (1972).

61. I.M. Cheshire, D.F. Gallaher and A.J. Taylor, J. Phys. B3,
     813 (1970).

62. D. Storm and D. Rapp, Phys. Rev. A8, 1784 (1973); J.C.Y. Chen
     and T. Ishihara, Phys. Rev. 186, 25 (1969).

63. T.J. Morgan, J. Geddes and H.B. Gilbody, J. Phys. B6, 2118
     (1973).

64. M.R. Flannery, J. Phys. B3, 798 (1970).

65. I.M. Cheshire, J. Phys. B2, 428 (1968).

66. C. Gaussorgues, R. McCarroll, P.D. Piacentini and A. Salin,
     J. Phys. B3, L138 (1970).

67. C.A. Coulson and A. Joseph, Proc. Phys. Soc. 90, 887 (1967).

68. P.D. Robinson, Proc. Phys. Soc. 71, 828 (1958).

69. J.E. Bayfield, Phys. Rev. Letters 20, 1223 (1968), and
     reference 11.

70. I.A. Poluektov and L.P. Presnaykov, Sov. Phys.-JETP 27,
     67 (1968).

71. J.E. Bayfield and G.A. Khayrallah, "Measurement of the Total
     Cross Section for Electron Transfer into the Metastable
     $He^+(2s)$ State for $^4He^{++}$ Collisions with Atomic Hydrogen,"
     Bull. Am. Phys. Soc. 18, 1516 (1973), and to be published.

72. R. McCarroll, Proc. Roy. Soc. London A264, 547 (1961).

73. D. Basu, D.M. Bhattacharya and G. Chatterjee, Phys. Rev. 163,
     8 (1967).

74. J.P. Coleman and S. Trelease, J. Phys. $\underline{B1}$, 172 (1968).

75. K.H. Berkner, R.V. Pyle, J.W. Stearns and J.C. Warren, Phys. Rev. $\underline{166}$, 44 (1968).

76. J.E. Bayfield and G.A. Khayrallah, Proc VIIIth ICPEAC, Belgrade, 1973, and to be published.

77. M.B. Shah and H.B. Gilbody, J. Phys. $\underline{B7}$, 256 (1974).

78. A.F. Ferguson and B.L. Moiseiwitsch, Proc. Phys. Soc. London $\underline{74}$, 457 (1959).

79. D. Basu, S.C. Mukherjee and N.C. Sil, in Atomic Collision Processes, edited by M.R.C. McDowell, Amsterdam, North-Holland (1964), page 769.

80. M.J. Fulton and M.H. Mittleman, Proc. Phys. Soc. $\underline{87}$, 669 (1966).

81. S.C. Mukherjee, K. Roy and N.C. Sil, J. Phys. $\underline{B6}$, 467 (1973).

82. Y. Band, Phys. Rev. $\underline{A8}$, 2857 (1973); Bull. Am. Phys. Soc. $\underline{19}$, 87 (1974), paper JE2.

83. D.S. Belkic and R.K. Janev, J. Phys. $\underline{B6}$, 1020 (1973).

84. G.A. Khayrallah and J.E. Bayfield, "Measurement of the Total Cross section for Production of $He^+(2S_{1/2})$ Metastable Ions in $^4He^{++}$ Collisions with He, H_2, Ar and N_2 Targets," Bull. Am. Phys. Soc. $\underline{18}$, 1516 (1973), and to be published.

85. J.T. Park and F.D. Schowengerdt, Rev. Sci. Instr. $\underline{40}$, 752 (1969).

86. J.H. Baynon and R.G. Cooks, J. Phys. $\underline{E7}$, 10 (1974).

87. M.W. Siegel, Y.H. Chen and J.W. Boring, Phys. Rev. Letters $\underline{28}$, 465 (1972).

88. J. Durup, J. Appell, F.C. Fehsenfeld and P. Fournier, J. Phys. $\underline{B5}$, L110 (1972).

89. B.H. Bransden and L.T. Sin Fai Lam, Proc. Phys. Soc. $\underline{87}$, 653 (1966).

90. T.A. Green, H.E. Stanley and Y.C. Chaing, Helv. Physica Acta $\underline{38}$, 109 (1965).

91. L.T. Sin Fai Lam, Proc. Phys. Soc. 87, 653 (1966).

92. R.K. Colegrave and D.B.L. Stephens, J. Phys. B1, 856 (1968).

93. H.H. Michels, J. Chem. Phys. 44, 3834 (1966).

94. R.L. Fitzwilson and E.W. Thomas, Phys. Rev. A3, 1305 (1971).

95. D.H. Crandall and D.H. Jaecks, Phys. Rev. A4, 2271 (1971).

96. R.H. Hughes, E.D. Stokes, S.S. Choe and T.J. King, Phys.
     Rev. A4, 1453 (1961).

97. I.A. Sellin, J.A. Biggerstaff and P.M. Griffin, Phys. Rev.
     A2, 423 (1970); J.W. Wooten and J.H. Macek, Phys. Rev. A5,
     137 (1972).

98. D. Jaecks, B. Van Zyl and R. Geballe, Phys. Rev. 137, A340
     (1965).

99. E.P. Andreev, V.A. Ankudinov and S.V. Bobashev, Sov. Phys.-
     JETP 23, 375 (1966).

100. J.E. Bayfield, Ph.D. Dissertation, Yale University, 1967,
     unpublished.

101. J.R. Hiskes, Phys. Rev. 137, A361 (1965); K. Omidvar, Phys.
     Rev. 153, 121 (1967).

102. S.T. Butler, R.M. May and I.D.S. Johnston, Phys. Letters 10,
     281 (1964); S.T. Butler and I.D.S. Johnston, Nuc. Fusion 4,
     196 (1964); R.M. May, Nuc. Fusion 4, 207 (1964); and also
     reference 49.

103. H.R. Dawson and D.H. Lloyd, Phys. Rev. A9, 166 (1974).

104. R.H. Hughes, C.A. Stigers, B.M. Doughty and E.D. Stokes,
     Phys. Rev. A1, 1424 (1970); R.H. Hughes, H.R. Dawson, B.M.
     Doughty, D.B. Kay and C.A. Stigers, Phys. Rev. 146, 53 (1966);
     R.H. Hughes, H.R. Dawson and B.M. Doughty, Phys. Rev. 164,
     166 (1967).

105. J.F. Williams and D.N.F. Dunbar, Phys. Rev. 149, 62 (1966);
     F.J. DeHeer, J. Schutten and H. Moustafa, Physica 32,
     1766 (1966),

106. J.C. Ford and E.W. Thomas, Phys. Rev. A5, 1694 (1972).

107. R.A. Mapleton, Phys. Rev. 122, 528 (1961); R.A. Mapleton,
      J. Phys. B1, 529 (1968); R.A. Mapleton, R.W. Doherty and
      P.E. Meehan, Phys. Rev. A9, 1013 (1974).

108. V. Dose, Helv. Physica Acta 41, 261 (1968).

109. R.H. McKnight and D.H. Jaecks, Phys. Rev. A4, 2281 (1971);
      and reference 95.

110. R.L. Fitzwilson and E.W. Thomas, Phys. Rev. A6, 1054 (1972).

111. J.E. Bayfield, G.A. Khayrallah and P.M. Koch, Phys. Rev.
      A9, 209 (1974).

112. H.C. Brinkman and H.A. Kramers, Proc. Acad. Sci. Amsterdam
      33, 973 (1930); also reference 15.

113. R.A. Mapleton, Proc. Phys. Soc. London 83, 895 (1964).

114. Y. Band, Phys. Rev. A8, 243 (1973).

115. P.J. Kramer, Phys. Rev. A6, 2125 (1972).

116. R.M. Drisko, thesis, Carnegie Institute of Technology,
      1955, unpublished.

117. I.M. Cheshire, Proc. Phys. Soc. 82, 113 (1963).

118. N. Urata and T. Watanabe, J. Phys. B4 L121 (1971).

119. I.M. Cheshire, Proc. Phys. Soc. 84, 89 (1964).

120. S. Geltman, J. Phys. B4, 1288 (1971).

121. J.C.Y. Chen and P.J. Kramer, Phys. Rev. A5, 1207 (1972);
      C.S. Shastry, A.K. Rajagopaland, J. Callaway, Phys. Rev.
      A6, 268 (1972).

122. A. Salin, J. Phys. B4, L125 (1971).

123. R.K. Janev and A. Salin, J. Phys. B4, L127 (1971).

124. R.A. Mapleton, Phys. Rev. 117, 479 (1960); Proc. Phys. Soc.
      85, 841 (1965).

125. G.W. McClure, Phys. Rev. 166, 72 (1968).

126. R.K. Janev and Dz. S. Belkic, J. Phys. B5, 1237 (1972).

127. A.H. Mousa and M.S. Abdel-Monem, Proc. Phys. Soc. 92,
      370 (1967).

128. Dz. S. Belkic and R.K. Janev, J. Phys. B6, 2613 (1973).

129. J.L. McKibben, G.P. Lawrence and G.G. Ohlsen, Phys. Rev.
      Letters 20, 1180 (1968).

130. W.E. Burcham, O. Karban, S. Oh and W.B. Powell, Nucl. Instr.
      and Methods 116, 1 (1974); O. Karban, S.W. Oh and W.B.
      Powell, Phys. Rev. Letters 31, 109 (1973).

131. J.E. Brolley, G.P. Lawrence and G.G. Ohlsen, "Proposal for
      a Lamb-shift Polarimeter", Los Alamos Laboratory Publication
      LA 4465-MS, 1970, pages 17.1 to 17.6.

132. M.O. Scully, W.H. Louisell and W.B. McKnight, Proc. VIII
      Inter. Quantum Electronics Conference, San Francisco, Calif.
      June 10-13, 1974, Paper M.9.

133. R.A. McCorkle, Phys. Rev. Letters 29, 982 (1972).

134. P.B. Kramer, S.R. Lundeen, B.O. Clark and F.M. Pipkin, Phys.
      Rev. Letters 32, 635 (1974); see also the paper by
      F.M. Pipkin in this volume, p. 119.

135. P.M. Koch and J.E. Bayfield, "Ionization of Highly Excited
      Atoms by Intense Electromagnetic Fields," Abstr. IV ICAP,
      Heidelberg, 1974, paper given at this conference.

136. L.S. Vasilenko, V.P. Chebotaev and A.V. Shishaev, JETP
      Letters 12, 113 (1970).

137. T.W. Hänsch, K.C. Harrey, G. Meisel and A.L. Schawlow, to
      be published.

138. J.E. Bayfield, Rev. Sci. Instr. 40, 869 (1969).

LOW ENERGY ELECTRON-MOLECULE SCATTERING

K. Takayanagi

Institute of Space and Aeronautical Science

University of Tokyo

Komaba, Meguro-ku, Tokyo, Japan

This paper is primarily concerned with the theoretical aspect of the electron-molecule scattering in the low-energy region (below about 10 eV).[1] Even in this limited range of energy, there are many kinds of processes involved. I shall confine myself to discussions of the elastic scattering and rotational transitions and a little bit of vibrational excitation in the electron-diatomic molecule scattering.

Before we understand any collision process, we have to know the nature of the relevant interaction. The electron-molecule interaction is usually divided into the short-range and the long-range parts. There is no definite way to separate these two parts, but this is a convenient way to start discussion. Very often the electrostatic interaction between the incident electron and the undeformed molecular charge distribution is adopted as the short-range part of the interaction. This of course joins smoothly with the long-range multipole interactions described below if the adopted charge distribution is sufficiently accurate. Even this interaction alone, theoretical calculation requires a large amount of numerical work unless we content ourselves with adopting a very rough estimation. (See, e.g., the calculations of potential for N_2 and CO. [2]) However, it is known in the study of rotational excitations of H_2 molecules that the electron exchange effect is essentially required to assure a quantitatively good agreement with experiment. This forces us to solve coupled integro-differential equations just as in the electron-atom collisions. It is a very hard task to do this in the multi-center problem under considera-

target molecules. Burke and his colleagues[3] introduced the
pseudo-potential method which approximately takes account of the
exchange effect. Furthermore, the deformation of the molecular
charge cloud, or more precisely the effect of correlation between
the incident electron and molecular electrons may be important.
At large distances, this is taken into account usually as the
polarization force. This adiabatic polarization force is often
used down to the closest distances. No investigation has been
done beyond this approximation, as far as I know, on the electron
correlation in electron-molecule scattering.

For an electron and a diatomic molecule, the asymptotic form
of the interaction potential is of the following form

$$V(\vec{r}, \hat{s}) \xrightarrow{r \to \infty} - \frac{De}{r^2} P_1(\hat{r} \cdot \hat{s}) - \frac{Qe}{r^3} P_2(\hat{r} \cdot \hat{s}) - \frac{(\alpha_\parallel + 2\alpha_\perp)e^2}{6r^4}$$

$$- \frac{2(\alpha_\parallel - \alpha_\perp)e^2}{6r^4} P_2(\hat{r} \cdot \hat{s}) - \dots \qquad (1)$$

where $\vec{r}$ is the position vector of the electron relative to the
center-of-mass of the molecule; $\hat{s}$ is the unit vector along the
molecular axis; D and Q are the dipole and quadrupole moments of
the molecule, respectively; and $\alpha_\parallel$, $\alpha_\perp$ are the electric polarizabilities of the molecule in the direction parallel and perpen-
dicular to its axis. The importance of these long-range forces
as compared with the short-range part of interaction in electron-
molecule scattering depends, of course, upon the magnitude of D,
Q, α's, etc. In the rotational transitions in the low-energy
electron scattering, it is often these long-range forces which
primarily determine the cross section as we shall see shortly. The
total cross section is usually determined by short-range forces.
But, in very low energy electron scattering from polar molecules,
the dipole interaction often gives the major contribution to the
total cross section.

In early days, only the short-range interaction was taken into
account in the theoretical study of electron-nonpolar molecule
scattering. In 1955, Gerjuoy and Stein[4] pointed out that the
rotational excitation due to the long-range quadrupole interaction
can account satisfactorily for the energy loss of thermal electrons
in gases such as N_2 and H_2. Their calculations are based on the
Born approximation. Since this approximation is usually not

applicable to low-energy scattering, it is interesting to investigate the reason for the success.   Thus, I made the partial-wave analysis.[5]   The Born cross section is decomposed into the partial cross sections, each term of which corresponds to a pair of definite orbital angular momenta for incident and outgoing electrons. It is found that the success of the Born approximation is, first of all, due to the long-range nature of the relevant interaction (quadrupole interaction in this case) and, secondly, due to the fact that the s wave, which is the only partial wave to be distorted considerably in the very low energy collision under consideration, does not contribute appreciably to the rotational transitions. The p wave, which interacts less strongly with the molecule, gives a dominant contribution.   This is due to the selection rule arising from the $P_2(\hat{r} \cdot \hat{s})$ factor associated with the quadrupole interaction.   In the scattering, the orbital quantum number ℓ of the electron should either change $\Delta \ell = \pm 2$ or remain unchanged ($\Delta \ell = 0$; $0 \to 0$ forbidden).   Thus, the s wave can couple only with the d wave which passes far outside the region of appreciable interaction.   The largest contribution to the rotational cross section thus comes from "p-wave to p-wave" scattering.   Since p wave passes the target at a distance where the interaction is still appreciable but not too large, so that the first-order perturbation treatment is applicable.   This is the reason why the Born approximation can be used in this problem.   It should perhaps be mentioned that the selection rule stated above is not applicable to a linear molecule with nonvanishing electronic angular momentum around its axis. In the following, the target molecule will be treated, for simplicity, as being in a $^1\Sigma$ state.

The momentum-transfer cross section for the electron scattering from a point quadrupole is approximately given by $\sigma_m = (16/45)$ $\pi a_0^2 (Q/\pi a_0^2)^2$ in the Born approximation.   For gases such as H_2, N_2 and O_2, this gives σ_m one or two orders of magnitude smaller than the values derived from electron-swarm experiments.   This fact indicates that the quadrupole interaction, which is most important in the rotational transitions by low-energy electrons is much less important in the elastic scattering.   Probably the s wave scattering from the short-range part of the interaction gives the most important contribution there.

As the collision energy increases, the p wave comes closer to the target, so that the distortion of this partial wave becomes appreciable.   Distorted-wave calculations[6] show that a large deviation from the Born result starts at about 0.1 eV in the case of N_2. For O_2, this occurs at a much lower energy, since the quadrupole moment of O_2 is much less than that of N_2.

One of the calculation methods which are expected to give more reliable cross sections is the close-coupling method.   In this method, the total wave function of the system is expanded as

$$\Psi = \sum_{\ell J \lambda \mu} \frac{1}{r} F_{\ell J}^{\lambda \mu}(r) \sum_{mM} (\ell \, J \, m \, M \mid \ell \, J \, \lambda \, \mu) \, Y_{\ell m}(\hat{r}) \, Y_{JM}(\hat{s}), \quad (2)$$

where the radial functions F's are determined by solving a set of
coupled differential equations.   If the molecular electrons are
explicitly taken into account, and Ψ is properly antisymmetrized
with respect to all the electrons in the system, we have to solve
a set of coupled integro-differential equations.   In any case, we
cannot solve a set of infinite number of coupled equations, so that
the truncation is inevitable.   For the electron scattering from H_2,
where the relevant interaction is fairly well known, close-coupling
calculations have been made by Henry and Lane[7] (earlier references
are found therein).   The results of their calculations are in close
agreement with experimental values.[8,9]   They also applied the same
method to calculate the rotational excitation of the D_2 molecule.[10]

The expansion (2) is appropriate especially for distant encounters, since the orbital angular momentum ℓ and the rotational
angular momentum J are approximately good quantum numbers there.
This is no more the case when the electron is in the vicinity of
the target molecule because of the strong interaction.   More
appropriate way to describe the electron motion relative to the
molecule in this region is to adopt the body-frame approach.
This is the treatment of scattering in the coordinate system fixed
to the target molecule.   In principle, this is equivalent to the
laboratory-frame formulation.   The wave function in the two frames
are related to each other by a unitary transformation.   Chang and
Fano[11] have developed a formalism where the laboratory frame is
used for large distances while the body frame is adopted at shorter
distances to take advantage of different simplifying factors in
different regions.

Henry and Chang[12] applied this frame-transformation theory to
electron scattering from H_2 and calculated pure vibrational and
simultaneous rotational-vibrational excitation cross sections.
The results are in good agreement with experiment.   In the calculation, however, they entirely neglected the rotational Hamiltonian.
Namely, since the vibrational transitions take place mainly in the
short electron-molecule distances, they connected the body-frame
solution in the inner region directly to the free electron wave
function outside the interaction region and obtained the scattering
amplitudes by an appropriate unitary transformation.   This approximation is equivalent to the adiabatic approximation (or fixed-nuclei
approximation), which was originally proposed by Chase[13] in a study
of neutron scattering in nuclear physics.   This approximation has

been used frequently in the low-energy electron-molecule scattering in recent years.

In the adiabatic approximation, the electron scattering from the target molecule fixed in space is first studied. The amplitude f for the scattering of electron from the incident wave vector $\vec{k}_0$ to the outgoing $\vec{k}$ naturally depends on the molecular orientation $\hat{s}$. The amplitude for the rotational transition $J_0 M_0 \rightarrow J M$ is then given by

$$f(J_0 M_0, \vec{k}_0 \rightarrow J M, \vec{k}) = \int Y^*_{JM}(\hat{s}) \, f(\vec{k}_0 \rightarrow \vec{k}, \hat{s}) \, Y_{J_0 M_0}(\hat{s}) \, d\hat{s}.$$

(3)

This approximation is valid provided that (i) the electron velocity is not too low and the effective collision duration is much shorter than the period of molecular rotation, and that (ii) the rotational level spacings are much smaller than the incident electron energy so that we may assume $k \simeq k_0$. Hara[14] has applied this method to electron-H_2 scattering and obtained rotational cross sections comparable to those of close-coupling calculation by Henry and Lane.[7]

The adiabatic approximation has many merits. For the electron scattering from a linear molecule, we can choose the molecular axis as the quantization axis. Then the component m of the electron orbital angular momentum ℓ along this axis will become a good quantum number, so that the set of coupled equations to be solved will be decomposed into smaller sets of equations. This greatly facilitates the numerical work. Secondly, in fixed-nuclei approximation, we can easily adopt methods other than the single-center treatment. In the close-coupling method in the laboratory frame, the interaction potential is usually expanded around the center-of-mass of the target molecule to make the calculation feasible. Since molecules have more than one nuclei, the single-center expansion of the potential requires a large number of terms in order to reproduce the actual interaction accurately. In the adiabatic approximation, on the other hand, we may use, without difficulty, the two-center coordinate system (spheroidal coordinates), which is especially suitable for diatomic targets.[15] Even for polyatomic molecules, one may apply the variational principle[16] to calculate the wave function of scattered electron in the fixed-nuclei approximation. Once the asymptotic form of the wave function is obtained in the body-frame, the elastic and rotational cross sections in the laboratory frame can be obtained easily.

Furthermore, the isotope effect can be studied easily in the adiabatic approximation. In the case of electron scattering from

HD, for instance, the interaction is nearly the same as in elec-
tron-H_2 system, except that the center-of-mass of the molecule is
shifted.   In the adiabatic approximation, the scattering amplitude
is obtained immediately from that of the electron-H_2 scattering as

$$f_{HD}(J_o M_o, \vec{k}_o \rightarrow J M, \vec{k})$$

$$= \int Y^*_{JM}(\hat{s}) \; \exp(-i\vec{K}\cdot\Delta\vec{R}) \; f_{H_2}(\vec{k}_o \rightarrow \vec{k}, \hat{s}) \; Y_{J_o M_o}(\hat{s}) \; d\hat{s}, \qquad (4)$$

where $\vec{K} = \vec{k}_o - \vec{k}$, and $\Delta\vec{R}$ is the vector representing the shift of
the center-of-mass from the midpoint of the two nuclei.[17]   This
relation has been used by Hara[18]   to calculate the rotational cross
sections for the system.

The adiabatic approximation can be applied also to the vibrational excitations.   For this purpose, the scattering amplitude in
the fixed-nuclei approximation is calculated at several internuclear
distances.   Then, the integral (3) is a function of the internuclear distance.   With appropriate initial and final vibrational
wave functions, we can easily obtain the amplitude for simultaneous
rotational-vibrational transitions.   The work by Henry and Chang
already mentioned is of this sort of calculation.   Faisal and
Temkin[19] also applied the adiabatic approximation (adiabatic-nuclei
theory according to their nomenclature) to the vibrational excitation of H_2.   The differential cross section (DCS) for the process
(i) v = 0 → 1; ΔJ = 0   and (ii) v = 0 → 1; J = 1 → 3   calculated
in this way are in close agreement with experimental values. [9,20]
The DCS for (i) has a minimum at a scattering angle near 90°, while
that of (ii) is nearly constant.   The same problem has been studied
also by Abram and Herzenberg,[21]   again in the adiabatic approximation (they call it impulse approximation).   They assume, however,
that only the scattering through the $^2\Sigma_u^+$ compound state contribute
significantly to vibrational excitation. [22]   They calculated the
ratio of the two DCS :   R(v) = DCS(v = 0 → 1;   ΔJ = 0) / DCS(v =
0 → 1; J = 1 → 3).   Their scattering amplitude consists of a
product of the vibrational and rotational factors.   Thus, their
ratio can be used also for the other final vibrational states v.
In a recent experiment by Wong and Schulz,[20]   the ratio R(v) has
been obtained for the three final states   v = 1, 2, and 3.   Since
the Abram and Herzenberg's theory neglects the direct component of
scattering and uses the resonance component only, their result is
probably more applicable to the cases of v = 2 and 3, since the
excitation to these states is alomst entirely due to the resonance
process.   The theoretical ratio agrees fairly well with the experimental values for v = 2.   However, the theory cannot explain that
the ratio for v = 3 is appreciably different from that for v = 2.

More refined theoretical study is thus desirable.

As to the vibrational excitation of H_2, there is another theoretical study based on the Glauber approximation. This approximation was originally introduced in nuclear physics and was designed to be   applied to high-energy collision processes.[23] However, it has been also applied to many electron-atom collision processes in recent years with considerable success, down to comparatively low energies.[24] Thus, Chang, Poe and Ray[25] applied this method to the vibrational excitation of H_2, assuming that the relevant interactions are the long-range polarization interaction and the short-range interaction (screened Coulomb interactions centered at the two nuclei). The straightforward calculations give the DCS in very good agreement with experiment[26] even at collision energy as low as 20 eV, to large scattering angles (up to $\theta \sim$ 80°). It is rather surprising that such a good agreement is obtained with the Glauber approximation using the approximate interaction chosen. The Glauber approximation usually is not applicable for large-angle scattering. Furthermore, in the rotational cross sections, many investigators (such as Henry and Lane,[7] and Hara[14]) have clearly shown the importance of the electron exchange effect. It is desirable, therefore, to further study the Glauber cross section to see whether the vibrational cross section depends sensitively on the interaction chosen.

We now turn to the rotational excitation in the electron-polar molecule scattering. Here, the Born approximation is valid in a wider energy range than in the electron-nonpolar molecule scattering. Of course, we find a considerable distortion in the inner partial waves especially as the collision energy increases, just as in the case of nonpolar targets. However, because of the very long-range nature of the dipole interaction, a considerable contribution comes from partial waves with large values of the orbital quantum number ℓ, so that the modifications in the inner partial cross sections do not give a drastic change in the total rotational cross section. Furthermore, unless the dipole moment D is very small, the rotational cross sections $\sigma(J \rightarrow J \pm 1)$ become very large in the sub-eV region and dominate the elastic cross section. According to a close-coupling calculation for the rotational cross section in electron-HCl system,[27] the rotational excitation cross section $\sigma(J = 0 \rightarrow 1)$ is one or two orders of magnitude larger than the elastic cross section $\sigma(J = 0 \rightarrow 0)$ for the collision energies below 1 eV. Since the Born approximation can be applied fairly accurately to the dipole transition cross section $\sigma(0 \rightarrow 1)$, this approximation also gives fairly satisfactory values of the total cross section in the low-energy region.

When the dipole moment D is increased, the wave function of the scattered electron is distorted not only by the short-range forces

but also by the dipole interaction itself.   However, at least up
to the dipole moment of about D = 1 atomic unit (ea$_o$ ∿ 2.54 debyes),
the difference between the close-coupling and the Born approximation
results is still small at small angles in the DCS for J = 0 → 1
excitation.   At large angles, there are appreciable differences,[28]
but the absolute magnitude of the DCS is small there.

Recently, Stern and his colleagues at Columbia University made
molecular-beam recoil experiments on the electron collision with
strongly polar molecules : CsF (D = 7.9 debyes), CsCl (D = 10.4
debyes) and KI (D = 11.0 debyes).[29]   They derived the DCS from
the observed data and then calculated the total scattering and the
momentum-transfer cross sections.   Their experiments have revealed
that the differential scattering cross section $d\sigma/d\Omega$ is considerably smaller for these molecules than the Born prediction not only
at large scattering angles but also at relatively small angles (θ =
10° ∿ 30°).   This must be the evidence for the breakdown of the
first-order perturbation theory for the strong dipole interaction.
The breakdown is probably because the region of appreciable field
strength is greatly widened and effective collision duration is
increased.   The incident electron can be scattered many times with
appreciable probability within the dipole field of a single target
molecule, while the multiquantum transition in the molecular rotation becomes important.   One of the simplest way to proceed beyond
the first-order theory is to apply the Glauber approximation.[23,24]
This approximation is expected to be applicable more favorably to
the present problem than in the electron-H$_2$ scattering or some of
the electron-atom collision processes, since the distant encounters
dominate here, and we are especially interested in the comparatively
small scattering angles.   Thus, we have recently applied the
Glauber approximation to the electron-polar molecule interaction.[30,31]

The Glauber approximation is a kind of the adiabatic approxi-
mation, where the scattering amplitude $f(\vec{k}_o \rightarrow \vec{k}, \hat{s})$ for fixed-
nuclei (fixed $\hat{s}$) calculation [see equation (3)] is given by

$$f(\vec{k}_o \rightarrow \vec{k}, \hat{s}) = \frac{ik_o}{2\pi} \int [1 - \exp(i\chi)] \exp(i\vec{K}\cdot\vec{b}) \, d^2b, \qquad (5)$$

where

$$\chi = -\frac{1}{\hbar v_o} \int_{-\infty}^{\infty} V(\vec{r}, \hat{s}) \, dz, \qquad \vec{K} = \vec{k}_o - \vec{k}, \qquad (6)$$

$v_o = \hbar k_o/m_e$ is the velocity of the incident electron, $\vec{b}$ is the two -
dimensional vector lying in the plane perpendicular to the z axis,
while the latter is chosen along the bisector of the two wave
vectors $\vec{k}_o$ and $\vec{k}$.   In the case of the point dipole interaction

$V = - eD \, (\hat{r} \cdot \hat{s}) \, / \, r^2$, it is easy to show that

$$\chi \;=\; \frac{2eD}{\hbar v_o b} \, (\hat{s} \cdot \hat{b}). \tag{7}$$

The straightforward calculation gives the DCS for total scattering as

$$\frac{d\sigma}{d\Omega} \;=\; k_o^2 \; (\frac{2eD}{\hbar v_o})^4 \; \sum_{JM} \frac{k}{k_o} \; G(J_o M_o \rightarrow J \, M; \; \xi), \tag{8}$$

where each G is the square of a definite integral which is a function of $\xi = 2eDK/\hbar v_o \; \cong \; (4eD m_e/\hbar^2) \, \sin(\theta/2)$. In the limit of vanishingly small D, only the transitions $J = J_o \pm 1$ are allowed and (8) becomes the usual Born approximation formula.

Although the molecules are distributed among various rotationally excited states under the experimental condition, we assume for simplicity that the target is initially in the rotationally ground state (J = 0). This is permissible since we know [Takayanagi (1967)[1]] that $d\sigma/d\Omega$, when summed up over all possible final states and averaged over the initial orientation of the molecule, does not depend much on the initial rotational quantum number J_0. Numerical calculation of (8) results in the DCS in very good agreement with experiment (4.77 eV electron scattering from CsCl, etc.) up to the scattering angle $\theta \sim 50°$. We also studied approximately the effect of spherical part of the short-range and polarization interactions, and also the effect of quadrupole interaction. These additional interactions modify the calculated DCS at larger angles, but the agreement with experiment for $\theta < 50°$ remains good. At larger angles, the Glauber approximation is not reliable. Also, the experimental DCS is not very accurate at these angles because of the limited flexibility of the model adopted in analysing the raw data from experiment. Therefore, the detailed comparison between theory and experiment is not possible at larger angles at present.

A drawback in the fixed-nuclei approximation for the electron-polar molecule scattering is that the differential cross section diverges in the forward direction and even the integrated scattering cross section diverges. This comes from the divergence of $G(J_0 M_0 \rightarrow J_0 \pm 1 \, M)$ $(\propto \xi^{-2})$ at $\xi \rightarrow 0$. For fixed-nuclei, the lower limit of ξ is 0. In reality, however, ξ ($\propto K$) has a finite lower limit, since we have $K \geq |k_0 - k|$, and $|k_0 - k|$ is small, but finite. If this is taken into account, we can obtain a definite theo-

retical cross section. The experimental total cross section for the
three polar molecules mentioned above varies roughly as $E^{-0.6}$ with
the incident energy E, and is about 50 - 70 % of the Born values
in the energy range of 3 - 5 eV for CsCl as well as for KI.   The
calculated total cross section is very close to the experimental
one in the same energy region, but the energy dependence is somewhat steeper (roughly proportional to $E^{-0.9}$).   Strictly speaking,
we have to calculate the total cross section for the actual rotational state distribution, since the lower limit of ξ (corresponding to $\theta = 0$) depends on the initial J_0.   Theoretical momentum-
transfer cross section σ_m, on the other hand, does not depend much
on J_0.   It varies as E^{-1} with energy and it is an order of magnitude smaller than the Born result.   This is in good agreement with
experimental findings for CsCl and KI at least for E < 5 eV.
Above 5 eV, however, the experimental σ_m does not decrease as E^{-1},
but starts to increase to a flat plateau.   This has not been
explained yet.   Since such an increase is hardly seen in the total
scattering cross section, it should be associated with large-angle
scattering.   Either short-range forces or electronic excitation
processes, or both, are probably responsible for the observed
behavior of σ_m.

     We shall now study the dependence of σ_m upon the dipole moment
D.   According to the exact calculation for the electron scattering
from a point dipole fixed in space,[32] there is no acceptable solution for $D \geq D_c = 1.625$ debyes.   For a finite dipole, consisting
of two point charges +Ze and -Ze fixed at a finite distance, we
have found resonance-like peaks in the σ_m curve as a function of
D.[33]   The first peak is at about 2 debyes and exceeds considerably the Born cross section.   In the energy region immediately
above the peak, the calculated cross section is less than the Born
values.   A similar behavior is found also in the close-coupling
calculation, where the singularity of the point-dipole interaction
at the origin is removed by multiplying an appropriate cut-off
factor.[28]   These peaks are not due to the long-range nature of the
dipole potential, but due to the presence of a bound state near
zero energy in the potential well, when the depth of the well,
characterized by D in the adopted models, takes appropriate values.
However,these peaks are pronounced only at very low energies.   As
the collision energy increases, the position of peaks move to higher
D values and at the same time become less pronounced.   Since the
Glauber approximation is designed for high-energy collisions, there
is no serious confliction between the Glauber and more accurate
calculations.

     Garrett[34] pointed out recently that the position of the peaks
in σ_m as a function of D depends considerably upon the magnitude of
other interactions (short-range interaction, polarization force,
etc.).   That it is so can be seen also in our early calculations,

where the peak moves to lower values of D when we add a short-range
attractive potential.   Because of the energy dependence and the
dependence on the other interactions, Garrett claims that there is
no definite dipole magnitude which corresponds to the resonance
effect in σ_m for thermal electrons.   This is true in the sense that
the critical dipole moment 1.625 debyes for the pure dipole interaction has nothing to do with real molecules.   However, as far as
the first peak (around 2 debyes in the model calculations) is
concerned, its position does not change so rapidly with energy.
If, therefore, we can select a group of molecules where the magnitude of polarizability and the effect of short-range forces do not
change drastically from molecule to molecule, we can still expect
some trend in the σ_m vs. D diagram, i.e., σ_m may tend to exceed
significantly the Born values for low D and to fall below the Born
values for somewhat larger D.   Compilation of observed σ_m by
Garrett[34] for a large number of molecules at 300 K seems to indicate
such behavior.

REFERENCES

1.   Some earlier review articles on the topic include :
     K. Takayanagi, Progr. Theor. Phys. (Kyoto) suppl. 40, 216
     (1967); K. Takayanagi and Y. Itikawa, Adv. Atom. Mol. Phys. 6,
     105 (1970); D. E. Golden, N. F. Lane, A. Temkin and E. Gerjuoy,
     Rev. Mod. Phys. 43, 642 (1971).

2.   D. G. Truhlar, F. A. Van-Catledge and Th. H. Dunning, J. Chem.
     Phys. 57, 4788 (1972); D. G. Truhlar and F. A. Van-Catledge,
     J. Chem. Phys. 59, 3207 (1973).

3.   P. G. Burke and N. Chandra, J. Phys. B5, 1696 (1972); P. G.
     Burke, N. Chandra and F. A. Gianturco, J. Phys. B5, 2212 (1972);
     N. Chandra and P. G. Burke, J. Phys. B6, 2355 (1973).

4.   E. Gerjuoy and S. Stein, Phys. Rev. 97, 1671 (1955); 98, 1848
     (1955).

5.   K. Takayanagi, Rept. Ionos. Space Res. Japan 19, 1 (1965);
     see also review articles in reference 1.

6.   S. Geltman and K. Takayanagi, Phys. Rev. 143, 35 (1966).

7. R. J. W. Henry and N. F. Lane, Phys. Rev. 183, 221 (1969).

8. R. W. Crompton, D. K. Gibson and A. I. McIntosh, Austr. J. Phys.
    22, 715 (1969).

9. F. Linder and H. Schmidt, Z. Naturforschg. 26a, 1603 (1971).

10. R. J. W. Henry and N. F. Lane, Phys. Rev. A4, 410 (1971).

11. E. S. Chang and U. Fano, VII ICPEAC, Abstracts of Papers
    (North-Holland, Amsterdam, 1971) p.1057; Phys. Rev. A6, 173
    (1972).

12. R. J. W. Henry and E. S. Chang, Phys. Rev. A5, 276 (1972).

13. D. M. Chase, Phys. Rev. 104, 838 (1956).

14. S. Hara, J. Phys. Soc. Japan 27, 1592 (1969).

15. See, e.g., K. Takayanagi in reference 1.   J. W. Darewych, P.
    Baille and S. Hara (to be published) have recently compared the
    one-center and two-center calculations for the positron scatter-
    ing from H_2 in the adiabatic approximation, and concluded that
    the two calculations give almost the same result.   This is
    probably because the H_2 molecule has extraordinarily small
    internuclear distance and the molecular electron cloud is nearly
    spherical and also because the low-energy positron cannot
    approach nuclei closely because of the repulsive force.
    Similar comparison for electron scattering from other molecules
    would be valuable.

16. K. Onda, J. Phys. Soc. Japan 36, 826 (1974).

17. K. Takayanagi, J. Phys. Soc. Japan 28, 1528 (1970).

18. S. Hara, J. Phys. Soc. Japan 30, 819 (1971).

19. F. H. M. Faisal and A. Temkin, Phys. Rev. Letters 28, 203
    (1972).

20. S. F. Wong and G. J. Schulz, Phys. Rev. Letters 32, 1089 (1974).

21. R. A. Abram and A. Herzenberg, Chem. Phys. Letters 3, 187 (1969).

22. The resonance scattering of electrons from molecules is not
    discussed here.   See, for instance, the review article (theore-
    tical) by J. N. Bardsley and F. Mandl, Rep. Progr. Phys. 31,
    471 (1968), and the review article (experimental) by G. J.
    Schulz, Rev. Mod. Phys. 45, 423 (1973).

23. R. J. Glauber, in Lectures in Theoretical Physics, vol. 1
    (Interscience, New York, 1959) p.315.

24. See, e.g., E. Gerjuoy, in Physics of Electronic and Atomic
    Collisions, VII ICPEAC (North-Holland, Amsterdam, 1972) p.247.

25. T. N. Chang, R. T. Poe and P. Ray, Phys. Rev. Letters 31, 1097
    (1973).

26. S. Trajmar, D. G. Truhlar, J. K. Rice and A. Kuppermann, J. Chem.
    Phys. 52, 4516 (1970).

27. Y. Itikawa and K. Takayanagi, VI ICPEAC, Abstracts of Papers
    (The MIT Press, Cambridge, 1969) p.144; J. Phys. Soc. Japan 26,
    1254 (1969).

28. Y. Itikawa, J. Phys. Soc. Japan 27, 444 (1969).

29. R. C. Slater, M. G. Fickes and R. C. Stern, Phys. Rev. Letters
    29, 333 (1972); R. C. Slater, M. G. Fickes, W. G. Becker and
    R. C. Stern, Scattering of alkali halides by electrons I. CsF;
    II. CsCl; III. KI (to be published).

30. K. Takayanagi, Progr. Theor. Phys. (Kyoto) 52,337(1974).

31. K. Takayanagi, O. Ashihara, I. Shimamura and Y. Itikawa,
    Fourth Intern. Conf. Atomic Physics in Heidelberg (July 1974),
    Abstracts of Papers.

32. M. H. Mittleman and R. E. von Holdt, Phys. Rev. 140A, 726 (1965).

33. K. Takayanagi and Y. Itikawa, J. Phys. Soc. Japan 24, 160 (1968).

34. W. R. Garrett, Molecular Phys. 24, 465 (1972).

ELECTRON-PHOTON COINCIDENCES AND POLARIZATION OF IMPACT

RADIATION

H. Kleinpoppen

Department of Physics, University of Stirling

Stirling, Scotland

I. INTRODUCTION

Two basic aspects are associated with the title of this talk: first, the physics of electron atom collisions in general, and secondly, as a special problem the angular distribution of photons and also their correlation with the angular distribution of electrons having excited atoms and from which photons are emitted. The collision process between unidirectional and monoenergetic electrons and atomic targets still represents an important chapter in our knowledge of fundamental scattering processes. Due to the recent success of new types of experiments which are different in their methods to conventional cross section measurements, the state of investigation of electron atom collision physics has changed considerably. On the one hand the situation can be characterised by the same remark as in Professor Bederson's[1] review paper on "Electronic Polarization Behaviour in Collisions" at the Third International Atomic Physics Conference in Boulder, namely, the subject of my report is part of the full story about the physics of electron atom collisions. On the other hand, the recent more ambitious and new types of experiments including electron spin and photon polarization analysis, and also the electron photon correlation measurements bring us to a much deeper understanding of the details of the electron atom collision process. 'Details' of the electron atom collision process means e.g. an analysis of the different types of interactions occurring during the collision, namely the Coulomb or the direct interaction, the exchange and the spin orbit interaction as investigated in the electron and atomic spin analysis experiments;[2] or details of the collision process means analysis of coherence effects or the atomic alignment and orientation parameters from electron photon coincidence measurements.

449

It appears that the new types of analysing experiments require
a complex theoretical and experimental analysis of the observables,
and the situation may be characterised by the fact that certain
papers published recently deal with aspects which can be called
theory of the measurement of electron atom collisions.[3] In these
theories links have been established between observables on the one
hand, and collision amplitudes and their phase differences and
alignment and orientation parameters on the other hand.

Of course, the relevant theoretical approximations for electron
atom collision processes provide predictions of these amplitudes,
their phase differences, and of atomic alignment and orientation
parameters. The normal differential cross section for example is
an averaging quantity consisting of a term like:

$$\sigma = \sum_{i,j} \ |a_i|^2 + |a_i - a_j|^2 \qquad (1)$$

with the excitation or scattering amplitudes $a_{i,j}$ describing a
given physical process of the electron atom collision.
Theoreticians normally carry out calculations of these amplitudes
based upon a given theoretical approximation and then 'hide' their
detailed information on interesting feature and structure of the
amplitudes by summing up the above terms and thereby averaging
over detailed information. It would certainly be advisable for
future comparisons between experimental and theoretical data to
have tables of amplitudes and their phase differences available.

The two main parts of my talk follow from its title. I will
start with the traditional measurements and then I will describe the
new type of electron photon angular correlation or coincidence
measurements for electron atom excitation processes.

II. THE SCHEME FOR ELECTRON IMPACT POLARISATION STUDIES AND ELECTRON PHOTON COINCIDENCE MEASUREMENTS

The two parts of Fig. 1 illustrate the basic schemes of the
experimental technique to be discussed in connection with this
section. In comparing these schemes we discuss the basic aspects
and their differences. The incoming unidirectional and monoenergetic
beam of electrons excites the atom and the subsequent emission of
photons is observed perpendicular to direction z (case (a) of Fig.1),
or in any direction in coincidence with the inelastically scattered
electron (case (b) of Fig. 1). Rotational symmetry about the beam
axis exists in case (a), whereas the electron photon coincidence
process has only reflection symmetry in the scattering plane as
defined by the incoming and outgoing beam of electrons. Simple
relations follow immediately from the geometry and the dynamics;
in case (a), a selection $\Delta m_L = 0$ exists for the excitation at

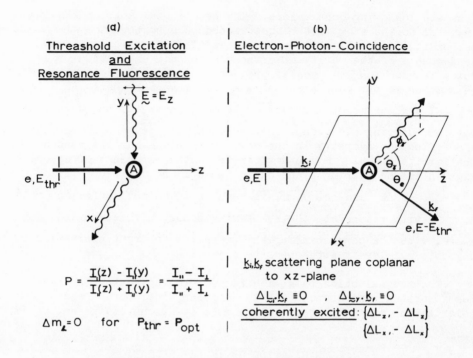

Fig. 1: Scheme for electron impact polarization studies (case a) and electron photon correlation measurements (case b).

threshold energy because the incoming electrons have no orbital angular momentum in their direction of travel. Therefore they cannot carry away any orbital angular momentum with respect to the z-axis because the electrons have lost all their energy and, thus, their ability to participate in any orbital angular momentum transfer to the atom.

The threshold excitation therefore is governed by the same selection rule $\Delta m_L = 0$ as with resonance fluorescence excitation by linearly polarized light. The polarization of line radiation of resonance transitions excited by electron impact bombardment at threshold (P_{thr}), or by linearly polarized resonance radiation (P_{opt}), should therefore be equivalent, if experimental arrangements are set up as indicated in Fig. 1(a). In Section III we present examples where the identity of $P_{thr} = P_{opt}$ is verified.

Fig. 1(b) illustrates simple consequences which result from the coincident detection of inelastic electrons having excited the atom and the photons from the decay of the excited atom. First we can draw the following conclusions. The net orbital angular momentum transfer to the atom must be restricted to a component perpendicular to the scattering plane. This restriction is a consequence of the parity invariance of atomic excitation processes which require a zero scalar product of an axial and a polar vector; this means in our case that $\overline{\Delta L_i} \bullet \overline{k_f}$ must vanish where $\overline{\Delta L_i}$ is the orbital angular momentum transfer vector, and $\overline{k_f}$ is the wave vector for the outgoing electrons. However, as will be seen in Section IV, the reflection symmetry in the scattering plane has the consequence that orbital angular momentum transfer with components of opposite sign in the scattering plane can coherently be excited (e.g. ΔL_z and $-\Delta L_z$ or ΔL_x and $-\Delta L_x$). This results in a 'net' selection rule $\Delta m_L = 0$ with respect to the scattering plane, but, as pointed out above, with the understanding that we allow coherent excitation of magnetic sub-levels with opposite signs.*

The net selection rule with respect to the scattering plane has important consequences for the angular distribution and polarization of the photons, which will be discussed in Section IV.

III. POLARIZATION OF IMPACT RADIATION

The revived interest in polarization studies of impact radiation since about the beginning of the sixties originates from the following facts. As pointed out in Fig. 1(a), the threshold polarization of electron impact line radiation is governed by the selection rule $\Delta m_L = 0$. Accordingly, threshold polarization can easily be calculated for many simple cases, e.g. the threshold polarization of the excitation/de-excitation process $^1S \rightarrow {}^1P \rightarrow {}^1S$ should be 100%, because only the $m_L = 0$ sub-level of the P state is excited, and subsequently only the π transition to the S state occurs. There were several experiments in the twenties, particularly that of Skinner and Appleyard[4], which were in complete disagreement with the threshold polarization as required from angular momentum conservation. Fig. 2 illustrates the situation with regard to several mercury lines: the crosses in Fig. 2 represent the threshold polarization of the various mercury lines and the full curves give the experimental results of Skinner and Appleyard[4] showing complete disagreement with what was to be expected theoretically for the polarization close to threshold. This startling discrepancy remained unexplained for several decades. The other important factor that stimulated renewed interest in

---

*The author gratefully acknowledges discussions with B. Bederson on this point.

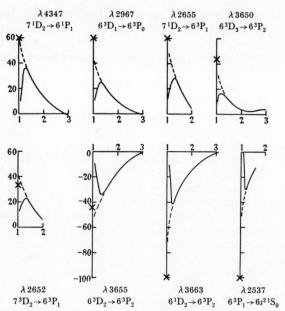

Fig. 2: Polarization of electron impact radiation of Hg as functions of $\sqrt{E/E_{thr}}$ (E electron impact energy, E_{thr} threshold excitation energy). The full lines represent experimental data of Skinner and Appleyard[4] (1927). Crosses give threshold polarization for mercury isotopes with zero nuclear spin and assuming LS coupling. Dashed lines represent extrapolations of full lines neglecting the experimental drop of polarization near threshold. (The above figure is taken from Percival and Seaton[5].)

polarization studies was the new theoretical approach to the problem mainly due to the papers by Percival and Seaton[5] and by Baranger and Gerjuoy[6] in 1958. The theory developed by Percival and Seaton[5] overcame the deficiencies of the previous Oppenheimer[7] and Penney[8] theory by allowing for radiation damping or what is equivalent to taking the finite level width of the excited fine and hyperfine structure states into account. The particular importance of the paper by Baranger and Gerjuoy[6] is often referred to the introduction of the atomic compound model for atomic collision processes. However, the most important applications and consequences of this model were related to the angular distribution and polarization of light emitted following excitation through compound states (see below).

Although there is a large amount of evidence that the existing theory on polarization of impact radiation describes the experimental data correctly, there are still problems left which might either call for extensions or modifications of certain aspects of the theory, or alternatively, for an improvement in the experimental conditions. It therefore appears reasonable to review selected examples which illustrate cases of agreement and disagreement between theory and experiment.

The theoretical basis for impact line polarization lies within the framework of quantum mechanics. The theory is well formulated for cases where spin orbit and spin spin interactions can be neglected.

The polarization of the line radiation depends on the anisotropic population of the magnetic sub-levels excited during the collision, and also on the relative transition probabilities for π and σ transitions of these sub-states into the magnetic sub-levels of the state into which the atom decays. In order to discuss recent results on line polarization, we concentrate on examining applications for atoms in which the ground state has no orbital angular momentum. The intensity components polarized parallel or perpendicular to the direction of the incoming beam (z direction, see Fig. 1(a)) is given by

$$I_{\|} \propto \sum_m \frac{A_m^\pi}{A} Q_m \quad , \quad I_\perp \propto \sum_m \tfrac{1}{2} \frac{A_m^\sigma}{A} Q_m \tag{2}$$

(A—total transition probability, A_m^π and A_m^σ refer to magnetic sub-states for π and σ transitions, Q_m— excitation cross section for magnetic sub-state with magnetic quantum m).

It then follows for the polarization of the line radiation (with $A = A_m^\pi + A_m^\sigma$; note that the total transition probability does not depend on m):

$$P = \frac{I_{\|} - I_\perp}{I_{\|} + I_\perp} = \frac{3K^\pi - K}{K^\pi + K} \tag{3}$$

$$K^\pi = \frac{1}{A} \sum_m A_m^\pi Q_m \quad \text{and} \quad K = \sum_m Q_m = Q \tag{4}$$

Note that in the above equation we did not specify the magnetic quantum number which could be taken for m_L, m_j or m_F. The quantities K^π and K are proportional to the production rates of the total linear polarized and total unpolarized line intensity of the emission respectively.

The recent emphasis on line polarization measurements according to the scheme as in Fig. 1(a) is mostly connected with some simple examples and arguments which can best be illustrated as

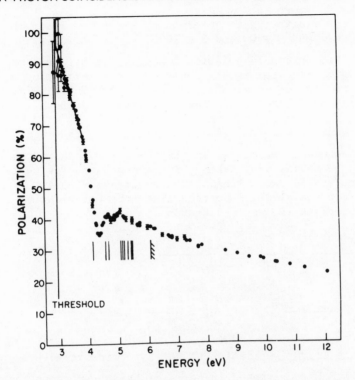

Fig. 3: Polarization of the $4^1P - 4^1S$, $\lambda = 4227\overset{\circ}{A}$ radiations for
low electron-bombardment energies (one-σ statistical error bars
arising from counting statistics, Ehlers and Gallagher[7]).

follows: let us consider the excitation/de-excitation process
$^1S \to {}^1P \to {}^1S$ with the total excitation cross sections
$Q(S \to P) = Q_{m_L = 0} + Q_{m_L = +1} + Q_{m_L = -1} = Q_0 + 2Q_1$.* Only the
Q_0 component results in the excitation of π light, therefore,
with $K^\pi \propto Q_0$ and $K \propto Q$, the polarization of the above excitation
process is given by

$$P = \frac{Q_0 - Q_1}{Q_0 + Q_1} = \frac{Q_0/Q_1 - 1}{Q_0/Q_1 + 1} \qquad\qquad (5)$$

---

*Parity invariance requires that the $m_L = \pm 1$ sub-states are
 excited with equal probability; note $Q_{m_L = \pm 1} = Q_1$ and
 $Q_{m_L = 0} = Q_0$.

At threshold the excitation is governed by the selection rule
$\Delta m_L = 0$ or $Q_1 = 0$ and $Q = Q_o$ and $P = 100\%$. A beautiful example
which proves that threshold polarization of the above process
should be 100% is revealed in the study of the polarization of the
first resonance line of calcium investigated as a function of the
electron energy by Ehlers and Gallagher[7] (Fig. 3). The high energy
limit (E_∞) of the polarization should approach
$P_\infty = -100\%$ because $Q_o (E_\infty)/Q_1 (E_\infty) \to 0$. Including fine and
hyperfine structure interactions in the $S \to P \to S$ transitions, the
theory has to be modified by taking into account the finite level
width and also the jI-recoupling of the excited states after the
excitation. For the alkali resonance lines the polarization is
given by the formula (Flower and Seaton[8])

$$P = \frac{3(9\alpha - 2)(Q_o - Q_1)}{12Q_o + 24Q_1 + (9\alpha - 2)(Q_o - Q_1)} \tag{6}$$

with

$$\alpha = \sum_{F,F^l} \frac{\zeta(I,F,F^l)}{1 + \epsilon^2_{F,F^l}} \quad , \quad \epsilon_{F,F^l} = \frac{2\pi \, \Delta\gamma_{F,F^l}}{A} \tag{7}$$

and $h\Delta\gamma_{F,F^l}$ the energy separation of the hyperfine structure
states and $\zeta(I,F,F^l)$ expressed in terms of Racah and vector coupling
coefficients. The value of α depends largely on the ratios of the
hyperfine structure separations to the natural line width which can
be calculated from the theory of Flower and Seaton[8] for cases of
interest and which are summarized in Table 1 below.

Table 1

$2^2P_{1/2,3/2}$ of atomic hydrogen $\lessdot$	with hfs	$\alpha = 0.441$
	without hfs	$\alpha = \frac{4}{9} = 0.445$
$2^2P_{1/2,3/2}$ of Li^6		$\alpha = 0.413$
$2^2P_{1/2,3/2}$ of Li^7		$\alpha = 0.326$
$3^2P_{1/2,3/2}$ of Na^{23}		$\alpha = 0.288$

Table 2 gives results for theoretical and experimental
threshold and optical line radiation of the first resonance line
of Li^6, Li^7 and Na^{23}, and also the Lyman-α impact polarization of

atomic hydrogen. Notice the excellent agreement between theory and experiment of the alkali resonance lines. This proves the relation $P_{opt} = P_{thr}$ is valid for the alkali resonance lines, at least for light alkalis. Very good agreement also exists for the experimental and theoretical polarization data of the first resonance lines of Li^6 and Li^7 in an energy range just above threshold, as seen in Fig. 4 and 5, while in the case of the sodium D line, the spread of experimental data precludes a clear comparison of them with a theory.

Table 2

Threshold and resonance fluorescence polarization (P^{opt}) of the first resonance line of Li^6, Li^7, Na^{23} and H. (P_{cal}, P_{exp}, theoretical and experimental data).

	$\underline{Na}^{23}$	$\underline{Li}^6$	$\underline{Li}^7$	$\underline{H}$
P_{cal}^{thr} =	14.1%	37.5%	21.6%	42.1%
P_{exp}^{thr} =	(14.8 ± 1.8)%	(39.7 ± 3.8)%	(20.6 ± 3.0)%	13%(?)[9]
P_{cal}^{opt} =	(14.2 ± 0.2)%	(37.7 ± 1.0)%	(21.7 ± 1.5)%	42.1%
P_{exp}^{opt} =	(14.0 ± 0.8)%	(37.5 ± 1.7)%	(23.1 ± 1.3)%	–

Contrary to the above examples, there are quite a number of atomic lines which appear not to satisfy the threshold polarization required by the $\Delta m_L = 0$ selection rule. Important examples are certain helium lines, the Lyman-α radiation (Table 2), the $3^2D \rightarrow 2^2P$ transition in lithium[10] and also certain mercury lines.[11] Most of these lines show a considerable drop and even oscillations in the polarization close to threshold. It has been suspected by several authors[3a, 6 and 11] that such a drop in the polarization near threshold is related to the existence of resonances in the inelastic electron atom cross section near threshold. Fano and Macek[3a] suggest that at a resonance the colliding electron and the one being excited in the atom remain strongly correlated for a time interval sufficient to allow extensive exchange of angular momentum between them with a decrease in alignment (see Section IV). The following examples to be discussed give clear evidence about the influence of resonances on the polarization of line radiation.

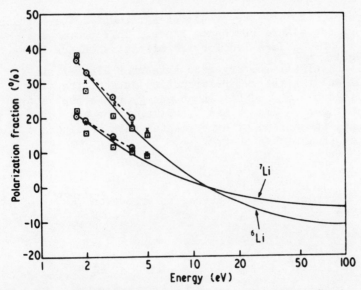

Fig. 4: Polarization of first resonance lines of Li[6] and Li[7].
_____ Glauber approximation[12]; xx close-coupling calculations
incorporating the dipole polarizability of atomic states[13]; ▣ close-
coupling calculations[14]; ☉···☉ experimental data[10].

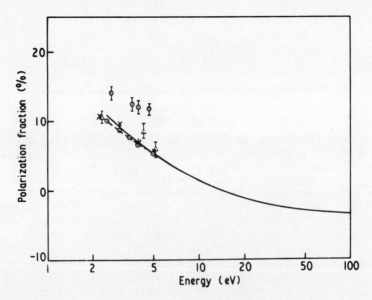

Fig. 5: Polarization of the first resonance of Na[23]; _____
Glauber approximation[12]; xx four state exchange close-coupling
approximation[15]; experimental data: ☉···☉ Enemark and Gallagher[16];
△ Gould[17]; ♀ Hafner and Kleinpoppen[10].

In this connection, we first discuss the mercury excitation/de-excitation processes $6^1S_0 \rightarrow 6^3P \rightarrow 6^1S_0$ ($\lambda = 2537$ Å) and $6^1S_0 \rightarrow 6^1P_1 \rightarrow 6^1S_0$ ($\lambda = 1849$Å). In the case of the threshold excitation of the 6^1P state, only the $M_J = 0$ is excited because of the selection rule $\Delta m_L = 0$; for the excitation of the 6^3P_1 state, only $M_J = \pm 1$ states can be excited at threshold because of the fact that the Clebsch–Gordan coefficient

$$C_{\substack{S = 1,\ L = 1,\ J = 1 \\ M_S = 0,\ M_L = 0,\ M_J = 0}} = 0, \tag{8}$$

while all the other angular momentum coupling coefficients are unequal to zero. This has the consequence that the threshold polarization of the above mercury lines should provide the largest possible difference ($\pm 100\%$) for the line polarization (P_{thr} ($\lambda = 2537$Å) $= -100\%$ and P_{thr} ($\lambda = 1849$Å) $= +100\%$.)

Figs. 6 and 7 show the latest results for the experimental studies with high resolution of the mercury 2537 line (Ottley[19]. The structure of the polarization curve above threshold coincides with resonance structure of an electron mercury transmission experiment carried out by Burrow and Michejda[18]. Baranger and Gerjuoy[6] first developed a theory in which they associated the failure of experimental data to approach the required threshold polarization with the formation of atomic compound states. They assumed that two negative ion compound states are formed just above the excitation threshold of the 2537Å line and assigned electron configurations corresponding to $^2P_{1/2,3/2}$ states. Baranger and Gerjuoy[6] also made predictions for the polarization of the 2537Å line in the resonance states and predicted a polarization of 60% and 0% in the centre of the $^2P_{3/2}$ and $^2P_{1/2}$ resonances respectively.

The calculations of Baranger and Gerjuoy[6] assumed zero orbital angular momentum of the outgoing electrons. The experimental data of the polarization curve do not confirm these theoretical predictions at the moment; one has, however, to bear in mind that the natural isotope composition and the remaining finite energy resolution of the electron beam still affect the polarization. As seen by comparing Fig. 6 and Fig. 7, the polarization peak at the first resonance close to the threshold rises from about 25% where the energy spread of the electron beam is about 140 meV to almost 40% where the energy spread of the electron was about 100 meV. A rough estimate shows that the predicted polarization of 60% in the $^2P_{3/2}$ should be reduced to approximately 45%, taking into account the hyperfine structure splitting for the normal isotope mixture of mercury. Fano

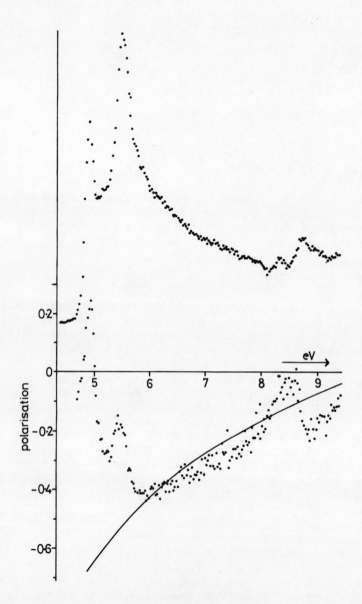

Fig. 6: Excitation function (upper part) and polarization curve (lower part) of the mercury intercombination line $6^3P_1 \rightarrow 6^1S_0$, $\lambda = 2537$ Å versus electron energy (Ottley[19]). Energy resolution of electron monochromator $\Delta E = 140$ meV. The full line represents the theoretical prediction by McConnell and Moiseiwitsch[21].

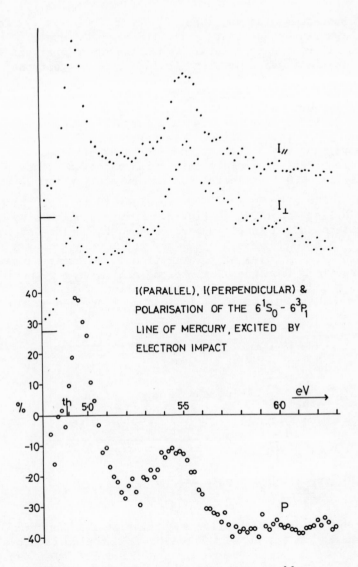

Fig. 7: Measured intensity components (Ottley[19]) polarized
parallel and perpendicular (upper part) to the exciting electron
beam for the mercury intercombination line λ = 2537 Å . The lower
part of the figure displays the polarization of the intercombination
line (Ottley[19]). Energy resolution of electron monochromator
ΔE = 100 meV. The error of the measured polarization at the peak
(4.92 eV) is ±7% with 90% confidence limits plus a small
uncertainty in the systematic correction.

and Cooper[20] have suggested the identification $^4P_{1/2,3/2,5/2}$ for the three lowest resonances associated with the compound state configuration 6S 6P^2. A more detailed configuration analysis will be necessary to distinguish between the different suggestions.

First polarization measurements of the $6^1P \rightarrow 6^1S$, $\lambda = 1850$ Å transition of mercury was recently reported by Ottley et al.[22] A quartz Rochon polarizer was used in this far ultra-violet spectral region. Clearly the polarization curve for the 1850 Å line as in Fig. 8 shows that the polarization near threshold does not approach the required threshold polarization based upon the selection rule $\Delta m_L = 0$. While resonances obviously affected the polarization of the 2537 Å, it is not so clear if this is the case for the 1850 Å line according to Fig. 8. Future studies with higher energy resolution should enable a more clear-cut conclusion.

High resolution studies of polarization curves near threshold for line transitions from the 4D states of helium were reported by Heddle, Keesing and Watkins[23]. The polarization of the line transition from the D states of helium can be expressed as follows:

$$P(4^1D \rightarrow 2^1P) = \frac{3Q_0 + 3Q_1 - 6Q_2}{5Q_0 + 9Q_1 + 6Q_2}$$

$$P_{thr}(4^1D \rightarrow 2^1P) = \frac{3}{5} = 60\%$$

$$P(4^3D \rightarrow 2^3P) = \frac{213Q_0 + 213Q_1 - 426Q_2}{671Q_0 + 1271Q_1 + 1058Q_2} \qquad (9)$$

$$P_{thr}(4^3D \rightarrow 2^3P) = \frac{213}{671} = 32\%$$

Figs. 9 and 10 show excitation and polarization functions lines associated with the above transitions. Although the total errors of the data in these figures is approximately 30% close to threshold, the measurements agree well with the threshold predictions. The above authors suggest that the sharpness of the fall in the polarization immediately above threshold is associated with a 1s4d^2 ^{2}S resonance, which strongly decays into $m_L = |2|$ sub-states of the D state. Previous studies reveal structure in the excitation functions of the above line transitions which coincide with the energy position of the polarization curves. The lowest possible polarization values occur if Q_2 is large compared with Q_0 and Q_1, which is -100% and -40% for the singlet and triplet transitions respectively. The initial fall to -22% in the triplet

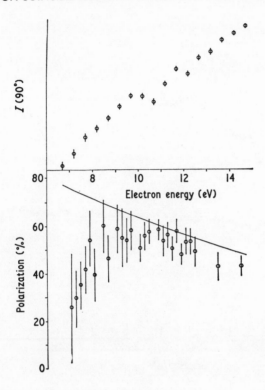

Fig. 8: Excitation function $I(90^\circ)$ (upper part) and polarization curve (lower part) of the mercury $6^1P_1 \rightarrow 6^1S_0$, $\lambda = 1850$ Å line as measured by Ottley et al.[22] The full line represents the theoretical prediction by McConnell amd Moiseiwitsch.[21] Error bars indicate 90% confidence limits plus a systematic error not exceeding 4%.

transition is followed by the return of a positive value. The true minimum polarization might be approached only with even higher energy resolution as demonstrated for example in the case of the mercury excitation as described above. The double feature around 23.8 eV in the polarization of the helium triplet transitions can also be associated with structure in the excitation function.

Recently Mumma et al.[24] determined the polarization of $n^1P \rightarrow 1^1S$ transitions of helium by measuring the angular intensity distribution of the photons. It was, however, not possible to separate the different lines of the $nP \rightarrow 1S$ transitions from each other in the vacuum ultraviolet spectral region. Fig. 11 shows the results of the above authors. It is clear that the lack of separation of the vacuum ultraviolet lines in the experiment of

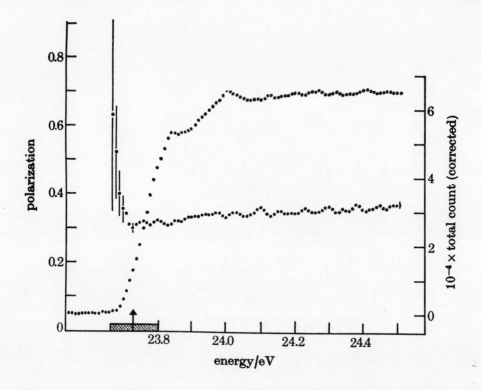

Fig. 9: The excitation function and polarization curve of the
$\lambda = 4922$ Å, $4^1D \rightarrow 2^1P$ line of helium as measured by Heddle et al.[23]
after 120 h.

Mumma et al.[24] makes a comparison with theoretical predictions
rather difficult, however, the polarization of the collective
$\sum n^1P \rightarrow 1^1S$ transition might become independent of n at higher
energies as indicated in comparing the experimental data with
the theoretical calculation of Vriens and Carriere.[25]

Similar oscillatory structure as seen in the polarization
curves of helium and mercury has also been detected in an electron
ion excitation process. Fig. 12 shows the polarization of the
4554 Å Ba$^+$ $2^2P_{3/2} \rightarrow 2^2S_{1/2}$ resonance line as a function of the
electron energy (Crandall et al.[26]).

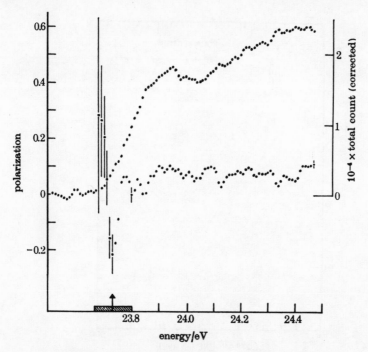

Fig. 10: The excitation function and polarization curve of the 4472 Å , $4^3D \rightarrow 2^3P$ line of helium as measured by Heddle et al.[23] after 170 h.

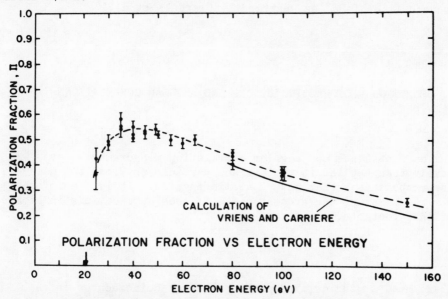

Fig. 11: Polarization curve of the collective $\sum_n n^1P \rightarrow 1^1S$ transition of helium as investigated by Mumma et al.[24] and compared with the calculation of Vriens and Carriere.[25]

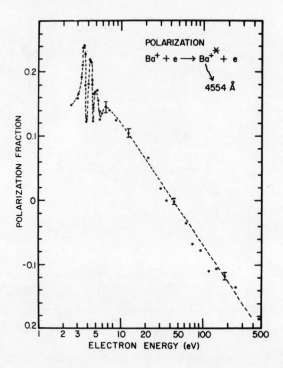

Fig. 12: Polarization of the $\lambda = 4554\mathring{A}$, $^2P_{3/2} \rightarrow {}^2S_{1/2}$ first
resonance line of Ba$^+$ as measured by Crandall et al.[26]

IV.   ELECTRON PHOTON COINCIDENCES AND ANGULAR CORRELATIONS

1.   Theory

     While coincidence and angular correlation measurements have
extensively been applied in nuclear and particle physics, the use
of these methods is rather new to atomic physics. It is, therefore,
possible to summarize the present state of such applications in atomic
physics in a footnote.*

─────────────────

*Imhof and Read[30] used delayed photon electron coincidences in the
measurement of lifetimes of excited states. Pochat, Rozvel and
Peresse (J. de Physique 34, 701, 1973) measured differential
inelastic cross sections for n = 4 and n = 5 states of helium. The

The method of electron photon coincidence techniques will be discussed for a special example of an electron atom excitation process. It will be demonstrated how it can be used as a tool for investigating the following parameters:

1) the collision parameters of the electron atom excitation process (i.e. differential inelastic cross sections, excitation amplitudes and their phase differences),

2) the "source parameters" of the excited atomic state (i.e. orbital angular momentum transfer to the atom, orientation and alignment parameters, and multipole states of the excited atom),

3) the characteristics of electron photon angular correlation (i.e. symmetry direction of the photon emission and its relation to the momentum transfer and to photon polarization data).

The theory of measurements in which photons are detected in delayed coincidences with scattered electrons has been developed in a form which relates the coincidence rates for electron photon angular correlations to scattering amplitudes (Macek and Jaecks[36])

---

excited levels with different ℓ values and multiplicity were identified by measuring coincidences between the scattered electrons with decay photons of selected wavelengths. King, Adams and Read (J. Phys. B, 5, 254, 1972) determined the threshold polarization of the helium $1^1S \rightarrow 3^1P \rightarrow 2^1S$ process by measuring in coincidence inelastically forward scattered electrons and the photons polarized parallel and perpendicular to the forward direction. Ehrhardt, Hesselbacher, Jung, Schultz and Willmann (J. Phys. B, 5, 2107, 1972), Weigold, Hood and Teubner (Phys. Rev. Lett. 30, 475, 1973) and Camilloni, Giardini, Guidoni, Tiribelli and Stefani (Phys. Rev. Lett. 29, 618, 1972) reported coincidence measurements on the electron atom ionization processes – the so-called (e,2e) experiments. El-Sherbini and van der Wiel (Physica, 62, 119, 1972) used coincidence measurements to determine oscillator strengths in inner-shell excitation processes. Backx, Klewer and van der Wiel (Chem. Phys. Lett. 20, 100, 1973) reported triple coincidence measurements between inelastically scattered electrons, ions and photons. Jaecks (Invited Paper at VIII ICPEAC, Beograd, 1973) reviewed coincidence techniques applied to particle–photon coincidences and summarized applications to atom–atom and atom–ion collisions. Hertel and Stoll (J. Phys. B, 7, 570 and 583, 1974) interpreted the electron scattering by laser excited atoms as a time reversed electron photon coincidence experiment, for which Hertel and Macek developed a theory (to be published; see also the Invited Paper by I V Hertel on Electron Scattering by Laser Excited Atoms in Well-Defined States to be published in these proceedings).

or to the orientation and alignment parameters (Fano and Macek[3a]) or
to multipole states (Fano and Macek[3a], and also Blum and
Kleinpoppen[3f]) We shall not derive these results but will apply
them to the following excitation/de-excitation process of helium:
$1^1S \rightarrow n^1P_1 \rightarrow {}^1S_0$. For that case, the links between the
coincidence rate and the excitation parameters follow from simple
and plausible arguments. According to what has already been
pointed out in Section I, the excitation into the excited states
can be described as a coherent superposition of excitation into
degenerate magnetic sub-levels neglecting spin orbit and spin spin
interactions. The atom is prepared into the P state of helium such
that the amplitudes a_{m_L} for magnetic sub-level excitation govern

the initial distribution of the sub-level excitation of the eigenstate
$\psi(^1P_1)$ of the excited atom:

$$\psi(^1P_1) = \sum_{m_L} a_{m_L} \psi_{m_L}(^1P_1) \qquad (10)$$

Mirror symmetry of the electron atom scattering process requires,
for the three excitation amplitudes a_0, a_1 and a_{-1} of the P-state,
the restrictions $a_1 = a_{-1}$. $\psi(^1P_1)$ can be normalized such that the
amplitudes are related to the inelastic differential excitation
cross section σ of the ^1P state:

$$|a_0|^2 = \sigma_0, |a_1|^2 = \sigma_1 \qquad (11)$$

$$\sigma = \sigma_0 + 2\sigma_1 \qquad (12)$$

σ_0, σ_1 are partial differential cross sections for exciting the
magnetic sub-levels $m_L = 0$ and $m_L = \pm 1$, respectively. The
amplitudes a_{m_L} are, in general, complex numbers depending on the

entire problem of the collision process. For any given excitation
process, determined by the excitation energy and the scattering
angle of the inelastically scattered electron, the amplitudes
a_1 and a_0 are expected to have a fixed phase relationship to each
other.

$$a_1 = |a_1|e^{i\alpha_1} \quad \text{and} \quad a_0 = |a_0|e^{i\alpha_0} \qquad (13)$$

$$\text{or} \quad \chi = \alpha_1 - \alpha_0 \qquad (14)$$

Taking $\lambda = \dfrac{\sigma_0}{\sigma}$ or $1 - \lambda = 2\dfrac{\sigma_1}{\sigma}$ the following coincidence rate N_c

for the simultaneous detection of the electron and the photon in
the scattering plane (see Fig. 1(b)) can be derived from the

general theory[3b] as

$$N_c \propto \lambda^2 \sin^2\theta_\gamma + (1 - \lambda)\cos^2\theta_\gamma + 2\sqrt{\lambda(1 - \lambda)}\cos\theta_\gamma \sin\theta_\gamma \cos\chi \quad (15)$$

whereby photons are counted without regard to polarization by a detector placed in any direction in the scattering plane and on the opposite side of the electrons scattered under a fixed angle. It can easily be verified that eq. (15) is identical with the eq.

$$N_c \propto \frac{1}{\sigma}\left| a_0 \sin\theta_\gamma + \sqrt{2}a_1 \cos\theta_\gamma \right|^2 \quad (16)$$

Of course this last equation calls for the picture that two classical oscillators with the relative phase difference χ and the amplitudes $|a_0|$ and $\sqrt{2}|a_1|$ are coherently excited in the z and x directions respectively.

The collision parameters λ and χ which, as shown below, can be determined from the coincidence experiment, can easily be linked with the expectation values of orbital angular momentum quantities of the atom in the excited P state (in units of $\hbar$):

$$\begin{array}{ll}
\langle L_x \rangle = 0 & \langle L_x^2 \rangle = \lambda \\
\langle L_y \rangle = -2\sqrt{\lambda(1 - \lambda)}\sin\chi & \langle L_y^2 \rangle = 1 \qquad (17) \\
\langle L_z \rangle = 0 & \langle L_z^2 \rangle = 1 - \lambda
\end{array}$$

with $L^2 = L_x^2 + L_y^2 + L_z^2 = L(L + 1) = 2$ for $L = 1$. Fano and Macek[3a] connected the anisotropic population of the magnetic sub-states with alignment and orientation parameters of the excited atom. These can also be linked with the above collision parameters λ and χ. The Fano-Macek alignment tensors for our example are given by

$$A_0^{col} = \langle 3L_z^2 - \underline{L}^2 \rangle \{L(L + 1)\}^{-1} = \tfrac{1}{2}(1 - 3\lambda)$$

$$A_{1+}^{col} = \langle L_x L_z + L_z L_x \rangle \{L(L + 1)\}^{-1} = \left[\lambda(1 - \lambda)\right]^{-\frac{1}{2}}\cos\chi \quad (18)$$

$$A_{2+}^{col} = \langle L_x^2 - L_y^2 \rangle \left[L(L + 1)\right]^{-1} = \lambda - 1$$

The orientation vector of the atom in the excited P state is defined by[3a]

$$O_{1-}^{col} = \langle L_y \rangle \left[L(L + 1)\right]^{-1} = -\left[\lambda(1 - \lambda)\right]^{\frac{1}{2}}\sin\chi \quad (19)$$

The polarization of the emitted line radiation with reference to the collision frame in Fig. 1(b) can also be linked to the amplitudes and the collision parameters λ and χ . The linear polarization P^{lin} of the emission from the $^1P \rightarrow {}^1S$ transition under consideration can be obtained as follows:

$$P_{zy}^{lin} = \frac{I_z - I_y}{I_z + I_y} = \frac{|a_0|^2}{|a_0|^2} = 1 \tag{20}$$

$$P_{zx}^{lin} = \frac{I_z - I_x}{I_z + I_x} = \frac{|a_0|^2 - 2|a_1|^2}{|a_0|^2 + 2|a_2|^2} = 2\lambda - 1 \tag{21}$$

$$P_{xy}^{lin} = \frac{I_x - I_y}{I_x + I_y} = 1 \tag{22}$$

(I_x, I_y and I_z are the intensity components polarized parallel to the x, y and z directions, respectively; direction of emission perpendicular to the polarization vectors of the polarized intensity components). The circular polarization of the line transition $^1P \rightarrow {}^1S$ observed perpendicular to the scattering plane is equal to the orbital angular momentum transfer:

$$P_{xz}^{circ} = -2\sqrt{\lambda(1-\lambda)}\sin\chi = \left\langle L_y \right\rangle = \frac{N^+ - N^-}{N^+ + N^-} = 2 Q_{1-}^{col} \tag{23}$$

(N^+, N^- are the number of photons with spin parallel and antiparallel to the direction of propagation.)

No circular polarization occurs if the direction of observation of the photons is within the scattering plane which is equivalent to the fact that no "net" orbital angular momentum transfer takes place with reference to the scattering place:

$$P_{xy}^{circ} = P_{yz}^{circ} = 0 \tag{24}$$

Wykes[27] calculated the photon polarization of alkali resonance radiation detected in coincidence with inelastically scattered electrons. The general equations derived by Wykes[27] describe the polarization for any arbitrary observation angle. Due to the fact that fine and hyperfine structure and also the finite level width of the excited state have to be taken into account, the polarization of the alkali resonance radiation leads to modified expressions compared to eqs. (20-24). The spin orbit and nuclear magnetic interactions depolarize the photon radiation with the effect that e.g. P_{xy}^{lin} and P_{zy}^{lin} become different and are no longer 100% as in the case of the helium $1^1S \rightarrow n^1P$ excitation process discussed above.

2. Experimental Method

The scheme of the experimental method is determined by the following requirements: a monoenergetic beam of electrons excites a well-defined target of atoms, e.g. a collimated beam of helium atoms; both the inelastically scattered electrons and the photons from the de-excitation process have to be "filtered", the unwanted elastic and inelastic electrons and the photons arise from those processes not being studied. The initial monoenergetic beam of electrons exciting the atoms can be taken from the exit slit of an electron monochromator, the inelastically scattered electrons having lost the threshold excitation energy can be filtered from unwanted electrons by means of a second electron monochromator tuned to the appropriate electron energy (Fig. 13). Photons with the frequency of the transition to be studied and the energy-selected electrons having excited the atom can be detected by observing them as delayed coincidences in a fast coincidence circuit (Bell[29], Imhof and Read[30], Eminyan et al.[28]). A typical energy-loss spectrum obtained by scanning the voltage on the entrance slit of the analyser is displayed in Fig. 14. Photons from the transition to be studied and the energy selected electrons are detected by appropriate electron and photon detectors and the corresponding pulses are fed into a time-to-amplitude converter (TAC). Correlated electron and photon pulses from true coincidence events arrive at

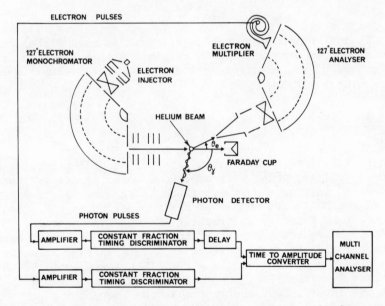

Fig. 13: Schematic diagram for electron photon coincidence measurements:[28] the atomic beam (helium) is crossed by the monoenergetic electron beam from the 127° electron monochromator. The electron and photon pulses from the excitation process are fed into the delayed coincidence electronics.

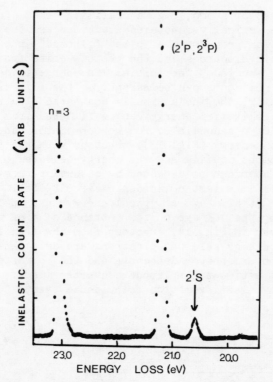

Fig. 14: Typical helium energy loss spectrum as used for electron photon coincidences from $2^1P \rightarrow 1^1S$ transitions (See Fig. 15): initial electron energy 80 eV, electron scattering angle $\theta_e = 16°$, energy resolution 0.25 eV (FWHM).

the TAC with a definite correlation in time. These "real" coincidences fall into a restricted interval Δt of channels of the multichannel analyser (MCA) whereas chance coincidences are randomly distributed and introduce only a uniform background on the delay-time spectrum of the MCA. The number of real coincidences collected in a given time can be found by subtracting the baseline measured outside Δt from the total number of coincidences, real and chance, within the coincidence peak confined to Δt. A delayed coincidence spectrum between photons from the $2^1P \rightarrow 1^1S$ transition (λ 584 Å) and the electrons from the $(2^1P, 2^3P)$ energy loss peak is shown in Fig. 15. Note that the insufficient energy resolution between the 2^1P and the 2^3P energy loss peak does not affect the number of real coincidences since the 2^3P state does not decay with the emission of vacuum ultra-violet light.

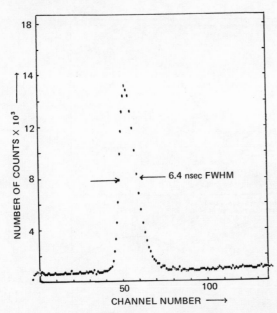

Fig. 15: Delayed coincidence spectrum from the excitation/de-
excitation process $1^1S \to 2^1P \to 1^1S$ of helium: initial electron
energy 80 eV, electron scattering angle $\theta_e = 16^O$, photon angle
$\theta_\gamma = 126^O$ (opposite to scattered electron beam), accumulation time
$\sim$ 12 hours, channel width 0.58 ns; inelastic electron rate $\sim$ 6 kHz,
photon rate 21 kHz.

3.   Results

The first measurements of Eminyan et al.[28] and Standage et al.[31]
for electron photon angular correlations of the helium $1^1S \to n^1P_1$
excitation process were restricted to the observation of the photon
angular distribution in the scattering plane. Fig. 16 shows angular
e-hν correlation curves for 1^1S_0-2^1P excitation with an incident
beam of electrons of 60 eV and for fixed electron scattering
angles of 16°, 35°, and 40°. The examples of Fig. 16 clearly
demonstrate the well-known fact that the Born approximation (dashed
curves in Fig. 16) approaches the experimental data for small momentum
transfer and sufficiently high energy; at the lowest scattering angle
in Fig. 16 ($\theta_e = 16^O$) the first Born approximation is close to the
experimental data, whereas it deviates substantially from reality
at larger angles ($\theta_0 = 35^O$ and 40°) at the same energy of 60 eV. It
might be useful to recall the fact that the Born approximation
requires the selection rule $\Delta m_L = 0$ with respect to the momentum

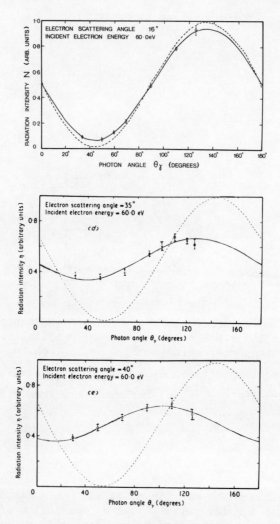

Fig. 16: Electron photon angular correlations for different electron scattering angles (θ_e) and for the $1^1S \to 2^1P \to 1^1S$ excitation/de-excitation of helium (Eminyan et al.[28]). One standard deviation. Full curve, least squares fit of equation (15) to the experimental data; dashed curve prediction of the first Born approximation.

transfer direction. Consequently, the electron photon angular correlation of our excitation/de-excitation process under consideration may be illustrated by the picture as shown in Fig. 17: for a fixed electron scattering angle the probability

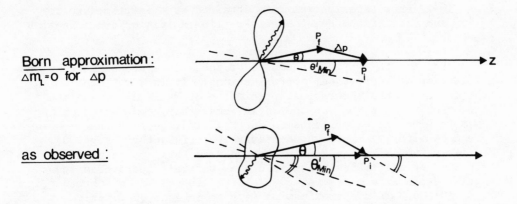

Born approximation:
$\Delta m_L = 0$ for Δp

as observed :

Fig. 17: Momentum analysis (P_i initial momentum of electron travelling in z direction, P_f final momentum of electron after collisional excitation, ΔP momentum transfer to atom) and photon angular distribution curve (arrow) of $^1S \to {}^1P \to {}^1S$ excitation/de-excitation process of helium; upper part: photon angular distribution according to Born approximation based upon selection rule $\Delta m_L = 0$ with reference to ΔP as axis of quantization, lower part: typical photon angular distribution as actually observed (See Fig. 16). Note the tilt between the symmetry directions of the two distribution curves.

of finding the photon is determined by the dipole characteristics of a classical oscillator parallel to the momentum transfer ΔP. Zero intensity of the photon emission occurs in the direction of the oscillating dipole or in the direction of the momentum transfer ΔP, which, together with the direction of the incoming electron beam, defines the angle θ_{min}. In the case as observed, and illustrated in the lower part of Fig. 17, the minimum photon intensity does not decrease to zero and does not occur in the direction of the momentum transfer. The tilt between the minimum angle as predicted by the Born approximation and the actual observed minimum angle can be as large as $40°$ according to the measurements (see Fig. 16).

In order to extract the collision parameters λ and χ and the minimum angle θ_{min} from the electron photon angular correlation, a least squares fit of the theoretical angular correlation function (eq. 15) to the data points, like those of Fig. 16, was carried out.

Figs. 18-22 show experimental data obtained by this procedure for
λ, $|\chi|$ and θ_{min} of the He, $1^1S \rightarrow 2^1P \rightarrow 1^1S$ excitation/de-excitation
processes.

Note that the various examples of these data are compared with
a selection of theoretical approximations. It is worth mentioning
that for 80 eV the theoretical prediction of Madison and Shelton[32]
(distorted wave approximation) is in remarkably good agreement in
$\lambda = \sigma_0/\sigma$ with the experimental data (Fig. 18), whereas the
agreement with $|\chi|$ is only qualitatively satisfactory (Fig. 20).
Similarly, the eikonal DWBA (distorted wave Born approximation)
with Glauber distorting potentials qualitatively approaches the
angular dependence of $|\chi|$ at 80 eV (Joachain and Vanderpoorten[33],
Figs. 19-20). This approximation is also in good qualitative
agreement with the experimental data for $|\chi|$ at 200 eV but shows
some discrepancy with λ (Fig. 21).

Clearly the data for λ and $|\chi|$ in Figs. 18-22 already
demonstrate how the collision parameters λ and $|\chi|$ can be used as
sensitive tests for collision theories. An interesting example for
the angular dependence of λ and $|\chi|$ of the Lyman-α excitation process
of atomic hydrogen has been calculated by several authors, by
McDowell, Morgan and Myerscough[35a], based upon the distorted wave
polarized orbital approximation, and also by Geltman[35b], based upon
the Coulomb projected Born approximations. An example for the
theoretical prediction of λ and χ for the Lyman-α line by
McDowell et al.[35a] is given in Fig. 23.

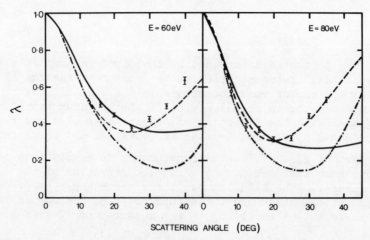

Fig. 18: Experimental data of $\lambda = \sigma_0/\sigma$ for the He, $1^1S \rightarrow 2^1P \rightarrow 1^1S$
excitation/de-excitation process[28] compared to theoretical predictions:
———— FBA (first Born approximation), ——— Madison and Shelton[32] and
—··—··— Csanak, Taylor and Tripathy[34].

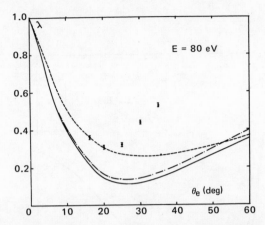

Fig. 19: Experimental data as in Fig. 18 compared to theoretical predictions: ---- FBA; _._._._eikonal DWBA (distorted wave Born approximation) with static distorting potentials and ———— eikonal DWBA with Glauber distorting potentials (Vanderpoorten and Joachain[33]).

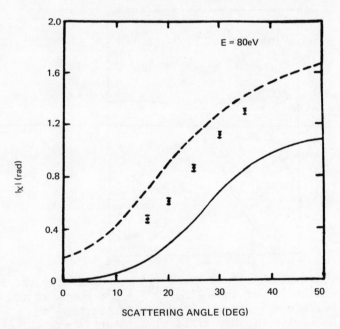

Fig. 20: Experimental data of $|\chi|$ for the He, $1^1S \rightarrow 2^1P \rightarrow 1^1S$ excitation/de-excitation process[28] at 80 eV compared to theoretical predictions: ---- Madison and Shelton[32] (78 eV), ———— eikonal DWBA with Glauber distorting potentials (Vanderpoorten and Joachain[34], 81.6 eV).

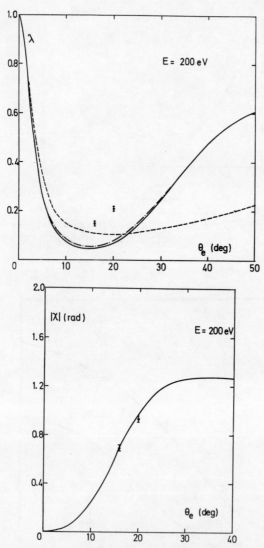

Fig. 21: Experimental data[28] of λ and $|\chi|$ at 200 eV for the He, excitation/de-excitation process $1^1S \rightarrow 2^1P \rightarrow 1^1S$ compared to theoretical predictions by Joachain and Vanderpoorten[33]: ———— eikonal DWBA with Glauber distorting potentials, —·—·— eikonal DWBA with static distorting potential, ----- FBA.

3^1P   ELECTRON   ENERGY   50 eV

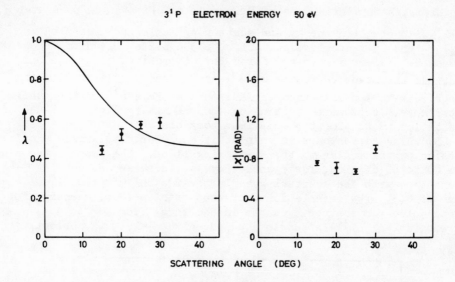

Fig. 22:   Experimental data of $\lambda = \sigma_0/\sigma$ and $|\chi|$ for the He, excitation/de-excitation process $1^1S^0 \rightarrow 3^1P \rightarrow 2^1S$ ($\lambda = 5016$ Å) at 50 eV (Standage et al.[31]). One standard deviation. ———FBA.

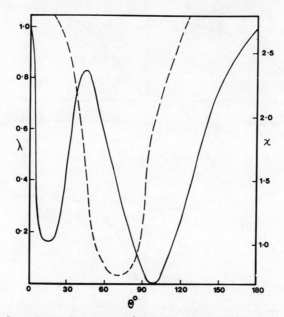

Fig. 23:   Angular dependence of λ and χ for the Lyman-α excitation of atomic hydrogen as predicted by McDowell, Morgan and Myerscough[35a] at 100 eV.

Examples for variations of θ_{min} and the orientation parameter $|O_1^{col-}|$ with electron energy and scattering angle are presented in Figs. 24 and 25 respectively. Notice the considerable difference between θ_{min} according to the first Born approximation on the one hand, and the observed values of θ_{min} and those resulting from the distorted wave approximation of Madison and Shelton[32] on the other. Furthermore, it is worth mentioning the remarkably high magnitude of orientation up to almost the maximum possible vale of $|0.5|$ (see eq. 23) which corresponds to completely circularly polarized photon emission perpendicular to the scattering plane. The data in the range from 40 to 80 eV show, for each energy, an increase of the orientation with the scattering angle towards the maximum allowed value of 0.5. The orientation increases with increasing energy for a fixed electron scattering angle in the range from 20° to 35°, whereas, at the smallest angle studied, at $\theta_e = 16°$, the orientation

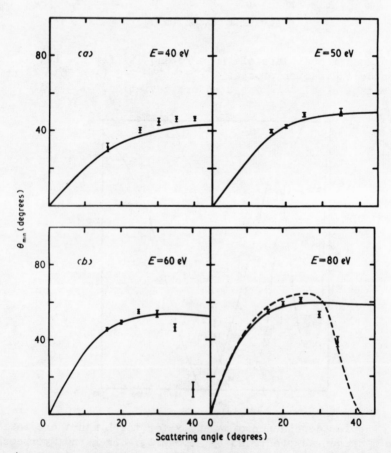

Fig. 24: Variation of θ_{min} with electron energy and scattering angle $1^1S \rightarrow 2^1P \rightarrow 1^1S$ excitation/de-excitation process[28]. First Born approximation, --- Madison and Shelton[32].

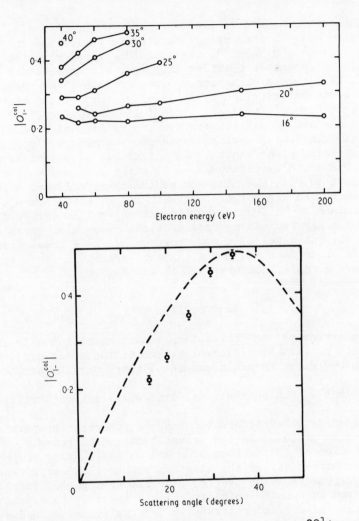

Fig. 25: Variation of the orientation parameter $|O_{1-}^{col}|$ for the excitation/de-excitation process $1^1S \rightarrow 2^1P \rightarrow 1^1S$ with electron energy at different electron scattering angles (upper part), and with electron angle at fixed energy of 80 eV. One standard deviation. ---- Madison and Shelton[32] at 78 eV.

is nearly constant from 40 eV to 200 eV.

Attention should also be drawn to the following fact. While the first Born approximation reproduces the experimental data for

λ at small scattering angles (see Fig. 18) reasonably well, it predicts zero orientation for all scattering angles and energies. According to Fano and Macek[3a], the momentum transfer vector ΔP is the only collision parameter on which alignment and orientation can depend in the Born approximation. Since ΔP is a polar vector, no orientation (to be described by an axial vector) of the atom can occur in this approximation. It then follows from eq. (23) that both the phase difference between the two amplitudes a_o and a_1 and also the circular polarization of the photons observed perpendicular to the scattering plane vanish. The magnitude of orientation for the $1^1S \rightarrow 2^2P \rightarrow 1^1S$ excitation/de-excitation process at 60 eV and 20° is almost $|0.25|$, i.e. the corresponding circular polarization is $\pm 50\%$ (note the sign of the circular polarization has not yet been determined). While a given approximation like the Born approximation may reproduce cross sections or partial cross sections (or $\lambda = \sigma_o/\sigma$) quite well, it may, however, fail in predicting phase differences or orientation and alignment parameters (as illustrated by the above example, compare Fig. 18 and Fig. 25).

V. CONCLUSIONS

The electron photon correlation measurements **not** only led to measurements of partial differential excitation cross sections σ_{m_L} , compared with the measurements of total partial excitation cross sections Q_{m_L} in conventional line polarization studies. They also provided detailed information on the coherence properties of the excitation process, on the orientation and alignment characteristics of the excited atom and on the angular momentum dynamics occurring during the electron atom collision process. Future developments might even combine electron photon correlation with electron and photon spin experiments in order to separate the exchange and direct processes(in the amplitudes a_o and a_1) from each other, as discussed by the author for the case of electron alkali excitation[3e]. Such developments might be initiated by the already successful spin flip experiments with polarized electrons[36] and polarized atoms[2], or by the electron scattering experiments with laser excited atoms[3h] by using polarized electrons, atoms and photons.

Acknowledgements

The author wishes to thank members of the Experimental Physics Research Group of the Physics Department of Stirling University for many inspiring discussions. The critical reading of the manuscript by I. McGregor is gratefully acknowledged.

REFERENCES

1)   B. Bederson, in Atomic Physics 3, Proceedings of 3rd Int.
     Conf. on Atomic Physics, S.J. Smith and G.K. Walters, eds.,
     Plenum Press, New York, 1973, p.401.

2)   Recent reviews in this field were given by B. Bederson,
     reference (1) of this article, also P.S. Farago, Reports on
     Progress in Physics, Vol. 34, No. 11, p.1055, 1971, and
     H. Kleinpoppen, Introductory Paper of Panel Discussion on
     Spin and Polarization Effects in Atomic Processes in
     "Fundamental Interactions in Physics", edited by B. Kursunoglu,
     A. Perlmutter et al., Plenum Press 1973.

3)   (a) U. Fano and J.H. Macek, Revs. Mod. Physics, Vol. 45,
     No. 4, p.553, 1973; (b) J.H. Macek and D.H. Jaecks, Phys.
     Rev. A4, p.2288, 1971; (c) P.G. Burke and J.F.B. Mitchell,
     J. Phys. B, Vol. 7, No. 2, p.229, 1974; (d) P.S. Farago,
     J. Phys. B, Vol. 7, No. 1, p.128; (e) H. Kleinpoppen, Phys.
     Rev. A3, p.2015, 1971; (f) K. Blum and H. Kleinpoppen, Phys.
     Rev. A9, p.1902, 1974; (g) D.L. Moores and D.W. Norcross, J.
     Phys. B, Vol. 5, p.1482, 1972; (h) I.V. Hertel and W. Stoll,
     J. Phys. B, Vol. 7, p.570, 1974; (i) J.H. Macek and I.V.
     Hertel, to be published; (j) K. Rubin, B. Bederson, M. Goldstein
     and R. E. Collins, Phys. Rev. 182, 201, 1969.

4)   H.W.B. Skinner and E.T. Appleyard, Proc. Roy. Soc. A, 117,
     224, 1927.

5)   I.C. Percival and M.J. Seaton, Phil. Trans. Roy. Soc.,
     London, Series A, No. 990, Vol. 251, p.113, 1958.

6)   E. Baranger and E. Gerjuoy, Proc. Phys. Soc. 72, 326, 1958.

7)   V.J. Ehlers and A.C. Gallagher, Phys. Rev. A7, 1573, 1973.

8)   D. R. Flower and M.J. Seaton, Proc. Phys. Soc., London, 91,
     59, 1967.

9)   W.R. Ott, W.E. Kauppila and W.L. Fite, Phys. Rev. Lett. 19,
     1361, 1963.

10) H. Hafner, H. Kleinpoppen and H. Krüger, Phys. Lett. 18, 270,
     1965 and also H. Hafner and H. Kleinpoppen, Z. Physik, 198,
     315, 1967.

11) H.G.M. Heideman, C. Smit and J.A. Smit, Physica 45, 305, 1969.

12) A.N. Tripathi, K.C. Mathur and S.K. Joshi, J. Phys. B, Vol.6, 1431, 1973.

13) N. Feautrier, J. Phys. B, 3, L152, 1970.

14) P.G. Burke and A.J. Taylor, J. Phys. B, 2, 869, 1969.

15) D.L. Moores and D.W. Norcross, J. Phys. B 5, 1482, 1972.

16) E.A. Enemark and A. Gallagher, Phys. Rev. A, 6, 192, 1972.

17) G. Gould, Ph.D. thesis, University of South Wales, 1970.

18) P.D. Burrow and J.A. Michejda, private communications, similar resonance structure was also observed in the angular dependence in low energy electron – mercury scattering by M. Düweke, N. Kirchner, E. Reichert and E. Staudt, J. Phys. B, 6, L208, 1973.

19) T.W. Ottley, Ph.D. thesis, University of Stirling, 1974.

20) U. Fano and J. W. Cooper, Phys. Rev. 138, A, 400, 1965.

21) J.C. McConnell and B.L. Moiseiwitsch, J. Phys. B, 1, 406, 1968.

22) T.W. Ottley, D.R. Denne and H. Kleinpoppen, J. Phys. B, Vol.7, L179, 1974.

23) D.W.O. Heddle, R.G.W. Keesing and R.D. Watkins, Proc. Roy. Soc. Lond. A, 337, 443, 1974.

24) M.J. Mumma, M. Misakian, W.M. Jackson, and J.L. Faris, Phys. Rev. A, 9, 203, 1974.

25) L. Vriens and J.D. Carriere, Physica 49, 517, 1970.

26) D.H. Crandall, P.O. Taylor and G.H. Dunn, quoted in U. Fano and J.H. Macek, Revs. Mod. Phys. 45, No. 4, 553, 1973.

27) J. Wykes, J. Phys. B, 5, 1126, 1972.

28) M. Eminyan, K.B. MacAdam, J. Slevin and H. Kleinpoppen, Phys. Rev. Lett. 31, 576, 1972 and also J. Phys. B, to be published 1974.

29) R.E. Bell, Alpha, Beta and Gamma Ray Spectroscopy edited by
     K. Siegbahn, Amsterdam, North Holland Vol. 2.

30) R.E. Imhof and F.H. Read, J. Phys. B, 4, 450, 1971 and
     Chem. Phys. Lett. 11, 326, 1971.

31) M.C. Standage, M. Eminyan, H. Kleinpoppen, K.B. MacAdam and
     J. Slevin, J. Phys. B, to be published.

32) D.H. Madison and W.N. Shelton, Phys. Rev. A7, 449, 1973.

33) C.J. Joachain and R. Vanderpoorten, private communication,
     1974.

34) G. Csanak, H.S. Taylor and D.N. Tripethy, J. Phys. B, 6,
     2040, 1973.

35) (a) M.R.C. McDowell, L.A. Morgan and Valerie P. Myerscough,
     private communication; see also Invited Papers by
     M.R.C. McDowell and (b) by S. Geltman in Proceedings of the
     International Symposium on Electron and Photon Interactions
     with Atoms in Honor of Ugo Fano, Stirling, 16th - 19th July,
     1974, to be published by Plenum Press 1975.

36) G.F. Hanne and J. Kessler, Phys. Rev. Lett. 33, 341, 1974.

THE NEGATIVE HYDROGEN ION AND ITS BEHAVIOR IN ATOMIC COLLISIONS

John S. Risley

Department of Physics, University of Washington

Seattle, Washington 98195 USA

I. INTRODUCTION

This paper reviews the structure of the negative hydrogen ion and its interaction with atomic particles.

The negative hydrogen ion has been of much interest in the past because of its simple structure. The two-electron system of H^- is one of the simplest physical systems to describe quantum mechanically. Although exact energy levels and analytic wave functions for H^- are not available, accurate approximate solutions can be obtained. Much attention has been directed toward using H^- as a proving ground for the validity of theoretical models and their extension to other more complicated systems. Important analogies also exist between the structure of H^- and resonances in e-H atom scattering. In many cases theoretical treatments have been prompted by the lack of definitive experimental measurements which unravel the mysteries of H^-.

The negative hydrogen ion has been of interest since the advent of quantum mechanics in connection with solid state theory. Bethe[1] and Hylleraas[2] both calculated the binding energy and radial distribution of H^- and applied their results to the crystaline structure of LiH in the late 1920's.

In the 1940's and 1950's the negative hydrogen ion was found to play an important role in the opacity of the atmospheres of the sun and other stars. H^- is formed by radiative recombination of low energy electrons and hydrogen atoms. Radiation from the surface of the sun is absorbed by photodetachment. The continual formation and destruction of H^- conserves the total energy being

radiated but modifies the characteristics of the light emitted
from the surface of the sun. Photodetachment occurs throughout
the visible region because of the weak binding energy of H^-.

In addition to basic theoretical interest, the study of H^- is
important for practical reasons. Knowledge of its behavior in
collisions with photons, electrons, ions and neutral atoms and
molecules is of crucial importance if we are to fully understand
astrophysical observations and to effectively utilize H^- for injection into high temperature plasmas. In addition, sharp lines in
the electron spectra of H^- serve as accurate calibration standards
for electron spectroscopy since the positions and widths are well-
known theoretically and experimentally. H^- is also important for
injection into high-energy ion accelerators.

We will see in the next few sections that our theoretical
ideas on the structure of H^- are well developed although they have
not been tested with sufficient accuracy by experiment. On the
other hand, experimental results on collision phenomena involving
H^- are much more advanced than our theoretical understanding.

II.   STRUCTURE

II.A.   Ground State

The two electrons are arranged around the proton so that the
extra electron experiences a small, attractive force. Although a
$(1s^2)$ electron configuration is frequently used to denote the ^1S
ground state of H^-, the two electrons are not equally spaced from
the proton. A simple product of radial, hydrogenic wave functions,
$\exp[-\alpha(r_1 + r_2)]$, yields, variationally, a negative value for the
binding energy.[3] A better approximation using a product of two
hydrogenic wave functions with different scaling factors,[4]
$\exp[-\alpha r_1 - \beta r_2]$, gives a positive binding energy of 0.36 eV with
$\alpha = 1.04 \, a_0^{-1}$ and $\beta = 0.28 \, a_0^{-1}$. H^- can thus be described as a
two-electron system in which one electron is weakly bound in the
field of a polarized hydrogen atom.

The Rayleigh-Ritz variational method has been applied to the
H^- ion using various expansions. Convergence is satisfactory considering the accessibility of high speed computers, e.g. a 3 parameter Hylleraas-type wave function[1] yields a binding energy of
0.688 eV; 6 parameter,[2] 0.719 eV; 11 parameter,[5] 0.7495 eV; 24
parameter,[6] 0.75405 eV and 100 parameter,[7] 0.754740 eV. The energy
of the ground state has also been calculated to be 0.7517 eV
using a three-state hydrogenic close coupling approximation with
16 correlation terms.[8]

The most definitive theoretical work of the energy of the

ground state of H⁻ is the variational calculations of Pekeris[9] and
Aashamar[10] in which the eigenenergies are accurate to one part in
10^9. The theoretical ionization energies are listed in Table I.
The lower accuracy (lower than $1:10^9$) shown in the table is due to
uncertainty in the energy conversion factor.[11] Pekeris, using
parametric coordinates, solved determinates up to order 444 while
Aashamer used the Hylleraas-Scherr-Knight variational perturbation
wave function through 20th order. In both treatments, mass-polar-
ization and relativistic corrections were included. These two
calculations are the most precise theoretical predictions for H⁻.

Table I. Energy of the ground state of H⁻. Number in parenthesis
gives uncertainty in last digits.

E(eV)	Method	Investigator
	Theoretical	
−0.754215(2)	Variational − 444 order deter.	Pekeris (1962)[9]
−0.754220(2)	Variational − 20 order deter.	Aashamar (1970)[10]
	Experimental	
−0.8(1)	Surface ionization	Khvostenko and Dukel'skii (1960)[12]
−0.77(2)	Extrapolation of photodetach-ment cross section	Weisner and Armstrong (1964)[14]
−0.756(13)	Stellar photodetachment spectrum	Berry (1969)[16]
−0.776(20)	Crossed-beam, photodetach-ment threshold	Feldmann (1970)[17]

Experimental measurements of the ground state energy are much
less accurate than theory. The first direct measurement of the
electron affinities of H⁻ was made by Khvostenko and Dukel'skii[12]
using the surface ionization technique. A value of 0.8 ± 0.1 eV
for the binding energy was obtained.

The photodetachment cross section of Smith and Burch[13] was
extrapolated to threshold, at about 1.66 µm, using a formula derived by Bethe, Peierls and Longmire.[14] At present, the most
precise experimental value comes from the spectral distribution
measurements of the star α-Taure recorded in a balloon flight by

Woolf, Schwarzschild, and Rose.[15]   The spectrum exhibits a distinct
knee at approximately 1.6 µm. By subtracting the background radiation a value accurate to 1.5% was obtained.[16]

A laboratory measurement of the threshold behavior of the H⁻
photodetachment cross section was made by Feldmann[17] in a crossed
beam experiment using an intense high pressure Xe lamp and prism
monochromator for the light source. The accuracy for the binding
energy in this experiment was about 3%.

The experimental determinations are listed in Table I. None
of the past measurements provide a real test to the remarkably
precise theoretical values. Hopefully a more definitive determination will be made in the near future.

II.B. Excited States

At present, no bound excited levels of H⁻ below the ionization
or detachment limit have been observed or calculated.[18-20] Doubly
excited states in which both electrons are promoted into outer
orbitals have been predicted theoretically and observed experimentally. These levels lie above the ionization limit of H⁻ and both
electron and photon de-excitation can occur. Because the coupling
between the excited H⁻ and the continuum state consisting of an H
atom plus free electron is usually strong, the autodetachment
transition rate is large, 10^{11} to $10^{14} sec^{-1}$, compared to optical
transition rates. One expects the excited state of H⁻ to decay
almost exclusively via electron emission, provided the transition
is allowed.

The energy of the excited states of H⁻ has been calculated
several ways. The Rayleigh-Ritz variational method is well suited
for providing an upper bound for the energy of the lowest lying
state of a given symmetry. For higher members of the series, the
Feshbach projection operator formalism has revealed additional
levels for an effective Hamiltonian representing the closed channels lying just below the excited states of the hydrogen atom. A
level shift of the energy must be calculated before one can compare
Feshbach eigenenergies with other calculations or experimental
results. The excited states of H⁻ can also be treated as resonance
in the elastic scattering of electrons on atomic hydrogen. Close
coupling and variational calculations for the phase shift of the
wave function of the scattered electron have uncovered many resonances, providing accurate values for the energy and width of the
H⁻ states. Several excellent reviews have been written dealing
with e-H scattering resonances.[21-25]

Theoretical studies of electron-hydrogen elastic scattering

resonances commenced with the 1s-2s close-coupling calculations of
Smith, McEachran and Fraser[26] and the 1s-2s-2p calculations of
Burke and Schey[27] in 1962. The earliest prediction of a doubly
excited state of H^- appears to be the variational calculation of
the $(2s2p)^3P$ level by Hylleraas in 1950.[28] Holøien[29] in 1958 used
a 20th ordered orthogonal basis set to find the energy of the
$(2s^2)^1S$ state.

     Precise calculations for excited Feshbach states below the
n=2 level have been made using the 3-state hydrogenic close-coupling
approximation plus 20 correlation terms by Burke,[30] the 6-state
close-coupling calculation for the $(2p^2)^1D$ by Ormonde, McEwen and
McGowan,[31] and the application of the variational technique using
the Feshbach projection operator formalism by Bhatia and Temkin[32]
and Chen and Chung.[33]

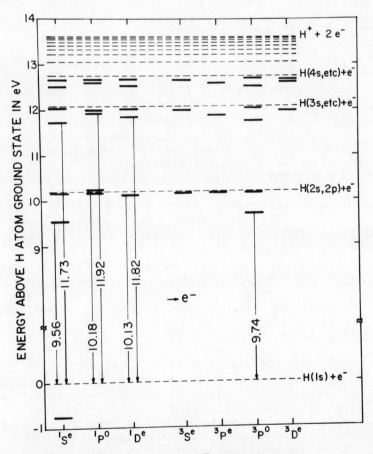

Fig. 1. Energy level diagram of H^- showing some of the autodetaching transitions observed experimentally.

For states lying below the n=3 level, calculations have been made using a six-state close coupling approximation by Burke, Ormonde and Whitaker.[34] Below the n=3 and n=4 levels Oberoi[35] has made calculations using the Feshbach formulism and Holøien and Midtdal[36] have calculated the $^3P^e$ energy using a multiconfigurational method.

The question of whether a $^3P^e$ state with even parity can exist below the n=2 level has been of great interest. L-S selection rules based on the e-e interaction prohibit the state from decaying via autodetachment. The decay scheme for autodetachment is

$$H^-(2p^2)^3P^e \rightarrow H(1s) + e^-.$$

From conservation of orbital angular momentum, the outgoing electron must be represented by a p wave after ejection. The parity of the <u>initial state</u> $(2p^2)^3P^e$ is even. The parity of the <u>final state</u> (hydrogen atom plus outgoing p electron) is odd. Thus, parity and angular momentum cannot be conserved simultaneously in electron decay. Optical emission is allowed.

The $^3P^e$ state is frequently referred to as the second bound state of H$^-$ because of its inability to autodetach via a Coulomb interaction. It is possible however for the state to eject an electron via a spin-orbit interaction with a characteristic transition probability 10^8 smaller than a typical autodetaching decay.[37,38] The energy of the $(2p^2)^3P^e$ state has been predicted by Drake,[39] Bhatia[40] (who also calculates the mass polarization correction, and by Aashamar[10] who, in addition, calculated the relativistic and radiative corrections. His value is 10.190003 eV.

A ^1P shape resonance above the n=2 level at 10.217 eV was predicted by Taylor and Burke[41,42] using 3-state close coupling plus 20 correlation terms. Shape resonances or potential resonances arise whenever the interaction potential has a sufficiently large repulsive barrier, sometimes caused by the centrifugal force, which temporarily traps the electron in the well.

Figure 1 is an energy level diagram based on the theoretical calculations described above. Included are transitions observed experimentally (see below).

As an example of the precision of theoretical efforts,[30,43-53] Table II lists the energy and width of the lowest $(2s^2)^1S$ excited state of H$^-$. All positions are within 10 meV of one another. The more recent calculations are within 3 meV. Less agreement is found in calculations of the width. Space does not allow a compilation of all the calculated values of the excited states of H$^-$. Currently over 40 papers have been published on theoretical predictions. Recent tabulations can be found elsewhere.[54,55,25]

Table II. Precision calculations of the $(2s^2)\,^1S$ autodetaching
state of H^-.

E(eV)	Γ(eV)	Method
9.555	0.048	Phase shift, 3 state close coupling with correlation[30,43]
9.553	0.051	Phase shift, 6 state close coupling[44]
9.555	0.039	Feshbach projection operator formalism[45]
9.552	0.050	Phase shift, modified close coupling[46]
9.545	0.052	Feshbach technique[47]
9.5522	0.047	Phase shift, Kohn variational method[48]
9.5517	0.0474	Rayleigh-Ritz variational method, complex coordinates[49]
9.5490	0.0411	Feshbach technique[50]
9.5487	0.0406	Feshbach technique[51,52]
9.5518	0.046	Rayleigh-Ritz variational method, complex coordinate rotation[53]

Excited states of H^- were first observed indirectly as resonances in the elastic scattering of electrons from atomic hydrogen
and as resonances in the emission cross section of Lyman alpha
produced by low energy electron impact. Schulz[56] in 1964 was the
first to detect the resonances in an electron transmission experiment. This technique was refined by Sanche and Burrow[57] in 1972
by using a high-resolution trochoidal monochromator and by modulating the voltage of the collision chamber and detecting the
derivative of the transmitted current. Their results are shown
in Figure 2. The solid line is the measured signal. The solid
circles are the results of an optimum fit to the experimental data
from which the resonance parameters were derived. The arrows
indicate theoretical predictions of the position of the resonances.

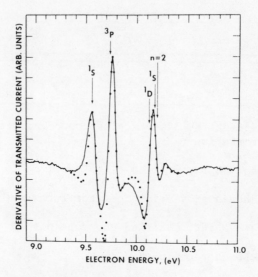

Fig. 2.   The solid line shows the derivative of the transmitted
current through atomic hydrogen measured as a function of the
electron impact energy.

Resonances in the differential elastic scattering cross section
for electrons on atomic hydrogen have been reported by Kleinpoppen
and Raible[58] and by McGowan and co-workers.[59-62]   Figure 3 shows
the electron signal observed at 90° versus electron energy.[59]
Subsequent to this publication the energy scale was corrected upwards by 0.010 eV.[60]

Several resonances below the n=3 level have been found in the
emission cross section of Lyman-alpha from H produced by low energy
electron impact. A large peak just above the n=2 threshold is
believed to be a [1]P shape resonance.[63] Feshbach resonances just
below n=3 were clearly resolved[64] although theoretical predictions
of the positions[34] are not in complete agreement. Additional structure was found just above the n=3 threshold and below the n=4.

The decay of excited states of H⁻ have now been observed
directly as sharp lines in electron spectra from collisionally
excited H⁻.[54] Different states of H⁻ can be excited by producing
fast H⁻ ions and allowing them to hit gas target atoms or molecules.

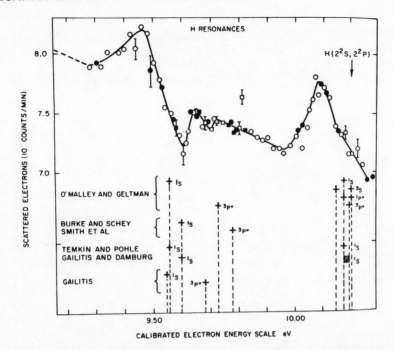

Fig. 3. Differential electron scattering signal at $90°$ observed
with atomic hydrogen. Indicated below the data are the best
theoretical values for H^- states prior to 1965. The large maximum
close to the n=2 threshold has now been identified as being due
primarily to the $(2p^2)^1D$ resonance.[31]

An example of an ejected electron spectrum is shown in Fig. 4. The
widths of the lines are predominantly instrumental due to the
analyzer resolution and kinematic broadening. The energy of the
lines in the lab frame is shifted owing to the high velocity of the
ion before decay.

In Table III the experimental values for the excited states
of H^- are listed along with theoretical results.[54] In general the
agreement is quite good within experimental uncertainty. Higher
accuracy in the experimental measurements is required before
any substantial conclusions can be drawn regarding the theoretical
approaches.

Table III. Experimental values for the energies of the autodetaching states of H^- in eV. Number in parentheses is uncertainty in last digits. References to theoretical values can be found in Ref. 54.

State	Risley et al.[54]	McGowan and co-workers[59-64]	Sanche and Burrow[57]	Kleinpoppen and Raible[58]	Schulz[56]	Theory
Below n=2 level						
$(2s^2)^1S$	9.59(3)	9.56(1)	9.558(10)			9.545–9.585
$(2s2p)^3P$	9.76(3)	9.71(3)	9.738(10)	9.73(12)	9.70(15)	9.731–9.767
$(2p^2)^1D$	10.18(3)	10.130(15)	10.128(10)			10.119–10.156
$(2s2p)^1P$						10.170–10.180
$(2p^2)^1S$						10.164–10.198
Below n=3 level						
$(3s^2)^1S$		11.65(3)				11.727
$(3s3p)^3P$		11.77(2)				11.759
$(3p^2)^1D$	11.86(4)	11.89(2)				11.813
$(3s3p)^1P$						11.910
$(3s4s)^1S$						12.031

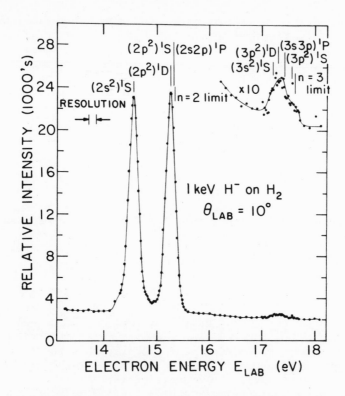

Fig. 4. Electron energy spectrum for 1-keV H⁻ on H₂ observed at 10⁰. The kinetic energy of the electrons from the autodetaching states is shifted to higher energies because of collision kinematics. The vertical lines above each peak indicate theoretical predictions of the energy, corrected for the lab reference frame.

II.C. Singly-Excited States

Recently attempts have been made to attribute certain inter-stellar absorption bands between 4400Å and 6200Å to resonances in the photodetachment spectrum of H⁻.[65-69] Calculations for the energies and widths of the $(1s2s)^{1,3}S$ and $(1s2p)^1P$ states were made.[67,69,70] A large discrepancy exists between the calculations of Ingemann-Hilberg and Rudkjøbing[67] and of Pekeris[9] for the 2^1S and 2^3S states.

Herzenberg and Mandl[71] noted some time ago that a resonance is a useful concept only if the width of the state is small compared to the level spacing of the series. The low lying ^1S, ^3S and ^1P resonances mentioned above could be members of an overlapping sequence.[72] Because of the importance of the problem more theoret-

ical work is needed in this energy range. It should be noted that
no laboratory experiments have demonstrated the existence of these
states. One would also expect to observe the absorption bands,
attributable to H⁻, in spectra of late-type stars including the sun
where continuous absorption by H⁻ is known to be important. No
diffuse absorption features have been observed in the solar spectrum.[73]

II.D. Decay of the $(2p^2)^3P^e$

In the Coulomb, e-e interaction approximation, two optical
decay channels are possible for the $(2p^2)^3P^e$ state of H⁻.   One
is to the $(2s2p)^3P^o$ state and the other to a continuum state consisting of a hydrogen atom in the ground state plus a photon and a
free electron.

$$H^-(2p^2)^3P^e \rightarrow H^-(2s2p)^3P^o + h\nu \ (29\ 000\overset{o}{A})$$

or

$$H^-(1skp)^3P^o + h\nu \ (\text{continuum distribution}).$$

The latter transition is much more probable than the former.
The calculated transition rate[74] for the $(2p^2)^3P^e$-$(2s2p)^3P^o$
decay is $2.5 \times 10^6 \text{sec}^{-1}$ while the transition rate for the $(2p^2)^3P^e$-
$(1skp)^3P^o$ reaction, integrated over energy, is $5.7 \times 10^8 \text{sec}^{-1}$.
Shown in Fig. 5 are two theoretical predictions[74,75] for the differential transition rate or optical emission spectrum expected in the
electron-photon decay of the $^3P^e$ state. Only small differences can
be seen between the two calculations, the latter one[75] showing that
the peak is closer to the maximum photon energy.

III. H⁻ COLLISIONS WITH PHOTONS

III.A. Photodetachment

The photodetachment spectrum of H⁻ is of great importance in
understanding the opacity in the visible and infra-red radiation
region in the atmosphere of certain stars including the sun.

$$H^- + h\nu \rightarrow H + e^-.$$

Much theoretical effort has been carried out on the bound-free
stimulated emission of H⁻ after Wilde[76] first suggested that negative hydrogen ions were responsible for this absorption. Special
theoretical interest in the photodetachment of H⁻ also arises from
the fact that the process is relatively easy to formulate, leading

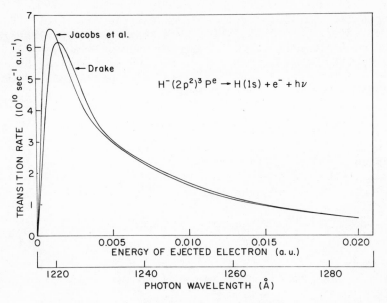

Fig. 5. Predicted differential spontaneous transition rates by Drake[74] and Jacobs et al.[75] for the $(2p^2)\,^3P^e-(1skp)\,^3P^o$ decay of H^- plotted as a function of the kinetic energy of the ejected electron and the wavelength of the emitted photon.

to clear interpretation of various theoretical approximations and methods.

The atomic absorption coefficient or photodetachment cross section can be evaluated equivalently using the "length," "velocity" and "acceleration" forms for the dipole matrix element.[77] If the wave functions for the bound ground state and the free continuum state of H^- were known exactly, the predicted cross section from each of the three matrix elements would be identical. However, since both the ground state and continuum wave functions are known only approximately and since the continuum wave function is, in general, difficult to parametrize, each form of the dipole matrix element is useful in our understanding of H^-. The dipole operator in the matrix element can be regarded as a weighting function for the continuum wave function. Chandrasekhar[78] found that, roughly speaking, the best result for the detachment cross section comes from the "velocity" form because the accuracy of the wave function for the bound and continuum states is highest for intermediate distances of the electrons from the proton. The "length" form weights the asymptotic region more strongly and the "acceleration" form weights the interior region or the electron-electron correlation of the ion.

Numerous calculations have been made for the detachment cross section (see Ref. 77). The most accurate theoretical results are probably those incorporating the 70-parameter wave function of Schwartz (unpublished) for the bound-state and a six-state, hydrogenic close coupled wave function included in a Hartree-Fock expansion for the continuum electron by Doughty, Fraser and McEachan[79] and a polarized orbital approximation for the continuum function by Bell and Kingston.[80] The results are shown in Fig. 6. The solid line was obtained using polarized-orbital free wave functions in the "length" matrix element, and the dashed in the "velocity" matrix element.[80] The + refers to the same calculation using close-coupling free functions in the "length" matrix element, the x refers to the "velocity" matrix element.[79]

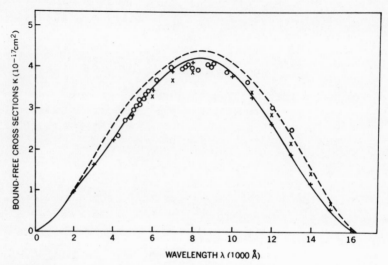

Fig. 6. Photodetachment cross section of H⁻. The open circles are experiment.[13]

Not included in Fig. 6 is some important, additional work on the photodetachment of H⁻ in the vacuum ultra-violet in the neighborhood of the n=2 threshold. Macek[81] found a significant resonance structure near 10.2 eV which he attributed to the ^1P shape resonance of H⁻. He also showed that above threshold the contribution from the n=2 partial cross section (hydrogen atom left in the n=2 state) is comparable to that from the n=1 partial cross section (hydrogen atom in the ground state).

$$H^- + h\nu \rightarrow H(n\ell) + e^-.$$

Hyman, Jacobs and Burke[82] extended the treatment and found that photodetachment into the 2p state is much more probable than to the 2s state close to threshold.

An experimental determination of the relative spectral dependence of the photodetachment cross section of H$^-$ has been made from 4000Å to 13 000Å with a resolution to 300Å by Smith and Burch.[13,83] Electrons produced in the interaction of a crossed beam of H$^-$ ions and photons were measured as a function of the frequency of the photons. Their relative measurements, accurate to 2%, were normalized to an earlier absolute measurement of the integrated cross section of Branscomb and Smith.[84] The experimental results are shown in Fig. 6 by the open circles. Good agreement is found between theory and experiment below about 10 000Å in the "length" formulation, but a considerable difference exists at larger wavelengths.

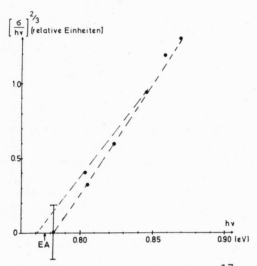

Fig. 7. Experimental measurement by Feldmann[17] of the threshold region of the photodetachment cross section of H$^-$.

A relative measurement of the detachment cross section for H$^-$ was extended to threshold by Feldmann[17] using a Xe lamp. The threshold behavior of the cross section is expected to follow a $(2\ell + 1)/2$ power law energy dependence[85] where ℓ is the angular momentum of the outgoing electrons. Since the H$^-$ ground state has two s electrons, the detached electron has one unit of angular momentum and thus the threshold behavior varies as $E^{3/2}$. The experimental results of Feldman[17] for the detachment cross section near threshold are shown in Fig. 7. A better determination of the threshold could be obtained with a light source of higher energy resolution.

III.B.   Photoattachment

Photoattachment is a process in which an electron, a photon and a hydrogen atom combine to form a $(2p^2)^3P^e$ state of H^-.

$$H(1s) + e^- + h\nu \rightarrow H^-(2p^2)^3P^e.$$

Drake[86] calculated the absorption coefficient and claims the process is efficient and should be considered as a source of continuum ultraviolet absorption in stellar atmospheres. He found that if the electron density is greater than 10% of the hydrogen atom density, then the photoattachment absorption coefficient is larger than that for Lyman alpha resonance broadening in the wavelength region from 1220Å to 1300Å. Independent calculations of the photoattachment cross section were also made by Jacobs, Bhatia and Temkin.[75] Using a 3-state close coupling continuum wave function, they found the maximum absorption to lie closer to the Lyman alpha line than predicted by Drake and in closer agreement to observations.

Recently Heap and Stecher[87] detected absorption features in the far ultraviolet spectrum of ζ Tau (B3pe) obtained in a rocket flight experiment. They attribute one of the absorption features to photoattachment of H^-. Their results are shown in Fig. 8. The absorption structure extends from 1218Å to 1233Å. In a note added to their paper, it was pointed out that water vapor has similar absorption features as observed in the rocket experiment. There is a possibility some water vapor was present in the apparatus. Further study of data from the Copernicus satellite is underway.

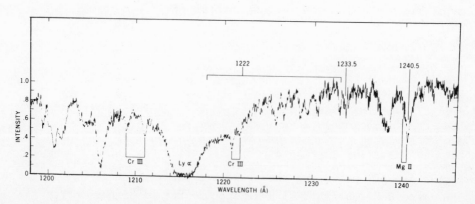

Fig. 8. Ultraviolet spectrum of ζ Tau obtained from spectrographs flown on a rocket experiment by Heap and Stecher.[87]

IV. H⁻ COLLISIONS WITH ELECTRONS

IV.A. Single-Electron Loss

Electron detachment of H⁻ by energetic electrons is an important simple atomic-collision system.

$$H^- + e^- \rightarrow H + 2e^-.$$

Initially, work in this area was stimulated in 1956 by Pagel[88] who suggested that if this process were rapid, H⁻ would not be in local thermodynamic equilibrium in the sun's atmosphere. Subsequent theoretical calculations, both classical and quantum mechanical, gave values for the cross section which differed widely both in energy dependence and in the maximum cross section by several orders of magnitude.[89]

Definitive, absolute experimental measurements were needed, and in 1964 Tisone and Branscomb[90] reported the first laboratory determination of the electron detachment cross section from 30 eV to 500 eV using sophisticated chopping techniques of both beams to recover the true signal from numerous spurious signals. In a later article they reported a more complex analysis of the experimental procedure along with an extension of the measurement down to 8.4 eV. At the same time, Dance, Harrison and Rundel[92] performed a similar experiment in the energy range from 9 eV to 500 eV. Although the two experiments were in agreement up to about 100 eV a significant difference was observed at high energies, in an energy region where the Bethe-Born approximation is believed accurate. In addition Dance et al. noted a possible resonance structure at 30 eV.

To clarify the situation, Peart, Walton and Dolder[93] undertook another experiment to independently measure the cross section with higher accuracy. Their results along with the previous measurements are shown in Fig. 9. It can be seen that above 50 eV remarkably good agreement is found between Dance et al.[92] and Peart et al.[93] and that the results of Tisone and Branscomb[91] are significantly higher. If one renormalizes the relative cross section of Tisone and Branscomb at 100 eV (lowers it by 14 percent, which is within their stated absolute accuracy), their cross section agrees very closely with that of Peart et al. below 100 eV. Peart et al. also observed no resonances near 30 eV.

By using intersecting beams of electrons and ions at 20°, another measurement of the detachment cross section was made by the Newcastle group[94] extending the cross section from 30 eV down to threshold 0.75 eV. The magnitude of the cross section agrees well with the 90° results of Peart et al.[93] for energies above 12.5 eV (see Fig. 12 below).

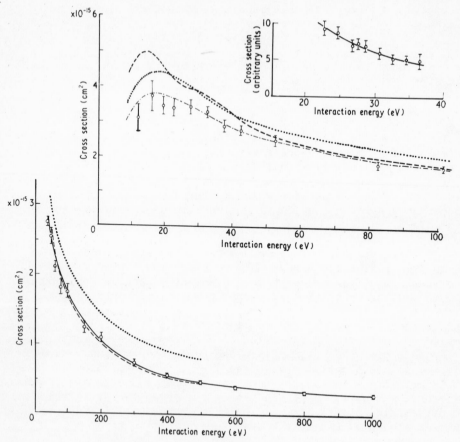

Fig. 9. Cross sections for detachment of H⁻ by electron impact.
Open circles, Peart et al.;[93] dashed curve, Dance et al.;[92] dotted
curve, Tisone and Branscomb;[91] dot-dash-dot curve, results of Tisone
and Branscomb renormalized to Peart et al. at 100 eV.

     Recent calculations of the detachment cross section by electron
impact are in much better agreement with each other and experiment.
A major source of error in previous calculations was the use of
nonorthogonal wave functions which increased the computed cross
section by a factor of 100. From 20 eV to 1000 eV, the Bethe-Born
approximation plus free-electron exchange calculation by Kim and
Inokuti[95] is somewhat higher than experiment, but it does exhibit
the correct energy dependence. Low energy predictions of the cross
section by Bely and Schwartz[96] using the impact-parameter method
for Coulomb trajectories are in good accord with experiment. Faisal
and Bhatia[97] also found that the detachment cross section for
leaving the hydrogen atom in the 2s state was two orders of magnitude smaller than that for leaving the H atom in the ground state.

The preceding theoretical results are compared with experiment in Fig. 10.

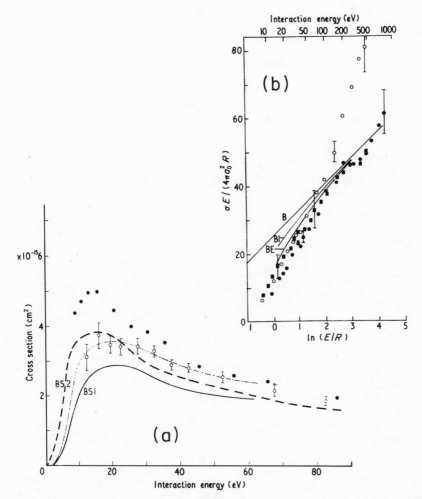

Fig. 10. Comparison of experimental and theoretical results of the H⁻ detachment cross section by electron impact. In part (a) open circles refer to the measurements of Peart et al.;[93] solid circles to Dance et al.;[92] solid curve BS1 and dot-dash-dot curve BS2 to calculations of Bely and Schwartz;[96] dashed curve to Faisal and Bhatia.[97] In part (b) open circles refer to Tisone and Branscomb;[91] solid circles to Peart et al.;[93] solid squares to Dance et al.;[92] solid curves are calculations of Kim and Inokuti;[95] B - Bethe's approximation, B1 - Born's approximation, BE - Born - exchange approximation.

IV.B. Double Electron Loss

The process of double detachment in H^- can be considered from several points of view.

$$H^- + e^- \rightarrow H^+ + 3e^-.$$

The "sudden approximation" assumes that one electron is knocked out in a time short compared with the orbital period. The remaining electron is left in a state which is not an eigenstate of the hydrogen atom and thus transitions to "good" quantum states, including the continuum are possible. Another suggestion is that the process is second order in that the fast projectile electron collides successively with both electrons either ionizing the first electron or forming a singly excited state before the second electron is knocked out. The unstable atom again can "shake off" the excited electron.

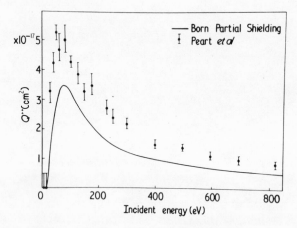

Fig. 11. Total cross section for double ionization of H^- by electron impact. Solid circles refer to experimental data of Peart et al.[99] The solid curve is a Born approximation calculation by Tweed.[98]

The calculation of multiple ionization cross sections is very sensitive to the choice of target wave function and especially to those terms involving electron correlation. Because of the loosely bound structure of H^- the negative hydrogen ion is a particularly good system to study the effects of electron correlation in the

A Born approximation calculation for double detachment of H^- by electrons has been made by Tweed.[98] His results, shown in Fig. 11, can be compared with the experimental determination from threshold 14.4 eV to 850 eV obtained in the crossed beam apparatus of Peart, Walton and Dolder.[99] The theoretical results lie significantly below experiment. Comparison of previous experimental results for single electron loss show that the ratio of double electron loss to single electron loss is approximately constant above 100 eV and equal to 0.026.

IV.C.   Resonance Structure in H^{2-}

In an attempt to extend experimental cross section measurements of single electron detachment of H^- by electron impact to lower energies, Walton, Peart and Dolder[100] discovered a significant, asymmetric structure in the cross section close to 14.2 eV.

$$H^- + e^- \rightarrow H^{2-*} \rightarrow H + 2e^-.$$

To achieve lower interaction energies between the electrons and H^- ions and maintain a sufficient signal to background ratio, the beams were inclined at 20° with each other. This technique allowed them to increase the electron energy, reducing losses due to space charge effects, and to increase the e-H^- interaction path length.

The results[101] from threshold to 30 eV are shown in Fig. 12. To check their measurements three different pairs of electron and ion beam energies were used. The broad structure at 14.2 eV has been identified by Taylor and Thomas[102] to be a short-lived resonant state of H^{2-} with a configuration $(2s^2 2p)^2 P^O$. The stabilization method with Hartree-Fock wave functions gave an energy of 14.8 eV for the state above the ground state of H^- with a width of roughly 1 eV.

Their calculation also showed the possibility of another resonance at higher energy, primarily due to a $2p^3$ configuration. Peart and Dolder[103] refined their apparatus by lowering the beam inclination angle to 10° and by reducing the residual gas pressure inside the interaction region. An increase of 25 in the signal to background ratio was obtained over the same ratio in their previous experiment.[101] The latest results are shown in Fig. 13. The 14.2 eV resonance is shown again with better statistics and at 17.2 eV another resonant structure is observed.

A new analysis[104] of the data calculated by Taylor and Thomas[102] yields an excited state of H^{2-} which is 60% $2p^3$ and 20% $2s^2 2p$ with an energy of 17.26 eV. Presumably the remaining 20% is distributed among other configurations.

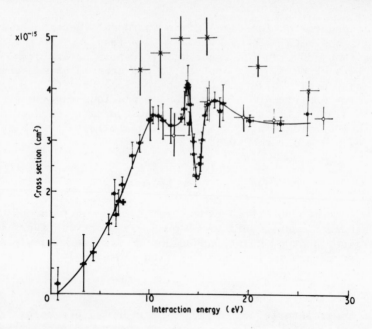

Fig. 12. Measured H⁻ detachment cross section for electron impact
vs interaction energy. Open circles 90° apparatus Peart et al.;[93]
crosses Dance et al.;[92] solid symbols 20° inclined beams Walton
et al.[101]

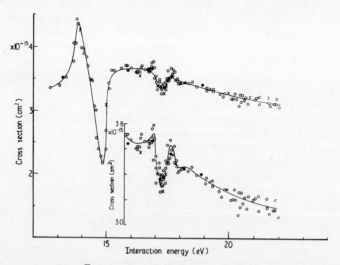

Fig. 13. Measured H⁻ detachment cross section for electron impact
vs interaction energy by Peart and Dolder.[103] Different symbols
refer to measurements with different ion beam energies.

V. H⁻ COLLISIONS WITH POSITIVE IONS

V.A. Neutralization with Protons

The ion-ion mutual neutralization reaction is particularly appealing for both theoretical and experimental reasons. The major theoretical emphasis in the past has been to consider the inelastic cross section in terms of the Landau-Zener curve-crossing model because of its physical simplicity and relative ease in computation. Neutralization cross sections can also be measured over a wide energy range owing to deamplification of the relative velocity in a merging or intersecting apparatus. The cross sections are important in understanding final ion removal in the upper atmosphere. The reader is referred to a recent comprehensive review on two-body ion-ion neutralization which covers in more detail the results and interpretations presented here.[105]

The charge transfer reaction of a proton and negative hydrogen is very suitable for an experimental test of the Landau-Zener theory because the atomic systems are simple.

$$H^+ + H^- \rightarrow H + H.$$

Potential energy curves as functions of internuclear separation for the $H^+ + H^-$ and H_2 systems are shown in Fig. 14. The H_2 molecular curves are flat at large internuclear separations. The attractive Coulomb potential curve of $H^+ + H^-$, however, crosses both the n=3 and n=4 curves of the H + H system at large internuclear separations. The curve crossing to n=4 occurs at very large distances where the exchange interaction is too weak to cause an appreciable transition probability. The crossing to n=1 occurs at very small internuclear separations and will contribute negligibly to the total charge transfer cross section.

An important characteristic of the neutralization reactions is that one of the product atoms will be left in an excited state. Also, because of the strong, long-range Coulomb force, the cross sections are very large, especially at thermal energies (exceeding $10^{-12} cm^2$).

The first experimental determination was made by Rundel, Aitken and Harrison[106] using a 20° intersecting apparatus for relative collision energies from 250 eV to 10 keV. These measurements were checked using a modified version of the same apparatus by Gaily and Harrison.[107] The results are essentially identical to within experimental uncertainty. In another experiment Moseley, Aberth and Peterson,[108] using the merging beam technique, extended the $H^+ + H^-$ cross section down to 0.15 eV. Their data compare well with the results of Rundel et al.[105] (see Fig. 15). The increase

in the cross section at low velocities is caused by the n=3 curve
crossing. The broad peak at higher velocities is caused by the
n=2 crossing. An interesting feature in this cross section is the
presence of two maxima around 150 eV and 1 keV plus some additional
"fine" structure near 300 eV.

The earliest Landau-Zener calculation by Bates and Lewis[109]
is indicated by curve "a." The behavior of the cross section below
10 eV is seen to agree in relative magnitude with the Landau-Zener
prediction but is too low by a factor of three. The discrepancy
at higher velocities can be expected since the reaction is not
described well by an adiabatic, molecular model for fast collisions.

Curve "b" is a Landau-Zener calculation by Dalgarno, Victor
and Blanchard based on more accurate potential curves for the H_2
molecule (see Ref. 105). Curve "c" is a Landau-Zener calculation
by Olson, Peterson and Moseley[110] using an interaction matrix element
evaluated by a method proposed by Smirnov. Good agreement exists
below 10 eV.

Curve "d" is a $2s-2p_0-2p_1$ close coupling calculation using the
impact parameter method by Roy and Mukherjee[111] in the energy range
from 500 eV to 8 keV. The calculated cross section using the close
coupled method appears to be much superior to Landau-Zener predictions in this energy range. The 30% disagreement at 500 eV may be
due to the lack of including n=3 states into the expansion of possible final states.

Curve "e" is another Landau-Zener calculation by Janev and
Tancic[112] with account taken for the exponential character of the
interaction and non-divergence of the potential curves at infinity.
Moseley et al.[105] point out a possible numerical error in Janev and
Tancic's calculation of the cross section arising from the n=2 state.
Their high energy cross section may be too large by a factor of three.

V.B. Neutralization with Other Positive Ions

Charge exchange cross sections for H^- on He^+ have been measured
from 200 eV to 8 keV by Gaily and Harrison[113] using modulated inclined ion beams and from 30 eV – 300 eV by Olson, Peterson and
Moseley[110] using merged beams. The later group actually used D^-
ions. No difference between D^- and H^- is expected since they
both have the same electronic structure. However, at very low
collision velocities an isotope effect is expected due to the
difference in center-of-mass energies.

The experimental results for H^- on He^+ are shown in Fig. 16
along with two predictions using the Landau-Zener model. The solid
curve is a calculation based on a semi-empirical formula for the

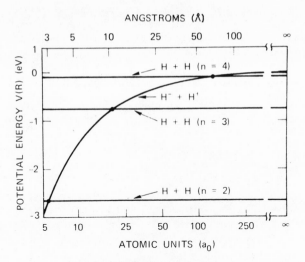

Fig. 14. Potential energy curves for the $H^+ + H^- \to H + H$
reaction.[105]

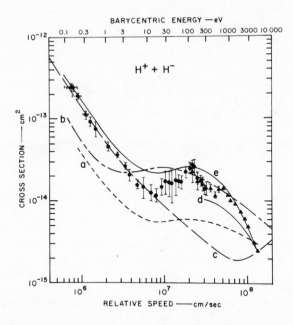

Fig. 15. Experimental and theoretical results for $H^+ + H^-$ neutralization. Circles refer to experiment of Moseley et al.,[108] triangles to Rundel et al.[106] and Gaily and Harrison.[107] See text for
references to theoretical results.

interaction matrix element[110] and the dashed curve refers to a calculation using matrix elements obtained more rigorously with an
asymptotic expansion for the wave functions of the negative ion
and excited atoms.[114]

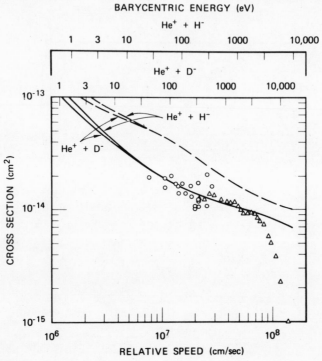

Fig. 16. Experimental results for $H^- + He^+$ by Gaily and Harrison[113]
open triangles, for $D^- + He^+$ by Olson et al.[110] open circles. Theoretical results for both systems using semi-empirical matrix
elements[110] (dashed line) and the Smirnov formula[114] (solid lines).

Charge exchange cross sections for D^- on a molecular ion H_2^+
have been measured by Aberth, Moseley and Peterson from 0.17 eV
to 50 eV (see Ref. 105). The two-state Landau-Zener theory is
difficult to apply to reactions involving molecules because of the
large number of excited electronic and vibrational states associated
with the molecule. Olson has modified the curve-crossing-model
for molecules by using a high density of crossing states.[115] He
assumes an absorbing-sphere model in which the neutralization
reaction goes with unit probability for some critical internuclear
distance. Using an empirical model for the critical distance based
on the detachment energy of the negative ion and the reduced mass
of the system, good agreement was found between experiment and the
absorbing-sphere calculation for the $D^- + H_2^+$ neutralization cross
section.[105]

VI. H⁻ COLLISIONS WITH ATOMS AND MOLECULES

VI.A.   Charge Transfer

The transfer of one electron from H^- to the target particle is an important inelastic reaction channel in H^- collisions with atomic and molecular targets which have a positive electron affinity.

$$H^- + H \rightarrow H + H^-.$$

The simplest system is the resonant charge transfer case in which the initial and final states are identical. The cross section for this process has been predicted by several authors[116-118] using a molecular expansion for a perturbed-stationary-state calculation and has been measured in a modulated crossed beam experiment in the energy range from 40 eV to 10 keV by Hummer, Stebbings, Fite and Branscomb.[119] The cross section σ can be written in the form

$$\sigma^{1/2} = A - B \ln v$$

where A and B are constants and v is the velocity. Below about 1 keV, theory and experiment are in good agreement. The cross section falls from about 85×10^{-16} cm^2 at 40 eV to 1×10^{-16} cm^2 at 10 keV.

Nonresonant charge transfer cross sections have been measured for H^- on O_2, NO_2 and atomic O in the energy range of a few eV to 4 keV.[120,121] In the latter case the inverse reaction was also studied.[121]

$$H^- + O \overset{\rightarrow}{\underset{\leftarrow}{}} H + O^-.$$

Using the principle of detail balancing, the transition probability is the same in both directions. When proper statistical weighting of the cross sections is made, the two cross sections can be compared. A broad maximum with a magnitude of 25×10^{-16} cm^2 is found for a relative collision velocity of 2×10^7 cm/sec.

VI.B.   Differential Scattering of H^-.

Total scattering cross sections for H^- on simple noble gas targets, H_2 and O_2 have been measured from 5 eV to 350 eV.[122-124] The effective cross sections, dependent on the scattering geometry of the apparatus, can be used to extract an interaction potential for the quasi-molecular system. Although the inversion procedure requires a simple form for the potential and yields a potential valid only over a short range of internuclear separations, the results obtained by this method are of some help to theorists in attempting to handle many-electron collision systems.

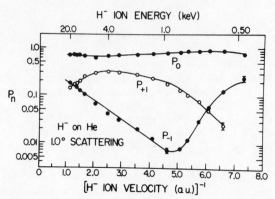

Fig. 17. Hydrogen charge fractions for a scattering angle of 1° vs reciprocal velocity.

Differential scattering measurements, to date, have been done only on the relative charge state fractions and on the energy loss spectra. Charge state fractions for H^- on He have been measured from 470 eV to 20 keV at scattering angles from 0.3° to 5°.[125] A large neutral fraction was found at all energies and all angles. In Fig. 17, the charge state fractions P_n are shown for a scattering angle of 1°.

In the energy range from 1 keV to 20 keV, the P_{-1} probability is related to the ion velocity v by the empirical formula

$$P_{-1} = 0.48 \exp(-v_0/v)$$

where v_0 is 2.0×10^8 cm/sec. McCaughey and Bednar[125] interpret this high velocity result in terms of a HeH^- molecular model.

Inspection of Fig. 17 shows that, except for very low collision energies, the positive charge fraction is much higher than the negative fraction. This same result also holds true for much higher scattering angles from 8° to 33° in roughly the same range of collision energies.[55]

Energy loss spectra for scattered H^- and doubly stripped H^+ ions at 8° are shown in Fig. 18 for 1 keV H^- incident on He in an experiment by Risley.[55] Only two "elastic" peaks are seen, no structure corresponding to excitation of optical or autoionization channels of He was detected. The difference in the energy of the two peaks is equal to the total binding energy of both electrons of H^-, 14.4 eV, plus an average of the sum of the kinetic energies of the two electrons, 8.4 eV in this case. The FWHM of the H^+ peak

is larger than the FWHM of the H⁻ peak. This result suggests that
the kinetic energy distribution of the detached electrons would be
broad.

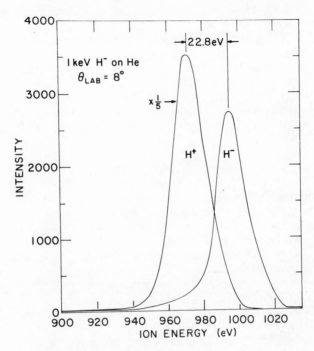

Fig. 18. Energy loss spectra of H⁻ and doubly stripped H⁺ from 1-
keV H⁻ on He observed at 8°.

VI.C. Optical Emission in H⁻ Collisions

The intensity of Lyman alpha radiation emitted during collisions
of H⁻ with rare gas target atoms has been measured from 5-40 keV.[126]

H⁻ + He → H(2ℓ)* + He + e⁻.

From the measurements the excitation cross sections for forming
hydrogen atoms in the 2s and 2p states were determined. The 2s
state was detected by quenching the metastable state with an electric field. For all target gases the total n=2 excitation cross
section was surprisingly large. It accounts for about 10% of the
total inelastic detachment channel.

The large emission cross section can be understood in part by
the structure of the H⁻ ground state. It was discussed earlier in
this paper that H⁻ can be described, in a certain sense, by consid-

ering one electron to be "tightly bound" to the proton with a wave
function similar to the ground state of a hydrogen atom and the
other electron to be "loosely bound" with a wave function corresponding to the n=2 excited state of hydrogen. In this framework,
Drukarev[127] has shown that the optical emission cross sections are
predicted quite well by considering the process as an expulsion of
the "tightly bound" electron followed by an adoption of the "loosely
bound" electron to the n=2 excited state of hydrogen. The cross
section for optical emission, then, would have the same energy
dependence as the cross section for ionization of hydrogen atoms.
Considering the simplicity of this model, the experimental results
are in good agreement.

VI.D. Electron Detachment.

The most important inelastic collision process with H^- ions
incident on neutral target atoms and molecules is electron loss. At
very low energies, much below the ionization energy of H^-, associative detachment, or recombination can occur if the target is chem-
ically active. In relatively fast collisions, however, the electron
is simply stripped from the ion.

The theory of associative detachment[128] is generally considered
in a molecular framework in which the H^- ion gradually combines with
the atomic target particle forming a negative molecular ion. The
electronic states of the quasimolecule may merge into autoionization
states of the negative molecular ion. Decays into states of the
neutral molecule can then occur. A laboratory measurement of the
associative detachment reaction

$$H^- + H \rightarrow H_2 + e^-$$

was made at $300^{\circ}K$ by Schmeltekopf, Fehsenfeld and Ferguson.[129] A
value of 1.3×10^{-9} cm^3/sec, accurate to within a factor of two, was
obtained for the rate constant. This reaction has special astrophysical interest because of its role in maintaining the H^- to H
ratio in stellar atmospheres.

Total electron detachment cross sections have been measured
over a wide energy range for many gas target species using a variety
of experimental methods. Because of the complexity in dealing with
a many-electron system, theoretical efforts have been limited. The
Born approximation has been used to calculate single electron loss
for H^- on H,[130] and for H^- on He with limited success.[131] There are
many reviews of past work done on electron detachment. The reader
is referred to two recent articles, Tawara and Russek[132] and Risley
and Geballe.[133]

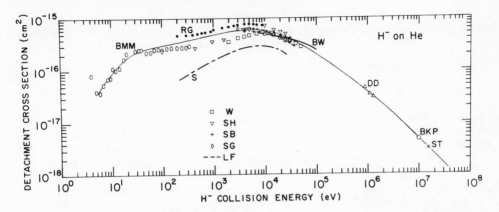

Fig. 19. Detachment cross section for H⁻ on He vs H⁻ collision
energy. Experimental: BMM - Bailey et al.,[134] RG - Risley and
Geballe,[133] W - Williams,[135] SH - Stedeford and Hasted,[136] SB -
Stier and Barnett,[137] SG - Simpson and Gilbody,[138] DD - Dimov and
Dudnikov,[139] BKP - Berkner et al.,[140] ST - Symthe and Toevs.[141]
Theoretical: BW - Bates and Walker,[142] LP - Lopantseva and Fir-
sov,[143] S - Sida.[131]

An example of the total cross section for electron detachment
of H⁻ incident on He is shown in Fig. 19. Measurements,[133-141]
extending from 4 eV to 14.7 MeV, reveal a slowly varying, relatively
structureless cross section with a maximum magnitude of about
8×10^{-16} cm². Some of the measurements plotted in Fig. 19 include
signals not only from single electron loss but also from double
electron loss. Measurements of the detachment cross section for
other target particles exhibit a similar energy dependence but
larger magnitude roughly in proportion to the number of electrons
in the target atom.

The Born approximation calculation by Sida[131] is seen to
grossly underestimate the cross section. The semi-empirical low-
energy-electron scattering model invoked by Bates and Walker[142]
and Lopantsev and Firsov[143] in their calculations is a much better
description of the detachment process. This model treats the
encounter as if it were an elastic collision of a free electron
(which is simply carried along by a hydrogen atom and which has a
mean velocity equal to the ion velocity) and a target atom. Calculations are dependent on measurements of the low-energy electron
scattering cross section for the target. In a recent article Lam,
Delos, Champion, and Doverspike[144] have discussed electron detach-

ment semi-classically using the complex potential method to describe detachment in terms of a quasimolecular decay. Their results
are probably applicable only for very slow collisions, below about
50-100 eV.

     Energy distributions of detached electrons from H⁻ have been
measured at different ejection angles for some simple atomic (He,Ar)
and molecular (H_2, N_2, O_2) gas targets by Risley using a beam-static
gas target system.[55] The electrons were analyzed with a rotatable
30^0 parallel plate analyzer.[145] An example of the doubly differential production cross section is shown in Fig. 20 for 200 eV H⁻
on He. The distributions are characterized by a fall of three
orders of magnitude for electron energies from 1 eV to about 10 eV.
The distribution probably peaks at much less than 1 eV. The intense
line at about 10 eV is due to several, unresolved autodetaching
states of H⁻ (see Fig. 4). Because of kinematic effects these lines
decrease in energy and intensity as the observation angle increases.
Energy distributions using other gas targets are quite similar to
those shown in Fig. 20.

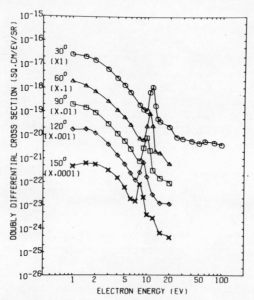

Fig. 20. Doubly differential cross section for secondary electron
production vs electron energy for 200 eV H⁻ on He.

     The energy distribution of all electrons emitted from H⁻ on
He is shown in Fig. 21 for selected collision energies. The differential cross section has been integrated over angle. The angular
distribution of electrons with the same kinetic energy is roughly

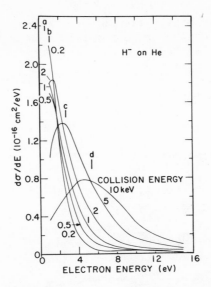

Fig. 21. Energy distribution of electrons ejected from collisions
of H⁻ on He. The number aside each curve is the collision energy
in keV. The vertical lines indicate the kinetic energy of an electron with the same velocity as the H⁻ ion. a - 1 keV, b - 2 keV,
c - 5 keV, d - 10 keV.

isotropic. The broad peak in the energy distribution is observed to
shift to higher electron energies as the collision energy is raised.

The vertical lines in Fig. 21 indicate the kinetic energy of a
free electron which has the same velocity as the H⁻ ion velocity.
The position of the peak is seen to agree quite well with this electron energy. This feature of the distribution can be interpreted in
terms of the free-electron-scattering model of electron detachment
where the lab velocity of the detached electron is equal to the
vectorial sum of the ion velocity and the binding energy velocity.

The angular distribution of the intensity of the autodetaching
lines of H⁻ has been measured for a few gas targets in the energy
range from 100 eV to 10 keV.[146] Special attention was focused on
the $(2s^2)^1S$ line produced in collisions of H⁻ with He. In a field
free region, neglecting any interference effects, the angular
distribution should be isotropic. In Fig. 22 the distribution is
seen to be isotropic (in the H⁻ rest frame) for high collision
energies. However, below 1 keV a significant asymmetric component is
observed which is peaked toward forward angles and which becomes more
pronounced as the collision energy is reduced. This asymmetric

distribution may be a manifestation of interference between continuum and discrete electron states or it may be due to a distortion of the electron cloud of the excited H⁻ ion by the target atoms before decay.[147]

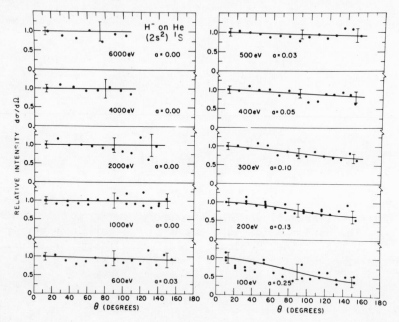

Fig. 22. Angular distribution of the intensity of the $(2s^2)^1S$ line of H⁻ in the rest frame of the H⁻ ion. The relative intensities are normalized to unity at 10^0. The collision energy for H⁻ on He is indicated for each curve along with the distortion parameter "a" determined by fitting the empirical formula for the intensity $I = A (1 + a \cos \Theta)$ to each set of data.

At high collision energies, above several keV, a $(2s2p)^3P$ line of H⁻ is observed close to the $(2s^2)^1S$ line.[55] Fig. 23 displays spectra observed at 10^0 for 8 keV H⁻ on several atomic and molecular gas targets. It should be noted that the ^3P line is missing from the He spectrum, whereas it is quite pronounced in the other spectra. If these collisions proceed via a Coulomb excitation, then one would expect the total spin of the two colliding partners to be conserved. To excite a triple state of H⁻, the target atom would have to be excited and a pair of electrons exchanged. The minimum excitation energy of He is more than twice that for the other target gases used.

In order to compare various excitation processes of H⁻, it is convenient to plot the ratio of the excitation cross section to the

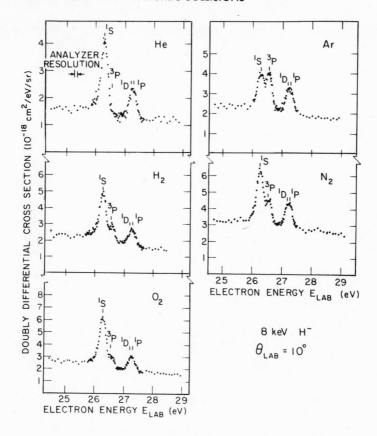

Fig. 23. Electron energy spectra for 8 – keV H⁻ collisions on atomic and molecular targets observed at 10°.

total inelastic or detachment cross section versus collision energy. Such a plot is shown in Fig. 24 for H⁻ on Ar. The optical excitation[126] of both the 2s and 2p states of atomic H accounts for more than 10% of the total inelastic channels. The autodetaching states account for much less than 1% of the total for collisions above 1 keV. The intensity of the ³P line is observed to increase sharply near 2 keV. One should be cautioned that the autodetachment cross section refers to the intensity of the line (area under the peak). The intensity is only indirectly proportional to the excitation cross section because of interference with the continuum states.

As a final example of some of the processes encountered in H⁻ collisions Fig. 25 and Fig. 26 show line spectra resulting from

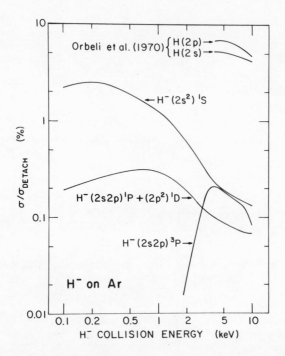

Fig. 24.   Ratio of selected excitation cross sections to the total
detachment cross section for H⁻ collisions on Ar.

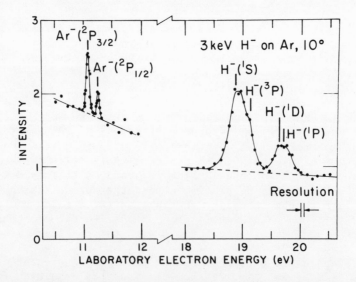

Fig. 25.   Electron energy spectra for 3-keV H⁻ on Ar observed at
10°.

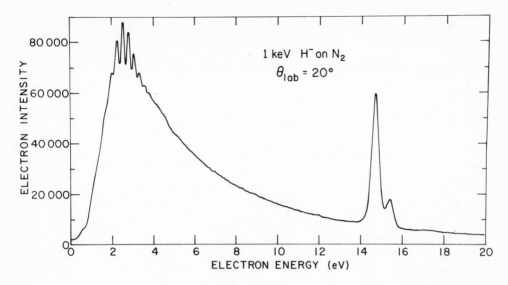

Fig. 26.   Electron energy spectra for 1-keV H^- on N_2 observed at 20°.

autodetaching states of the negative ion of the target particle
produced in a charge transfer reaction. The energy of the Ar^- lines
is well known[25] and, in fact, were used to calibrate the energy scale
for the H^- determinations.[54] The resonance structure in Fig. 26 was
only recently discovered[148] in 1 keV H^- collisions with N_2. Preliminary analysis shows that the energy and spacing of the lines are
in excellent agreement with electron scattering experiments. It is
surprising that the shape resonances of N_2^- are so pronounced compared with other possible excitation channels.

VII.   ACKNOWLEDGEMENTS

The author wishes to thank Prof. R. Geballe for continued
support and encouragement and the organizing committee of the
Fourth International Conference on Atomic Physics for giving me
the opportunity to present this paper.

REFERENCES

1. H. Bethe, Z. Physik 57, 815 (1929).
2. E. A. Hylleraas, Z. Physik 63, 291 (1930).
3. H. A. Bethe and E. E. Salpeter, in "Handbuch der Physik." (S. Fluegge, ed.), Vol. 35, Springer-Verlag, Berlin, 1957, p. 88.
4. J. N. Silverman, O. Platas and F. H. Matsen, J. Chem. Phys. 32, 1402 (1960).
5. L. R. Henrich, Astrophys. J. 99, 59 (1944).
6. E. Hylleraas and J. Midtdal, Phys. Rev. 103, 829 (1956).
7. C. W. Scherr and R. E. Knight, Rev. Mod. Phys. 35, 436 (1963).
8. P. G. Burke and A. J. Taylor, Proc. Phys. Soc. Lond. 88, 549 (1966).
9. C. L. Pekeris, Phys. Rev. 126, 1470 (1962).
10. K. Aashamar, Nucl. Instrum. Methods 90, 263 (1970).
11. B. N. Taylor, W. H. Parker and D. N. Laugenberg, Rev. Mod. Phys. 41, 375 (1969).
12. V. I. Khvostenko and V. M. Dukel'skii, Sov. Phy.-JETP 10, 465 (1960).
13. S. J. Smith and D. S. Burch, Phys. Rev. 116, 1125 (1959).
14. J. D. Weisner and B. H. Armstrong, Proc. Phys. Soc. (London) 83, 31 (1964).
15. M. J. Woolf, M. Schwarzschild, and W. K. Rose, Astrophys. J. 140, 833 (1964).
16. R. S. Berry, Chem. Rev. 69, 533 (1969).
17. D. Feldmann, Z. Naturforsch. 25a, 621 (1970).
18. A. Wallis, D. L. S. McElwain, and H. O. Pritchard, J. Chem. Phys. 50, 4543 (1969).
19. L. M. Branscomb, Adv. Electron Phys. 9, 43 (1957).
20. L. M. Branscomb, in Atomic and Molecular Processes, edited by D. R. Bates, (Academic Press, New York, 1962), p. 100.
21. P. G. Burke, Adv. Phys. 14, 521 (1965).
22. P. G. Burke, Advan. At. Mol. Phys. 4, 173 (1968).
23. K. Smith, Rep. Prog. Phys. 29, 373 (1966).
24. J. C. Y. Chen, Nucl. Instrum. Methods 90, 237 (1970).
25. G. J. Schulz, Rev. Mod. Phys. 45, 378 (1973).
26. K. Smith, R. P. McEachran, and P. A. Fraser, Phys. Rev. 125, 553 (1962).
27. P. G. Burke and H. M. Schey, Phys. Rev. 126, 147 (1962).
28. E. Hylleraas, Astrophys. J. 111, 209 (1950).
29. E. Holøien, Proc. Phys. Soc. Lond. A71, 357 (1958); 88, 538 (1966).
30. See Ref. 22.
31. S. Ormonde, J. McEwen and J. Wm. McGowan, Phys. Rev. Letters 22, 1165 (1969).
32. A. K. Bhatia and A. Temkin, Phys. Rev. 182, 15 (1969); A6, 562 (1972).
33. J. C. Y. Chen and K. T. Chung, Phys. Rev. A2, 1892 (1970).

34. P. G. Burke, S. Ormonde, and W. Whitaker, Proc. Phys. Soc. (London) 92, 319 (1967).
35. R. S. Oberoi, J. Phys. B5, 1120 (1972).
36. E. Holøien and J. Midtdal, J. Phys. B6, 1992 (1973).
37. The author is indebted to Dr. J. Macek for pointing out the relative importance of this decay scheme.
38. P. Feldman and R. Novick, Phys. Rev. 160, 143 (1967).
39. G. W. F. Drake, Phys. Rev. Lett. 24, 126 (1970).
40. A. K. Bhatia, Phys. Rev. A. 2, 1667 (1970).
41. A. J. Taylor and P. G. Burke, Proc. Phys. Soc. (London) 92, 336 (1967).
42. J. Macek and P. G. Burke, Proc. Phys. Soc. (London) 92, 351 (1967).
43. See Ref. 8.
44. See Ref. 34.
45. See Ref. 33.
46. J. J. Matese and R. S. Oberoi, Phys. Rev. A4, 569 (1971).
47. J. C. Y. Chen, K. T. Chung and A. L. Sinfailam, Phys. Rev. A4, 1517 (1971).
48. I. Shimamura, J. Phys. Soc. Jap. 31, 852 (1971).
49. J. N. Bardsley and B. R. Junker, J. Phys. B5, L178 (1972).
50. K. T. Chung and J. C. Y. Chen, Phys. Rev. A6, 686 (1972).
51. A. K. Bhatia and A. Temkin, Phys. Rev. A8, 2184 (1973).
52. A. K. Bhatia and A. Temkin, Phys. Rev. A (to be published).
53. G. D. Doolen, J. Nuttall and R. W. Stagat, Phys. Rev. A (to be published).
54. J. S. Risley, A. K. Edwards and R. Geballe, Phys. Rev. A9, 1115 (1974).
55. J. S. Risley, Ph.D. dissertation (University of Washington, 1973) (unpublished).
56. G. J. Schulz, Phys. Rev. Lett. 13, 583 (1964).
57. L. Sanche and P. D. Burrow, Phys. Rev. Lett. 29, 1639 (1972).
58. H. Kleinpoppen and V. Raible, Physics Letters 18, 24 (1965).
59. J. Wm. McGowan, E. M. Clarke and E. K. Curley, Phys. Rev. Lett. 15, 917 (1965).
60. J. Wm. McGowan, E. M. Clarke and E. K. Curley, Phys. Rev. Lett. 17, 66(E) (1966).
61. J. Wm. McGowan, Phys. Rev. 156, 165 (1967).
62. J. Wm. McGowan, Phys. Rev. Lett. 17, 1207 (1966).
63. J. F. Williams and J. Wm. McGowan, Phys. Rev. Lett. 21, 719 (1968).
64. J. W. McGowan, J. F. Williams and E. K. Curley, Phys. Rev. 180, 132 (1969).
65. M. Rudkjøbing, Astrophys. Space Sci. 3, 102 (1969).
66. M. Rudkjøbing, Astrophys. Space Sci. 5, 68 (1969).
67. C. Ingemann-Hilberg and M. Rudkjøbing, Astrophys. Space Sci. 6, 101 (1970).
68. M. Rudkjøbing, Astrophys. Space Sci. 6, 157 (1970).
69. M. Rudkjøbing, J. Quant. Spectrosc. Radiat. Transfer. 13, 1479 (1973).

70. W. Van Rensbergen, J. Quant. Spectrosc. Radiat. Transfer. $\underline{11}$, 1125 (1971).

71. A. Herzenberg and F. Mandl, Proc. Roy. Soc. (London) $\underline{A274}$, 253 (1963).

72. A. Herzenberg, private communication.

73. T. P. Snow, Jr., Astrophys. J. $\underline{184}$, 135 (1973).

74. G. W. F. Drake, Astrophys. J. $\underline{184}$, 145 (1973).

75. V. L. Jacobs, A. K. Bhatia, and A. Temkin, Astrophys. J. (to be printed 1 August 1974).

76. R. Wildt, Astrophys. J. $\underline{89}$, 295 (1939).

77. S. Geltman, Astrophys. J. $\underline{136}$, 935 (1962).

78. S. Chandrasekhar, Ap. J. $\underline{102}$, 395 (1945).

79. N. A. Doughty, P. A. Fraser and R. P. McEachran, Mon. Not. R. Astr. Soc. $\underline{132}$, 255 (1966).

80. K. L. Bell and A. E. Kingston, Proc. Phys. Soc. $\underline{90}$, 895 (1967).

81. J. Macek, Proc. Phys. Soc. $\underline{92}$, 365 (1967).

82. H. A. Hyman, V. L. Jacobs and P. G. Burke, J. Phys. $B\underline{5}$, 2282 (1972).

83. S. J. Smith and D. S. Burch, Phys. Rev. Lett. $\underline{2}$, 165 (1959).

84. L. M. Branscomb and S. J. Smith, Phys. Rev. $\underline{98}$, 1028 (1955).

85. L. M. Branscomb, D. S. Burch, S. J. Smith and S. Geltman, Phys. Rev. $\underline{111}$, 504 (1958).

86. G. W. F. Drake, Astrophys. J. $\underline{189}$, 161 (1974).

87. S. R. Heap and T. P. Stecher, Astrophys. J. $\underline{187}$, L27 (1974).

88. B. E. J. Pagel, Mon. Nat. R. Astron. Soc. $\underline{116}$, 608 (1956).

89. D. F. Dance, M. F. A. Harrison and R. D. Rundel, Proc. Roy. Soc. (London) $\underline{A299}$, 525 (1967).

90. G. C. Tisone and L. M. Branscomb, Phys. Rev. Letters $\underline{17}$, 236 (1966).

91. G. C. Tisone and L. M. Branscomb, Phys. Rev. $\underline{170}$, 169 (1968).

92. See Ref. 89.

93. B. Peart, D. S. Walton and K. T. Dolder, J. Phys. $B\underline{3}$, 1346 (1970).

94. D. S. Walton, B. Peart and K. T. Dolder, J. Phys. $B\underline{4}$, 1343 (1971).

95. Y. K. Kim and M. Inokuti, Phys. Rev. A. $\underline{3}$, 665 (1971).

96. O. Bely and S. B. Schwartz, J. Phys. $B\underline{2}$, 159 (1969).

97. F. H. M. Faisal and A. K. Bhatia, Phys. Rev. $A\underline{5}$, 2144 (1972).

98. R. J. Tweed, J. Phys. $B\underline{6}$, 270 (1973).

99. B. Peart, D. S. Walton and K. T. Dolder, J. Phys. $B\underline{4}$, 88 (1971).

100. D. S. Walton, B. Peart and K. T. Dolder, J. Phys. $B\underline{3}$, L148 (1970).

101. See Ref. 94.

102. H. S. Taylor and L. D. Thomas, Phys. Rev. Lett. $\underline{28}$, 1091 (1972).

103. B. Peart and K. T. Dolder, J. Phys. $B\underline{6}$, 1497 (1973).

104. L. D. Thomas, J. Phys. $B\underline{7}$, L97 (1974).

105. J. T. Moseley, R. E. Olson and J. R. Peterson, Case Studies in Atomic Physics (submitted March, 1974).

106. R. D. Rundel, K. L. Aitken and M. F. A. Harrison, J. Phys. $B\underline{2}$, 954 (1969).

107. T. D. Gaily and M. F. A. Harrison, J. Phys. B3, L25 (1970).
108. J. Moseley, W. Aberth and J. R. Peterson, Phys. Rev. Lett. 24, 435 (1970).
109. D. R. Bates and J. T. Lewis, Proc. Phys. Soc. (London) A68, 173 (1955).
110. R. E. Olson, J. R. Peterson and J. Moseley, J. Chem. Phys. 53, 3391 (1970).
111. K. Roy and S. C. Mukherjee, Phys. Rev. A7, 130 (1973).
112. R. K. Janev and A. R. Tancic, J. Phys. B5, L250 (1972).
113. T. D. Gaily and M. F. A. Harrison, J. Phys. B3, 1098 (1970).
114. B. M. Smirnov, Sov. Phys. Dokl. 10, 218 (1965); 12, 242 (1967).
115. R. E. Olson, J. Chem. Phys. 56, 2979 (1972).
116. A. Dalgarno and M. R. C. McDowell, Proc. Phys. Soc. (London) A69, 615 (1956).
117. B. M. Smirnov and O. B. Firsov (Zh. Eksp. Teor. Fiz. 47, 232 (1964). [Sov. Phys. - JETP 20, 156 (1965)].
118. D. M. Davidovic and R. K. Janev, Phys. Rev. 186, 89 (1969).
119. D. G. Hummer, R. F. Stebbings, W. L. Fite and L. M. Branscomb, Phys. Rev. 119, 668 (1960).
120. T. L. Bailey and P. Mahadevan, J. Chem. Phys. 52, 179 (1970).
121. W. R. Snow, R. D. Rundel and R. Geballe, Phys. Rev. 178, 228 (1969).
122. T. L. Bailey, C. J. May and E. E. Muschlitz, Jr., J. Chem. Phys. 26, 1446 (1957).
123. E. E. Muschlitz, Jr., T. L. Bailey and J. H. Simons, J. Chem. Phys. 24, 1202 (1956); 26, 711 (1957).
124. C. E. Baker, J. M. McGuire and E. E. Muschlitz, J. Chem. Phys. 37, 2571 (1962).
125. M. P. McCoughey and J. A. Bednar, Phys. Rev. Lett. 28, 1011 (1972).
126. A. L. Orbeli, E. P. Andreev, V. A. Ankudinov, and V. M. Dukelskii, Zh. Eksp. Teor. Fiz. 58, 1938 (1970) [Sov. Phys. - JETP 31, 1044 (1970).
127. G. F. Drukarev, Sov. Phys. - JETP 31, 1193 (1970).
128. J. C. Y. Chen, Phys. Rev. 156, 12 (1967).
129. A. L. Schmeltekopf, F. C. Fehsenfeld and E. E. Ferguson, Astrophys. J. 148, L155 (1967).
130. M. R. C. McDowell and G. Peach, Proc. Phys. Soc. (London) 74, 463 (1959).
131. D. W. Sida, Proc. Phys. Soc. London A68, 240 (1955).
132. H. Tawara and A. Russek, Rev. Mod. Phys. 45, 178 (1973).
133. J. S. Risley and R. Geballe, Phys. Rev. A9, 2485 (1974).
134. See Ref. 122.
135. J. F. Williams, Phys. Rev. 154, 9 (1967).
136. J. B. H. Stedeford and J. B. Hasted, Proc. Roy. Soc. (London) A227, 466 (1955), and J. B. Hasted, Proc. Roy. Soc. (London) A212, 235 (1952).
137. P. M. Stier and C. F. Barnett, Phys. Rev. 103, 896 (1956).
138. F. R. Simpson and H. B. Gilbody, J. Phys. B5, 1959 (1972).

139. G. I. Dimov and V. G. Dudnikov, Sov. Phys. - Tech. Phys. 11, 919 (1967).

140. K. H. Berkner, S. N. Kaplan, and P. V. Pyle, Phys. Rev. 134, A1461 (1964).

141. R. Smythe and J. W. Toevs, Phys. Rev. 139, A15 (1965).

142. D. R. Bates and J. C. G. Walker, Proc. Phys. Soc. London 90, 333 (1967).

143. G. B. Lopantseva and O. B. Firsov, Zh. Eksp. Teor. Fiz. 50, 975 (1966) [Sov. Phys. - JETP 23, 648 (1966)].

144. S. K. Lam, J. B. Delos, R. L. Champion and L. D. Doverspike, Phys. Rev. A9, 1828 (1974).

145. J. S. Risley, Rev. Sci. Instrum. 43, 95 (1972).

146. J. S. Risley, A. K. Edwards and R. Geballe, Phys. Rev. Lett. 29, 904 (1972).

147. J. S. Risley and R. Geballe, Phys. Rev. (submitted July, 1974).

148. J. S. Risley, to be published.

INTRAMULTIPLET MIXING AND DEPOLARIZATION IN ATOMIC COLLISIONS

E.E. Nikitin

Institute of Chemical Physics, USSR

Academy of Sciences Moscow, V-334

I. INTRODUCTION

Intramultiplet mixing and depolarization of atomic states in collisions are the simplest processes occuring in collisions of heavy particles. These processes are of interest in phenomena of redistribution of resonance radiation [1], sensitized fluorescence [2,3], optical pumping, and level-crossing spectroscopy [4-7]. They are also often responsible for kinetics in gas lasers [8] and in interstellar media [9,10].

However, the theoretical interest in mixing and depolarization is explained not only by the practical importance but also by the inherent simplicity which makes it possible to obtain cross sections using a b i n i - t i o calculations. The pecularity of these processes is that transitions occur between atomic states of one multiplet so that - at sufficiently low energy - the discussion can be confined to electronic states emerging from a single pair of terms of colliding atoms. As the energy transfer ΔE is low, interatomic interactions outside atomic cores are important and the spin-orbit coupling can be considered - at least to the first approximation - as uneffected by a perturbation of the "tails" of atomic wave functions.

Though the characteristic range of energy ΔE is small compared to that of electronic excitation the mechanisms of these processes vary tremendously from resonant ones

529

for depolarization (cross sections up to 10^2 $\mathring{A}^2$) to almost
adiabatic ones for mixing (cross sections down to 10^{-6} A^2).
Such a large variation in cross sections raises a number
of questions concerning the interatomic interaction, the
solutions of dynamical equations for the scattering matrix,
and the derivation of relaxation equations for physical
properties describing mixing and depolarization in relaxing atoms.

This presentation is not intended to be a comprehensive review on the subject: the discussion will be centered on problems which emerged recently in connection
with numerous works in this field.

II. RELAXATION OF ATOMIC STATES IN IMPACT APPROXIMATION

Let $\rho\, {}^{j_2 j_1}_{m_2 m_1}$ be the density matrix of an atom A inter-

acting with a gaseous heat bath.* The relaxation equation
reads:

$$d\, \rho\, {}^{j_2 j_1}_{m_2 m_1} / dt = -\sum_{j_2', m_2', j_1', m_1'} \Gamma\, {}^{j_2 j_1, j_2' j_1'}_{m_2 m_1, m_2' m_1'}\ \ \rho\, {}^{j_2' j_1'}_{m_2' m_1'} \qquad (2.1)$$

According to Omont[11] and D'yakonov and Perel'[12] the
relaxation matrix Γ has a number of important properties
which will be reviewed now.

The matrix $\Gamma\, {}^{j_2 j_1, j_2' j_1'}_{m_2 m_1, m_2' m_1'}$ is diagonal in q, q' with

$q = m_2 - m_1$ and $q' = m_2' - m_1'$. Uncoupled matricies

$\Gamma\, {}^{j_2\ \ j_1,\ \ j_2\ \ j_1}_{m_2 m_2 - q,\ \ m_2' m_2' - q}$ with different q values have indentical

eigenvalues.

Eigenvectors $\ell_m^q (j_2 j_1 \varkappa)$ of diagonal blocks

$\Gamma\, {}^{j_2\ \ j_1,\ \ j_2\ \ j_1}_{m, m-q,\ \ m' m'-q}$ satisfy a standard equation:

---

*In what follows j stands for any angular momentum which
characterizes an atomic state; S,L,J,I,F, have their usual spectroscopic meaning.

$$\sum_{m'} \Gamma^{j_2 j_1, \ j_2 j_1}_{m \ m-q, \ m' m'-q} \ \ell^q_{m'}(j_2 j_1 \mathcal{X}) = \Gamma^{j_2 j_1, j_2 j_1}_{\mathcal{X}} \ell^q_m(j_2 j_1 \mathcal{X}). \tag{2.2}$$

Eigenvectors are independent of the nature of interaction and are given by

$$\ell^q_m(j_2 j_1 \mathcal{X}) = (-1)^{j_1 - m + q} \begin{pmatrix} j_2 & \mathcal{X} & j_1 \\ -m & q & m-q \end{pmatrix} (2\mathcal{X} + 1)^{1/2} \tag{2.3}$$

Eigenvectors are orthonormal:

$$\sum_m \ell^q_m(j_2 j_1 \mathcal{X}) \ \ell^{q'}_m(j_2 j_1 \mathcal{X}') = \delta_{qq'} \delta_{\mathcal{X}\mathcal{X}'},$$

$$\sum_{\mathcal{X}} \ell^q_m(j_2 j_1 \mathcal{X}) \ \ell^{q'}_{m'}(j_2 j_1 \mathcal{X}) = \delta_{mm'} \delta_{qq'}, \tag{2.4}$$

and display a symmetry property

$$\ell^q_m = (-1)^{\mathcal{X}} \ell^{-q}_{-m} = (-1)^{\mathcal{X}+q} \ell^q_{q-m}. \tag{2.5}$$

Equations (2.2) and (2.4) yield

$$\Gamma^{j_2 j_1, j_2 j_1}_{\mathcal{X}} = \sum_{m,m'} \ell^q_m (j_2 j_1 \mathcal{X}) \Gamma^{j_2 j_1, j_2 j_1}_{m \ m-q, m' m'-q} \ell^q_{m'}(j_2 j_1 \mathcal{X}) \tag{2.6}$$

with any q in the rhs of (2.6).

The diagonalization of submatrices corresponds to a transformation from components $\rho^{j_2 j_1}_{m_2 m_1}$ to spherical components $\rho^{j_2 j_1}_{\mathcal{X}}(q)$ given by

$$\rho^{j_2 j_1}_{\mathcal{X}}(q) = \sum_m \ell^q_m(j_2 j_1 \mathcal{X}) \rho^{j_2 j_1}_{m \ m-q}. \tag{2.7}$$

This transformation accomplishes the diagonalization also of non diagonal boxes $\Gamma^{j_2 j_1, \ j_2 j_1}_{m \ m-q, m' m'-q}$ to diagonal elements $\Gamma^{j_2 j_1, j_2 j_1}_{\mathcal{X}}$. These elements and those given by Eq.(2.2) which are called relaxation constants, can be expressed by the corresponding cross sections in the usual relation

$$\Gamma^{j_2 j_1, j_2 j_1}_{\mathcal{X}} = \left\langle \sigma^{j_2 j_1, j_2 j_1}_{\mathcal{X}}(v) \cdot v \right\rangle. \tag{2.8}$$

In semiclassical approximation the cross sections are expressed by probabilities $P^{j_2 j_1 j_2 j_1}_{\mathcal{X}}(v,b)$ depending on the velocity v and the impact parameter b:

$$\sigma_{\chi}^{j_2 j_1, j_2' j_1'} = 2\pi \int_0^\infty b\,db\, P_{\chi}^{j_2 j_1, j_2' j_1'}. \qquad (2.9)$$

The semiclassical transition probability is

$$P_{\chi}^{j_2 j_1, j_2' j_1'} = \delta_{j_2 j_2'}\, \delta_{j_1 j_1'} -$$

$$- (2\chi+1)^{-1} \sum_{m,n,q} \ell_m^q(j_2 j_1 \chi)\, \ell_n^q(j_2 j_1 \chi)\, S^*{}_{m\,n}^{j_1 j_1'}\; S_{m-q\;n-q}^{j_2 j_2'}, \qquad (2.10)$$

where S is the scattering matrix.

This expression is invariant under rotation of a space fixed reference frame. This can be used to facilitate actual calculations by an appropriate choice of the coordinate system. This completes the general simplification which is due to the isotropy of the perturbation exerted by heat bath particles on a relaxing atom.

Now let us consider further simplifications for specific conditions often met in practice.

The actual time dependence of the density matrix is controlled by joint dynamical and relaxational evolution. The former is characterized by atomic frequencies $\omega_{j\,j'}$ and the latter by two characteristic times: the mean time between collisions τ_0 ($1/\tau_0$ often provides a measure of matrix elements of Γ) and the collision time τ_c (in impact approximation $\tau_c \ll \tau_0$). From these parameters we construct two dimensionless parameters which are small and large, respectively for the relaxation of atomic states in the gas phase. This permits to develop a more detailed relaxation theory.

The small parameter is a product of the hyperfine frequency $\omega_{FF'}$ and the collision time τ_c, i.e., $\omega_{FF'}\tau_c \ll 1$. The large parameter is a product of electronic frequency $\omega_{JJ'}$ and the time between collisions τ_0, $\omega_{JJ'}\tau_0 \gg 1$. These conditions make possible the approximations of fixed nuclear spin during collision[13] and the so called secular approximation[14]

a. Approximation of the Fixed Nuclear Spin. This approximation which is in fact the Frank-Condon principle in the collision process states that in a sudden process which affects the electron configuration (a molecular collision) the orientation of the nuclear axis remains unchanged[13]. It follows that $S_{FF'}^{FF'}$ can be constructed from $S_{MM'}^{JJ'}$ and 3-j symbols. In particular for hyperfine states belonging to one J-state the following relation holds:[15]

$$\Gamma \, {}_{\varkappa}^{F_2 F_1, F_2' F_1'} = \sum_{\varkappa'} P_{\varkappa \varkappa'} \; \Gamma_{\varkappa'}^{JJJJ}$$

$$P_{\varkappa \varkappa'} = \frac{(2\varkappa'+1)}{(2\varkappa+1)} \, (-1)^{F_1 - F_1'} \left[(2F_1+1)(2F_1'+1)(2F_2+1)(2F_2'+1) \right]^{\frac{1}{2}}$$

$$\cdot \begin{bmatrix} \varkappa' & I & I & \varkappa \\ J & J & F_1 & F_1' \\ J & F_2 & J & F_2' \end{bmatrix}$$

where $\begin{bmatrix} : & : & : & : \\ \cdot & & & \cdot \end{bmatrix}$ is a 12-j symbol of second kind [16] .

   b. Secular Approximation for Electronic Transitions.
Any matrix element $\rho_{\varkappa}^{J_2 J_1}$ will generally oscillate with
frequency $\omega_{J_2 J_1}$ and approach its asymptotic value with
characteristic time τ_0 . Using this property of ρ we can
simplify the relaxation part of the problem by stating
that in a slow process which affects the density matrix
due to molecular collisions the relaxation couples matrix
elements of ρ possessing the same electronic frequency
 $\omega_{J_2 J_1}$[14]. This approximation allows to study the relaxation
and transfer of multipole moments of atoms in which case
 $\omega_{J_2 J_1} = 0$) independently of the relaxation of coherences
(or lines) in the optical region of spectra (as for this
region $\omega_{J_2 J_1} \tau_0 \gg 1$).

   From now on we shall concentrate on the relaxation and
transfer of multipole moments assuming in the beginning
that there is no nuclear spin (I=0). For this case the ma-
trix $\rho_{\varkappa}^{JJ}(q)$ at fixed $\varkappa$ and q can be represented as a vector $X_J(\varkappa,q)$ obeying the relaxation equation

$$\dot{X}_J(\varkappa,q) = - \sum_{J'} \gamma_{\varkappa}^{JJ'} X_{J'}(\varkappa,q) \qquad (2.12)$$

with

$$\gamma_{\varkappa}^{JJ'} = \Gamma_{\varkappa}^{JJ,J'J'} \qquad (2.13)$$

and $\Gamma_{\varkappa}^{JJ,J'J'}$ defindes by (2.7) and(2.10). The kinetic coefficients $\gamma_{\varkappa}^{JJ'}$ obey the detailed balancing

$$\gamma_{\varkappa}^{JJ'} / \gamma_{\varkappa}^{J'J} = \exp \left(-\frac{E_J - E_{J'}}{kT} \right) \qquad (2.14)$$

To provide a simple macroscopic interpretation for $\gamma_{\varkappa}^{JJ'}$,
we note that for q = 0

$$X_J(\chi, 0) = \sum_M \ell_M^0(JJ\chi) \, N_M^J \tag{2.15}$$

where N_M^J are populations of the Zeeman states $|J,M\rangle$. Now, a vector component $\ell_M^0(JJ\chi)$ is proportional to the mean value of the $q = 0$ projection of a tensor T^χ in the state $|J,M\rangle$:

$$\ell_M^0(JJ\chi) = \left[G(\chi,J)\right]^{-1/2} (T_0^\chi)_{MM}^{JJ} \tag{2.16}$$

with a factor $G(\chi,J)$ defined by the normalisation

$$G(\chi,J) = \sum_M |(T_0^\chi)_{MM}^{JJ}|^2. \tag{2.17}$$

Thus we have from (2.15)

$$X_J(\chi,0) = (T_0^\chi)_J \left[G(\chi,J)\right]^{-1/2}$$

where $(T_0^\chi)_J$ is the average value of a tensor component T_0^χ in an atomic state with the angular momentum J.

In particular for $\chi = 0, 1$ and 2 these average values are the total population N_J, the total z-projection M_J of the angular momentum J and the total z-projection of the quadrupole moment of an atom in the state J:

$$N_J = \sum_M N_M^J , \qquad \text{i.e., } (T_0^0)_{MM}^{JJ} = 1,$$

$$M_J = \sum_M M N_M^J, \qquad \text{i.e., } (T_0^1)_{MM}^{JJ} = M, \tag{2.19}$$

$$Q_J = \sum_M \left[M^2 - \frac{J(J+1)}{3}\right] N_M^J, \quad \text{i.e., } (T_0^2)_{NN}^{JJ} = M^2 - \frac{J(J+1)}{3}$$

Using the indicated values of $(T_0)_{MM}^{JJ}$ and calculating $G(\chi,J)$ from equation (2.17) we arrive at the following equations which describe the collision induced relaxation and transfer of populations and z-components of angular magnetic dipole and quadrupole moments:

$$\dot{N}_J = -\sum_{J'} \gamma^{JJ'} \sqrt{\frac{2J+1}{2J'+1}} \, N_{J'}$$

$$\dot{M}_J = -\sum_{J'} \gamma_1^{JJ'} \sqrt{\frac{2J(2J+1)(2J+2)}{2J'(2J'+1)(2J'+2)}}\, M_{J'}$$

$$\dot{Q}_J = -\sum_{J'} \gamma_2^{JJ'} \sqrt{\frac{(2J-1)2J(2J+1)(2J+2)(2J+3)}{(2J'-1)2J'(2J'+1)(2J'+2)(2J'+3)}}\, Q_{J'}.$$

(2.2o)

On account of the isotropy of the relaxation, the same equations are valid for any projection of a multipole moment.

Consider now transitions within one atomic multiplet $^{2S+1}L_J$ assuming that there is no quenching. For each J there will be 2J+1 coefficients $\gamma_\varkappa^{JJ}$ ($\varkappa = o,1,\dots 2J$); one describing the depletion due to transitions to neighbouring fine structure states. The total number N_{LS} of different coefficients with L≥S taking into account the symmetry relation (2.14) is given by

$$N_{LS} = (1/6)(2S+1)(2S+2)(6L-2S+3).$$

For 2S+1 fine structure states there are 2S+1 extra constraints due to unitarian property of the S-matrix. This leaves $N_{LS}-(2S+1)$ rate constants for describing the relaxation. In particular for states $^2P_J (J = 1/2, 3/2)$ there are 6 coefficients as was first found by Grawert[17].

Now, in the case of non-zero nuclear spin, the relaxation of diagonal and off diagonal elements of the density $\rho_{M_2M_1}^{F_2F_1}$ will be coupled. However, the number of independent relaxation constants is the same as if there was a nuclear spin: all relaxation coefficients $\Gamma_\varkappa^{F_2F_1,F_2'F_1'}$ can be expressed by $\gamma_\varkappa^{JJ'}$ in an equation similar to (2.11).

On condition that $\omega_{F_2F_1}\tau_o \gg 1$, the secular approximation for hyperfine transition can be applied and relaxation equations take the form of (2.12) with J replaced by F. In this approximation Bulos and Happer[12] observed an appreciable influence of nuclear spin I on the shape of the Hanle signal for the $^2P_{1/2}$ state of the alkalis. Franz and Sooriamoorthi[19] discussed similar effects of I on the depolarization.

For intermediate values of $\omega_{F_2F_1}\tau_o$ it is necessary to solve the complete problem, taking into account the coupling of hyperfine populations and coherences. In this connection reference is made to a recent paper by Rebane, Rebane and Cherenkovskii[20] who studied the depolarization of the $6^2D_{3/2}$ state of Tl over a broad range of parameters

$\omega_{F_2F_1}\tau_0$ They found that effective cross sections for
collisional depolarization of resonance fluorescence defined according to Born[21] depended on the pressure. This
dependence is due to competition of radiative decay and
multi-exponential relaxation and is complicated by transition from the general to the secular case.

If, at the other extreme, $\omega_{F_2F_1}\tau_0$ is small, it is
possible to describe the relaxation as proceeding at a
fixed orientation of the nuclear spin. Experimentally
this case has been studied by Elbel and Schneider[22] for
the sensitized Hanle signal of sodium in the $^2P_{3/2}$ state.
For an inert gas pressure of about 2o Torr, the Hanle
curve shows no pecularities which would be expected for
the usual case of hyperfine coupling; for pressures two
order of magnitude lower the signal exhibits a characteristic shape which reflects the decoupling of electron
and nuclear spins with increasing magnetic field.

Thus the nuclear spin essentially effects the relaxation of atomic states. But this is due exclusively to the
influence of the nuclear spin on the dynamics of an atom
in the time between collisions. For a correct   interpretation of experimental data in terms of electronic relaxa-
tion constants $\gamma_{\varkappa}^{JJ}$ and a particular interatomic interaction, it is necessary either to take properly into account
the hyperfine interaction between collisions or to eliminate it by applying a strong external magnetic field.

In concluding this section, two more approximations
should be mentioned, which are not so general in nature.
They are based on either a large or a small value of the
collision time τ_c as compared to the inverse frequency
$(\omega_{JJ'})^{-1}$ for fine-structure transitions.

c. Adiabatic Approximation. At $\omega_{JJ'}\cdot\tau_c \gg 1$ the non-diagonal elements of the S matrix are much smaller than the
diagonal ones. Then $\gamma_{\varkappa}^{JJ'} \ll \gamma_{\varkappa}^{JJ}$, and, as a zero approximation, the multipole relaxation for each fine structure
level can be considered to be uncoupled from the relaxation of other levels. The first order non-adiabatic perturbation theory should be used to calculate $\gamma_{\varkappa}^{JJ'}$.

d. Approximation for Quasidegenerate States. For
$\omega_{JJ'}\cdot\tau_c \ll 1$ the orientation of the electron spin can be considered as fixed during the collision[23,24]. In this case
the relaxation constants $\gamma_{\varkappa}^{JJ'}$ for all states of the term
$^{2S+1}L_J$ are expressed via $2L$ relaxation constants

$$\gamma_{\varkappa}^{LL} \quad (\varkappa = 1, \ldots, 2L)$$

of a spinless electronic term with electron orbital angular momentum equal to L (compare Eq.(2.11)).

The four approximations, a),b),c) and d) qualitatively classify the relaxation and transfer of atomic multipole moments with respect to the relation between characteristic times of intra- and interatomic processes. This classification is summarized in Fig. 1 where different regions of the $\omega\tau_c,\omega\tau_0$-plane correspond to the different approximations for relaxation and transfer. The figure presents also regions of different approximations for relaxation of coherence which show up in the shapes of corresponding lines. This completes the description of the relaxation. We turn now to the problem of collision dynamics which is the basis for calculation of the kinetic coefficients $\gamma_{\varkappa}^{JJ}$

III. QUALITATIVE DESCRIPTION OF THE COLLISION DYNAMICS

Exact calculation of the S-matrix for a collision process is based on the solution of quantal or semi-classical coupled differential equations. Time consuming numerical calculations are justified in cases when the interatomic interaction is reasonably well known. Unfortunately, this is not always the case. Only for resonance processes with rather large cross sections the long-range part of the interaction which can be reliably estimated can be safely assumed to be responsible for the process. In the next section some examples for such processes are discussed.

If the interaction is not known in detail or, moreover, if some sort of correlation between the interaction and the cross section is looked for, it would be helpful to use approximative methods for the construction of the S-matrix for intramultiplet transitions. The idea that will be discussed now is to separate the interaction region into smaller regions and to take into account only the strongest interaction in each case. In terms of the adiabatic electronic representation of a colliding pair there are only two mechanisms of interaction between states, the radial and the rotational non-adiabatic coupling. Both are due to the relative motion of the atoms. The coupling depends on the ratio of the electrostatic interatomic interaction to the spin-orbit interaction. The different coupling cases can be conveniently classified in terms of the Hund coupling cases which were originally formulated for stable molecules[25,26]. For colli-

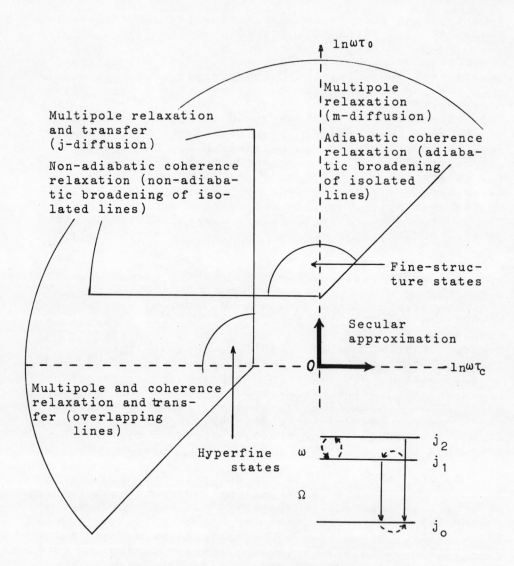

Fig. 1 Different domains of the relaxation of atomic
        states in the impact approximation

sion problems a similar approach has been used in earlier
papers on intramultiplet mixing[27] and has recently been
accepted as a standard description of processes which
lead to depolarization[28-29] and broadening of spectral
lines[33-36]. An analogous approach has been formulated in
the theory of electron-atom collisions[37].

a. The Hund Coupling Cases for Diatomic Molecules.
The coupling in question involves three angular momenta,
the electronic orbital angular momentum $\underline{L}$, the spin $\underline{S}$
and the relative angular momentum $\underline{N}$ of nuclei, which give
the total angular momentum of a diatomic molecule. A particular coupling scheme provides a rule for construction
of the zero-order wave functions which are used to calculate the first and higher order corrections for energy
levels. If this approach fails, i.e., if the breakdown of
a particular coupling case occurs, intermediate schemes
must be considered, which take into account simultaneously three relevant interactions: the electrostatic one V_e
(coupling of $\underline{L}$ to the molecular axis $\underline{n}$), the magnetic one
V_m (coupling of $\underline{L}$ and $\underline{S}$) and the rotational interaction
V_r (decoupling of $\underline{L}$ and $\underline{S}$ from $\underline{n}$ due to the rotation of
the axis).

The different Hund coupling cases and their correlations with the interaction V_e, V_m and V_r are illustrated
in Table I which essentially coincides with that shown in
Mulliken's paper[26] concerning the a,b,c,d, and e cases.
For completeness, we add here the case e' which comprises
all possible combinations (3! = 6) of the values for the
interactions in question. As we confine ourselves to the
semiclassical approximation, only the electronic quantum
numbersare given in Table I: Λ is the projection of the
electronic orbital angular momentum on the molecular axis
$\underline{n}$, Ω is that of the total angular momentum, L,S, and J
are, restrictively, the total orbital, the spin and the
total angular momenta of the atom (we consider, for simplicity, only one atom in the dimer having an L and S
value different from zero), S_n is the projection of $\underline{S}$ on
$\underline{n}$, S_N is the projection of $\underline{S}$ on the angular momentum of
the dimer, $\underline{N}$. In the semiclassical approximation, $\underline{N}$ is directed along the total angular momentum vector and thus de-
fines the axis of the space-fixed coordinate system. Table
I also shows the particular cases b' and d'[26] and the type
I case c discussed by Mulliken[38].

We note now that the case e', which has not been discussed by Mulliken, is physically identical to the case e.
This "degeneracy" of two cases is due to the impossibility

Table I. Hund coupling cases

Hund coupling case	Ve, coupling of $\underline{L}$ to $\underline{n}$	Vm, coupling of $\underline{S}$ to $\underline{L}$	Vr, effect of rotation on $\underline{L}$ or $\underline{S}$	Electron quantum numbers	Limitations and remarks
a	large	moderate	small	Λ, S, S_n	$\Lambda > 0, S > 0$
b	large	small	moderate	Λ, S, S_N	case b' if $S = 0$
c	moderate	large	small	Ω	type I case c if J is a good quantum number
d	moderate	small	large	L, S, L_N, S_N	case d' if $S = 0$
e	small	moderate	large	J, J_N	both cases are
e'	small	large	moderate	J, J_N	physically identical

of the direct coupling (different from indirect coupling
via $\underline{L}$) of $\underline{S}$ to $\underline{n}$. Technically the cases e and e' differ
because in the case of e $\underline{L}$ and $\underline{S}$ are first quantized on
$\underline{N}$ (thus giving quantum numbers L_N, S_N, and $J_N=L_N+S_N$, but
still not forming J) and then L and S are coupled to give
J; in the case of e' the vectors $\underline{L}$ and $\underline{S}$ first form $\underline{J}$
(the quantization axis at this stage is still unspecified),
and then $\underline{J}$ is quantized on $\underline{N}$, giving J_N. Designating both
the e and e' cases as e, we shall consider now a dimer with
the fixed interatomic distance b and relative nuclear ve-
locity $\underline{v}$ (evidently $\underline{vn}$= 0). Different coupling cases correspond to different regions of the (v/ω, b)-plane with
$\omega=V_m/\hbar$. If the scale is given by the distance R_0, which
is defined by the condition $V_e(R_0)=V_m(R_0)$, the two dimensionless parameters $\nu=v/(\omega R_0)$ and $\beta =b/R_0$ (which can be
thought to arise from two other dimensionless parameters
coming from the three interactions V_e, V_m, and V_r) will be
convenient for the classification of the coupling. The result of such an approach is presented in Fig. 2, the boundaries between regions being defined by the conditions
$V_e=V_m$ (straight line II, $\beta=1$), $V_r=V_m$ with $V_r=\hbar v/b$ and $V_m=$
$\hbar\omega$ (part I of the straight $\beta=\nu$), and $V_e=V_r$ (curve III,
$\beta=\beta_m(\nu)$). Pure coupling cases correspond to the regions
well inside the boundaries. The intermediate cases are
those at the boundaries. Those, which most commonly occur
and which are discussed by Herzberg[39] are indicated by
the arrow 1 (spin-orbit decoupling) and by the arrow 2
(decoupling of $\underline{L}$ from $\underline{n}$ which leads to Λ-doubling when it
is weak).

   b. Transient Coupling Cases. We shall discuss now a
three-dimensional extension of the coupling regions allowing the relative velocity $\underline{v}$ of the dimer to be nonorthogonal with respect to $\underline{n}$. In line with the usual parametri-
zation of a trajectory describing the relative motion of
two colliding atoms, we introduce the impact parameter b
and interatomic distance R which define the linear coordinate $z=(R^2-b^2)^{1/2}$. At a fixed R an adiabatic molecular term
is characterized by "static" interactions $V_e(R)$ and $V_m(R)$,
and also by the Coriolis (rotational) interaction $V_r(R,b)=$
$\hbar vb/R^2$. Using the already defined β and ν and defining
$\xi=z/R_0$, we draw in the space ξ,ν,β various surfaces confining transient coupling cases, which again are specified
by relative values of V_e, V_m, V_r given in Table I. For
Λ≠ 0, S≠0 and Λ=0, S≠0 these surfaces are shown in Fig.3.
If Λ≠0, S=0, only one surface (III) separates the coupling
cases b' and d'. If the surfaces are cut by the ν,β-plane,
the picture obtained will be that in Fig. 2 for Λ≠0, S≠0
and Λ=0, S≠0, or similar to that for the case, Λ≠0, S=0.
The other cuts, however, which will be discussed now are

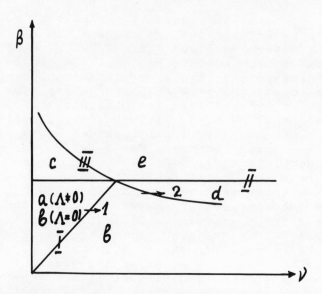

Fig. 2   Two dimensional picture of the Hund coupling
         cases for diatomic molecules

of greater interest for the collision problem.

For slow atomic collisions, when overlapping of the
atomic functions is small, the spin-orbit interaction is
to a good approximation independent of R. This means that
for a given collision the parameter ν does not depend on
R. The collision itself can then be described by the motion of a representative point in a plane $\nu=$ const. Along
the trajectory (not necessarily being a rectilinear one)
there will be changes for the different coupling cases.
This can be used for an approximative calculation of the
scattering matrix by means of the matching method[40]. For
such an approximation transitions between the zero-order
states which are just those corresponding to the particular Hund cases are localized near the boundaries. Correspondingly, the problem of the S-matrix calculation re-
duces to the calculation of the matrix $\underline{N}_{kl}$ which describes
the non-adiabatic transition at the boundary between the
cases k and l. As follows from the theory of non-adiabatic transitions, the important parameter in $\underline{N}_{kl}$ which
governs the transition probabilities is the switching rate

of the cases k→l. If the switching rate is high the ma-
trix $\underline{N}_{kl}$ is completely defined by projection of the basis
|k> on |l> so that the known basis transformations are of
immediate use (for the switching d→ b, see[41], for the
switching a→b, see[42]). If the switching rate is low, one
must solve the problem of non-adiabatic coupling in the
vicinity of the boundaries.

It follows from Fig. 3 that the space ξ, ν, β is devi-
ded into two subspaces $\nu < 1$ and $\nu > 1$. In the first subspace,
$\nu < 1$, typical switching sequences are: e→c→a→b→a→c→e (small
impact parameters), e→c→a→c→e (intermediate impact parameters) and e→c→e (large impact parameters). For the second
subspace, $\nu > 1$, the sequences are e→c→a→b→a→c→e (small impact parameters), e→d→b→d→e (intermediate impact parameters), e→d→e (large impact parameters). If there are special reasons to neglect the contributions of the one
region of impact parameters, the variety of sequences
diminishes, and often the main contributions to the cross
sections are due to only one switching sequence. This is
the very reason for the transient coupling cases beeing
of use for interpreting the mechanism of an inelastic
collision event.

The simplest possible case which illustrates the
above approach is a depolarizing collision of a $^2S_{1/2}$
ground state alkali atom with a noble gas atom. The depolarization is due to spin-orbit coupling which can be calculated with the wave function that corresponds to a mixture of the d and b Hund coupling cases. The main contribution of the depolarization probability comes from the
distance of the closest approach. The cross section which
is very small depends critically on the distortion of the
atomic wave function, the mixture coefficient which is proportional to the angular velocity of rotation of the molecular axis, and the spin-orbit coupling constant in the
nearest excited state[43,44]. This explains why for the two
isoelectronic systems, $Sr^+(5^2S_{1/2})$ and $Rb(5^2S_{1/2})$, the
depolarization cross-sections differ by two to three orders of magnitude (the larger values are those for the
ion[45]).

More examples are provided by the intramultiplet
mixing and the depolarization in the excited states of the
alkalis which have been reviewed recently[46,47]. For these
processes the region of impact parameters which mainly
contribute to the cross sections are designated in Fig.3
by arrows.

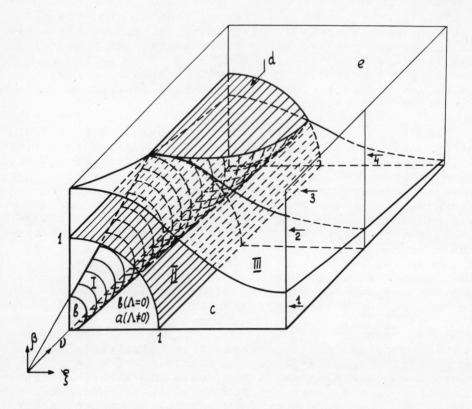

Fig. 3   Three-dimensional picture of the transient
cases for two colliding atoms

1. Intramultiplet mixing $^2P_{1/2}-^2P_{3/2}$ in the atoms
Rb*(5^2P) and Cs*(6^2P) induced by collision with noble
gases. A typical switching is c→a→b→a→c[48]. For the heavy
noble gases the switchings c→a, a→c are almost adiabatic,
so that the main contribution comes from a→b→a. The specific of the collisions is that the core-core interaction
bends the trajectory thus making it impossible to accomp-
lish the complete switching a→b. The beginning of breakdown in the pure case a, however, is essential for the
explanation of the experimental data.

2. Depolarization of the $^2P_{1/2}$ states of Rb*(5^2P)
and Cs*(6^2P) induced by collisions with noble gases . A
typical switching is c→a→c[49]. The specific of the process
is that substantial depolarization cross sections can be
expected only if the atomic state $^2P_{1/2}$ adiabatically

correlates with the molecular state $^2\Pi_{1/2}$, but not with $^2\Sigma$. Since any correlation scheme could be excepted a priori for the different noble gases, the rather large experimental values of the cross sections[50] would imply that only the first possibility is realized for all the noble gases.

3. Depolarization of the states $^2P_{3/2}$ of Rb*$(5^2P_{3/2})$ and Cs*$(6^2P_{3/2})$, induced by collisions with noble gases. The typical switching is thought to be $e{\to}c{\to}e$[51]. The specific of the process is that this switching sequence makes the main contribution only if the depolarization cross sections(four altogether) are substantially larger than πR_o^2. Otherwise, one must consider the sequence $e{\to}c{\to}a{\to}c{\to}e$. Since the switching $e{\to}c$ for the approach and $c{\to}e$ for the reseparation is complete, one would not expect a strong dependence of the depolarization cross section on subtle features of the interaction. This is substantiated by the comparison of the depolarization cross-sections of the atom Rb$(5^2P_{3/2})$ and of the isoelectronic ion Sr$^+(5^2P_{3/2})$[52].

4. Multipole relaxation and mixing in the sates $^2P_{1/2}$ and $^2P_{3/2}$ of Na*(3^2P) induced by collisions with He. The typical switching is $e{\to}b{\to}e$[29,31,32]. The specific of the events is that the switching is almost instantaneous and the bending of the trajectories occurs well inside the region of the pure case b. As a consequence, the approximation of straight line trajectories is expected to be reasonable.

These examples show that the interpretation of the collision processes, for which the Coriolis and spin orbit interactions are of the same order of magnitude, can be given in terms of transient Hund coupling cases. It is hoped, that more work along this line will help interpreting a large variety of available experimental results[2,3]. One more comment is appropriate. For the semiclassical picture of a collision the common procedure is to use one rectilinear trajectory describing the relative motion of colliding partners on several electronic terms. This approach cannot be accepted for thermal collisions at its face value since interatomic potentials are of the order of or larger than kT at separations contributing to the transitions. The use of different trajectories for different terms requires that the transitions be localized. This is just the case for the electronic basis set constructed according to the Hund prescription. Only for regions of intermediate coupling, the one rectilinear trajectory approach fails. But this weakness of the semiclassical method is probably the lowest price that has to be paid for simplicity.

IV. CALCULATIONS FOR MODEL INTERACTIONS

Exact calculations of the S-matrix for model potentials are of interest because of two reasons: first, they
can be applied for the interpretation of selected processes with interactions that can be simulated by a model;
second, an exact result is an important standard case
which should be referred to in any calculation involving
more or less complicated interactions.

Below a short review is given of the most important
model calculations within the framework of the semiclassical approximation using rectilinear trajectories.

a. Resonant (First Order) Dipole-Dipole Interaction.
For a collision of two identical atoms A_1 and A_2 one of
which is in the excited (e) and the other in the ground
(g) state, this interaction is the most important contribution to the interatomic potential at large distances
provided the oscillator force for the transition $e \rightarrow g$ is
not too weak.

The most detailed case studied is that of S -- P
excitation transfer ($J_g = 0$, $J_e = 1$) (for review see[47]).
The research aimed at the calculation of the excited transfer cross section $\sigma_{ex. trans}$[47,52], of the cross section
for transitions between hyperfine states[54] and the optical manifestation of the resonance interaction[53,55]. For
this case as well as in the general case for other J_e and
J_g, $\sigma_{ex. trans}$ is proportional to $\sigma_o^{J_g J_e} = \frac{\pi}{3} \frac{e^2}{h\nu} (J_e \| z \| J_g)^2$.
Satisfactory approximation to exact results can be achie-
ved if one considers the Hund coupling case d(for S = 0)
or e(for S $\neq$ 0), i.e., the so called "fixed atom approximation"[56].

Recently for this type of interaction all cross sections of relaxation and transfer have been calculated
for J_g, $J_e = 0,1/2,1,3/2$[57]. The results are tabulated as
parameters $_{ab}g(A_k)$ (a,b = e,g, $A_k = A_1, A_2$) which give the
corresponding cross sections in terms of $\sigma_o^{J_g J_e}$.

b. Quasiresonant Dipole-Dipole Interaction. In the
collision of an excited atom A and an unexcited atom B,
the leading interaction at large distances is the second
order dipole-dipole interaction $\propto R^{-6}$ which turns into the
first order interaction $\propto R^{-3}$ at $R < R_p$ such that
$d_{J_e J_g}^A d_{J_g J_e}^B / R_p^3 > \Delta E$ where ΔE is the mismatch of the resonance
for excitation transfer and $d_{J_e J_g}^A$ is the matrix element of
the atom A.

The simple case $J_e^A = 1$, J_g^B $0 \to J_g^A = 0$, $J_e^B = 1$ is treated in a number of papers (see, e.g.[47]). Particular attention is paid to the interpolation formulae[58] which would provide an approximate dependence of $\sigma_{ex,trans}$ on ΔE between exactly treated limiting cases of low ($\Delta E R_p/\hbar v \ll 1$)[59] and high ($\Delta E R_p/\hbar v \gg 1$)[60] values of transferred energy ΔE. Though it has been noted quite a while ago[61] that the reported formulae (see, e.g.[62]) would give grossly incorrect estimates of the excitation transfer cross sections, a numerical study using the approximation of a non-rotating molecular axis (coupling case b) provided a better approximation only recently[63]. However more work is needed to take into account the rotation of the molecular axis (switching $d \leftrightarrow b \leftrightarrow d'$).

Practical application of the theory of non-resonant energy transfer is limited by the existence of a number of close lying multiplet states which strongly interact under the perturbing effect of an incoming atom. Much is to be done in this respect. Even non-resonant energy transfer in collisions of alkali atoms which is probably the simplest process of that kind has not yet been satisfactorily interpreted[47]. The reason is that with increasing fine structure splitting $\Delta E(^2P_{3/2} - {}^2P_{1/2})$ the distance R_p markedly diminishes so that the dipole-dipole part of the interaction ceases to represent the main part. Of all alkalis only the Na*-Na collisions leading to intramultiplet mixing ($\Delta E = 17cm^{-1}$) can be interpreted in terms of the dipole-dipole interaction; for heavier alkalis the exchange interaction is substantial[47].

c. Polarization (Second Order Dipole-Dipole) Interaction. For non-resonant collisions the leading term in the interaction potential most often is the second order dipole-dipole interaction proportional to R^{-6}. An example discussed in many papers is a collision of an atom A in an adiabatically isolated state J with a spherically symmetrical noble gas atom $X(^1S_0)$.

The only case which can be treated in a closed form for any potential is that with $J = 1/2$[49,64].

For $J > 1/2$ the calculation of depolarization cross sections would require a solution of coupled equations describing switching $d \leftrightarrow b \leftrightarrow d'$ (if $S = 0$) or $e \to c \to e$ (if $S \neq 0$). This has been done for $J = 1$[65,66], $J = 3/2$[49,51] and $J = 2$[67]. All cross sections σ_χ^{JJ} are proportional to $\sigma_0^{JJ} = [(C_\parallel^J - C_\perp^J)/\hbar v]^{2/5}$ with $C_\parallel^J$ and $C_\perp^J$ being interaction coefficients for adiabatic molecular terms with a maximal

and minimal projection of J onto the molecular axis.
These exact results were used in a recent discussion[68] on
different approaches to the calculation of depolarization
cross sections and on the appraisal of approximative methods. Hence, it will not be repeated here.

There are many papers dealing with the comparison between
theory and experiment. If the polarizability of X is
large (X = Ar,Kr,Xe) experimental results on depolarization can be usually explained by taking into account only the polarization interaction. A more strict test of
the theory would require interpretation of the experimental data in terms of the same interaction model, both on
the depolarization of a state and on the broadening (and
possibly the shifts) of a spectral line originating from
that state. At this level, as well, it is often possible
to arrive at satisfactory results[69] for heavier noble gases though it was most recently reported that there was
found no such agreement for the depolarization of the
$^3P_1$ state and broadening of the $^3P_1-^1S_0$ line of Hg[70].

For the light noble gases it is not possible to explain experimental findings in terms of the long-range
polarization attraction. That called forth the appearance
of empirical models in which a kind of repulsive interaction was added to supplement the attraction. This approach
with varying parameters of the repulsive part makes the
experimental data fit into a consistent scheme, but the
theory still remains on an empirical level. A new move
would require the use of independent information on the
short-range interaction of colliding partners.

d. Exchange Interaction. The a b i n i t i o calculation of a full interaction potential at large distances is difficult because the variational approach using a limited basis fails. Though reasonable results are
sometimes obtained by simple superposition of the exchange and polarization parts of the interaction[71] the
general situation is most simple in cases when the R^{-6}
interaction can be completely neglected compared to the
exponentially decaying exchange interaction. The sufficient condition for this is that the main contribution
to the scattering phase-shift δ at the cut-off impact parameter b^* (say, $\delta(b^*) = 1/\pi$, according to Firsov[72]) is
due to exchange forces. This would be the case with He
which represents a system convenient for studying because it consists of a small number of electrons (hence
the a b i n i t i o calculation is easy) and has a
large ionization potential (hence it is possible to apply

the asymtotic type of calculation[73] using as a measure
for the interaction the scattering length for slow electrons). At least, for He, with a>0, those two approaches
do not disagree in the sign of the potential (for alkali-
helium interaction both approaches give exchange repulsion) as they do for heavier X atoms with a<0.

The asymptotic method of calculation was recently
applied to systems M-He(M=Rb,Cs) and the potentials obtained were used for the semiclassical calculation of the
scattering matrix[35]. The calculated cross sections and
broadening agree resonably well with the available experimental findings[35]. This provides a clear indication
that the exchange interaction usually described as a
short-range one, prevails the "long-range" polarization
interaction.

V. SODIUM-HELIUM COLLISIONS

Among all systems for which relaxation and transfer
have been studied, the case of $Na^*(3^2P_J)$ + $He(^1S_0)$ is
probably the best understood. All cross sections but one
(that corresponding to the octupole relaxation in the
$^2P_{3/2}$ state) were measured[22,74-76] and these data are
complemented by the recently reported cross section
σ_{opt} for broadening of the resonance lines $^2P_{3/2}$ - $^2S_{1/2}$
and $^2P_{1/2}$ - $^2S_{1/2}$ [34]. The potential curves have been calculated by different methods[34,77,78,79,80,81] and the
collision dynamic studies range from crude semiclassical
calculation[24] to exact quantal treatment[82]. The discussion
will be based now on what we believe to be the best semiclassical approximation[31,32].

In spite of some earlier[83-85] and even recent[82,86]
calculations using the van der Waals or the Lennard-Jones
type potentials with a large well these models are not
discussed here in view of the unrealistic nature of the
interaction. The pattern of adiabatic molecular terms of
the Na*-He system is shown in Fig. 4. Qualitatively, there
are two parameters R_1 and R_2 which define the range of the
interaction in the $B^2\Sigma$ and $A^2\Pi$ states, and two parameters
α_1 and α_2 which characterize the steepness of the repulsion in these states. The attraction which for heavier
noble gases would lead to the potential well in the $A^2\Pi$
state, can be neglected for He. The spin-orbit coupling
can be allowed for by diagonalizing the interaction in
the basis of atomic states, i.e., by assuming that the
electrostatic interaction Na-He does not perturb the spin-

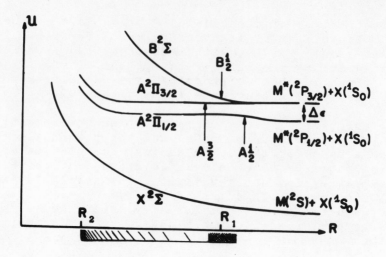

Fig. 4 Qualitative pattern of the adiabatic mole-
        cular terms for an alkali-helium pair. The
        shaded regions on the R-axis correspond to
        regions of non-adiabatic coupling.

orbit interaction in Na[78,81].

        The mechanism of relaxation and transfer can be des-
cribed in terms of the relative dephasing of the degenerate $^2P_{3/2}$ state and the breakdown of the adiabatic
approximation for the pair of terms, A1/2 and B1/2, due
to radial motion and for the pair of terms, A1/2 and A3/2,
due to rotation. The first region of non-adiabatic coupling is localized near R_1 (c→a switching for the approach
and a→c switching for the separation). The second region
extends from R_2 to R_1 (a→b→a switching for the approach
and separation). For almost adiabatic collisions, with
$\Delta ER_1/\hbar v \gg 1$, the second coupling tends to be localized
near R_2.

        The calculated splitting of $^2\Sigma$ and $^2\Pi$ terms for the
system Na -He is shown in Fig. 5 in the region compris-
ing R_1 and neglecting the spin-orbit interaction. The results which differ most of all from the others are obtained by the asymptotic method[34,77].

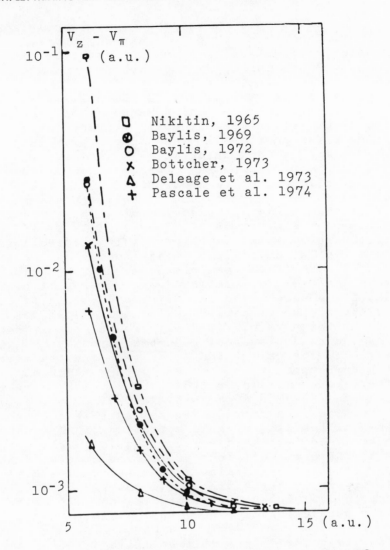

Fig. 5 Calculated energy difference between $^2\Sigma$ and
$^2\Pi$ terms of the Na(^{2}P)-He system neglecting
the spin-orbit coupling

Empirical[27], pseudopotential[78,80,82], and model[79]
calculations give values for R_1 ranging from 13.2-13.8
a.u., whereas the asymtotic method gives 1o.7a.u. The
question is whether this difference in the potentials
is large enough to result in an appreciable difference
in the cross sections and whether this difference lies

outside possible errors of the dynamical calculation. As
for the term $A^2\Pi$, it is expected that any possible uncertainty of R_2 will effect the computed cross section less
than the variation in the $B^2\Sigma$ potential.

Now, the ratio $\Delta ER_1/\hbar v$ for Na^*-He collisions is less
than 1 and the general switching sequence of the transient Hund cases can be approximately replaced by the $e \to b \to e$
switching which provides a reasonable approximation for
the exact semiclassical calculation[29]. However the latter
was done for rectilinear trajectories which neglect the
distorting influence of the potential in the relative motion of the colliding pair. Though this approximation is
valid for large impact parameters, $b \approx R_1$ where the $B^2\Sigma$ potential is of the order of ΔE, it certainly fails for
smaller b values. With the corresponding corrections allowing for different curvilinear trajectories on the $A^2\Pi$ and
$B^2\Sigma$ adiabatic terms the semiclassical approximation seems
to attain its best formulation[31,32]. The result of such a
calculation is summarized in the second column of Table II.
The results of the primitive semiclassical calculation[24]
based on the picture of fixed electronic spin at $R < R_1$,
i.e., on the sudden switchings $e \to b \to e$, are also shown in
the Table II. The calculated relaxation and transfer cross
sections correspond to potentials with $R_1 = 13.8$ a.u. For
any other value of R_1 close to 13.8 a.u. they vary proportional to $(R_1/13.8)^2$. With $R_1 = 13.8$ a.u., the calculated cross section shows a general agreement with the
experimental results. A small variation of R_1 would probably make it better. This value of R_1 is roughly consistent with one group of theoretical predictions, but dis-
agrees with asymtotic calculations which give $R_1 = 10.7$
a.u. It is the latter vaule of R_1, however, which recon-
ciles the calculated and experimental optical cross sections for broadening of the D_1 and D_2 lines of Na[34].

Thus even for this simple collision process there is
not as yet a satisfactory overall picture of interaction
which describes all phenomena encountered in the resonance doublet of Na. The semiclassical approximation can
hardly be blamed for the discrepancy: there is little doubt
about its applicability to the calculation of σ_{opt}. Its
estimated inherent error in the calculation of the relaxation and transfer cross sections (by direct comparison
with quantal results) has been shown to be small[32]. More
work is needed to resolve this inconsistency. Nevertheless
the example of sodium-helium collisions demonstrates a
rather high degree of our understanding of the mechanisms
operating in the depolarizing and mixing collisions.

Table II. Relaxation, transfer, and optical cross
sections for $Na^*(^2P_j)$ perturbed by He

Cross sections ($Å^2$)	Best semiclassical calculation	Primitive semiclassical calculation	Experiment
$\sigma_{depol}^{(1/2)} =$ $\frac{1}{2}\left[\sigma_1^{(1/2\ 1/2)} + \sqrt{2}\sigma_0^{(1/2\ 3/2)}\right]$	12.8^a	16.8^c	15.6^e
$\sigma_1^{(3/2\ 3/2)}$	99^a	97^c	128^h
$\sigma_2^{(3/2\ 3/2)}$	147^a	159^c	146^f
$\sigma_{poltrans}^{(1/2\to 3/2)} = -\sqrt{10}\,\sigma_1^{(1/2\to 3/2)}$	-55^a	-74^c	$-64^e, -71^e$
$\sigma_{sens}(1/2\to 3/2) =$ $= -\sqrt{2}\,\sigma_0^{(1/2\ 3/2)}$	92^a	112^c	$89^e, 86^g$
$\sigma_{opt}(D_1)$	$148^b, 98^{b*}$		90^b
$\sigma_{opt}(D_2)$	$154^b, 95^{b*}$		96^b

Theoretical with $R_1=13.8$ a.u.: a- [31,32]; b - [34], c - [24]

Theoretical with $R_1=10.7$ a.u.: b^* [34]

Experimental: b- [34], e- [75], f - [22], g - [74], h - [76]

(The ref. are given by the numbers in brackets)

REFERENCES

1. A.Omont, E.W.Smith, J.Cooper, Astrophys.J.175,185(1972).
                                 Astrophys.J.182,283(1973).
2. C.Th.J.Alkemade, P.J.Th.Zeegers in:*Spectrochemical Methods of Quantitative Analysis of Atoms and Molecules,*
   p.3, ed.by J.D.Winefordner, John Wiley and Sons,Inc.1971.

3. L.Krause in:*The Physics of Electronic and Atomic Collisions,* VII.ICPEAC invited papers and progress reports,
   p.65,ed.by T.R.Govers and F.J.de Heer, North Holland
   Publ.Co.,1972.

4. W.Happer, Rev.Mod.Phys. 44,169 (1972).

5. G.zu Putlitz in:*Atomic Physics,*Proceedings of the First
   International Conference on Atomic Physics, ed.by
   B. Bederson, V.W.Cohen, F.M.J.Pichanick, Plenum Press,
   N.Y. 1969.

6. E.B.Aleksandrov, Uspekhi Fiz.Nauk 1o7, 595 (1972).

7. V.G.Pokazaniev, G.V.Skrotzki, Uspekhi Fiz.Nauk 1o7,
   623 (1972).

8. C.H.Wang, W.J.Tomlison, Phys.Rev. 181, 125 (1969).

9. N.H.Dieter,W.M.Gross, Rev.Mod.Phys. 38, 125 (1966).

1o. J.N.Bahcall, R.A.Wolf, Astrophys.J. 152, 7o1 (1968).
11. A.Omont, J.Phys. 26, 26 (1965).
12. M.I.D'yakonov, V.I.Perel,Zhurn.Eksp.Teor.Fiz.48,345
    (1965).
13. C.Cohen-Tannoudji, Comm.Atom.Mol.Phys. 2,24(197o).

14. J.P.Barrat, C.Cohen-Tannoudji,J.Phys. 22,443 (1961).

15. V.N.Rebane, T.K.Rebane, Optika i Spekt. 33,4o5(1972).

16. P.A.Yutzis, A.L.Bandzaitis,*Teoria momenta kolichestva
    dvizheniya v kvantovoi mekhanike,* Vilnus, 1965.

17. G.Grawert, Z.Physik 225, 283 (1969).

18. B.R.Bulos, W.Happer, Phys.Rev. A4, 849 (1971).

19. F.A.Franz, C.E.Sooriamoorthi, Phys.Rev. A8,239o(1971).

2o. V.N.Rebane, T.K.Rebane, V.A.Cherenkovki, Optika i
    Spektr. 33,616(1972).

21. M. Born, E.Wolf, *Principles of Optics*,Pergamon Press
    N.Y.1959.

22. M.Elbel, W.B.Schneider, Physica 68, 146(1973)

23. M.Elbel, Z.Physik 248,375(1971).

24. E.I. Dashevskaya, N.A.Mokhova,Chem.Phys. Lett.2o,454
                                                 (1973).

25. F. Hund, Z.Physik 36,637(1926).

26. R.S.Mulliken, Rev.Mod.Phys.2,6o(193o).

27. E.E.Nikitin, J.Chem.Phys. 43,744(1965).

28. F. Masnou-Seeuws, J.Phys. B3,1437(197o).
29. F. Masnou, E.Roueff, Chem.Phys.Lett.16,593(1972)-

3o. F.H.Mies, Phys.Rev. A7,942(1973),A7,957(1973).

31. E.I.Dashevskaya, F.Masnou, R.McCarroll, E.E.Nikitin
    Optika i Spektr. 37,2o9(1974).

32. F.Masnou-Seeuws, R.McCarroll,J.Phys.B(to be published)

33. E.L.Lewis, L.F.McNamara, Phys.Rev. A5,2643(1972).

34. J.P.Deleage, D.Kunth,G.Testor, F.Rostas, E.Roueff,
    J.Phys. B6,1892(1973).

35. E.Roueff,A.Suzor, J.Phys.B 3 5 , 727 (1974).

36. E.I.Dashevskaya, E.A.Kobzeva, Optika i Spektr. 3o,8o7
                                                   (1971).

37. U.Fano, Comm.Atom.Mol.Phys. 2,47(197o).

38. R.S.Mulliken, Phys.Rev. 57,5oo(194o).

39. G.Herzberg, *Molecular Spectra and Molecular Structure*,
    I.Spectra of Diatomic Molecules, 2-nd ed., VanNostrand,
    Princeton, N.Y.195o.

4o. J.E.Bayfield, E.E.Nikitin, A.I.Reznikov, Chem.Phys.Lett.
    19,471(1973).

41. E.S.Chang, U.Fano, Phys.Rev. A6,173(1972).

42. Y.Chiu, J.Chem.Phys. 58,722(1973).

43. R.M. Herman, Phys.Rev. 136A,1576(1964).

44. E.I.Dashevskaya,E.A.Kobzeva, Optika i Spektr. $\underline{3o}$,436
    (1971).

45. E.W.Weber, H.Ackermann, N.S.Laulainen, G.zu Putlitz,
    Z.Physik $\underline{259}$,371(1973).

46. P.L.Lijnse, Review of Literature on Quenching, Exci-
    tation and Mixing Collision Cross-Sections for the
    First Resonance Doublets of the Alkalies, Report 398,
    Fisisch Laboratorium, Rijksuniversiteit Utrecht,1972.

47. E.E.Nikitin, Adv.Chem.Phys. $\underline{28}$(to be published).

48. E.I.Dashevskaya, E.E.Nikitin, A.I.Reznikov, J.Chem.
    Phys. $\underline{53}$,1175(197o).

49. E.P.Gordeev, E.E.Nikitin, M.Ya. Ovchinnikova, Optika
    i. Spektr. $\underline{3o}$,189(1971).

5o. J.Guiry, L.Krause, Phys.Rev. $\underline{A6}$,273(1972).

51. A.I.Okunevich, V.I.Perel., Zhurn.Eksp.Teor.Fiz. $\underline{58}$,
    667(197o).

52. E.W.Weber, H.Ackermann, N.S.Laulainen, G.zu Putlitz,
    Z.Physik $\underline{26o}$,341(1973).

53. P.K.Berman, W.E.Lamb, Phys.Rev. $\underline{187}$,221(1969).

54. M.I.Chibisov, Doklady Akademii Nauk $\underline{186}$,1o63(1969).

55. *Voprosy Teorii Atomnykh Stolknovenii*, ed. By Yu.A.Vdo-
    vin, Atomizdat, Moskva,197o.

56. M.Mori, T.Watanabe, K.Katsuura, J.Phys.Soc.Jap. $\underline{19}$,
    38o(1964), $\underline{19}$,15o4(1964).

57. C.G.Carrington, D.N.Stacey,J.Cooper., J.Phys. $\underline{B6}$,417
    (1973)

58. I.I.Sobel'man, *Introduction to the Theory of the Atom-
    ic Spectra*, Pergamon, §46,1972.

59. V.Ermatchenko, C.R.Acad.Sc . $\underline{277}$,B475(1973).

6o. E.E.Nikitin, Chem.Phys.Lett. $\underline{2}$,4o2(1968).

61. A.Gallagher, Phys.Rev. $\underline{179}$,1o5(1969).

62. N.F.Mott, H.S.W.Massey, *The Theory of Atomic Colli-
    sions*, Oxford University Press, London, 1965.

63. J.C.Gay, A.Omont, J.Phys. 35,9(1974).

64. E.A.Gordeev, E.E.Nikitin, M.Ya.Ovchinnikova, Can J.Phys. 47,1819(1969).

65. J.P.Faroux, J.Brossel, C.R.Acad.Sc . 263,B612(1966).

66. V.N.Rebane, Optika i Spektr. 24,296(1968).

67. A.G.Petrashen', V.N.Rebane, T.K.Rebane, Optika i Spektr. 35,4o8(1973).

68. E.E.Nikitin, Comm,Atom.Mol.Phys. 3,7(1971).

69. V.N.Rebane, Optika i Spektr. 26,673(1969).

7o. J.Butaux, Schuller F., R. Lennuier, J.Phys. 35,361 (1974).

71. A.F.Wagner, G.Das, A.C.Wahl, J.Chem.Phys. 6o,1885 (1974).

72. O.B.Firsov, Zhurn.Eksp.Teor.Fiz. 32,1oo1(1951).

73. B.M.Smirnov, *Asimptoticheskiye metody v teorii atomnykh stolknovenii*, Atomizdat, Moskva, 1973.

74. J.Pitre, L.Krause, Can.J.Phys. 45,2671(1967).

75. W.B.Schneider,Z .Physik 248,387(1971).

76. M.Elbel, A.Koch, W.B.Schneider, Z.Physik 255,14(1972).

77. E.Roueff, J.Phys. B5,L79(1972).

78. W.E.Baylis, J.Chem.Phys. 51,2665(1969).

79. C.Bottcher, Chem.Phys.Lett. 18,457(1973).

8o. W.E.Baylis, as cited in ref. 32.

81. J.Pascale, J.Vandeplanque, J.Chem.Phys. 6o,2278(1974).

82. R.H.G.Reid, J.Phys. B6,2o18(1973).

83. J.Callaway, E.Bauer, Phys.Rev.14o,A1o72 (1965).

84. J.Callaway, P.S.Laghos, Phys.Lett. A26,394(1968).

85. L.Kumar, J.Callaway, Phys.Lett.A28,385(1968).

86. R.H.G.Reid, A.Dalgarno, Phys.Rev.Lett.22,1o29(1969).

THE SPECTRA OF COLLIDING ATOMS

Alan Gallagher

Joint Institute for Laboratory Astrophysics[†]

Boulder, Colorado   80302

A Tribute

   While the purpose of this paper is to describe some historical
and current developments in the interpretation of collisional radiation, we wish to draw particular attention to the important role
that Edward Condon played in these developments. The classical and
quantum-mechanical Franck-Condon principles are the foundation of
both modern and historical understandings of atomic collisional radiation, and the "new" developments we will describe here come
largely from application of these historical principles. Dr.
Condon and many others have long recognized that these principles
apply to bound-free and free-free molecular continua as well as
to the more familiar bound-bound transitions. The "physics" of
pressure broadening is the same topic as molecular continuum radiation; while there are some regions of collisional lines for
which a non-adiabatic Born-Oppenheimer-approximation treatment is
required, the basic concepts are still those elucidated by Edward
Condon 40-50 years ago.

[†]Operated jointly by the National Bureau of Standards and
University of Colorado.

INTRODUCTION

The shapes of collisionally broadened atomic lines have been studied for more than fifty years. The basic theoretical concepts were developed in the 1920's and 1930's, but very few definitive measurements or calculations for foreign-gas broadening are yet available. Progress has been difficult in large measure because the line shape depends critically on the electronic interaction energies, or molecular adiabatic states and energies, and these have seldom been known. Attempts to infer interaction energies from line shapes have been hampered by a lack of uniqueness and the use of oversimplified models. On the other hand line-shape data can be very powerful tools for learning these interaction energies, as we will demonstrate shortly. Whereas most colli-sional processes depend only on the scattering matrix, or the state of the system before and after the collision, the radiative interaction in the line wings can be localized to a brief instant during a collision. Thus it is possible to directly detect the interaction-energy surface with a radiative spectrum, whereas this must be inferred by modeling to fit scattering or cross-section data. But as we will see below, radiation and scattering measure-ments on alkali-noble gas molecules very beautifully complement each other. The experimental and theoretical scattering studies that have been carried out at a variety of laboratories are very sensitive to the region of the Van der Waals minimum whereas our spectral data yield interaction potentials at closer radii.

We will discuss here what can be learned from line shapes and the basic ideas of how they are interpreted with theoretical ap-proximations. Only "pressure" or neutral-gas broadening will be considered, not "Stark" or plasma broadening. We will give exam-ples of recent theoretical and experimental work, but this is not intended as a review and we will concentrate on our measurements in the far wings of the lines.

It is convenient to divide the line shape into five regions, as indicated in Fig. 1. These are:

(1) The impact-theory region corresponding to the Lorentzian-shaped center of the line.

(2) The far-wing region where the quasi-static theory is appli-cable.

(3) The very-far or "extreme" wing where the quasi-static theory is applicable and major pressure and temperature dependences also occur. These three regions, at progressively increasing frequency shifts from the atomic line, are associated with in-creasing collisional interaction strength and decreasing atomic separations.

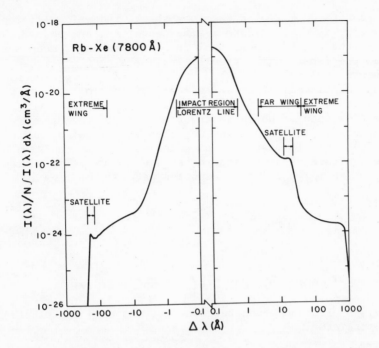

Fig. 1. The normalized emission spectrum of Rb $(5^2P_{3/2})$ in xenon of density N. The data are from Refs. 1, 2, and 3. $T \cong 315°K$.

(4) Satellite-line regions, characterized by a maximum in the spectrum.

(5) Transition regions, where finite collision-time dynamics dominate the spectrum but cannot be treated by the impact approximation.

LINE SHAPE THEORY

The simple semiclassical models of the 1930's are sufficient to explain the distinction between these different regions. These models represent the atom as an oscillator whose resonant frequency ω_0 is perturbed to $\omega(t)$ during a collision, so that one may consider $\hbar[\omega(t)-\omega_0]$ as the perturbation in the spacing of the atomic energy levels during the collision.[4] The absorption or emission at ω' is proportional to the squared Fourier amplitude of the time-varying oscillator:

$$I(\omega') \propto \left| \int dt\ \exp\left[-i(\omega'-\omega_0)t + i \int^t [\omega(t')-\omega_0]dt'\right] \right|^2 \quad . \quad (1)$$

The typical duration τ_c of these long-range, thermal collisions is $\sim 10^{-12}$ sec. If one considers a line-center region where $\omega - \omega_0 \ll 1/\tau_c$ the second exponential in Eq. (1) changes much more rapidly than the first during a collision. The oscillator can then be approximated as unperturbed except for abrupt phase shifts $\Delta\phi$ from each collision. Fourier analysis of this oscillator [Eq. (1)] yields a Lorentzian line of width 2Γ where $\Gamma = 1/\Delta T$ is the effective rate of phase destruction and ΔT the average time between collisions. The line may also be shifted by an amount Δ from ω_0 if most $\Delta\phi$ have the same sign; the resulting average rate of oscillator phase advance or delay is equivalent to a shift in ω_0. The Lorentzian part of the line is

$$I(\omega') \propto \frac{\Gamma}{(\omega - \omega_0 - \Delta)^2 + \Gamma^2} \qquad (2)$$

where

$$\Gamma + i\Delta = N \int dv \; v \; f(v) \int db \; 2\pi b (1 - \exp[-i\Delta\phi(b,v)])$$

and N, $f(v)$, b are the perturber density, velocity distribution, and impact parameter.

In quantum mechanical theories the $[1 - \exp(-i\Delta\phi)]$ is replaced by $(1 - S_{ii}S_{ff}^{-1})$ where S is the scattering matrix suitably averaged over allowed m_J state combinations and i, f refer to the two atomic levels, assumed isolated.[5]

There have been very few realistic calculations of Γ and Δ for non-resonant atomic collisions due to the difficulty of obtaining accurate interatomic potentials and of evaluating the necessary S-matrix terms. Two calculations of H broadening of the Na D lines, one by Lewis, McNamara and Michels[6] and the other by Roueff,[7] are outstanding exceptions which demonstrate the complexities of a full treatment. Here an _ab initio_ calculation of the adiabatic molecular potentials has been combined with collisional S-matrix calculations utilizing these potentials. We reproduce in Fig. 2a the phase shifts given by Lewis et al., using straight-line trajectories, for the various excited states originating in Na(3p) + H(1s). The broadening-rate contribution at each impact parameter resulting from a weighted average of these phase shifts is shown in Fig. 2b, along with the effect of the non-adiabatic rotational mixing. Roueff found an additional 20% correction in a fully non-adiabatic calculation. It is clear that most real problems will generally involve a similar number of complications; simple two-level models with adjustable potential parameters are conceptually appealing but quite unrealistic. Conversely, a measurement of Γ or Δ may be of value in an astro-

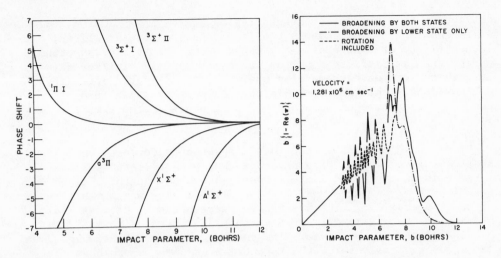

Fig. 2. (a) Phase shifts versus impact parameter for Na + H col-
lisons. The various adiabatic molecular states associated with
Na*(3p) are indicated. (b) Contributions to the Na D-line broadening cross section versus impact parameter. Both figures are reproduced from Ref. 6.

physical or plasma problem, but it yields no direct information
about the atomic interactions.

      The far-wing region is fixed by the inverse condition,
$\omega-\omega_0 >> 1/\tau_c$. In this case the integrand in Eq. (1) oscillates
rapidly at all times except when $\omega(t') \cong \omega'$ during a collision.
The major contribution to the intensity at ω' is thus very small
except when $\omega(t') \cong \omega'$ and the fraction of the total line intensity in the ω' to $\omega' + d\omega'$ integral becomes the ensemble averaged
fraction of time in which $\omega' \leq \omega(t') \leq \omega' + d\omega'$. If the oscillator
frequency $\omega[R(t)]$ is identified as $[V*(R) - V(R)]/\hbar$ this is the same
as the probability that a perturber is between the appropriate internuclear spacing R to R + dR. Here V^* and V represent one
excited and one ground-state molecular adiabatic potential. Thus:

$$I(\omega') \propto N(R)\ 4\pi R^2(\omega')\ dR(\omega')/d\omega' \qquad (3)$$

where N(R) is the perturber density at R. This very simple formula
is obtained because the rate of change of $\omega(t')$ in (1) is ignored; it is assumed that intensity occurs only at $\omega' = \omega(t')$ as
if the perturbers were in a static distribution. Thus the name
"Quasi-Static Theory" (QST).[4]

      The cause of a satellite in the far wing can then be identified
as due to an extremum in V*(R) - V(R) so that a large ΔR interval

contributes to a narrow wavelength region [$dR/d\omega' \to \infty$ in (3)]. The finite collision times clearly cannot be ignored for satellites, since the rate of change of $\omega(t')$ spreads out the Fourier spectrum whereas the QST predicts a narrow intensity spike.[8] In fact anywhere in the far or extreme wing where the quasi-static intensity changes rapidly the finite collision time must be taken into account; for example on both red and blue edges of the spectrum in Fig. 1.

The finite collision time must also be considered at an ω' in the far-wing region which is never reached by $\omega(t')$; the QST predicts no intensity at ω' in this case. The blue wing of a line for which $V^* - V = \omega_0 - C_n R^{-n}$ is the classic case which was first solved by Holstein.[9] The $n = 6$ case representing a Van der Waals interaction that is larger for V^* than V is expected to apply to the blue wings of many lines. The theory predicts a blue wing that drops gradually below the Lorentzian as $\omega'-\omega_0$ approaches $1/\tau_c$, and finally reaches a rate of decrease that is exponential in $\omega-\omega_0$ for $\omega' - \omega_0 \gg 1/\tau_c$. The -1 to -10 Å portion of the blue wing in Fig. 1 is expected to be largely due to such a collision-time effect, but the shape does not conform to the theory for a $C_6 R^{-6}$ interaction because the real interactions only follow this form at very long range, whereas the larger $\omega' - \omega_0$ are in general due to close-range collisions. The extreme-blue-wing spectrum in Fig. 1 is due to a $V^* - V > \hbar\omega_0$ at close range, which will be discussed later.

Finally, we note that in the transition zones, where $\omega'-\omega_0 \sim 1/\tau_c$, the finite collision time is an essential ingredient and neither the QST nor impact theory can be used to simplify the integral in Eq. (1); a "unified" theory is necessary. Here, also, very little can be learned directly about the atomic interactions, and so far the adequacy of theories is also untested due to uncertainties in the interactions.

The far and extreme wings offer much greater possibilities for potential determinations for here each radius yields in the QST a single ω'; the spectrum should allow us to map out V^*-V using Eq. (2). But there are several problems that have so far prevented definitive results in the far wing region. First, there are often more than one $V_i^*-V_j$ combination that can radiate at the same ω' since each atomic state splits generally into several molecular levels (i,j). Next, the various V_i^* (or V_j) can be non-adiabatically mixed, invalidating the QST, if they are not separated by much more than $\hbar/\tau_c$ where τ_c corresponds to the duration of a close collision. Further, the QST spectrum depends on $V_i^*-V_j$ and neither potential separately. Finally, we are aware of only one published quantitative measurement of a far wing intensity distribution,[10] as opposed to a spectral shape or only the position of features such as satellites.

The fraction of the total emission or absorption in a $\omega \rightarrow \omega + d\omega$
interval determines the volume of the corresponding $R \rightarrow R+dR$ region
through Eq. (2), so that an actual fractional intensity measurement severely constrains the possible set of $V_i^*-V_j$. Such measurements in the far wing can be used to test theoretical or assumed
potentials, but they cannot be used to uniquely infer $V_i^*-V_j$. We
will now demonstrate how the extreme wing spectrum can be used to
directly obtain V_i^* and V_j at small R.

INTERACTION POTENTIALS FROM EXTREME-WING DATA

We distinguish the extreme wing from the far wing region by
the dependence of the spectrum on temperature and perturber pressure. The far-wing region corresponds to $N(R) \cong N$, the perturber
density, in Eq. (2), and as a result the spectrum is simply proportional to N. In the extreme-wing region $N(R)/N$ is a function
of temperature, of the initial-state potential V_j or V_i^* and sometimes of N. Figure 3 gives an example of this behavior in the
emission spectrum of Rb perturbed by Xe. Note that opposite temperature dependences are observed on the red and blue wings. For an
explanation of this behavior we refer to Fig. 4, where the same
emission spectrum versus photon energy is shown on the left and the
adiabatic states responsible for the red-wing are on the right. The
classical Franck-Condon Principle (CFCP) requires that the nuclear
velocity remain unchanged during the electronic radiative transition, as indicated in Fig. 4. Thus the photon energy E_i-E_j equals
$V_i^*(R)-V_j(R)$ if the transition occurs while the nuclei are at separation R, exactly as in the QST.[11] (In Fig. 4 V_i^* and V_j are the
$A^2\Pi_{3/2}$ and $X^2\Sigma_{1/2}$ states.) A single initial state, such as the free
collision state indicated in Fig. 4, thus yields a contribution to
the spectrum at each $\omega \rightarrow \omega + d\omega$ proportional to the time, $2dR/v_R$
(R,b), spent at the appropriate $R \rightarrow R + dR$ (v_R is the radial velocity component and b the impact parameter). The total intensity
at ω is obtained by summing the ensemble average density at the
appropriate R.

If only free perturber collisions occur, this classical density for the adiabatic molecular state i is:

$$N_i(R) = N \frac{g_i}{g_a} \int_0^\infty dv v f(v) \int_0^{b_{max}(v,R)} db 4\pi b \left(\frac{2}{\mu} [U-V_i(R)-\mu v^2 b^2/2R^2]\right)^{-1/2} \quad (4)$$

where the last factor is the classical $1/v_R$, $U = \mu v^2/2 + V_i(\infty)$, μ is the
nuclear reduced mass, $f(v)$ is the normalized collisional velocity
distribution, and g_i and g_a are the molecular and atomic statistical weights. If $V(R)$ is repulsive, then b_{max} is the value at which
the bracketed expression is zero and for a Maxwellian $f(v)$ one
obtains $N_i(R) = N \exp\{-[V_i(R)-V_i(\infty)]/kT\}$ from (4). If $V(R)$ is

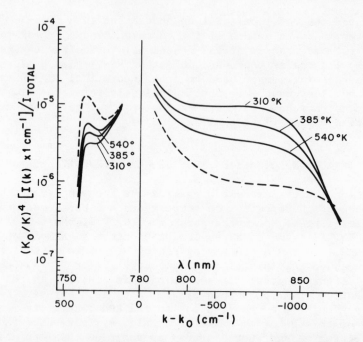

Fig. 3. Normalized emission spectrum of the Rb 7800Å D-line perturbed by xenon at 10^{19}/cm^3 density. The gas temperatures in °K
are indicated. From Ref. 2.

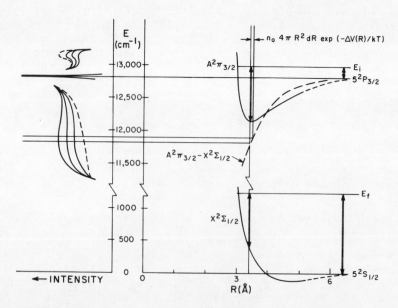

Fig. 4. The model used to interpret the extreme-wing spectra in
Fig. 3, based on the classical Franck-Condon principle.

attractive, all bound and quasi-bound states inside the rotational barrier are unpopulated in this purely collisional distribution. Thus $b_{max}(v,R)$ is the b for which the rotational barrier height equals E; N(R) then depends on the shape of V(R') at all R' > R rather than simply on V(R).

The complication associated with rotational barriers arises only when all bound and quasi-bound states inside the angular-momentum barrier are unpopulated. This situation applies only in the zero perturber-density limit to an emission spectrum.[12] For an absorption spectrum or emission at high perturber densities the collision processes are more rapid than the radiative rate and the vapor is in a Boltzmann distribution of bound and free states. Since we are using a classical approximation (the QST or CFCP) for the radiation it is appropriate to use also the classical canonical distribution for the resultant populations:

$$dN(\underset{\sim}{P},\underset{\sim}{R},T) = Nh^{-3}Z^{-1}d^3\underset{\sim}{P} \; d^3\underset{\sim}{R} \; \exp(-E/kT) \qquad (5)$$

with $E = P^2/2u + V(R)$ and Z the partition function. For repulsive potentials this is identical to the distribution in Eq. (4) since then the classical collisional distribution is identical to the canonical distribution. Integrating Eq. (5) over $d^3\underset{\sim}{P}$, one obtains for all forms of V(R):

$$N_i(R) = N\frac{g_i}{g_a} \exp(-[V_i(R)-V_i(\infty)]/kT). \qquad (6)$$

Note that this Boltzmann factor results from summing the densities of states of all nuclear kinetic energy $P^2/2u$, including bound, quasi-bound, and free collision states. The major contribution to the sum is at $P^2/2u \cong kT$, not the classical turning point energy $E = V(R)$. If this $N_i(R)$ density is evaluated quantum mechanically the $N_i(R)$ distribution is close to the classical distribution when kT is much greater than the vibrational spacing, but the quantum mechanical $N_i(R)$ is more spread out at replusive barriers and it may retain some undulations.[13] In Eqs. (3) and (6) we now have a method of separately determining the two potentials. The photon energy equals $V_i^*-V_j$ at an R and the temperature dependence determines, at the same R, V_i^* in emission or V_j in absorption (see Fig. 4).

The other complications in interpreting the far-wing spectra are also alleviated in the small R region responsible for the extreme wing: The various $V_i(R)$ and $V_j(R)$ are usually far from atomic energies $V_i(\infty)$ and $V_j(\infty)$ and from each other so that nonadiabatic mixing is negligible. The large separation of the various V_i (and V_j) also tends to spectrally separate the radiation from various V_i^*, V_j combinations. However, one new problem appears. For the strong perturbations of atomic energies corresponding to these extreme-wing spectra, the transition dipole moment may vary considerably from the atomic value, adding an unknown transition-probability factor $A[R(\omega)]/A(\infty)$ to the right side of Eq. (3). In our work on alkali-noble gas molecules involving strongly-allowed dipole moments we believe the dipole moment is constant within typically 10% and uncertainty in $A_i(R)$ is only a minor problem. But in a case like $Hg(6^3P) + Hg$ the atomic and some of the associated molecular-state radiation is forbidden, whereas other molecular states can undergo allowed dipole radiation. In addition, rapid radial variations in molecular-state coupling schemes are expected, accompanied by variations in the associated $A_i(R)$. These $A_i(R)$ uncertainties have so far prevented definitive interpretation of the blue Hg_2 bands.

In our experiments we detect the extreme wing emission intensity as a function of the total line emission, as indicated in Fig. 1. Combined with the assumption of a constant transition dipole moment, Eqs. (3) and (6) yield for this case:[14]

$$\frac{I(\omega')}{\int I(\omega')d\omega'} = N \frac{g_i^*}{g_a^*} \left(\frac{\omega_o}{\omega'}\right)^4 4\pi R^2(\omega') \frac{dR(\omega')}{d\omega'} e^{-[V^*(\omega')-V^*(\infty)]/kT} \quad (7)$$

where g_i^* and g_a^* are the statistical weights of the molecular and parent atomic state. Absorption coefficients using the same models can be found in Ref. 14. As an example of this behavior, the temperature-dependent RbXe spectrum from Ref. 2 is shown in Fig. 3. Note that the $A^2\Pi_{3/2} \rightarrow X^2\Sigma_{1/2}$ radiation produces the red extreme wing and $B^2\Sigma_{1/2} \rightarrow X^2\Sigma_{1/2}$ the blue wing. The $T=\infty$ limit shown corresponds to Eq. (7) with the exponential factor equal to one, so $R(\omega')$ or $V_i^*-V_j$ versus R is established from this red-wing curve. Then V_i^* versus $V_i^*-V_j$ obtained from the temperature dependence fixes V_i^* versus R, and by subtraction of $\hbar\omega'$ from V_i^* we obtain V_j versus R. The resulting potentials are shown in Fig. 5. The blue wing includes a satellite, so we expect $V_i^*-V_j$ to yield the same ω' at an $R > R_{sat}$ and an $R < R_{sat}$. The major contribution at each ω' comes from the $R > R_{sat}$ due to the last three factors in Eq. (7); thus as can be seen in Fig. 5 we do not infer much about the $B\Sigma_{1/2}$ potential at $R < R_{sat}$.

The QST predicts an intensity from ω_1' to ω_2' proportional to the volume [Eq. (7)] from $R(\omega_1')$ to $R(\omega_2')$, but the theory cannot be extended to line center corresponding to $R \rightarrow \infty$. Thus the starting separation is undetermined in this analysis; we do not know where to place $R(\omega_1')$. This is equivalent to bound-bound molecular spectra in which rotational lines are unresolved so that R_e is not determined. Our procedure for fixing a starting point is to utilize alkali-noble gas atomic-beam-scattering data. We compare our ground $X^2\Sigma_{1/2}$ state potential to the potentials inferred from

such data, or to the semiempirical potentials of Baylis[15] and
Pascale[16] that are parameterized to fit the scattering data. Once
this starting point is fixed, the $B^2\Sigma_{1/2}$, $A^2\Pi_{3/2}$, and $X^2\Sigma_{1/2}$ potentials are uniquely reconstructed from the data. In Fig. 5 this
starting R is 3.3 Å, where $X^2\Sigma_{1/2}$ is 500 cm^{-1} repulsive. A
fractional variation of the dipole moment from the assumed constant value will alter the radial scale of the potentials in Fig.
5 by a similar fraction. Our data uncertainties are typically at
the 5-10% level, yielding a 20-40 cm^{-1} vertical uncertainty and a
5-10% uncertainty in the radial scale of the potentials.

We have constructed interaction potentials for the Cs-noble
gas and Rb-noble gas molecules using this technique, but for Na and
Li which rapidly destroy windows we have found it convenient to
take pressure measurements at a single low temperature. Such data
are shown for the NaXe case in Fig. 6; measurements on the Li-noble
gas system are currently underway. These data are analyzed by
noting that in the high-density limit $N_i(R)$ in Eq. (3) is given
by Eq. (6) whereas in the zero density limit $N(R)$ is given by
Eq. (4) with b_{max} determined from the rotational barrier height.
Thus V_i^* versus ω' or $V_i^*-V_j$ can be obtained from the ratio
of the two limiting intensities (Fig. 6), although uncertainty in
the unknown long-range $V_i(R)$ add about 30 cm^{-1} uncertainty to

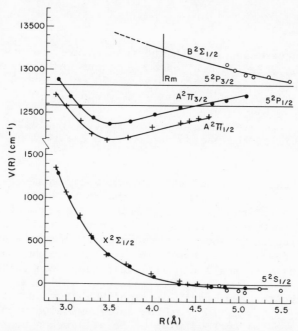

Fig. 5. The Rb-Xe adiabatic potentials inferred from the data in
Fig. 3. The classical satellite radius R_{sat} is labeled R_m.
From Ref. 2.

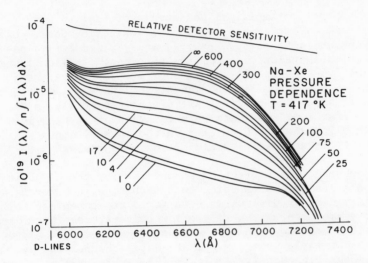

Fig. 6. Normalized emission spectrum of the Na D lines perturbed
by Xe. The Xe pressures in Torr are indicated. From Ref. 12.

$V_i[R(\omega')]$.[12] We have measured both the temperature and pressure
dependences for the RbXe (7948 Å) case, and obtained agreement of
V_i^* within the experimental uncertainty of ~50 cm^{-1} across a
V_i^* range of –200 to –630 cm^{-1} (Refs. 2, 12, and 17).

The pressure dependence measurements yield, in addition to
interaction potentials, the recombination rate coefficient for

$$M^* + 2X \to MX^* + X \quad ,$$

where M^* is the excited alkali and X the noble gas. This recombination rate competes with the radiative rate, so the growth
of the bound portion of the spectrum yields the rate coefficient.
In fact different portions of the spectrum grow at different rates
so that the distribution of bound vibrational populations has been
inferred from the data. Note, however, that the pressure-dependence
measurements yield quite small signal levels since the extreme-
wing intensity decreases more rapidly than N, while N
must be lowered about three orders of magnitude from the value
shown in Fig. 1. Typically 10^{-8} of the total emission occurs in
a 1 cm^{-1} interval on the extreme wing of Na at a xenon density of
10^7/cm^3, and one must detect fluorescence to avoid obscuring these
weak continua by other lamp radiation. A typical signal level is
1-10 photoelectrons/sec in our apparatus.

SATELLITES

Physicists have been fascinated by satellites since the earliest days of line-shape studies. These wing intensity maximas are generally attributed to an extremum in V^*-V, for which the QST yields an intensity spike at the extremum photon energy.[8] The actual satellite shape is broadened by the finite collision time effects,[18] or equivalently by the finite nuclear de Broglie wavelength in the Franck-Condon-factor approach.[19] Until recently only the satellite positions were generally noted, but several satellite shapes have now been measured quantitatively and several calculations have been made. In the far wing region a red satellite of the second K doublet broadened by Kr has been observed to change from a well resolved satellite at low temperature to a mere shoulder at high temperature, demonstrating the expected strong dependence on collision velocities.[20] In the extreme wing, satellites of Cs-noble gas and Rb-noble gas have been measured and a semi-classical calculation of Franck-Condon factors, using WKB nuclear wavefunctions, has been applied to the Cs-Ar case.[21] Finally, a Franck-Condon-factor calculation using exact Born-Oppenheimer nuclear wavefunctions has been carried out for H_2 for what would be the extreme red wing of Lyman α.[19]

These calculations and observations have clarified the basic features of satellite shapes. Three characteristic features can be seen in the Rb (7800 Å) + Kr satellite in Fig. 7. These are: an approximately symmetric central maximum, an exponential intensity decrease beyond the maximum, and a series of undulations between the satellite and parent line. As noted above, the frequency breadth of the central maximum is due to the finite time of traversing the satellite radius. The exponential decrease beyond the satellite is characteristic of Fourier analysis at a frequency that is never reached by the oscillator, and is predicted by the semi-classical Franck-Condon-factor theory (Ref. 19). The undulations are attributed to interference between contributions to the radiation amplitude at the two radii where $V^*-V = \hbar\omega$, one on either side of the satellite radius.[19,22] As ω moves away from the satellite frequency ω_S these two radii separate. The accumulated phase between the two radii is different for motion in V^* versus V so the relative phase of the two contributions grows as $\omega-\omega_S$ grows, yielding alternate in and out of phase superpositions. The resulting intensity undulations are not of very large amplitude because the relative phases are averaged over a thermal collisional velocity distribution.

While the causes of these basic features are now understood, there is so far no theoretical prediction of all the observed features or a method for uniquely reconstructing V^* and V from a measured satellite shape.

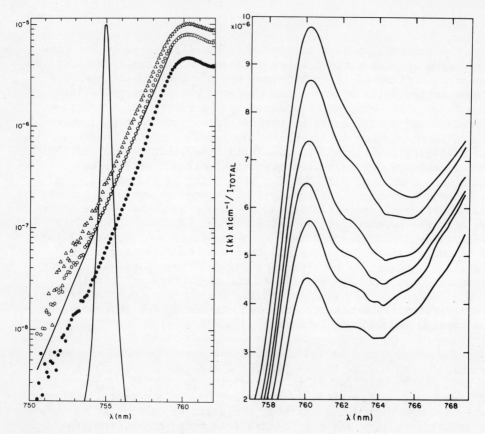

Fig. 7. Normalized emission spectrum of the blue satellite of
Rb 7800 Å line perturbed by $10^{19}/cm^3$ of Kr: (a) Logarithmic plot
at temperatures, from top to bottom, of 740, 530, and 330°K. The
spectrometer instrument function, centered at 755 nm is also
shown. (b) Linear plot of the peak region at temperatures, from
top to bottom, of 706, 603, 488, 423, 377, and 323°K.

LOOKING FORWARD

Extreme-wing data can thus be used to almost uniquely recon-
struct adiabatic interaction potentials in the small R, strong-
interaction regions. (The A(R) and starting-R uncertainties
prevent uniqueness.) Future analysis of intensity undulations
and satellite shapes should further improve this knowledge. Semi-
empirical[15],[16] and ab-initio[23] potentials are now becoming available for many cases. These can be tested in the larger R regions
with far wing data, and perhaps adjusted or improved to yield
agreement with quantitative far wing data of the type shown in
Fig. 1.

There is another extremely powerful technique available for determining some of these intermediate-range potentials. Note that in Fig. 5 there is a radial and spectral region in which bound $A^2\Pi_{1/2}$ to bound $X^2\Sigma_{1/2}$ transitions occur. This bound-bound spectra is an unresolved portion of the continuum we have so far studied. Our fluorescence signals are too weak to allow high resolution measurements of these lines, but tunable laser absorption detected as fluorescence could yield this bound-bound spectrum. This portion of the potentials could then be determined with the typical 0.1 cm^{-1} accuracy of bound-bound spectra, and in addition R_e could be accurately determined.

If the interaction energies can thus be established at close and intermediate range, it should be possible to connect to perturbative or model potentials at long range. Then impact-approximation and other dynamical calculations of the line-center spectrum can be definitively tested against observations. The effect of temperature and high pressures on the line center can make these comparisons more definitive; this is already being utilized in some line-center analyses.

There are also a variety of interesting applications of these line-broadening ideas to chemical physics. We have already mentioned measurements of the vibrational distribution of bound molecules formed by three-body recombination; a crucial but previously unobserved feature of recombination theories. The absorption or emission spectrum of colliding atoms can be used to map out interaction surfaces for many types of processes using the principles applied above to the extreme wing. A velocity-selected atomic beam is just as effective as a thermal distribution for mapping the initial-state potential [Eq. (4) without the velocity integration]. One example is chemiluminescence, where a continuum will be emitted as the dissociating fragments separate (normally the bound-bound decay of the luminescing molecule is detected after separation). Another example is collisional excitation by heavy particles; part of the decay radiation will occur before the particles separate. In these spontaneous emission cases the dissociating continuum will be a small fraction ($\sim\Gamma/\tau_{dissociation}$) of the total emission. In laser absorption by interacting ground-state species signal levels should generally not be a problem.

References

1. C. G. Carrington and A. Gallagher, "Blue satellite bands of Rb broadened by noble gases," Phys. Rev. A (in press).

2. D. L. Drummond and A. Gallagher, J. Chem. Phys. $\underline{60}$, 3426 (1974).

3. C. Ottinger, R. Scheps, G. York, and A. Gallagher, "Noble-gas broadening of the Rb resonance lines," to be published.

4. S. Y. Chen and M. Takeo, Rev. Mod. Phys. 29, 20 (1957).

5. J. Cooper, Rev. Mod. Phys. 39, 167 (1967).

6. E. L. Lewis, L. F. McNamara, and H. H. Michels, Phys. Rev. A 3, 1939 (1971).

7. E. Roueff, J. Phys. B 7, 185 (1974).

8. J. Cooper, "Comments on the theory of satellite bands," JILA Report 111, University of Colorado, Boulder.

9. T. Holstein, Phys. Rev. 79, 744 (1950); see also G. N. Plass and D. Warner, Phys. Rev. 86, 138 (1952) and Ref. 10.

10. C. L. Chen and A. V. Phelps, Phys. Rev. A 7, 470 (1973).

11. A Jablonski, Phys. Rev. 68, 78 (1945).

12. R. E. Roberts, R. B. Bernstein, and C. F. Curtiss, J. Chem. Phys. 50, 5163 (1969). In the present application: G. W. York, R. Scheps, and A. Gallagher, to be published.

13. C. G. Carrington, D. L. Drummond, A. Gallagher, and A. V. Phelps, Chem. Phys. Lett. 22, 511 (1973).

14. R. E. M. Hedges, D. L. Drummond, and A. Gallagher, Phys. Rev. A 6, 1519 (1972).

15. W. E. Baylis, J. Chem. Phys. 51, 2665 (1969).

16. J. Pascale and J. Vanderplanque, J. Chem. Phys. 60, 2278 (1974).

17. C. G. Carrington and A. Gallagher, J. Chem. Phys. 60, 3436 (1974).

18. E. A. Andreev, Opt. Spektrosk. 34, 603 (1973) [Opt. Spectrosc. 34, 346 (1973)].

19. K. M. Sando and J. C. Wormhoudt, Phys. Rev. A 7, 1889 (1973).

20. D. G. McCarten, J. M. Farr, and W. R. Hindmarsh, J. Phys. B 7, 208 (1974).

21. K. M. Sando, Phys. Rev. A 9, 1103 (1974).

22. W. H. Miller, J. Chem. Phys. 52, 3563 (1970).

23. M. Krauss, P. Maldonado, and A. C. Wahl, J. Chem. Phys. 54, 4944 (1971).

COLLISION STUDIES WITH ION STORAGE TECHNIQUES

G. H. Dunn

Joint Institute for Laboratory Astrophysics[†]

Boulder, Colorado 80302

INTRODUCTION

Ion trap technology has been with us for over a decade now, and has provided us with a powerful tool for atomic physics advances. Dehmelt[1] has reviewed the mechanics of traps and their applications in the rf spectroscopy of ions. Significant contributions have been made[2-4] since these reviews -- particularly in electron spectroscopy -- and reported current work[5] is exciting indeed.

It has been clear for nearly the same decade that traps have been with us, that these devices can as well be a new weapon in the collisions study arsenal. Despite this recognition, it is only very recently that ion traps have been successfully applied to collision measurements. This is notwithstanding the fact that many of the spectroscopic studies[1] utilize collisions as a key part of the measurement technique. Because of the infancy and near non-existence of published literature on ion-trap collision measurements, it is not reasonable to try to "review" the field; although brief mention will be made of other collision experiments of which the author is aware. It is more the purpose of this paper to give

[†]Operated jointly by the National Bureau of Standards and University of Colorado.

a progress report on recombination measurements made by studying
electron-ion collisions in ion traps.

We will proceed by briefly reviewing two-body recombination
processes between electrons and ions and the means by which they
have previously been measured. The trapping scheme and its realization for collision work will then be discussed, after which
results for some recombination measurements will be given. We
will then note work at other laboratories on collision studies
using ion traps.

TWO-BODY RECOMBINATION MECHANISMS AND THEIR MEASUREMENT

The first recombination process normally coming to mind is
that of radiative recombination. The excess energy of the electron
is simply radiated, leaving the resultant neutral atom in an
excited state. The electron passes the ion in about 10^{-15} sec,
and radiative times are $\sim 10^{-8}$ sec; so the process has low probability with cross sections about 10^{-20} cm^2 which decrease with
electron energy E roughly as 1/E. The process has not been
measured directly, although radiative recombination data can sometimes be inferred from measurements of the reverse process of
photoionization.

The second process is dielectronic recombination which involves multiple excited states of the neutral which lie in the
ionization continuum. If the incident electron has just the right
energy to excite an electron in the ion core, and itself be left
in a very high Rydberg state, then there is a reasonable chance
that the core will radiate before autoionization occurs, thus
resulting in an excited neutral. This process can have cross
section values comparable to the ion excitation cross section --
10^{-14} cm^2 to 10^{-15} cm^2 -- but they are sizeable over only a very
narrow energy range (perhaps a few meV). The process has been
invoked[6] to explain ionization balance in high temperature plasmas,
but has not been more directly observed.

The third process is dissociative recombination[7] which occurs
only with molecular ions. Here again, we have excited neutral
states which lie in the continuum -- but this time the energy
varies with R, and may be below the continuum at large R. In this
case the excess energy the electron brought in can be carried away
as kinetic energy of the dissociating partners. This takes place
in 10^{-15} to 10^{-13} sec; so the process can be very efficient with
cross sections at low energies greater than 10^{-13} cm^2. One notes
that again the mechanism is resonant; but the resonance is broad,
since the initial molecular ion exists over a range of R.

When densities become high enough for three-body collisions to be frequent, we have three-body recombination as well as third body enhancement of the above mechanisms. In a way, dissociative recombination is a process in which the third body is carried around with the ion. It is usually dominant when molecular gases are present.

Though radiative and dielectronic recombination have not yet been directly measured, dissociative recombination has been. Until very recently the only measurements on the process were those in plasmas[7] in which the quantity measured is the rate coefficient -- the cross section times the velocity averaged over a Maxwellian velocity distribution. The most extensive and successful experiments have been those of Biondi and his coworkers who have studied these things over a 25-year period. In their experiments a plasma is generated in a microwave cavity, the excitation is turned off, and the time decay of conductivity (equivalent to electron density) and mass selected ion species are observed. There is provision for microwave heating of the electrons, so that the process can be studied as a function of electron temperature up to a few thousand degrees. The cavity can be physically heated so that recombination can be studied over a range of total temperature (up to 450°K). The electron and ion density decay rates -- corrected for diffusion -- are quantitatively related to recombination rates.

Other methods used to obtain rate coefficients include heating of a plasma by a traveling shock, and observing the electron density as a function of distance (time), using electrostatic probe pairs placed along the shock tube. Observations for total temperatures up to a few thousand degrees can be made with this technique.

Despite the success of the plasma measurements, there are limitations: 1) the energy range is limited; 2) the energy resolution is limited, obscuring details, and not allowing calculations of rates for other than Maxwellian distributions; 3) the species one can study are limited; and 4) the internal energy states are not controllable.

Colliding beam measurements can overcome to some degree each of the listed limitations except the last, where beams are even more limited. In such experiments one collides a beam of electrons with a beam of ions and observes the appearance of neutrals, the disappearance of ions, or the appearance of photons from excited dissociation fragments. The experiments are few in number[8-11]; though the very recent experiments of Peart and Dolder[10,11] indicate the method is capable of high accuracy and reliability subject to the limitation noted above.

ION TRAPS

This brings us to the focus of this paper, the use of ion
traps, introducing a third and complementary means to measure
electron recombination with ions. Most limitations of the other
methods are overcome, though new limitations arise.

Ion traps depend primarily on an electric saddle potential V
given by

$$V = A(r^2-2z^2) , \tag{1}$$

where A is a constant, and r and z are the normal cylindrical
coordinates. Ions in the r = 0 plane will tend to oscillate in a
parabolic well along z; but if they are at finite r, or have
finite v_r, they will move out in the r direction.

Now, in the radio frequency trap, the potential is made to
change sign before the ion can escape along r, change again before
escaping along z, etc.; if the amplitude and frequency are properly
chosen, the ion will be trapped. This containment of ions by means
of alternating electric field gradients grew out of the work of
W. Paul and collaborators[12,13] on electric mass filters; and the
trap mechanism and the rf spectroscopy of ions using these traps
have been reviewed by Dehmelt.[1] This active trap, however, keeps
the ions hot; and the fields in the trap make it unsuitable to try
to study collisions of the trapped ions with monoenergetic
electrons.

To make a trap more suitable for collision studies, a strong
magnetic field is introduced along z with a dc potential given by
Eq. (1); so that, as the ions start to escape along r, they are
turned back in cyclotron orbits. Since the ions see a crossed
$\vec{E}$ and $\vec{B}$ field, they experience an $\vec{E} \times \vec{B}$ drift, which results in a
precession about the z axis. Thus, the ions undergo harmonic
motion in the parabolic well along z, cyclotron motion in the
r - θ plane, and precession about the z axis. In the absence of
collisions they will be trapped indefinitely.

Tracing the history of the development of the static trap is
not straightforward, since there seems to have been a considerable
amount of unpublished work. The basic configuration is that used
by Penning[14] in the study of discharges. Pierce[15] noted the possibility of trapping in the potential given by Eq. (1). Solutions to
the equations of motion were published by Farago,[16] with references
to unpublished work of Frisch (1954), Bloch (1953), and Dicke
(1965). Dehmelt[1] has reviewed the technique and noted unpublished
work of Dehmelt (1961). Gräff and co-workers[17-19] studied the
mechanics of these traps and reported successfully trapping
electrons for several minutes in their electron spectroscopy work.
However, the trap awaited the work of Dehmelt and Walls[20] to demon-

strate its ability to give long trapping times; they achieved trapping times for electrons of about 3×10^6 sec, or five weeks. Furthermore, and of as much importance, they demonstrated a non-destructive bolometric detection technique useable with the trap.

Though the dc trap is not suitable for some kinds of ion spectroscopy because of the large magnetic fields and associated Zeeman splittings, it seems precisely suited for collision studies. First, the passive nature of the trap allows one to cool the ions and obtain energy resolution of tens of milli-electron volts; second, long lifetimes make it possible to prepare the ions in the desired state; and third, one can make repeated non-destructive measurements of the numbers of various ion species that may be present. We now examine the trapping scheme and realization of these features in more detail.

Figure 1 shows a cross sectional view of a dc ion trap and a block schematic of the heterodyne detection system used. The trap electrodes are hyperbolae of revolution which form equipotential surfaces of the distribution given in Eq. (1), where now $A = V_0/(a^2+2b^2)$. Here 2a is the equatorial diameter of the ring, 2b is the axial distance between end caps, and V_0 is the potential applied to the ring electrode with respect to the end cap electrodes. The dimensions of our trap are 2b = 0.76 cm and 2a = 1.25 cm.

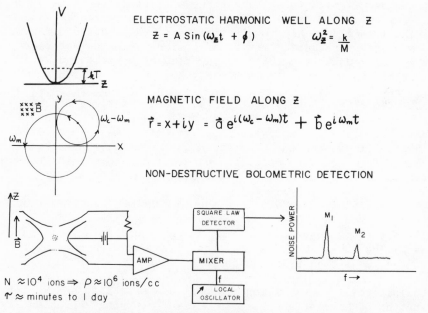

Fig. 1.   Schematic illustration of ion trap motions (upper, middle), and of the trapping and detection schemes (lower).

Ions can be introduced to traps in a variety of ways, such as photoionization, colliding ion beams, and ion beam-residual gas collisions; but the most common means is by electron impact on gas of a select species.

The axial motion of ions in the trap is harmonic oscillation at a frequency

$$\nu_z = \left[\left(\frac{qV_0}{\pi^2 m}\right)\middle/(a^2+2b^2)\right]^{\frac{1}{2}} \qquad (2)$$

where q/m is the charge-to-mass ratio of the ions. As noted the ions additionally move in small circles at a frequency $\nu' = \nu_c-\nu_m$ near the cyclotron frequency $\nu_c = qB/2\pi m$, and the centers of the circles precess about the z axis at a frequency of ν_m. The frequencies are related to one another by $\nu_z^2-2\nu_c\nu_m+2\nu_m^2 = 0$. Note that due to the harmonic nature of the potentials and the assumed uniformity of $\vec{B}$, the frequencies are independent of energy and position. For O_2^+ with $V_0 = -1$ V and B = 1.15 T (1 Tesla = 10^4 Gauss), $\nu_c = 552$ kHz, $\nu_z = 67$ kHz, and $\nu_m = 4.1$ kHz.

The conditions for long-term ion storage are: 1) the cyclotron radius is much smaller than "a", 2) the radial magnetic force, Bqv_r, is larger than the radial electric force, qE_R, and 3) the magnetic and electric fields are cylindrically symmetric with coinciding axes. With these conditions satisfied, the loss of ions is due solely to collisions with other species (background gas). The losses due only to elastic scattering induced diffusion can be studied by measuring lifetimes of ions which do not react with the background gas. The trapping time of Dehmelt and Walls for electrons (3×10^6 sec) has already been noted. Measurements on NH_4^+ with B = 1.15 T, V_0 = 1.8 V, and a background pressure of $\sim 1 \times 10^{-10}$ Torr (presumably mostly H_2), yield lifetimes of about 10^5 sec. Both indicate that diffusion effects can be made small in attainable vacuum. Typically, however, lifetimes are determined by reactions with the background gas, such as charge transfer, clustering, ion-atom interchange, etc.; and this leads to more usual lifetimes of about 10^3 sec at the noted pressure.

These trapping times are long enough that most excited ions will have relaxed to the ground state. An exception is vibrationally excited homonuclear molecular ions whose lifetimes are $\sim 10^7$ sec. The times are also long enough to envision pumping the ions to excited states using suitable radiation.

Coulomb collisions between trapped ions establish a Maxwellian distribution of kinetic energies in a time[21] varying approximately as $T_i^{3/2} m^{1/2} n^{-1} Z^{-4}$. For T_i = 300°K, m = 30 a.m.u., n = 10^6 ions/cm^3, and Z = (q/e) = 1, then $\tau_c \sim 0.02$ sec.

The fact that the trapping times are so much longer than τ_c allows one to speak of an ion gas at a definite temperature T_i, and makes possible the use of a variation of the non-destructive detection technique of Dehmelt and Walls.[20]

The axial motion of the ions causes an image current to flow from one end cap to the other. If a resistor, R, is inserted in this current path, the Joule heating losses will cause the ions to approach the noise temperature of the resistor with a time constant of $\sim 8mZ_0^2/q^2R$, assuming no other external coupling to the ion motion. The collective motion of N_i ions is damped N_i times faster. Note that this process can be used either to heat or to cool the ions. Evaporation from the trap will also efficiently cool the ions to a mean energy about 1/10 of the well depth. Beyond this, there are not sufficient ions high enough in the tail of the distribution to have much effect. The ions are non-destructively detected by measuring the noise power in the image currents to one end cap of the trap. The heterodyne detection circuit shown in Fig. 1 is very broad band, allowing non-destructive detection of ions from mass 1/1836 to several hundred a.m.u.'s. Individual ion species are manifest as peaks in the noise power spectrum at the ion's z oscillation frequency given by Eq. (2). The area under a peak is proportional to ion number times temperature, $N_i T_i$.

If N_i is constant, one can investigate processes that raise the ions' kinetic temperature. For example, introducing power at the cyclotron frequency raises the ions' temperature, which is manifest as an increase in the area of the noise power peak. Similar heating of internal degrees of freedom which couple to the ions' kinetic motion can also be so detected.[4,20] Resonant cyclotron heating is used to measure ion q/m ratios with a resolution of about 2 parts in 10^4; it has been used, for example, to distinguish between H_2O^+(m = 18.020) and NH_2^+(m = 18.044).

When T_i is constant, one can investigate processes which change N_i. The wideband, non-destructive nature of the detection makes it possible to repeatedly measure N_i for each ion species of interest, and so keep track of both primary and competing processes. Recombination is a process in which T_i is constant, and one is interested in the depletion of N_i in the presence of electrons.

The number of ions that can be stored in a trap of the cited dimensions and a well depth of 1 V is limited by space charge to about 3×10^6. In recombination studies the need for a well defined interaction energy between electrons and stored ions reduces the radial extent of useful trap volume and correspondingly the maximum number of stored ions is reduced to about 10^5, with a density of about $10^7/cm^3$.

RECOMBINATION MEASUREMENTS

Electrons from either of two guns are introduced into the trap through holes in the end caps. A simple "on axis" gun gives a beam of FWHM of about 120 meV with currents about 5×10^{-8} A; the other is similar to the trochoidal design of Stamatović and Schulz[22] and gives a beam of FWHM about 35 meV and currents about 2×10^{-9} A.

Gas is introduced into the system to about 3×10^{-10} Torr, ions are formed by bombardment with electrons from the low resolution gun. After formation the ions are allowed to cool for about five minutes by coupling to an external resistor as described above. The well depth is then reduced from about 0.8 V to 0.4 V and the ions cool further by evaporation so that the mean energy in their z motion is about 15 meV.

The area under the noise power peak for a given ion is then observed as a function of time to establish the natural residence time in the trap. As noted above, this period is generally determined by ion-molecule reactions with the background gas. Electrons of a chosen energy and current i then bombard the ion cloud for a measured time t_e, chosen to give a measureable ion loss; the natural period is then measured again. This measurement sequence is demonstrated in Fig. 2. The cross section at a given energy is then calculated from the expression

$$\sigma(E) = \frac{eA}{i \, t_e} \left[\ell n(n_1/n_2) - \ell n(n_1'/n_2') \right] . \tag{3}$$

Here A is a geometric quantity, which in a simple model is just the area of the ion cloud as seen by the electron beam. The number of ions before and after an electron "kill" are given by n_1 and n_2 respectively, and n_1' and n_2' are the ion numbers before and after a period of <u>natural</u> decay only, and where the natural decay is observed for the same time t_e as an electron "kill."

Quantities in Eq. (3) are readily measured except for the overlap A, and estimates of this quantity are accurate only to within about a factor of two. A procedure is adopted to measure A indirectly by measuring quantities in Eq. (3) where σ is known. The process chosen for this determination is electron-O_2^+ recombination where, in fact, σ is not known; but rather the recombination rate coefficient $\alpha = \int_0^\infty \sigma(v) v \, F(v) \, dv$ has been accurately measured over a broad range of temperatures with good agreement among investigators. Since we know σ to within a factor A from the trap measurements and Eq. (3), α/A can be calculated. Comparison to the known value of α determines A, and this serves as a trap calibration for measurements on all other species.

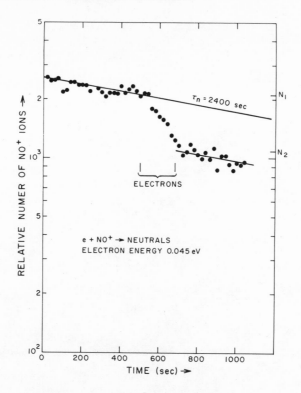

Fig. 2. Illustration of measurement sequence. Ions decay from the trap due to interactions with background gas between 0 and 500 sec and between 700 and 1000 sec. Between 500 and 700 sec there is additional loss due to recombination of electrons and ions.

Figures 3 and 4 show measured[23,24] cross sections for NO^+ and H_3O^+ respectively. Measurements in Fig. 4a were made with the high resolution electron gun and those in Fig. 4b were made with the low resolution gun. The solid curve is our best estimate of the cross section deduced from the data, and the dashed curves are convolutions of the respective electron energy distributions with the deduced (solid curve) cross section. Rate coefficients calculated from the cross sections agree well with published measured rate coefficients. Structures in the cross section curves indicate contributions from several final states. Measurements have also been made for NH_4^+, HCO^+, and N_2H^+.

Recombination studies using the trapped ion technique uncover details that will help in guiding theoretical efforts to calculate recombination cross sections. The measurements provide a heretofore unavailable means to calculate rate coefficients in practical

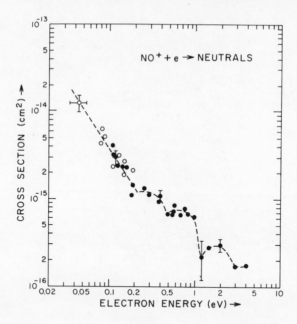

Fig. 3. Measured cross section for recombination of electrons and NO$^+$ as a function of electron energy (from Ref. 23).

plasmas where the energy distribution of electrons may be different from Maxwellian. The technique extends the range of energies accessible with other methods, and provides an environment in which one can more accurately define the internal state distribution of the reacting ions.

Primary limitations of the technique applied to recombination are 1) limited storage times imposed by ion-molecule reactions on the background gas, 2) indirect measurement of A, and 3) a lower limit on the electron energy of about 0.040 eV.

OTHER COLLISION STUDIES

Examples of other collision measurements using traps include the work of McGuire and Fortson[25] who store electrons in a dc trap and study elastic, inelastic, and spin dependent collisions with selected atoms and molecules. These measurements make use of the dependence of the trapped electrons' lifetimes and temperatures upon elastic and inelastic collisions. Details of the method and numerical results have not been published. Ensberg[26] has studied

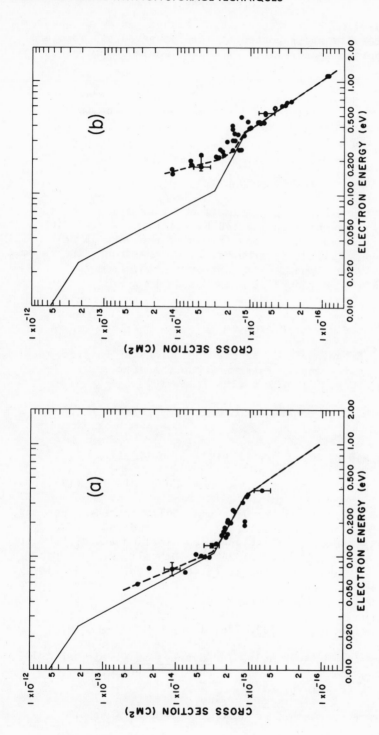

Fig. 4.   Points show measured cross sections for recombination of electrons and H_3O^+ as a function of electron energy (from Ref. 24):   (a) measurements using an electron beam of FWHM 0.040 eV,   (b) measurements with a FWHM of 0.120 eV.   The solid line is deduced from the measurements, and the dashed lines are convolutions of the energy distributions with the solid line.

photodissociation of CH_4^+, SiH_4^+, and SiH^+ trapped in an rf quadrupole trap. His results are reported at this conference. Todd and
Lawson[27] have reported measurements of ion-molecule reactions
using an rf trap.

Photoionization, photodissociation, electron-ion recombination,
and electron impact ionization and dissociation all seem to lend
themselves to measurement (for both positive and negative ions)
with the trapped ion method.

ACKNOWLEDGMENTS

The dc ion trap used for recombination measurements discussed
here was primarily assembled and tested by Dr. F. L. Walls. The
recombination measurements have been primarily the work of Dr. Walls
and Dr. R. A. Heppner with help from Mr. W. T. Armstrong. The recombination work has been supported in part by a grant from the
Aeronomy branch of the National Science Foundation.

References

1. H. G. Dehmelt, in Advances in Atomic and Molecular Physics,
    Vol. 3 (D. R. Bates and I. Esterman, eds., Academic Press,
    New York, 1967), p. 53; Vol. 5 (1969), p. 109.

2. D. Wineland, P. Ekstrom and H. Dehmelt, Phys. Rev. Lett. 31,
    1279 (1973).

3. D. A. Church and B. Mokri, Z. Physik 244, 6 (1971).

4. F. L. Walls and T. S. Stein, Phys. Rev. Lett. 31, 975 (1973).

5. H. Dehmelt, P. Ekstrom and D. Wineland, Bull. Am. Phys. Soc. 19,
    572 (1974); H. Dehmelt, P. Ekstrom, D. Wineland and R. Van Dyck,
    Bull. Am. Phys. Soc. 19, 572 (1974); D. Wineland and H. Dehmelt,
    Bull. Am. Phys. Soc. 19, 642 (1974); H. Dehmelt, R. Van Dyck,
    P. Ekstrom and D. Wineland, Bull. Am. Phys. Soc. 19, 643 (1974);
    D. Wineland and H. Dehmelt, Fourth Int. Conf. on Atomic Physics,
    Heidelberg, July, 1974 (Abstract).

6. A. Burgess, Astrophys. J. 139, 776 (1964).

7. J. A. Barsley and M. A. Biondi, in Advances in Atomic and Mole-
    cular Physics, Vol. 6 (D. R. Bates and I. Esterman, eds.,
    Academic Press, New York, 1970), p. 1.

8. G. Hagen, Air Force Cambridge Research Laboratory Report
    AFCRL-68-0649 (1968).

9. M. K. Vogler and G. H. Dunn, Bull. Am. Phys. Soc. $\underline{15}$, 417 (1970).

10. B. Peart and K. T. Dolder, J. Phys. B $\underline{6}$, L359 (1973).

11. B. Peart and K. T. Dolder, J. Phys. B $\underline{1}$, 236 (1974).

12. W. Paul and H. Steinwedel, Z. Naturforsch. $\underline{8a}$, 448 (1953).

13. W. Paul and M. Raether, Z. Physik $\underline{140}$, 262 (1955).

14. F. M. Penning, Physica $\underline{3}$, 873 (1936).

15. J. R. Pierce, Theory and Design of Electron Beams (Van Nostrand, Princeton, N.J., 1954), 2nd ed., p. 40.

16. J. Byrne and P. S. Farago, Proc. Phys. Soc. (London) $\underline{86}$, 801 (1965).

17. G. Gräff and E. Klempt, Z. Naturforsch. $\underline{22a}$, 1960 (1967).

18. G. Gräff, F. G. Major, R.W.H. Roeder and G. Werth, Phys. Rev. Lett. $\underline{21}$, 340 (1968).

19. G. Gräff, E. Klempt and G. Werth, Z. Physik $\underline{222}$, 201 (1969).

20. H. G. Dehmelt and F. L. Walls, Phys. Rev. Lett. $\underline{21}$, 127 (1968).

21. L. Spitzer, Physics of Fully Ionized Gases (Interscience, New York, 1956), p. 76.

22. A Stamatović and G. J. Schulz, Rev. Sci. Instr. $\underline{41}$, 423 (1970).

23. F. L. Walls and G. H. Dunn, J. Geophys. Res. $\underline{79}$, 1911 (1974).

24. R. A. Heppner, F. L. Walls, W. T. Armstrong and G. H. Dunn (submitted to Phys. Rev. A).

25. M. D. McGuire and D. N. Fortson, Bull. Am. Phys. Soc. $\underline{18}$, 710 (1973).

26. E. Ensberg, Fourth Int. Conf. on Atomic Physics, Heidelberg, July, 1974 (Abstract).

OPTICAL PUMPING WITH LASERS

C. COHEN-TANNOUDJI

Ecole Normale Supérieure and Collège de France, Paris

Optical pumping is a method for transferring to an ensemble
of atoms a fraction of the angular momentum carried by a beam of
polarized resonance radiation ([1]). The observation of the characteristics of the light absorbed or reemitted by these atoms gives
a lot of information about atomic structure, relaxation phenomena,
interaction processes between atoms and photons ...

The light sources which have been used so far in optical pumping experiments are ordinary spectral lamps excited by RF discharges. The emitted light has in general a very broad spectral width,
of the order of a few gigahertz. Even the brightest lamps have a
relatively low intensity : this means that the pumping time, i.e.,
the mean time between 2 successive absorptions of pumping photons
by the same atom, is much longer than the radiative lifetime of
the excited state. In other words, the absorption and induced
emission processes associated with the light beam are weak compared to spontaneous emission.

The spectacular development of lasers has now changed the situation. We have at our disposal light sources with very interesting
characteristics : high intensity which permits one to easily saturate an atomic or molecular transition-tunability over large spectral ranges which gives the possibility of studying more levels-
monochromaticity which makes it possible to get rid of the Doppler
width for optical lines - pulsed operation which opens the way to
a time resolved spectroscopy ...

I would like in this paper to analyse some of the new effects
which can be observed in optical pumping experiments and which are

a consequence of the improvement of light sources. I will put emphasis on the theoretical problems and, more particularly, on the modification of the equations describing the optical pumping cycle. Some consequences of these modifications will be illustrated by a few examples, chosen as simple as possible.

More complete and more detailed treatments may be found in references $(^2)$, $(^3)$. This paper will be restricted to a simple presentation and discussion of some physical ideas.

OPTICAL PUMPING EQUATIONS

The central problem is to find, and eventually to solve, the equations which describe the evolution of the atomic density matrix. This matrix has the following form :

$$
\begin{array}{|c|c|}
\hline
\sigma_e & \sigma_{eg} \\
\hline
\sigma_{ge} & \sigma_g \\
\hline
\end{array}
\qquad (1)
$$

σ_e and σ_g are the density matrices describing the ensemble of atoms inside the excited state e and the ground state g. Their diagonal elements are the populations of e and g sublevels, their off diagonal elements, the "hertzian" coherences between these sublevels. σ_{eg} and $\sigma_{ge} = \sigma_{eg}^+$ contain only off diagonal elements connecting a sublevel of g to a sublevel of e; they evolve at optical frequencies and are called for that reason "optical" coherences.

Let us give, as an example, the density matrix corresponding to a $J_g = 0 \leftrightarrow J_e = 1$ transition excited by a light beam having a σ linear polarization, i.e., perpendicular to the axis Oz of quantization. Only sublevels m = ±1 of the $J_e = 1$ excited state and sublevel m = 0 of the $J_g = 0$ ground state have to be considered (we can forget the m = 0 excited sublevel; see fig. 1), so that (1) takes the form :

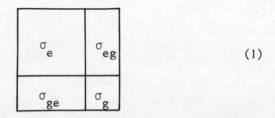

$$
\begin{array}{|c c|c|}
\hline
\sigma_{++} & \sigma_{+-} & \sigma_{+o} \\
\sigma_{-+} & \sigma_{--} & \sigma_{-o} \\
\hline
\sigma_{o+} & \sigma_{o-} & \sigma_{oo} \\
\hline
\end{array}
\qquad (2)
$$

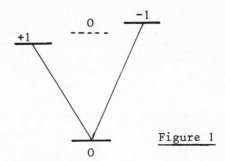

Figure 1

Generally, for an atomic vapour or for a gas, one has to introduce such a density matrix for each velocity group. However, under certain circumstances, it may happen, as we shall see later, that the internal state of an atom and its velocity are uncorrelated. In this case, σ refers to internal variables only. For an atomic beam perpendicular to the light beam, such a problem does not exist as the Doppler effect disappears.

The light beam is supposed to result from the superposition of parallel plane waves having all the same polarization $\vec{u}$, but different (complex) amplitudes $\mathcal{E}_\mu$ and frequencies ω_μ. The matrix describing the interaction between the atom and the light beam is purely non-diagonal and can be written (in the so-called "rotating-wave approximation") :

$$
\begin{array}{|c|c|}
\hline
0 & -\sum_\mu D_{eg}\,\mathcal{E}_\mu\,e^{-i\omega_\mu t} \\
\hline
-\sum_\mu D_{ge}\,\mathcal{E}_\mu^\star\,e^{i\omega_\mu t} & 0 \\
\hline
\end{array}
\tag{3}
$$

We have put $D_{eg} = (\vec{u}.\vec{D})_{eg}$ where $\vec{D}$ is the electric dipole moment operator.

To simplify the discussion, we have treated the light beam as a classical field. Such an approach is not correct for describing the effect of spontaneous emission. This process may be taken into account by just adding to the equations giving the rate of variation of the components of σ the following terms (see references ([4])([5])) :

$$
\left\{
\begin{array}{ll}
\dot{\sigma}_e = -\,\Gamma\,\sigma_e & \text{(4-a)} \\[2mm]
\dot{\sigma}_{eg} = -\dfrac{\Gamma}{2}\,\sigma_{eg} & \text{(4-b)} \\[2mm]
\dot{\sigma}_g = \mathcal{C}\,(\sigma_e) & \text{(4-c)}
\end{array}
\right.
$$

where Γ is the natural width of the excited state (equal to the reciprocal of the radiative lifetime τ of this state). Equations (4-a) and (4-b) give the damping of σ_e and σ_{eg} by spontaneous emission while (4-c) describes the transfer from e to g associated with such a process. For each element of σ_g , the right hand side of (4-c) is a linear combination of the matrix elements of σ_e. Due to the spherical symmetry of spontaneous emission, this coupling appears only between quantities having the same symmetry (populations are coupled only to populations, coherences to coherences). In the example considered above, the last equation (5) can be written :

$$\dot{\sigma}_{oo} = \Gamma(\sigma_{++} + \sigma_{--}) \tag{5}$$

(σ_{oo} is not coupled to σ_{+-} and to σ_{-+}).

Finally the rate of variation of the components of σ, including the effect of the atomic Hamiltonian H_o (free evolution), the coupling with the incident light beam, and the spontaneous emission, is given by (we take $\hbar = 1$) :

$$\begin{cases} \dot{\sigma}_e = -i\left[H_o,\sigma_e\right] - \Gamma\sigma_e + i\sum_\mu \left[D_{eg}\sigma_{ge}\mathcal{E}_\mu e^{-i\omega_\mu t} - \sigma_{eg}D_{ge}\mathcal{E}_\mu^\star e^{i\omega_\mu t}\right] & \text{(6-a)} \\[2mm] \dot{\sigma}_g = -i\left[H_o,\sigma_g\right] + \mathcal{C}(\sigma_e) + i\sum_\mu \left[D_{ge}\sigma_{eg}\mathcal{E}_\mu^\star e^{i\omega_\mu t} - \sigma_{ge}D_{eg}\mathcal{E}_\mu e^{-i\omega_\mu t}\right] & \text{(6-b)} \\[2mm] \dot{\sigma}_{eg} = -i\left[H_o,\sigma_{eg}\right] - \dfrac{\Gamma}{2}\sigma_{eg} + i\sum_\mu \left[D_{eg}\sigma_{gg} - \sigma_{ee}D_{eg}\right]\mathcal{E}_\mu e^{-i\omega_\mu t} & \text{(6-c)} \end{cases}$$

Some algebraic manipulations can be done on equations (6). We can integrate equation (6-c) (and its hermitian conjugate) and insert the expression so obtained for σ_{eg} and σ_{ge} in the right hand side of (6-a) and (6-b). One gets in this way a system of 2 integro-differential equations involving only σ_e and σ_g. This is more easily done in interaction representation with respect to H_o. If we put :

$$\begin{cases} \tilde{\sigma}(t) = e^{iH_o t}\,\sigma(t)\,e^{-iH_o t} & \text{(7-a)} \\[2mm] \tilde{D}(t) = e^{iH_o t}\,D\,e^{-iH_o t} & \text{(7-b)} \\[2mm] \mathcal{E}(t) = \sum_\mu \mathcal{E}_\mu\, e^{-i\omega_\mu t} & \text{(7-c)} \end{cases}$$

we get after some simple calculations :

$$\begin{aligned} \dot{\tilde{\sigma}}_g(t) = {}& e^{iH_o t}\mathcal{C}(\sigma_e)\,e^{-iH_o t} \\ & - \int_0^t dt'\left[\tilde{D}_{ge}(t)\tilde{D}_{eg}(t')e^{-\frac{\Gamma}{2}(t-t')}\mathcal{E}^\star(t)\mathcal{E}(t')\tilde{\sigma}_g(t') + \text{hermit.conjug.}\right] \\ & + \int_0^t dt'\left[\tilde{D}_{ge}(t)\tilde{\sigma}_e(t')\tilde{D}_{eg}(t')e^{-\frac{\Gamma}{2}(t-t')}\mathcal{E}^\star(t)\mathcal{E}(t') + \text{hermit.conjug.}\right] \end{aligned} \tag{8}$$

plus a similar equation for $\overset{\sim}{\overset{\bullet}{\sigma}}_e$. The first integral in the right
hand side of (8) involves only $\overset{\sim}{\sigma}_g$: it describes how the ground
state is affected by the absorption process. The second integral,
which involves only $\overset{\sim}{\sigma}_e$, describes the transfer from e to g by
induced emission.

The problem is now to solve equations (6) or (8) which are
equivalent. A first idea would be to get a solution of (6) in the
form of a perturbation expansion in the field amplitudes $\mathcal{E}_\mu$. Reference ([6]) gives an example of such a calculation up to 4th order.

These types of calculations are quite laborious. I prefer here
to discuss some situations where it is possible to by-pass perturbation treatments and to get non-perturbative solutions of equations (6) or (8). Such a possibility depends of course on the type
of light beam which is used, i.e., on the properties of the amplitudes $\mathcal{E}_\mu$. I will consider first the case of an ordinary spectral lamp
or of a free-running multimode laser for which it is more convenient
to start from (8), and then the case of a single mode laser where
equations (6) are simpler.

BROAD-LINE EXCITATION

Figure 2 shows the intensities $|\mathcal{E}_\mu|^2$ of the various waves
forming the light beam

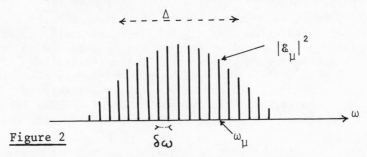

Figure 2

In the case of a spectral lamp, the frequencies ω_μ of these waves
form a continuum. If we have a laser beam, we suppose that the
laser oscillates on a great number of modes. In both cases, we will
assume that the width Δ of the spectral interval covered by the
frequencies ω_μ (see fig. 2) is very large compared to the Doppler
width $\Delta\nu_D$ and the natural width Γ of the atomic line, and that the
spacing $\delta\omega$ between 2 successive modes is small compared to Γ :

$$\left\{ \begin{array}{l} \Delta \gg \Delta\nu_D, \ \Gamma \\ \delta\omega < \Gamma \end{array} \right. \tag{9}$$

In this case (see reference ([2])), the different "Bennett holes"
burnt by the various modes in the Doppler profile overlap, and it

is easy to understand that the response of an atom does not depend on its velocity, so that σ refers to internal variables only.

The relative phases of the various modes are assumed to be random : we have a "free-running" multimode laser and not a "phase-locked" one. The instantaneous electric field $\mathcal{E}(t)$ of the light wave (see equation (7-c)) may be considered as a stationary random function. The correlation function $\overline{\mathcal{E}(t)\mathcal{E}^*(t-\tau)}$ of $\mathcal{E}(t)$ only depends on τ and tends to zero when τ is larger than the correlation time τ_c which is of the order of $1/\Delta$.[19] When the ω_μ's form a discrete set of equidistant frequencies, there is another correlation time $\tau'_c = 1/\delta\omega$, which is much longer.

The strength of the coupling between the atom and the light wave may be characterized by a parameter v which is the product of the atomic electric dipole moment d by an electric field amplitude and which gives an order of magnitude of the evolution frequency of $\tilde{\sigma}$.

$$v = d.\left[\,\overline{|\mathcal{E}(t)|^2}\,\right]^{\frac{1}{2}} = d.\left[\,\sum_\mu |\mathcal{E}_\mu|^2\,\right]^{\frac{1}{2}} \tag{10}$$

Let us come back to equation (8), and more precisely to the 2 integrals over t'. $\overline{\tilde{D}(t)\tilde{D}(t')}e^{-\Gamma(t-t')/2}$ and $\overline{\mathcal{E}^*(t)\mathcal{E}(t')}$ are the correlation functions of the atomic dipole moment and of the electric field. They have correlation times respectively equal to $1/\Gamma$ and to $1/\Delta$. The product of the 2 correlation functions has a memory which is determined by the shorter correlation time, i.e., by $1/\Delta$ (the other correlation time $\tau'_c = 1/\delta\omega$ of the electric field does not play any role as it is longer than $1/\Gamma$, according to condition (9)). One therefore expects that the contributions of $\tilde{\sigma}_g(t')$ $\left[\text{ or } \tilde{\sigma}_e(t')\right]$ with $t-t' \gg 1/\Delta$ will be cut down. This gives the key for a very convenient method for solving equations (8). If $1/v$, which is the characteristic time of evolution of $\tilde{\sigma}$ under the influence of the coupling with the light, is much longer than $1/\Delta$, i.e., if

$$v \ll \Delta \tag{11}$$

we can replace, to a very good approximation, $\tilde{\sigma}_g(t')$ and $\tilde{\sigma}_e(t')$ by $\tilde{\sigma}_g(t)$ and $\tilde{\sigma}_e(t)$. We transform the system of integro-differential equations (8) into a set of differential equations, or "rate equations" describing the coupled evolutions of σ_g and σ_e. The coefficients γ appearing in these rate equations are given by the integral over t' of the product of the 2 correlation functions of $\tilde{D}$ and $\mathcal{E}$. As soon as $t \gg 1/\Delta$, the result of this integration, i.e., γ, is independent of t : this is due to the stationary character of the random function $\mathcal{E}(t)$. The order of magnitude of γ is found to be :

$$\gamma \sim v\,\frac{v}{\Delta} \tag{12}$$

So, the evolution frequency of σ under the influence of the coupling with the light is reduced from the expected value, v, by a factor v/Δ much smaller than 1. This result may be compared with the motional narrowing appearing in the theory of relaxation in liquids and gases [7] : if the effect of the perturbation is weak during the correlation time, the net effect of this perturbation is reduced even more.

Due to (9), the value of γ does not depend on Γ or $Δν_D$. It does not depend also on the static magnetic field if we assume that the Zeeman splittings in e and g are weak compared to Δ. γ has in general an imaginary part which describes the so-called light shifts, i.e., the displacements of atomic energy levels produced by the light irradiation [4]. I will assume here that the distribution of figure 2 is centered on the atomic frequency, in which case γ is real. The angular part of the coefficients appearing in the rate equations giving $\dot{σ}_e$ and $\dot{σ}_g$ are determined by the polarization $\vec{u}$ of the light and by the values J_e and J_g of the angular momenta. Let us give, as an example, these rate equations for the case considered above ($J_e = 1$, $J_g = 0$, $\vec{u}$ being a linear polarization perpendicular to the axis of quantization Oz). We assume in addition that a magnetic field is applied along Oz and we call $ω_e$ the Larmor frequency in level e.

Free evolution	Spontaneous emission	Absorption	Stimulated emission
$\dot{σ}_{++} =$	$-Γσ_{++}$	$+γσ_{oo}$	$- γ (σ_{++}+σ_{--}-σ_{-+}-σ_{+-})/2$
$\dot{σ}_{--} =$	$-Γσ_{--}$	$+γσ_{oo}$	$- γ (σ_{++}+σ_{--}-σ_{-+}-σ_{+-})/2$
$\dot{σ}_{-+} = 2iω_eσ_{-+}$	$-Γσ_{-+}$	$-γσ_{oo}$	$+ γ (σ_{++}+σ_{--}-2σ_{-+})/2$
$\dot{σ}_{oo} =$	$+Γ(σ_{++}+σ_{--})$	$-2γσ_{oo}$	$+γ(σ_{++}+σ_{--}-σ_{-+}-σ_{+-})$

$$(13)$$

Each vertical column contains the terms corresponding to a given process : free evolution (in this case, the Larmor precession), spontaneous emission, absorption and stimulated emission.

We see from equations (13) that the spontaneous and stimulated emission terms contain both the matrix elements of $σ_e$ and in the first case Γ, the second one γ.

If γ << Γ, i.e., if according to (12)

$$v << \sqrt{ΓΔ} \qquad (14)$$

one can drop the stimulated emission terms which are negligible compared to the spontaneous emission ones (note that we must keep the absorption terms as they involve $σ_g$ which may be much greater than $σ_e$). Omitting the last column of equations (13), we get the

usual optical pumping equations (see references (^{4})) derived for
thermal sources (γ is the inverse of the pumping time T_p).

Suppose now that γ is of the order of Γ, or larger than Γ
(the condition of validity (11) being conserved) :

$$\sqrt{\Gamma\Delta} \lesssim v \ll \Delta \tag{15}$$

In this case, we must keep the terms contained in the last column
of (13). We will discuss later some of the new effects which come
from these terms and which can be observed only if optical pumping
is performed with a laser source. Let us already note that these
terms are anisotropic, contrary to those of column 2. For example,
they couple the Zeeman coherence σ_{-+} to the population σ_{oo}. This is
due to the fact that the polarization $\vec{u}$ of the light is a preferential direction introduced by the interaction processes.

We will see later that the physical quantities calculated from
equations (13) can be expanded in powers of γ, all powers of γ appearing in these expansions. This shows that the solutions obtained
from (13) are non-perturbative.

It is clear however that these solutions do not correspond to
a summation of all the terms of the perturbation series because an
approximation has been done when replacing t' by t in some terms of
equations (8). The physical meaning of this approximation is the
following : the correlation time $1/\Delta$ of the light wave is so short
that each interaction process may be considered as uncorrelated
with the previous ones if (11) is satisfied. In other words, the
rate equations (13) take into account the effect of an indefinite
number of uncorrelated one-photon processes, but they neglect all
the possible interferences between 2 successive interactions.

One can now ask the question of what happens if (11) is not
satisfied, i.e., if $v \gtrsim \Delta$. We will come back to this problem after
having discussed the case of a monochromatic excitation.

MONOCHROMATIC EXCITATION

We now consider a single mode laser of frequency ω irradiating an atomic beam perpendicular to it, so that there is no
Doppler effect.

The correlation time of the perturbation is, in this case, very
long and it becomes impossible to neglect the correlations between
successive interactions of the atom with the light wave. We cannot
consider that the atom undergoes, from time to time and without
any phase memory, transitions from g to e or from e to g. We have

now a coherent oscillation between e and g, analogous to the Rabi
nutation in magnetic resonance, and which proceeds at a frequency
of the order of v [where v = d& is the coupling parameter analogous to (10)] . Furthermore, the optical coherence σ_{eg} becomes significant and oscillates at the same frequency, in quadrature with
the populations of e and g. In the previous case, σ_{eg} was negligible as the coherence time $1/\Delta$ was too short for permitting σ_{eg} to
build up appreciably. This was the real justification for trying
to eliminate the optical coherences from equations (6) and replacing them by equations (8). It is now better to start directly
from equations (6) which, since there is only one frequency $\omega_\mu = \omega$,
look like Bloch's equations in magnetic resonance. The only difference is that the 2 levels e and g are not simple but have a structure.

A transformation analogous to the transformation to the rotating frame can be performed. If we put :

$$\sigma_{eg} = \rho_{eg} \, e^{-i\omega t} \tag{16}$$

we get a set of differential equations with time independent coef-
ficients, coupling σ_e , σ_g , ρ_{eg} and $\rho_{ge} = \rho_{eg}{}^+$. Let us give these
equations for the same transition and for the same polarization $\vec{u}$
as for equations (13) :

Free evolution	Spontaneous emission	Coupling with the laser
$\dot\sigma_{++} =$	$-\Gamma\sigma_{++}$	$-iv(\rho_{o1} - \rho_{1o})$
$\dot\sigma_{--} =$	$-\Gamma\sigma_{--}$	$+iv(\rho_{o-1} - \rho_{-1o})$
$\dot\sigma_{-+} = 2i\omega_e\sigma_{-+}$	$-\Gamma\sigma_{-+}$	$+iv(\rho_{o1} + \rho_{-1o})$
$\dot\sigma_{oo} =$	$\Gamma(\sigma_{++}+\sigma_{--})$	$-iv(\rho_{1o}-\rho_{-1o}-\rho_{o1}+\rho_{o-1})$
$\dot\rho_{o1} = -i(\omega-\omega_o-\omega_e)\rho_{o1}$	$-\frac{\Gamma}{2}\rho_{o1}$	$-iv(\sigma_{++} - \sigma_{oo} - \sigma_{-+})$
$\dot\rho_{o-1} = -i(\omega-\omega_o+\omega_e)\rho_{o-1}$	$-\frac{\Gamma}{2}\rho_{o-1}$	$+iv(\sigma_{--} - \sigma_{oo} - \sigma_{+-})$

$$\tag{17}$$

We have already defined the coupling parameter v. ω_0 is the e-g
separation in zero magnetic field.

Here again, the solution of (17) corresponds to a summation
of the perturbation series as all powers of v appear in the expressions calculated from (17).

Let us note first that it is also possible for a monochromatic excitation to get, in some cases, rate equations coupling only
σ_e and σ_g . The correlation function of the atomic dipole moment

$\tilde{D}(t)\tilde{D}(t')\,e^{-\Gamma(t-t')/2}$, which appears in equation (8) and which has now the shortest correlation time (as $\Delta = 0$), cuts the contributions of $\tilde{\sigma}_g(t')$ [or $\tilde{\sigma}_e(t')$] when $t-t' \gg 1/\Gamma$. If the nutation frequency v is small compared to Γ, it is justified to replace, as before, t' by t in these quantities. One can say that spontaneous emission suppresses all phase memory between 2 successive interactions of the atom with the light beam. The rate equations so obtained are however less general than equations (17) which are valid even if $v \gg \Gamma$.

Finally, we come back to the case of a broad line excitation so intense that $v \gg \Delta \gg \Gamma$. From the previous discussion, we see that the nutation frequency v is so high that several nutations occur during the correlation time $1/\Delta$ of the light wave. It is therefore no longer possible to introduce rate equations coupling only σ_e and σ_g. Moreover one cannot consider that the light wave appears as a monochromatic excitation for the atom, since the amplitude of this wave fluctuates appreciably within the decay time $1/\Gamma$ of the dipole moment. We are in a difficult intermediate situation which presents some analogies with the problem of thermal relaxation when the motion narrowing condition is not satisfied.

ILLUSTRATION ON A VERY SIMPLE CASE : THE HANLE EFFECT OF A $J_g = 0 \leftrightarrow J_e = 1$ TRANSITION

In order to discuss some of the new effects which appear in optical pumping experiments performed with laser sources, we will consider the simplest possible transition $J_g = 0 \leftrightarrow J_e = 1$, and a resonance which does not require the use of any RF field, the Hanle zero-field level crossing resonance. We will suppose that the light beam, propagating along Oz, is linearly polarized along Ox, and that a magnetic field $\vec{B}_0$ is applied along Oz [situation considered for equations (13) and (17)]. One measures the variation versus B_0 of the total fluorescence light L_f , emitted along Oy with a linear polarization parallel to Ox. One can show that such detection signals are linear combinations of the matrix elements of σ_e. In the present case :

$$L_F \sim \sigma_{++} + \sigma_{--} - \sigma_{-+} - \sigma_{+-} \tag{18}$$

In some experiments, where g is not the ground state, but the lower level of a pair of excited levels, the observation of the fluorescence light emitted from g, with for example a π-polarization gives a signal I_π proportional to the population σ_{oo} of g

$$I_\pi \sim \sigma_{oo} \tag{19}$$

[In this case, equations (13) and (17) have to be slightly modified to introduce the rate of preparation of atoms in levels e

and g, the spontaneous decay of g, the spontaneous decay of e to
levels other than g ([2]). But this does not modify the physical
results .⌋

We will solve equations (13) or (17) according to the type of
light irradiation which is considered. This will give us a quantitative expression for the 2 signals (18) and (19).

Let us first briefly recall what the situation is when an ordinary thermal source is used (broad line excitation with $\gamma \ll \Gamma$).
Neglecting the terms in the last column of (13), we readily get the
steady state solution of these equations. To lowest order in γ, this
solution is :

$$\begin{cases} \sigma_{++} = \sigma_{--} = (N_o - \sigma_{oo})/2 = \gamma N_o / \Gamma & \text{(20-a)} \\[2em] \sigma_{-+} = -\dfrac{\gamma N_o}{\Gamma - 2i\omega_e} & \text{(20-b)} \end{cases}$$

where $N_o = \sigma_{++} + \sigma_{--} + \sigma_{oo}$ is the total number of atoms ⌈ N_o is a
constant of motion, as can be seen by adding the 2 first equations
(13) to the last one ⌋ . We see that the Zeeman coherence σ_{-+} exhibits a resonant behaviour when the Larmor frequency ω_e is varied
around 0, by sweeping the magnetic field B_o. This is the origin of
the Hanle zero field level crossing resonance appearing on the
fluorescence light (18) :

$$L_f = \frac{2\gamma N_o}{\Gamma} \left(1 + \frac{\Gamma^2}{\Gamma^2 + 4\omega_e^2} \right) \qquad (21)$$

and which has a Lorentzian shape and a width Γ independent of γ,
i.e. of the light intensity. On the other hand, no resonances appear
on the populations σ_{++} , σ_{--} , σ_{oo} which are independant of ω_e.

What are the modifications which appear when we use a much more intense broad-line source (for example, a free running multi-mode
laser) ? We now have to keep the last column of equations (13). The
calculations are a little more difficult, but it remains possible
to get analytical expressions for the steady state solution of
these equations.

We find for the populations :

$$\sigma_{++} = \sigma_{--} = (N_o - \sigma_{oo})/2 = \frac{\gamma N_o}{\Gamma + 3\gamma} \left[1 - S \frac{\Gamma'^2}{\Gamma'^2 + 4\omega_e^2} \right] \qquad (22)$$

where :

$$S = \frac{\gamma}{\Gamma+4\gamma} \qquad \Gamma' = \left[\frac{\Gamma(\Gamma+\gamma)(\Gamma+4\gamma)}{(\Gamma+3\gamma)} \right]^{1/2} \qquad (23)$$

The populations now exhibit a resonant behaviour near $\omega_e = 0$. The corresponding resonances are called "saturation resonances". They have a Lorentzian shape, a contrast S, and a width Γ' (see figure 3).

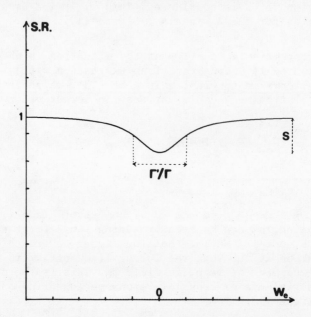

Figure 3 : Saturation resonance S.R. normalized to 1 in high field
($\omega_e \gg \Gamma'$). S is the contrast of the resonance, Γ' its
width. $W_e = 2\omega_e/\Gamma$ is a normalized Larmor frequency.

The saturation resonance appearing on σ_{00} may be interpreted in the following way. A first interaction with the laser (absorption process) removes the atom from the ground state and puts it in a coherent superposition of the -1 and +1 sublevels of e (figure 4-a). The combined effect of Larmor precession and spontaneous emission gives rise to the well known resonant behaviour of the Zeeman coherence σ_{-+}. A second interaction with the laser (induced emission process) brings back the atom to the ground state (figure 4-b) and partially confers to the population σ_{00} of this state the resonant behaviour of σ_{-+}. Such a process cannot occur for spontaneous emission which is an isotropic process and which, on the average, does not couple σ_{-+} to σ_{00}.

Another way of interpreting the saturation resonance is to take the axis of quantization along the direction Ox of the laser polarization. In zero magnetic field, the energy levels may be taken as the eigenstates of J_x and the polarization of the laser is

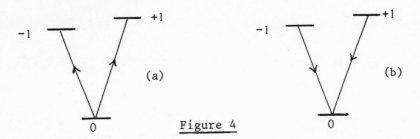

Figure 4

a π-polarization (figure 5). The $0 \leftrightarrow 0$ transition is saturated by the laser and the populations of these 2 sublevels tend to be equalized. The application of a magnetic field along Oz induces transitions between the 3 upper sublevels of figure 5. The corresponding changes in the populations of the m=0 upper sublevel are transferred to σ_{oo} through stimulated emission processes.

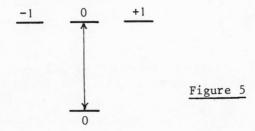

Figure 5

   The variations with γ of the contrast S and the width Γ' of the saturation resonance are represented on figures 6 and 7 (C is the dimensionless quantity γ/Γ). If we put $\gamma_0 = \Gamma/4$, the expression (23) of S may be written for $\gamma < \gamma_0$:

$$S = \frac{1}{4} \sum_{p=0}^{\infty} (-1)^P \left(\frac{\gamma}{\gamma_0} \right)^{p+1} \qquad (24)$$

We obtain a perturbation expansion containing all orders of γ and which is not convergent for $\gamma > \gamma_0$. This clearly shows that the solution (22) of equations (13) is non perturbative. Γ', which is equal to Γ for $\gamma = 0$, increases linearly with γ for $\gamma \ll \Gamma$. This may be interpreted as a radiative broadening proportional to the laser intensity. For $\gamma \gg \Gamma$, Γ' increases only as $\sqrt{4\gamma\Gamma/3}$, i.e., as the amplitude of the light wave. This shows that some care must be taken when extracting atomic data from experimental results. Plotting the width Γ' of a saturation (or Hanle) resonance as a function of the laser intensity, and extrapolating linearly to zero light intensity, may lead to wrong results if the majority of experimental points do not fall in the linear range of figure 7 ($C = \gamma/\Gamma \ll 1$).

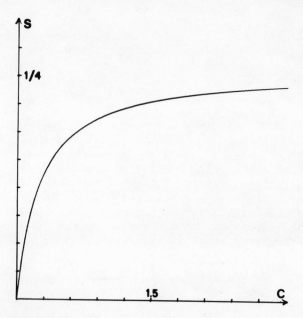

<u>Figure 6</u> : Variations with C = γ/Γ of the contrast S of the satu-
ration resonance.

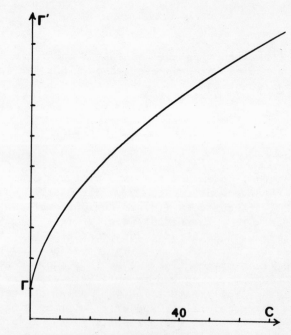

<u>Figure 7</u> : Variations with C = γ/Γ of the width Γ' of the satura-
tion resonance (Γ' is also the width of the Hanle resonance). For
C << 1, Γ' increases linearly with C. For C >> 1, Γ' increases as $\sqrt{C}$.

The steady state solution for σ_{-+} may also be calculated from equations (13), and included with (22) in the expression (18) of L_f. We get for L_f :

$$L_f = \frac{2\gamma N_o}{\Gamma+3\gamma} + \frac{2\gamma N_o\ (\Gamma+2\gamma)}{(\Gamma+3\gamma)(\Gamma+4\gamma)}\ \frac{\Gamma'^2}{\Gamma'^2 + 4\omega_e^2} \tag{25}$$

i.e., the sum of a constant and of a Lorentzian curve having the same width Γ' as the saturation resonance. Figure 8 shows a set of Hanle curves corresponding to different values of the dimensionless parameter $C = \gamma/\Gamma$. One clearly sees the radiative broadening of the resonances. For large values of γ/Γ, the shape of the resonances does not change when γ increases, provided that the scale of the horizontal axis is contracted proportionally to $\sqrt{\gamma}$.

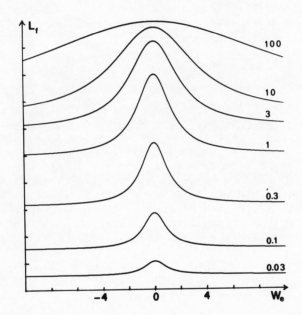

Figure 8 : Set of Hanle resonances detected on L_f . The excitation is broad-line. Each curve corresponds to a value of γ/Γ indicated on the figure. $W_e = 2\omega_e/\Gamma$ is a normalized Larmor frequency.

A detailed experimental verification of all the above results has been done on the $2s_2 \leftrightarrow 2p_1$ transition of Neon (λ = 1.52 μ) [8].

Let us summarize these new results which appear when the light source is a free running multimode laser : saturation resonances observable on the populations of the Zeeman sublevels, radiative broadening of these resonances (and also of the Hanle resonances), which
is not a simple linear function of the laser intensity.

We consider now the case of an atomic beam irradiated perpendicularly by a single mode laser. We have therefore to use equations (17). We suppose that the illuminated portion of the beam is
sufficiently long so that each atom reaches a steady state regime
when passing through this zone. As before, $\sigma_{++} + \sigma_{--} + \sigma_{00} = N_0$ is
a constant of motion and represents the total number of atoms in
the illuminated zone. To simplify the discussion, we will suppose
that $\omega = \omega_0$, i.e., that the laser frequency ω is tuned at the center of the atomic line.

The steady state solution of equations (17) may be found in
an analytical form after some simple algebra, and we get for the
Hanle signal :

$$L_f = 16 \ v^2 \ N_o \ (\Gamma^2 + 4v^2) \ / \ D \tag{26}$$

where :

$$D = 16\omega_e^4 + (8\Gamma^2 + 16 \ v^2) \ \omega_e^2 + (\Gamma^2 + 4 \ v^2) \ (\Gamma^2 + 16 \ v^2) \tag{27}$$

Figure 9 shows a set of such curves corresponding to different
values of the dimensionless parameter $4v^2/\Gamma$. One sees clearly the
radiative broadening of the resonance when the laser intensity,
i.e., v^2, increases, but the shape is no more Lorentzian and the
signal does not tend to a non zero value (as in figure 8) when ω_e
is very large. This is due to the fact that, when ω_e increases,
the frequencies $\omega_0 \pm \omega_e$ of the 2 optical lines $0 \leftrightarrow +1$ and $0 \leftrightarrow -1$
are out of resonance with the laser frequency ω.

When $v \to 0$, expression (26) takes the simple form :

$$L_f = 16 \ v^2 \ N_o \ \frac{\Gamma^2}{(4\omega_e^2 + \Gamma^2)^2} \tag{28}$$

We find the square of a Lorentz curve [20] which is easy to understand:
a first Lorentz denominator describes, as in the previous case, the
decrease of the Zeeman coherence due to the Larmor precession, the
second one comes from the Zeeman detuning of the 2 components of
the optical line with respect to the laser frequency. Expression
(28) may also be obtained from the Born amplitude for the resonant
scattering [9]. The initial state corresponds to the atom in the
ground state in the presence of an impinging ω photon. The atom
can absorb this photon and jump to one of the 2 excited sublevels

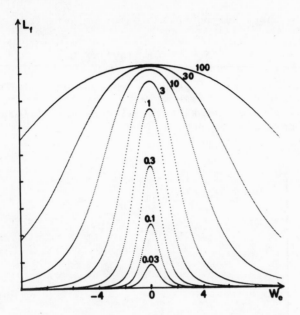

Figure 9 : Set of Hanle resonances detected on L_f. The excitation is monochromatic. Each curve corresponds to a value of $4v^2/\Gamma$ indicated on the figure. $W_e = 2\omega_e/\Gamma$.

±1 of energies $\omega_0 \pm \omega_e$, and then fall back to the ground state by emitting the fluorescence photon. As there are 2 intermediate states for the scattering process, the scattering amplitude A is the sum of 2 terms which are, respectively, proportional to $1/(\omega-\omega_0-\omega_e + i\frac{\Gamma}{2})$ and to $1/(\omega-\omega_0+\omega_e + i\frac{\Gamma}{2})$. As we assume $\omega = \omega_0$, we get :

$$A = \frac{1}{-\omega_e+i\frac{\Gamma}{2}} + \frac{1}{\omega_e + i\frac{\Gamma}{2}} = -\frac{4i\Gamma}{4\omega_e^2 + \Gamma^2} \qquad (29)$$

The cross section is proportional to $|A|^2$ and has the same ω_e and Γ dependance as expression (28).

For $v^2 \gg \Gamma$, the shape of the curve giving L_f does not change any more when v increases, provided that the scale of the horizontal axis is contracted proportionally to v.

Let us also study the variations with the magnetic field of the quantity $\sigma_{-+} + \sigma_{+-}$ (which may be experimentally observed by opposing the σ polarized fluorescence light emitted in 2 directions perpendicular to the magnetic field, and, respectively, parallel and

perpendicular to the laser polarization). One gets from equations
(17) :

$$\sigma_{-+} + \sigma_{+-} = -8v^2 N_o \ (\Gamma^2 - 4\omega_e^2 + 4v^2) \ / \ D \tag{30}$$

where D is given by (27). The shape of such signals is more complicated and is represented in figure 10 for different values of $4v^2/\Gamma$
(in the broad line case, the same signal consists of Lorentz curves tending to zero when $\omega_e \to \infty$). [21]

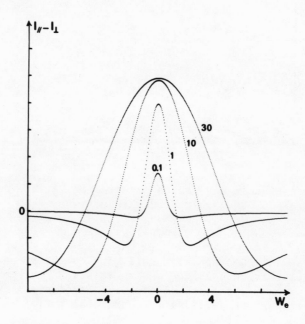

<u>Figure 10</u> : Set of curves giving the difference between the σ-pola-
rized light emitted in directions perpendicular to the magnetic
field and,respectively,parallel and perpendicular to the laser polarization. Each number gives the corresponding value of $4v^2/\Gamma$.
$W_e = 2\omega_e/\Gamma$.

We will not consider the variations of σ_{oo}. They result from
the combination of different factors : coupling between σ_{-+} and
σ_{oo} as in the broad line case, Zeeman detuning of the atomic lines.

To summarize, we see that the essentially new results obtained
in the absence of Doppler effect (single mode laser and atomic beam)
come from the Zeeman detuning of the atomic lines. The zero-field
level crossing resonances have more complicated shapes (non-Lorentz-
ian), but they still have a width which is of the order of Γ at low
laser intensity and which increases with the laser intensity.

MORE COMPLICATED SITUATIONS

In the examples studied above, no structure was existing in level g. There was only one Zeeman coherence in level e, and the Hanle effect was only observable in this level.

We would like now to give an idea of what happens when Zeeman coherences and Hanle effects exist in both levels e and g, and, for that purpose, we take the simplest possible example of such a situation, the case of a transition $J_g = 1/2 \leftrightarrow J_e = 1/2$.

We restrict ourselves to a broad-line excitation. The light beam is supposed to be σ^+ polarized and to propagate along Oz, the magnetic field being applied along Ox. The relaxation of the ground state (which was absent in the previous case as $J_g = 0$) is supposed to be produced by the leakage of atoms from the cell through a small hole (the probability per unit time of escaping from the cell is $1/T$). A balance is provided by an entering flux of n_0 atoms per unit time, all in the ground state and completely unpolarized. If the collisions with the inner walls of the cell are not disorienting, the relaxation time is simply T.

$$\text{Relaxation} \qquad\qquad \text{Larmor precession}$$

$$\dot{\sigma}^e_{++} = \overbrace{-\sigma^e_{++}/T}^{} \quad \overbrace{+i\omega_e(\sigma^e_{+-}-\sigma^e_{-+})/2}^{}$$
$$-\Gamma\sigma^e_{++} \qquad\qquad +2\gamma\sigma^g_{--}/3 \qquad -2\gamma\sigma^e_{++}/3$$

$$\dot{\sigma}^e_{--} = -\sigma^e_{--}/T \quad -i\omega_e(\sigma^e_{+-}-\sigma^e_{-+})/2$$
$$-\Gamma\sigma^e_{--}$$

$$\dot{\sigma}^e_{-+} = -\sigma^e_{-+}/T \quad -i\omega_e(\sigma^e_{++}-\sigma^e_{--})/2$$
$$-\Gamma\sigma^e_{-+} \qquad\qquad\qquad\qquad -\gamma\sigma^e_{-+}/3$$

$$\dot{\sigma}^g_{++} = n_0/2 - \sigma^g_{++}/T \quad +i\omega_g(\sigma^g_{+-}-\sigma^g_{-+})/2$$
$$+\Gamma\sigma^e_{++}/3 + 2\Gamma\sigma^e_{--}/3$$

$$\dot{\sigma}^g_{--} = n_0/2 - \sigma^g_{--}/T \quad -i\omega_g(\sigma^g_{+-}-\sigma^g_{-+})/2$$
$$+\Gamma\sigma^e_{--}/3 + 2\Gamma\sigma^e_{++}/3 \qquad -2\gamma\sigma^g_{--}/3 \qquad +2\gamma\sigma^e_{++}/3$$

$$\dot{\sigma}^g_{-+} = -\sigma^g_{-+}/T \quad -i\omega_g(\sigma^g_{++}-\sigma^g_{--})/2$$
$$\underbrace{-\Gamma\sigma^e_{-+}/3}_{\substack{\text{spontaneous} \\ \text{emission}}} \qquad \underbrace{-\gamma\sigma^g_{-+}/3}_{\text{absorption}} \qquad \underbrace{}_{\substack{\text{stimulated} \\ \text{emission}}} \qquad (31)$$

Equations (31) are the rate equation for the various matrix elements σ^e_{++}, σ^e_{--}, σ^e_{-+}, and σ^g_{++}, σ^g_{--}, σ^g_{-+} of σ_e, and σ_g, and are derived using the same method as above. ω_g is the Larmor precession in g. As before, γ is the reciprocal of the pumping time. We note that the absorption and stimulated emission terms are simpler than in (13), whereas the Larmor precession terms are a little more complicated. This is due to the choice of the axis of quantization Oz which is parallel not to the magnetic field, but to the direction of propagation of the σ^+ polarized laser beam.

In a Hanle experiment performed on a $J = 1/2$ level, one detects components of the atomic orientation $\vec{J}$ perpendicular to the magnetic field $\vec{B}_0$. As $\vec{B}_0$ is along Ox, we are interested here in J^e_z (z component of the orientation of level e), i.e., in $\sigma^e_{++} - \sigma^e_{--}$ (one can for example measure the difference between the σ^+ and σ^- fluorescence light reemitted along Oz). To study these Hanle signals, we have to find the steady state solution of (31). Putting

$$\left\{ \begin{array}{l} \Gamma_e = \Gamma + \dfrac{1}{T} \\[2mm] \Gamma_g = \dfrac{1}{T} \end{array} \right. \qquad \left\{ \begin{array}{l} \Gamma'_e = \Gamma_e + \dfrac{\gamma}{3} \\[2mm] \Gamma'_g = \Gamma_g + \dfrac{\gamma}{3} \end{array} \right. \tag{32}$$

we get for $\sigma^e_{++} - \sigma^e_{--}$

$$\sigma^e_{++} - \sigma^e_{--} = \frac{n_o T}{2} \; \frac{\Gamma_e \Gamma'_e}{\omega_e^2 + \Gamma_e \Gamma'_e} \; \frac{1}{D} \tag{33}$$

where :

$$D = 1 + \frac{3\Gamma_e}{2\gamma} + \frac{1}{2} \frac{\Gamma_e \Gamma'_e}{\omega_e^2 + \Gamma_e \Gamma'_e} + \frac{1}{2} \frac{\Gamma_e}{\Gamma_g} \frac{\Gamma_g \Gamma'_g}{\omega_g^2 + \Gamma_g \Gamma'_g} +$$

$$+ \frac{1}{3} \frac{\Gamma}{\Gamma_g} \frac{(\omega_e \omega_g - \Gamma'_e \Gamma'_g) \Gamma_e \Gamma_g}{(\omega_e^2 + \Gamma_e \Gamma'_e)(\omega_g^2 + \Gamma_g \Gamma'_g)} \tag{34}$$

Figures 11, 12, 13 show the variations with the magnetic field of the Hanle signal computed from (33). The various curves correspond to different values of the dimensionless parameter $\gamma/3\Gamma$, indicated on the figures. We have supposed that $\Gamma T = 100$, i.e., that the relaxation time T of the ground state is 100 times longer than the radiative lifetime $1/\Gamma$ of the excited state.

Let us first interpret the results at very low intensities ($\gamma \ll \Gamma$). We see on figure 11 that the Hanle signal appears as a superposition of 2 curves. The broad resonance is the Hanle effect of the excited state. It has a width equal to Γ (as $\gamma \ll \Gamma$, the radiative broadening is negligible). The amplitude of this resonance

increases proportionally to γ. The narrow resonance is the Hanle
effect of the ground state which has a width determined by $1/T$
(similar resonances have been observed in the ground state of ^{87}Rb
atoms optically pumped by a discharge lamp. Relaxation times as long
as 1 second may be obtained, so that the width of the resonance
may be as low as 10^{-6} gauss) [10]. For the last curve of figure 11
($\gamma/3\Gamma = 0.01$), γ is of the order of $1/T$, and the radiative broadening becomes visible. The intensity of the resonance increases as
γ^2 and not as γ. This is due to the fact that we detect this resonance indirectly on the fluorescence light (and not on the absorbed
light, as this is done usually). We need at least 2 interactions
with the pumping beam in order to get the resonance : the first one,
to create an atomic orientation in the ground state which gives rise
to the Hanle effect of this state, the second one, to transfer this
orientation to the excited state from which it is detected on the
fluorescence light. This explains why the ground state resonance is
so small for the first curve of figure 11 ($\gamma/3\Gamma = 0.001$).

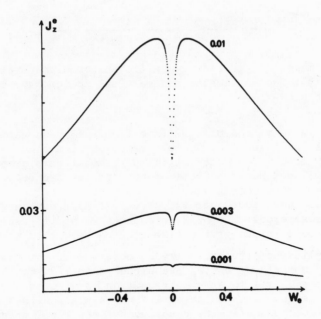

Figure 11 : Set of curves giving the Hanle effect of a $J_g = 1/2 \leftrightarrow$
$J_e = 1/2$ transition. J_z^e is proportional to $\sigma_{++}^e - \sigma_{--}^e$ and is in
arbitrary units. Each curve corresponds to a value of $\gamma/3\Gamma$ indicated on the figure. ΓT is equal to 100. To simplify, levels e and
g are supposed to have the same Landé factor ($\omega_e = \omega_g$). $W_e = \omega_e/\Gamma$
is a dimensionless Larmor frequency.

When γ increases the width of the ground state resonance increases more rapidly than the width of the other one (figure 12).

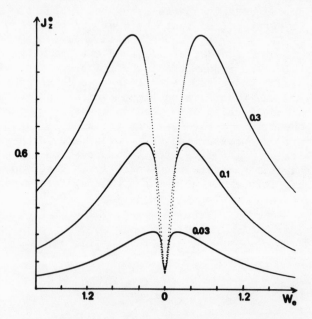

Figure 12 : Same signal as for figure 11 but for higher value of
γ/3Γ. The horizontal and vertical scales have been
changed.

For very large values of γ/3Γ (see figure 13), we get the
same result as for figures 8, 9, 10. The shape of the curve does
not change any more if we contract the horizontal axis proportionally to √γ . The 2 Hanle effects of both states are completely
mixed in a time short compared to 1/Γ (and of course to T).
This mixing does not smooth out the structure apparent on figure
11. We get the superposition of 2 resonances with different
widths and opposite signs giving rise to a curve with 2 maxima.

More complicated structures may be observed if the values
of the angular momenta J_e and J_g are higher than 1/2. For example,
in the case of a J_e = 1 ↔ J_g = 2 transition, and for a σ linearly
polarized excitation, one can observe Hanle signals with 3 maxima.
As in the previous example, the coupling between the 2 transverse
alignments of e and g (perpendicular to the magnetic field) gives
rise to a structure similar to that of figures 11, 12, 13. But
as J_g > 1, there is also in the ground state g a "hexadecapole"
moment (hertzian coherence $\sigma^g_{-2,+2}$) which can be induced in this
state after 2 interactions with the laser, one absorption and one
induced emission processes (see figure 14-a). A third interaction

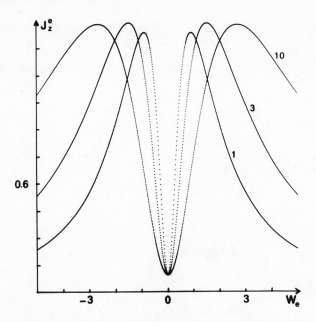

Figure 13 : Same signal as for figures 11 and 12, but for still
            higher values of $\gamma/3\Gamma$. The horizontal and vertical
            scales have been changed.

with the laser (absorption) can couple this hexadecapole moment to
the transverse alignment of e ($\sigma^e_{-1,+1}$) (see figure 14-b). As the
Hanle resonance associated to $\sigma^g_{-2,+2}$ has a smaller width (the resonant denominator is $\Gamma^2_g + 16\omega^2_g$), the total result of these various
couplings is to give for some values of γ a structure with 3 maxima.
This effect has been observed on the $3s_2 \leftrightarrow 2p_4$ transition of Ne
($\lambda = 6328$ Å) and interpreted quantitatively ([11]). It would be interesting to see if other recent observations ([12]) could be explained in the same way.

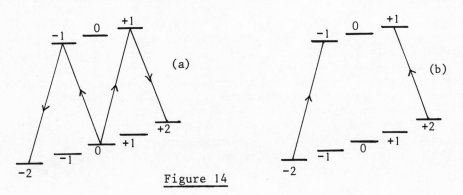

Figure 14

Optical pumping of molecules also provide several examples of Hanle resonances observed in levels having very high angular momentum ([13]). Some efforts are being done to write the optical pumping equations in a basis of quasiclassical states well adapted to the high values of J ([14]).

To summarize the results of this paragraph, we see that one can observe on the fluorescence light emitted from e, level crossing resonances having a width much smaller than the natural width of e. This is not related to the broad-line or narrow-line character of the pumping light (as it appears already on the results of the previous paragraph). These narrow resonances must be attributed to the other state g of the optical line which has a longer lifetime or a higher J value.

OTHER PROBLEMS

In the previous paragraphs, we have discussed some new effects appearing in optical pumping experiments performed with lasers. This review is far from being complete. Let us just mention a few other interesting problems.

We have restricted ourselves to the study of the intensity of the total fluorescence light, a signal which is proportional to some observables of the excited state (see expression (18) for example). Another possibility would be to study the spectral distribution of this fluorescence light, which gives information on the correlation function of the atomic dipole moment driven by the incident light wave. The lowest order theory predicts that a monochromatic wave is scattered elastically by an atom moving perpendicularly to it (it is the same type of theory which leads to the Born expression (29) for the scattering amplitude). At high laser intensities non linear processes take place which of course conserve the total energy but change the spectrum of the scattered light. Experimental evidence for such effects has just been obtained (see ref. ([15]) and references in).

We have also considered only steady state processes corresponding to stationary light beams. This does not mean of course that short light pulses, such as those delivered by mode locked lasers, are not interesting. In particular, the observation of the quantum beats appearing in the light spontaneously emitted by an atom which has been prepared by a short light pulse in a coherent superposition of different excited sublevels ([16]), provides a very powerful method for determining various Zeeman, hyperfine or fine structures ([17]). Furthermore, the spectral range covered by pulsed dye lasers and the peak power of such sources are much larger than for c.w. operation, which increases considerably the number of atomic or molecular transitions which can be optically pumped.

Let us finally mention a very interesting possibility, which has recently be demonstrated independently by several groups [18], of getting rid of the Doppler effect in 2 photon absorption processes. This possibility rests on the high intensity and monochromaticity of the light delivered by dye lasers. It opens the way to a high resolution study of atomic or molecular transitions connecting 2 levels with the same parity. This would be interesting for example for determining the absolute position of metastable levels.

[1] Some review articles on optical pumping :
J. BROSSEL, Quantum Optics and Electronics
Les Houches Summer School, 1964, p. 187 (Gordon and Breach, New York, 1964)

C. COHEN-TANNOUDJI and A. KASTLER,
Progress in Optics (North Holland, Amsterdam, 1966), vol 5, p.1

A. KASTLER, Science 158, 214 (1967)

G.W. SERIES, Quantum Optics (edited by S.M. Kay and A. Maitland) p. 395, Academic Press, 1970

W. HAPPER, Rev. Mod. Phys. 44, 169 (1972)

[2] M. DUCLOY, Phys. Rev. A 8, 1844 (1973)
                        9, 1319 (1974)

[3] M. DUMONT, Thesis, University of Paris, 1971
(C.N.R.S., A.O. 5608), chapter III

[4] J.P. BARRAT and C. COHEN-TANNOUDJI,
J. Phys. Rad. 22, 329 (1961)
              22, 443 (1961)

C. COHEN-TANNOUDJI,
Ann. de Phys. 7, 423 (1962)
              7, 469 (1962)

[5] M. DUCLOY and M. DUMONT,
J. Phys. (Paris) 31, 419 (1970)

[6] M. DUMONT,
J. Phys. (Paris) 33, 971 (1972)

[7] A. ABRAGAM,
The Principles of Nuclear Magnetism, chapter VIII
Oxford University Press, 1961

[8] M. DUCLOY,
Optics Commun. 3, 205 (1971)
Ann. de Phys. 1974, to be published

(^{9}) C. COHEN-TANNOUDJI,
Cargese lectures in Physics, vol. 2, edited by M. Lévy (Gordon and Breach, New York, 1968)

(^{10}) J. DUPONT-ROC, S. HAROCHE, and C. COHEN-TANNOUDJI,
Phys. Lett. 28A, 638 (1969)

(^{11}) M. DUCLOY, M.P. GORZA, and B. DECOMPS,
Opt. Commun. 8, 21 (1973)

(^{12}) W. GAWLIK, J. KOWALSKI, R. NEUMANN, F. TRÄGER,
Phys. Lett., to be published

(^{13}) M. BROYER and J.C. LEHMANN,
Phys. Lett. 40A, 43 (1972)

(^{14}) M. DUCLOY, To be published

(^{15}) F. SCHUDA, C.R. STROUD Jr., and M. HERCHER,
J. Phys. B 7, L 198 (1974)

(^{16}) A. CORNEY and G.W. SERIES,
Proc. Phys. Soc. 83, 207 (1964)

(^{17}) S. HAROCHE, J.A. PAISNER and A.L. SCHAWLOW,
Phys. Rev. Lett. 30, 948 (1973)

(^{18}) F. BIRABEN, B. CAGNAC, and G. GRYNBERG,
Phys. Rev. Lett. 32, 643 (1974)

M.D. LEVENSON and N. BLOEMBERGEN,
Phys. Rev. Lett. 32, 645 (1974)

T.W. HÄNSCH, K.C. HARVEY, G. MEISEL, and A.L. SCHAWLOW,
Opt. Commun. 11, 50 (1974)

(^{19}) For coherence functions to represent light fields
see:
J. DODD, G.W. SERIES, Proc. Roy. Soc. A 263, 353
(1961).
See also supplementary note published in the above
paper reprinted in:
R.A. BERNHEIM,
Optical Pumping, Benjamin, New York 1965.
Later treatment: A. V. DURRANT, B. LANDHEER,
J. Phys. B 4 L 36 and 1200 (1971)

(^{20}) G.W. SERIES,
Proc. Phys. Soc. 89, 1017 (1966)

(^{21}) For recent experimental work done with monochromatic
excitation, see: W. RASMUSSEN, R. SCHIEDER, and
H. WALTHER,
Abstracts of 4th International Conference on Atomic
Physics, Heidelberg 1974, p. 295

HIGH RESOLUTION SPECTROSCOPY BY MEANS OF TUNABLE LASERS

AND ATOMIC BEAMS

Pierre Jacquinot

Laboratoire Aimé Cotton - C.N.R.S. II -

F 91405 Orsay

The subject to be treated here is the intersection of three
wider domains and we shall restrict ourselves to what is really
specific to this intersection. Excellent review articles have been
published on the broader aspects (1) (2) (3) (4) (5) (6) and most of
their content will be supposed to be known, for instance the general
properties of dye lasers or the principles of non linear spectroscopy.
As a rule the references cited in these articles will not be repeated
here, except for publications referring specifically to our discussion.

I

In this paper we refer by "High Resolution" to resolution beyond
the normal Doppler limit, and by "Spectroscopy" to spectroscopy by
means of optical transitions. Because of their valuable properties,
lasers and atomic beams are obviously very useful tools for this type
of spectroscopy, but they are not always necessary. High resolution
can be obtained with lasers in gases or vapors without the help of
atomic beams, with the use of such methods as saturated absorption,
fluorescence line narrowing or two-photon transitions; moreover it
is not even necessary to use lasers if one makes use of the fact that
different levels may be excited coherently by a modulated wave. One
should also remember that resolution beyond the normal Doppler limit
was obtained before the laser era by analyzing the light absorbed or
emitted by atomic beams with interferometers of high resolving power
(Jackson and Kuhn, Meissner).

In Table I, we compare two classes of experiments using lasers.
We indicate in the first line methods in which the Doppler broadening

in gases or vapors is either suppressed (or reduced) by a selective
action on atoms belonging to a particular class of velocities (saturated absorption spectroscopy, S.A.S.; fluorescence line narrowing,
F.L.N.; 2 step optical-optical excitation) or cancelled in a transition produced by two photons of opposite directions. The second line
lists experiments with an atomic (or molecular) beam; here the Doppler broadening is suppressed or at least reduced by the fact that
the atoms have velocities all approximately in the same direction.
A distinction should be made between purely spectroscopic experiments
where the aim is to measure with precision the position of levels,
and some experiments where the aim is to selectively populate a definite level in order to separate from the others (by photoionization
or more complex processes) the atoms having reached this level, for
isotope separation for instance: different signs are used in the table
to indicate that a method can be applied in the first ($\bullet$) or in the
second case ($\blacktriangle$). For instance saturated absorption spectroscopy
would not be able to prepare only one isotopic species in an excited
level if the isotopic shift were less than the Doppler width; in that
case the excited state would be populated by atoms belonging to both
species but of different velocities; on the contrary in a two-photon
process, since there is an exact cancellation of the Doppler shift
for all atoms (7), this preparation would be possible.

It can be seen in the table that a number of results are obtainable only with an atomic beam: some of them will be discussed in
this paper. Of course any of the methods applicable to vapors can
also be used with atomic beams: but usually the latter would not
present fundamental advantages except in very special cases. For
instance, it has been proposed to use saturated absorption with a
fast atomic beam in order to study the relativistic second order Doppler shift (8). Another case, 2 step excitation, will be discussed
later.

Compared to other methods, e.g. saturated absorption, the use
of atomic beams has several particular features:

a) The recorded spectrum is free from spurious lines due to
cross resonances which occur when two components share a common
level. As these spurious components involve atoms belonging to
different classes of velocity, they cannot appear with an atomic
beam: in a relatively simple structure these cross-over signals
can be identified and may even be helpful; but in a structure with
(true) components very close to each other, the result may be quite
ambiguous. An example is given in Fig. I: in the S.A.S. pattern of
this narrow hyperfine structure of the $6s6p\ ^1P_1 - 6s5d\ ^1D_2$ (1.50 μm)
line of Ba I (9) the spurious components, the calculated positions
of which are designated by arrows, cannot be separated from the
others.

TABLE I

Different methods applicable to: ● level determination beyond the optical Doppler limit; ▲ selective excitation of atoms allowing operations such as isotope separation; the empty symbols correspond to less frequent uses.

	Resonant levels				Non resonant or highly excited levels			
	Direct absorption		F.L.N.	S.A.S.	2 step excit. Opt+Opt	2 step excit. El.+Opt	2 phot. trans.	El. or opt. exc. + SAS, FIN...
	Opt. Det.	Non Opt. Det.						
GAS or VAPOR	●	● ▲	●	●	○		● ▲	●
ATOMIC BEAM				○	● ▲	● ▲	○ △	

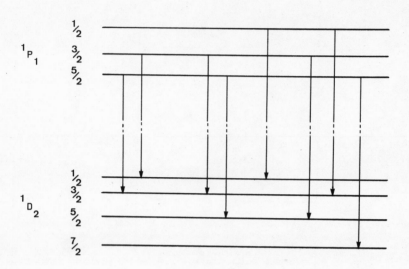

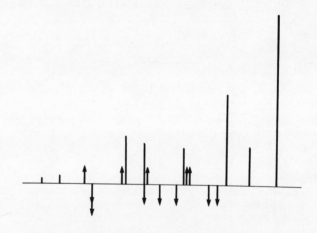

<u>Fig. 1</u> - Theoretical structure of the 6s6p $^1P_1$ - 6s5d $^1D_2$
(λ = 1.50μm) line of Ba in saturated absorption: arrows indicate
cross resonances between transitions sharing a common upper level.
From (9).

b) At low laser power, the homogeneous width of the lines is very close to the natural one since collision effects are negligible in the beams normally used; pressure shifts, always very difficult to evaluate or to extrapolate (10) are also absent. This property will be especially valuable for the study, which is now planned in several laboratories, of highly excited states, since these are very sensitive to collisions. This absence of spurious phenomena will also be of particular interest in the study of small effects that depend on the power of the light, more or less related to the dynamic Stark effect (Autler-Townes). For these studies it is essential to operate under very clean conditions that only atomic beams can provide.

c) Since purely linear processes can be used with the atomic beam, the power of the laser beam can be kept small enough so that power broadening is avoided. One should beware of this type of broadening since power densities as small as a few mW/mm^2 are sufficient to saturate the lowest resonance transition of many atoms. The same is true for the influence of optical pumping effects which may change dramatically the relative populations of the hyperfine sublevels of the ground level and, therefore, the relative intensities of the components. These effects can be avoided more readily with linear processes, that can be achieved with atomic beams, than with S.A.S.

d) At low power the relative intensities are proportional to the oscillator strengths; in contradistinction, with non-linear methods like S.A.S. or two-photon transitions, they are proportional to the square of the oscillator strengths.

e) When only a very small quantity of atoms is available it is necessary to use very sensitive methods of detection of the transitions and atomic beams lend themselves to methods of non-optical detection of the optical transition that may afford a greater sensitivity. In S.A.S. the signal is relatively weak and it would probably be difficult to work with unstable atoms produced in small quantities by nuclear reactions. In addition, particularly in an "on line" experiment, it may be much easier to handle very few atoms in a beam than in a cell.

The atomic beam technique has been known for long and benefits from the extensive experience of physicists who have used it in conjunction with magnetic (or electric) resonance (11): it is now mastered for various materials, from gases to refractory elements. In the simple cases it is no more difficult to use a beam than a cell, and the operating conditions are under much better control. The Doppler width is reduced, compared to the gas at the same temperature, by a factor equal to the colimation ratio C: for instance, in order to reach a residual Doppler width of 2 MHz with Na atoms, C should be as high as 10^3. Of course, with such a high C the density of the beam is low (typically of the order of 10^5 at.cm^{-3}, which corresponds to about 10^{10} at.cm^{-2}s^{-1}, and the number of atoms that interact with the laser beam is small. It is possible to escape this compromise between density and reduction of the Doppler width by using multiple hole collimators.

A very strong reduction of the Doppler width is not always necessary for the problem to be treated, and in this case a low collimation ratio would be sufficient. However, some methods of non-optical detection of optical resonances, like the Rabi-type or the deflection by recoil, require a high C: it may then become necessary to look for other methods such as the detection by photoionization from the excited state or that of magnetic filtering explained further on; these do not require highly collimated beams.

Tunable lasers appear to have reached by now almost all the qualities required for work with atomic beams. In particular, a high degree of sophistication has been achieved with CW single-mode dye lasers, which are now available even commercially. Frequency and intensity stability, and power output are satisfactory: an effective band width of 1 MHz is now commonly obtained with a power of more than 10 mW. The range of fine tunability around a given frequency is still small, typically 2,000 MHz, and in the best case 15,000 MHz (12). The only point which is still unsatisfactory with CW dye lasers is that their wavelength coverage is too narrow: Rhodamine 6G, which is by far the best dye, works only between 570 nm and 640 nm and, although it is possible to extend the domain by using other dyes, for instance toward the blue with coumarin, much progress still remains to be done. When a pulsed laser can be used it is much easier to cover the domain from near UV to IR. Semiconductor diode lasers can also meet the requirements for working with atomic beams if they are used under appropriate conditions. For instance, by controlling independently the temperature and the injection current of a GaAs diode laser it has recently been found possible (13) to obtain single-mode operation at the wavelength of the cesium resonance line (852 nm). The long term stability of the free running laser was of the order of 30 MHz, and the power was sufficient to completely empty one of the hyperfine levels.

The scanning of the laser frequency is usually made by applying a linear ramp voltage to the element which controls the length of the cavity. This gives approximately a linear frequency scale, but it is not sufficient for precise calibrations. The most common way

of calibration consits of simultaneously recording the output signal of a long Fabry-Perot etalon the bandwidth of which may be as
narrow as 1 MHz. It may be preferable to use the signals provided
by a Michelson interferometer with two channels $\pi/2$ out of phase.
These two signals contain, respectively, terms in $\sin 2\pi\sigma\Delta$ and $\cos$
$2\pi\sigma\Delta$ (σ wave number, Δ fixed path difference) from which it is
possible to extract, by appropriate electronic treatment, either a
measurement of σ (Δ being known) or an error signal to lock the
laser to a given frequency. Such a system has been developed and
used at Laboratoire Aimé Cotton and will be published soon (14).
Direct reading can be made down to 2 MHz and this should be possible even with pulsed lasers. The system can be controlled by electric pulses so as to scan a given frequency range step by step, the
size of the steps being chosen at will.

This device is now included in an optical experiment under way
at Laboratoire Aimé Cotton for the measurement on line, of the hyperfine structure and isotope shift in the D lines of ^{25}Na ($\tau \simeq 1$ min);
the sodium atoms are produced by a spallation reaction in a bombardment of ^{27}Al with 150 MeV protons accelerated by the Orsay synchrocyclotron.

For very precise measurements of frequency differences one can
use the beat frequency between a fixed-frequency stabilized laser
and the scanning laser. One can also use the frequency offset
locking of two dye lasers, the difference between the two frequencies
being determined by a tunable RF generator. Another possibility
is to shift the frequency of a stabilized laser by electro-optical
or acousto-optical (15) devices driven by a tunable RF generator.
These methods, which rely upon radiofrequency devices have, at
present, a somewhat limited range of the order of a few GHz; the
scanning over this range presents, of course, the usual problems of
wide-band RF and microwave coverage.

II

In the recent few years a number of publications have appeared
on high resolution spectroscopy by means of tunable lasers and
atomic beams. Most of them still deal with simple cases and their
aim is essentially to demonstrate the feasibility of new methods.
In atomic spectroscopy the hyperfine structure of sodium lines is
their common test and the line width is limited essentially by the
natural width (of the order of 10 MHz). With molecules (mostly I_2),
the excited states of which have a much longer life time, the linewidth obtained was smaller than 100 MHz in the best cases (15). The
tunable lasers used in the experiments on sodium and in some of the
experiments with I_2 were single-mode CW dye lasers.

The detection of absorption transitions is most often made by observing the fluorescence from the upper state. It should be noted that each atom can make several transitions during its interaction with the laser beam so that it can give rise to several fluoroescence photons. In the limiting case where the upper level is saturated, the number of fluorescence photons would be equal to the number of lifetimes contained in the interaction time, thus leading to a noticeable gain in detection sensitivity. But at the required laser power, optical pumping effects and power broadening would occur. One of the best results (16) obtained on Na D lines is shown in Fig. 2. In order to obtain this, it was necessary to attenuate the beam of the CW dye laser to less than 1 mW/mm^2: this demonstrated clearly effects of optical pumping and of power broadening.

Some experiments have been published on the study of the light scattered by atoms excited with radiation narrower than the natural width in order to test the predictions of different existing - but not concordant - theories. An atomic beam is of course the ideal medium for such experiments where each atom can be considered as isolated. In an experiment on the sodium D_2 line (17), the scattered light was analyzed by means of a Fabry-Perot interferometer: two stark-effect side bands, the splitting of which increases with the light intensity, were clearly seen. Another experiment has been reported on sodium and barium resonant states, where the scattered light is analyzed by photon correlation with the incoming beam in order to measure its spectral width (18).

Several methods of non-optical detection of absorption transitions have also been reported and tested on the same hyperfine structure (Na):

a) A "Rabi type" method (19) applicable to paramagnetic atoms: the transition to be studied produces optical pumping of the magnetic sub-levels of the ground state: the resulting change in the orientation of the magnetic moments is detected by a change in the trajectory of the beam in an inhomogeneous magnetic field. The device is exactly similar to the normal Rabi experiment, except that it does not require a magnet in the C region and that the radio-frequency is replaced by the laser beam.

b) A method making use of the recoil of the atoms due to transfer of momentum of the photon to the atom when the absorption occurs (20). The deflection of the beam is very small but sufficient to be used with a highly collimated beam. The atoms are detected, as in the preceding experiment, by different methods depending on the particular ones under study: ionization on a hot wire or by their radioactivity in the case of unstable atoms.

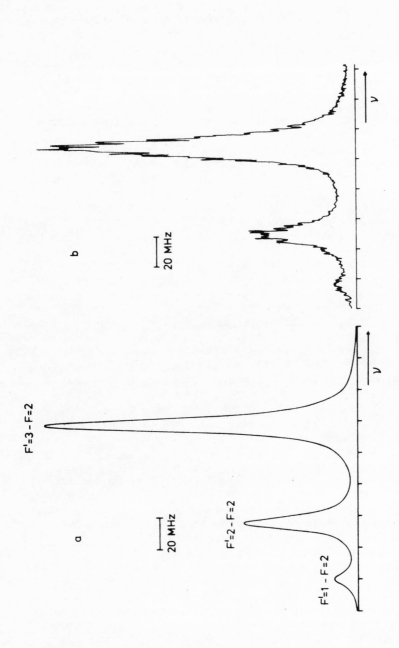

Fig. 2 - Portion of the hyperfine structure of the D$_2$ line of Na, obtained by recording the fluorescence of an atomic beam of Na excited by a tunable CW dye laser. a) calculated signals with natural width; b) experiment, sweep time 8s, 3 sweeps averaged. From (16).

c) Photoionization from the upper level of the transition by a second light beam (21): if the photons of this beam have just enough energy to photoionize the atom from this level, the number of collected ions is proportional to the number of transitions induced by the tunable laser. Ion counting is very efficient, but of course the overall efficiency of the method depends on the photoionization cross section of the atomic state under study and on the light source available for photoionization. Advances with this method will depend strongly on the extension of the spectral range of tunable dye lasers.

Atomic beams and tunable lasers are also being proposed for isotope separation of Ba (22), Ca (23) and U (24). The principle of these methods is very similar to the experiments mentioned above in b) and c). If the isotope shift of the transition used is smaller than the Doppler width, the use of an atomic beam is indispensable (except in the case of a two-photon transition): for this type of application methods such as saturated absorption spectroscopy could not work, as has been stated in the introduction. But even in the case of larger isotope shifts it may be advantageous, for practical reasons, to use an atomic beam (uranium).

Some experiments have also been made with atomic beams, although they could, in principle, have given the same result with a vapor (25) (26). It is, for instance, the case of an experiment on Zeeman quantum beats in ytterbium (27): this is a good example of a case where an atomic beam was, for practical reasons, easier to use than a vapor.

Three experiments performed recently at Laboratoire Aimé Cotton, not yet published or still under way, will now be reported briefly.

1) <u>A new method of detection of optical resonances by magnetic filtering of an atomic beam (28).</u>
As already stated, it may be useful, at times, to sacrifice somewhat the narrowing of the Doppler width to the gain of greater sensitivity, by decreasing the collimation ratio, since, we recall, the density of the beam is proportional to C^{-2}. The Rabi-type method has been modified so that it can work with a much higher aperture of the atomic beam. The arrangement is shown in Fig. 3. It uses a six-pole magnet that focusses the atoms with $m_J = +\frac{1}{2}$ and defocusses those with $m_J = -\frac{1}{2}$, thus acting as a magnetic filter. The action of the laser beam on the atomic beam takes place before the magnet in a region of weak field. In the case of Na, when the laser is tuned to a hyperfine component of one of the D lines, the relative populations of the hyperfine ground levels $F = 1$ and $F = 2$ are changed: this, in turn, produces a change in the relative population of the states $m_J = +\frac{1}{2}$ and $m_J = -\frac{1}{2}$ (Fig. 4) in the magnet, and, because of the magnetic filtering action, a change in the number of atoms that arrive at a small detector placed after the magnet.

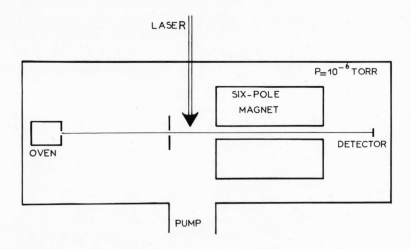

Fig. 3 - Experimental set up for the detection of optical
resonances by magnetic filtering (28).

Due to the fact that all atoms with F = 1 end in the $m_J = -\frac{1}{2}$ state
in the strong field and 80% of the atoms with F = 2 end in $m_J = +\frac{1}{2}$,
there is a background: consequently, some hyperfine transitions
give a positive signal relative to this background whereas others
give a negative signal. Figure 5 shows the result of an experiment
made on [23]Na designed to test the method. A detailed study of the
signal and of the S/N ratio has been made. This method of detection
will be used in the experiment on [25]Na to which we referred above:
a S/N ratio of 10 with a counting time of 1 minute is expected with
a production of 10^8 atoms per minute.

   2) Measurement of the hyperfine structure of the 5 $^2S_{1/2}$ state
of [23]Na by two-step excitation (29).
   The S and D states of alkali - and more generally states that
differ in angular momentum ℓ by 0 or 2 from that of the ground state -
can be reached only by two-photon transitions, or by two-step excitation via an intermediate level.

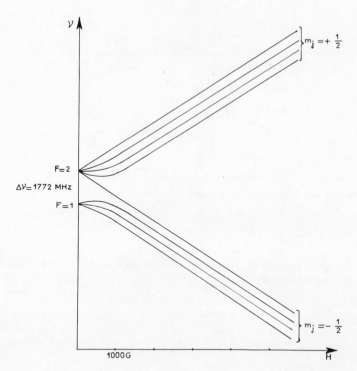

Fig. 4 – Breit-Rabi diagram for the ground hyperfine levels of
^{23}Na (I = 3/2).

The first method gives a complete cancellation of the Doppler
effect for all atoms even in a gas and has been applied with great
success recently. It is also interesting to explore the possibilities of the first method since the probabilities for the two one-
photon transitions are much higher than for the direct two-photon
transition. The principle of the experiment is very simple, but it
requires two tunable lasers: the first one must be tuned very stably
to the intermediate level whereas the second scans the structure of
the higher one. The experiment has been made (33) on the $3S_{1/2} \rightarrow$
$3P_{1/2} \rightarrow 5S_{1/2}$ transition of sodium (λ_1 = 589.6 nm, λ_2 = 615.4 nm),
the excitation of the $5S_{1/2}$ state being detected by observing the
fluorescence (Fig. 6, 7, 8).

In the preliminary experiment the linewidth is 24 MHz and the
measured hyperfine splitting of $5\,^2S_{1/2}$ state is 159 ± 6 MHz.

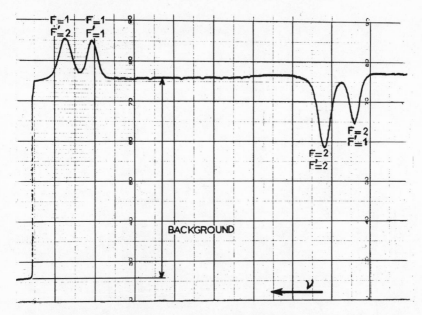

Fig. 5 - Experimental result with the set up of Fig. 3:
the "negative" signals correspond to the transitions resulting
from optical pumping from F = 2 to F = 1.

Cancellation of the Doppler broadening should be obtained in
a vapor if the two beams are colinear: in that case the first
beam populates the intermediate level with atoms of only one class
of velocity and only these atoms are available for the second transition. But spurious resonances occur if the intermediate level
has several components within the Doppler width since in that case
the different sublevels are populated by atoms belonging to different classes of velocity. It is easy to show that if there are
two intermediate levels separated by $\delta\nu$, $(\delta\nu < \delta\nu_{Doppl})$, there
appears a spurious resonance at a frequency $\nu'_2 = \nu_2 + \delta\nu \ [\nu_2/\nu_1 -1]$.
These spurious signals would disappear at $\nu_2 = \nu_1$, which is precisely the case of the two-photon transition. Indeed, in that case,
the intermediate level, which is virtual, can be considered as a
continuum of components spread over much more than the Doppler
width so that all atoms contribute to the signal without adding unwanted components.

With the two-step process, since ν_2 is never equal to ν_1, it
is preferable to use an atomic beam in order to avoid these spurious components.

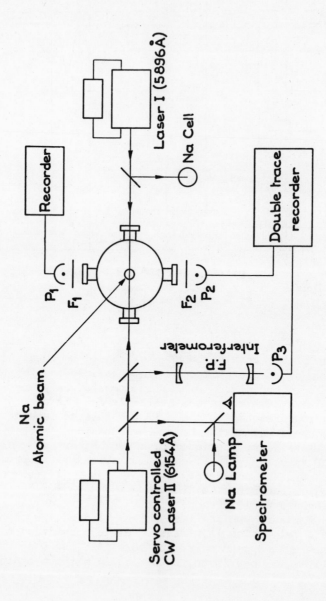

<u>Fig. 6</u> - Experimental setup for two-step excitation in an atomic beam of Na (33).

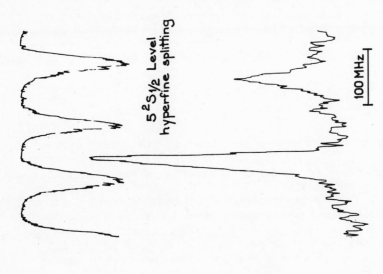

Fig. 8 - Recording of the hyperfine structure of the $5\ ^2S_{1/2}$ state obtained by scanning across the 615.4 nm line. The upper trace gives the calibration with the use of a spherical Fabry-Perot (33).

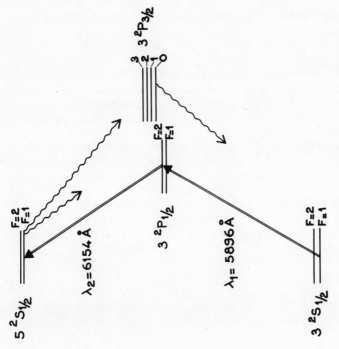

Fig. 7 - Transitions involved in the two-step excitation of the $5\ ^2S_{1/2}$ state of Na.

3) <u>Measurements on rare gases with an atomic beam excited by</u>
<u>an electron beam.</u>

In these experiments metastable states of a rare gas are produced by electron excitation and transitions from these metastables
can be studied. In the experiment underway the hyperfine structure
of several levels of the $2p^53p$ configuration of ^{21}Ne will be studied.
The detection is made by fluorescence.

III

We have now to ask the question: what is the future of this
type of spectroscopy and, namely, what are the problems in atomic
physics which could now be studied? Of course, a vast field of
high resolution spectroscopy can be covered without the use of
atomic beams if S.A.S. or two-photon transitions are used. On the
other hand, tunable lasers and atomic beams can be used simultaneously in experiments which are not high resolution spectroscopy.
This is the case for studies of photoionization from an excited
level where the use of an atomic beam provides many advantages;
the high resolution available may be useful in cases where there
is an intermediate autonionizing level (33) and, chiefly, in the
case of two-photon ionization if there is an approximately resonant
intermediate level below the continuum.

But there are a number of areas where all the advantages of
the atomic beam-tunable laser combination can be fruitfully exploited, for example:

1) <u>Isotope shifts and hyperfine structure in atomic spectra.</u>
Many of the experiments which were made by conventional spectroscopic techniques can now be made with a much better precision,
and some of them which were impossible can now be undertaken. The
interest will be mostly in isotope shifts since these quantities can
be measured only by means of optical transitions, whereas hyperfine
structures in excited as well as in ground states are measured with
much higher precision by radiofrequency spectroscopy. Measurements
of isotope shifts for atoms far from the stable valley will be accessible since the number of atoms required for the experiment can
be made much lower than by conventional spectroscopy.

In addition, isotope separation, that relies on the isotope
shift, is expected to become an importnat application.

2) <u>Structure of highly excited atomic levels.</u>
This is an important point for the theory of atomic structure.
In the present state of precision of the existing experimental data
a perturbed hydrogen-like model is sufficient to understand the

structure of states with a high principal quantum number. But when
these structures become known with a higher precision it will be
possible to study the connection between the two descriptions of
the atom valid in the limiting cases corresponding to an outer
electron close or very far from the core. In particular, very
little is known on the evolution of "electronic correlations" along
a series, and just about nothing is known on that of fine and hyperfine structures for increasing values of n. Hyperfine structure
will be particularly interesting since it involves the simplest
atomic operators and interactions. Measurements of Landé g factors and of the specific isotope mass shift will also be of great
value.

Of course, most of these quantities can also be measured by
other methods without atomic beam, such as two-photon transitions
or by radiofrequency spectroscopy of levels populated by cascades
(34) or by direct or stepwide laser excitation (35). But as these
states are very sensitive to collisions it may often be useful to
operate in such conditions that the atoms under study can be considered as isolated. It could be objected that the most interesting
property of these states is precisely their behaviour when they
are not isolated: but even from this point of view it is necessary to know their properties in the limiting case which is realized
in a beam.

3) Interaction of atoms with strong light fields.
As already mentioned, some studies on this fundamental question
have been undertaken with atomic beams and tunable lasers. Of
course, here also, it is possible to observe such effects under
different conditions, for instance in S.A.S. (36); however as the
phenomenon of saturated absorption is rather complicated by itself,
the interpretation of the power effects, e.g. optical nutation, is
more difficult. With atomic beams and tunable lasers all the
requisites for the observation of power effects under very well
controlled conditions are now met.

Since the laser line may be made much narrower than the natural
width, the effects produced by a strong monochromatic optical field
can be measured by means of optical transitions: this is quite different from the experiments where shifts produced by strong non-
monochromatic optical fields were measured by means of radiofrequency experiments. Some of the effects studied many years ago
entirely in the radiofrequency domain can now be studied entirely
in the optical domain.

REFERENCES

1. H. H. Stroke, Tunable wavelength high resolution spectroscopy,
    Comments At. Molec. Phys. <u>3</u>, 69, (1972) and <u>3</u>, 167 (1972).

2. W. Demtröder, High resolution spectroscopy with lasers,
    Physics Reports <u>7</u>, 223-277 (1973).

3. R. G. Brewer, Non linear spectroscopy, Science <u>178</u>, 247 (1972).

4. W. Lange, J. Luther, A. Steudel, Dye lasers in atomic spectro-
    scopy, Advances in atomic and molecular physics <u>10</u>,   (1974).

5. P. Toschek, General survey of saturation spectroscopy, Interna-
    tional Conference on "Spectroscopie sans largeur Doppler de
    systèmes moléculaires simples", Aussois 1973, C.N.R.S. Paris.

6. H. Walther, Recent advances in dye lasers,   Ibid.

7. B. Cagnac, G. Grynberg, F. Biraben, J. Phys. (Paris) <u>34</u>, 845,
    (1973).

8. J. J. Snyder, J. L. Hall, M. S. Sorem, VIII° Intern. Quantum
    Electronics Conf., San Francisco (1974).

9. Ph. Cahuzac, Physica, <u>67</u>, 567 (1973).

10. R. Vetter and D. Reymann, J. Phys. B <u>7</u>, 323 (1974).

11. N. F. Ramsey, Molecular Beams, Clarendon, Oxford (1956).

12. S. Liberman and J. Pinard, Appl. Phys. Lett. <u>24</u>, 142 (1974).

13. J.-L. Picqué, S. Roizen, H. H. Stroke, O. Testard, Submitted to
    Applied Physics.

14. P. Juncar and J. Pinard, to be published.

15. R. E. Grove, L. A. Hackel, F. Y. Wu, D. G. Youmand, S. Ezekiel,
    VIII° Intern. Quantum Electronics Conf., San Francisco (1974).

16. W. Lange, J. Luther, B. Nottbeck, H. W. Schröder, Opt. Comm. <u>8</u>,
    157 (1973).

17. F. Schuda, C. R. Stroud Jr., M. Hercher, J. Phys. B <u>7</u>, L198
    (1974).

18. H. Walther, VIII° Intern. Quantum Electronics Conf.,
     San Francisco (1974).

19. H. T. Duong, P. Jacquinot, S. Liberman, J.-L. Picque, J. Pinard,
     J.-L. Vialle, Opt. Comm. 7, 371 (1973).

20. P. Jacquinot, S. Liberman, J.-L. Picqué, J. Pinard, Opt. Comm.,
     8, 163 (1973).

21. H. T. Duong, P. Jacquinot, S. Liberman, J. Pinard, J.-L. Vialle,
     C. R. Acad. Sci. Paris B276, 909 (1973).

22. A. Bernhardt, D. Duerre, J. Simpson, L. Wood, J. Goldsborough,
     C. Jones, VIII° Intern. Quantum Electronics Conf., San Francisco
     (1974).

23. U. Brinkmann, W. Hartig, H. Telle, H. Walther, Ibid.

24. S. A. Tuccio, J. W. Dubrin, O. G. Peterson, B. B. Snavely,
                                                        Ibid.

25. S. Haroche, J. A. Paisner, A. L. Schawlow, Phys. Rev. Lett. 30,
     948 (1973).

26. R. Wallenstein, J. A. Paisner, A. L. Schawlow, Phys. Rev. Lett.
     32, 1333 (1974).

27. W. Gornik, D. Kaiser, W. Lange, J. Luther, H. H. Schulz,
     Opt. Comm. 6, 327 (1972).

28. H. T. Duong and J.-L. Vialle, Submitted to Opt. Comm.

29. D. Pritchard, J. Apt, T. W. Ducas, Phys. Rev. Lett. 32, 641
     (1974).

30. F. Biraben, B. Cagnac, G. Grynberg, Phys. Rev. Lett. 32, 643
     (1974).

31. M. D. Levenson and N. Bloembergen, Phys. Rev. Lett. 32, 645
     (1974).

32. T. Haensch, K. Harvey, G. Meisel, A. Schawlow, Opt. Comm. 11,
     50 (1974).

33. H. T. Duong, S. Liberman, J. Pinard, J.-L. Vialle, Submitted to
     Phys. Rev.

34. R. Gupta, W. Happer, L. K. Lam, S. Svanberg, Phys. Rev. A 8, 2792 (1973).

35. P. Tsekeris, R. Gupta, W. Happer, G. Belin, S. Svanberg, Phys. Lett. 48A, 101 (1974).

36. I. S. Shahin and T. W. Haensch, Opt. Comm. 8, 312 (1973).

QUANTUM-BEATS AND LASER EXCITATION IN FAST-BEAM SPECTROSCOPY

H.J. Andrä

Institut für Atom- und Festkörperphysik

Freie Universität Berlin, Germany

INTRODUCTION

When Prof. Bashkin presented his paper in 1970 on "Recent Advances of Beam-Foil Spectroscopy" at the Atomic Physics Conference in Oxford[1] he included as one of the advances the measurements of periodic intensity variations in the decay of hydrogen-like atoms (H, He$^+$) exposed to external electric fields. Although this phenomenon had been observed in 1965 for the first time[2] by beam foil spectroscopy[3] (BFS) it was only by 1969/70 that this phenomenon was understood as just one possible example of the interference occuring when the time-resolved decay of close-lying levels is observed after pulsed excitation[4].

The basic principle of such interferences, known as quantum beats, is easily explained by a two-level system decaying to a common ground state (Fig. 1a). According to Franken[5] such atoms may be described after pulsed excitation in a simplified way (pulse length $\Delta t < 1/\omega_{21}$) by a coherent superposition state

$$|\psi(t)> = a|1>e^{-(i\omega_1 + \frac{1}{2}\gamma_1)t} + b|2>e^{-(i\omega_2 + \frac{1}{2}\gamma_2)t}, \qquad (1)$$

where a and b are expansion coefficients, $\hbar\omega_1$ and $\hbar\omega_2$ are the energies of the two levels, and γ_1 and γ_2 are their decay constants. The time-dependent intensity distribution of photons emitted from such atoms after the excitation pulse is given by

$$I(t) \propto |<g| \hat{\varepsilon} \cdot \hat{r} |\psi(t)>|^2 \qquad (2)$$

where $\hat{\varepsilon}$ is the polarization vector of the emitted photon and $\hat{r}$ is

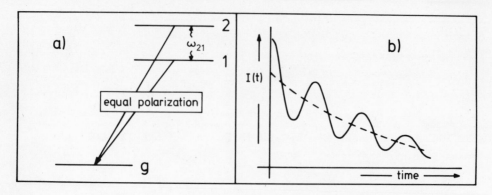

Fig. 1:   a) Illustration of two levels with energy difference $\hbar\omega_{21}$, which decay to the same groundstate g.   b) Expected intensity versus time after impulsive excitation of the two levels 1 and 2 in a)

the dipole length operator. Assuming an observation time resolution $\delta t < 1/\omega_{21}$ one obtains in the case $\gamma_1 = \gamma_2 = \gamma$ a single exponential with a superimposed oscillation at the frequency ω_{21} of the level separation

$$I(t) \propto e^{-\gamma t} \left[A+B \cos \omega_{21} t\right] \tag{3}$$

(Fig. 1b). B is non-zero if both matrix elements $\langle g|\hat{\varepsilon}\cdot\hat{r}|1\rangle$ and $\langle g|\hat{\varepsilon}\cdot\hat{r}|2\rangle$ are non-zero for the transition to the same final state. (It is this condition which requires electric fields for the observation of beats between fine structure levels of different parity in hydrogen-like atoms).

The obvious possibility of measuring small level separations by such quantum beat observations has been experimentally verified since 1962[6] . However, the lack of appropriate equipment for time resolution measurements limited these experiments only to Zeeman effect studies after optical or electron excitation[7] where the level splittings could be adjusted by the magnetic field to a few MHz. Therefore a wider use of quantum beats was not possible until BFS offered unprecedented experimental conditions which are summarized in Table 1.

As is well known these outstanding features A-E are used after foil excitation for the main applications in BFS[3] - namely spectroscopic and lifetime studies. In addition they are also ideally suited for quantum beat measurements not only on hydrogen-like atoms but

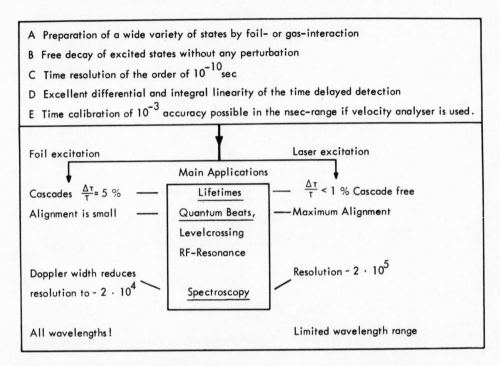

A Preparation of a wide variety of states by foil- or gas-interaction

B Free decay of excited states without any perturbation

C Time resolution of the order of 10^{-10} sec

D Excellent differential and integral linearity of the time delayed detection

E Time calibration of 10^{-3} accuracy possible in the nsec-range if velocity analyser is used.

Foil excitation   Laser excitation

                         Main Applications

Cascades $\frac{\Delta \tau}{\tau} \approx 5$ % ——   Lifetimes   —— $\frac{\Delta \tau}{\tau} < 1$ % Cascade free

Alignment is small   ——   Quantum Beats, ——Maximum Alignment

                                 Levelcrossing

                                 RF-Resonance

Doppler width reduces   Resolution ~ $2 \cdot 10^5$

resolution to ~ $2 \cdot 10^4$ Spectroscopy

All wavelengths !   Limited wavelength range

Table 1:   The basic properties of fast beam spectroscopy.

also on non-hydrogen-like atoms if sufficient alignment by the foil excitation is obtained. The importance of alignment will be pointed out later.

The cascades, the Doppler broadened spectral lines and particularly the small alignment characteristic of the foil excitation have led us to replace the foil by selective laser excitation. This modification solves some of these intrinsic problems of foil excitation for such transitions where laser wavelengths are available. As shown on the lower right of Table 1 one expects then to have excellent conditions for precision lifetime and quantum beat measurements, and even for high resolution spectroscopy.

Therefore I want to focus in the first part of this report on the development of zero-field quantum beats after foil excitation as a high resolution tool. In the second part I want to show then a number of experimental results which will clearly demonstrate the progress which has been made by using laser beams instead of foils for the excitation of fast atomic beams.

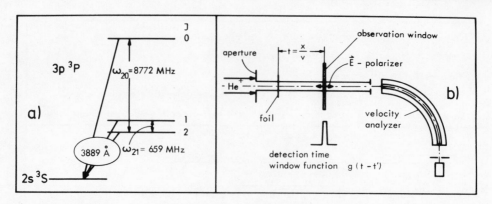

Fig. 2: a) Fine-structure of the HeI-3p^{3}P-state with the two possible beat frequencies ω_{21} and ω_{20} which can be observed in the λ 3889 Å emission. b) Schematic experimental BFS setup as used for quantum-beat observations.

QUANTUM BEATS AFTER FOIL EXCITATION

In our simplified picture of the beat phenomenon in Fig. 1 one can imagine that the levels 1 and 2 represent two fine-structure (fs) levels of an atom. As a result one would expect to observe fs quantum beats in a time-resolved measurement on the light emitted by such impulsively-excited atoms[8]. This indeed occurs and allows one to measure directly fs and hyperfine structure (hfs) separations of excited atomic states in the GHz range without applying any external field. However, the picture is not quite as simple when one studies a realistic example like the HeI 3p^{3}P fine structure (Fig. 2a). For a complete description of this case a total of (2S+1)(2L+1) = 9 magnetic sublevels and certain symmetry assumptions for the excitation mechanism by the foil have to be taken into account[9]. For such calculations an elegant formalism using the density matrix and its expansion into irreducible tensor components has been developed[10].

Applying it to our example for an experiment where the light is observed perpendicular to the beam with a polarization parallel to the beam (Fig. 2b) a distribution of intensity versus time

$$I(t) \propto \int_{-\infty}^{+\infty} dt' g(t-t') e^{-\gamma t'} \ x$$

$$x \ \{14\sigma_0 + 13\sigma_1 + (\sigma_0 - \sigma_1) \ [9\cos\omega_{21}t' + 4\cos\omega_{20}t']\} \quad (4)$$

is obtained[11] where the initial cross sections for the excitation of the orbital angular momentum sublevels are given by:

$$\sigma_o = \sigma(M_L=0) \quad \text{and} \quad \sigma_1 = \sigma(M_L=+1) = \sigma(M_L=-1). \qquad (5)$$

This intensity formula clearly shows why the aforementioned alignment, that is the anisotropic excitation $\sigma_o \neq \sigma_1$, is of crucial importance for the observation of any beat signal since with $\sigma_o = \sigma_1$ the oscillating terms in eq. 4 would vanish.

That alignment is indeed produced by the foil interaction was shown for the first time in 1970 by observing the ω_{21}=659 MHz beats in HeI[12,13] (Fig. 2a) with the experimental set up in Fig. 2b (without a velocity analyzer, however). A beam of 475 keV He$^+$ was excited by a foil and its time delayed (t=x/v) emission at 3889 Å ($3p\,^3P$ - $2s\,^3S$) was observed (Fig. 3). Light polarized parallel to the beam was detected and the foil was moved relative to the fixed observation window. The timewindow function $g(t-t')$ and the spatial reproducibility in this experiment did not allow for the resolution of ω_{20}.

Since this experiment a lot of effort has been invested, principally by groups in Aarhus, Lyon, Tucson and Berlin, in order to improve the quality of such quantum beat measurements[16-24]. High-precision target chambers with overall reproducibilities of foil positions relative to the detection optics of better than 0.05 mm had to be constructed and effective normalization systems had to be developed in order to reduce the influence of fluctuations in beam current. These experimental improvements are best demonstrated by a new measurement of the fs-beats of HeI-3p^{3}P at 250 keV, shown

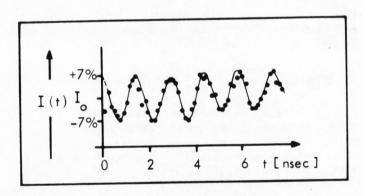

Fig. 3: First zero-field quantum beat observation. Only the ω_{21} fs beats of HeI-3p^{3}P are resolved and the nonoscillating intensity I_o has been subtracted.

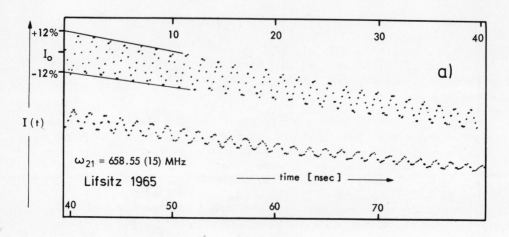

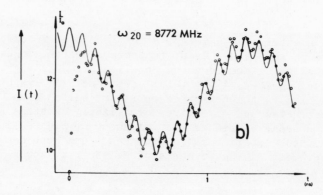

Fig. 4: a) HeI-3p^{3}P ω_{21} fs beats. The lower part is the continuation of the upper one. b) HeI-3p^{3}P ω_{20} fs beats superimposed on
one period of the ω_{21} fs beats of a).

in Fig. 4a,b[11,20]. The excellent statistics and the good stability
of the apparatus for a total delay-time of 80 nsec (300mm) in the
measurement of ω_{21} in Fig. 4a and the high time resolution obtained
for the measurement of ω_{20} in Fig. 4b prove the high quality possible
with quantum beat measurements.

However, such beautiful curves are of no great use unless a
highly accurate time-scale calibration (that is, an accurate velocity determination) is performed during the beat measurement. Many of
the experiments performed so far on zero-field quantum beats (Table
2) have suffered considerably in accuracy, because of the lack of
such information. Therefore it has become an absolute necessity to

		Ref.
H I	$2p^2P, 3d^2D, 4p^2P, 4d^2S$-fs	12-16
He^{3}I	$3p^3P$-hfs	17
He^{4}I	$3\text{-}8p^3P, 3\text{-}6d^3D$-fs	11,18-20
Li^{6}II	$2\text{-}5p^3P$-fs-hfs	21
Li^{7}II	$2,4,5p^3P$-fs-hfs	21
Be^{9}II	$4,5d^2D, 4,5f^2F$-fs	22
Be^{9}III	$2p^3P$-hfs	23
N^{14}III	$2s2p3p^4D, 2s2p3d^4F$-hfs	24

Table 2: Zero-field quantum beat measurements performed so far on fast beams with foil excitation.

use high-precision velocity analyzers after the foil. Such an instrument has been installed in Berlin and has been calibrated to 0.1% by using the He $3p^3P$ ω_{21} fs beats as an atomic clock[25] which itself is known with an accuracy of $2 \cdot 10^{-4}$.

Equipped with this instrument (Fig. 2b) W. Wittmann was able to perform a quantum beat measurement on the Li^{7}II $2p^3P$ hfs which is presented here as the state-of-the-art experiment in Fig. 5. The upper part shows the time-resolved intensity distribution of the $2p^3P\text{-}2s^3S$ λ 5485 Å transition obtained from a 335 keV Li$^+$ beam over a total length of 8 nsec. The step width was about 16 psec and the nonoscillating intensity I_0 has been subtracted. The beat amplitudes reach a maximum of ± 4% at a statistical uncertainty of ± 0.3% for the data points. The quality of this measurement is best indicated by its Fourier transform in the lower part of Fig. 5. The beat frequencies were extracted with uncertainties of less than ± 0.3% except for the lowest frequency with ± 0.55%. A recent theoretical calculation by Jette et al.[26] is compared with these results and shows excellent agreement.

From this experiment one might foresee a bright future for the general use of quantum-beat measurements after foil excitation. Unfortunately, however, the alignment seems to decrease at typical beam energies of a few hundred keV with increasing mass of the beam atoms and with increasing charge of the excited atoms[11-23]. Until systematic alignment studies are done for the clarification of this point, this imposes a serious limitation on the use of the zero-field quantum-beat method with foil excitation.

In order to solve this problem we have started a series of experiments with the ultimate goal of replacing the foil by a polarized laser beam which could yield maximum alignment.

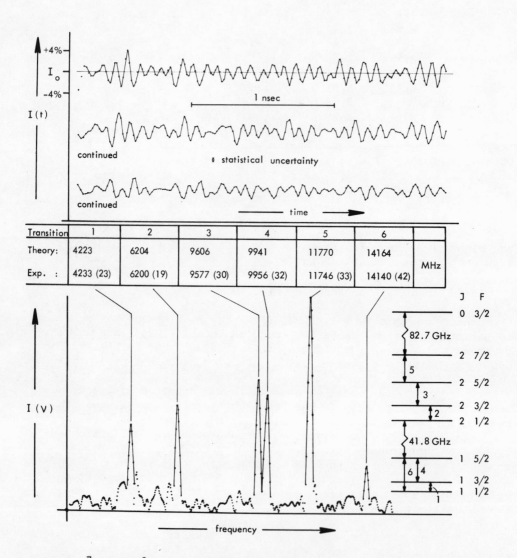

Fig. 5: ^{7}LiII–2p^{3}P (fs) hfs measurement with the zero-field quantum beat method. The upper part shows the twice continued beat pattern in the 2p^{3}P–2s^{3}S λ 5485 Å emission after subtraction of the nonoscillating intensity I_0. The line connects the data points only. The lower part shows the Fourier transform of the beat pattern. The zero point of the frequency has been suppressed. In the center six experimentally determined frequencies as indicated in the level diagram are compared with the theory of ref. 26.

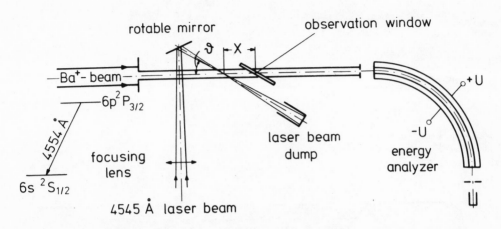

Fig. 6: Schematic experimental set up used for laser excitation of fast Ba$^+$ beams with a fixed wavelength argon laser via Doppler tuning. The observation window is moved along the beam axis relative to the fixed laser excitation region.

LASER EXCITATION OF FAST BEAMS

Using an argon-ion laser line at 4545 Å with a power of 20 mW it was found possible[27,28] to excite the BaII D_2 resonance line at 4554 Å with a probability per beam atom of the order of 10^{-2} to 10^{-3}. In this experiment the laser beam had to intersect the ion beam at a particular angle ϑ, adjusted by a rotatable mirror for fine tuning of the resonance via the Doppler effect. In order to achieve good time resolution the laser beam is focused by a lens onto the ion beam and the observation window is also tilted at the same angle ϑ (Fig. 6). A time-resolved measurement is then performed exactly as in BFS by measuring the emitted light intensity from the ion beam as a function of the distance x from the excitation region.

Remembering now the excellent properties of fast beam spectroscopy from Table 1 it becomes obvious that with such selective laser excitation (which eliminates completely the cascade problem of BFS) an unprecedented opportunity has been obtained for measuring precision lifetimes.

By choosing the isotope Ba-138 with nuclear spin I=0 we can indeed push the accuracy of lifetime measurements in atomic physics beyond the magic barrier of ±1%. However, the ultimate limit of what can be achieved is not clear yet if one compares for instance two independent sets of measurements in fall 1973 and spring 1974 (Fig. 7) which deviate by 0.5%. It may be too early to give definite error bars until the total analysis of systematic errors is completed.

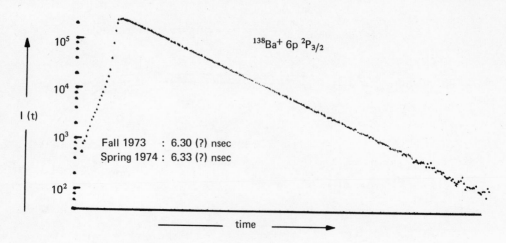

Fig. 7: Typical intensity decay curve of the ^{138}BaII $6p^2P_{3/2}$-$6s^2S_{1/2}$ λ 4554 Å emission after laser excitation. The averaged results of $\tau(6p^2P_{3/2})$ from two independent sets of measurements are compared.

By using polarized laser light with the $\vec{E}$-vector perpendicular to the image plane the ultimate goal of observing quantum beats was easily achieved by choosing the isotope Ba-137 with nuclear spin I=3/2. A great surprise appeared during the Doppler-tuning of the resonance (Fig. 8). As compared to the isotope Ba-138 where an estimated resonance linewidth of about 5 GHz was found a much wider resonance was obtained with the isotope Ba-137. The explanation is the large ground state hfs-splitting of 8.04 GHz[29] which was partly resolved. Consequently one has selective excitation of either the upper three or the lower three hfs-components of the excited level and expects therefore only beats among the upper three or the lower three hfs-components depending on which angular setting (a or b in Fig. 8) of the rotatable mirror has been chosen. (Very similar conditions have been obtained by Haroche et al.[30] in a pulsed laser experiment on CsI with less electronic time re-solution.) This is exactly what has been observed in the experiment; in the upper part of Fig. 9 two low-frequency beats from the lower three hfs-components appear and in the lower part one low frequency plus two high frequency beats appear from the upper three hfs-com-ponents, each with their own Fourier transforms. The results ob-tained from this measurement is preliminary and gives hfs-coupling constants A and B with accuracies which will be certainly refined by a detailed analysis of the data but which already compare favor-able with an earlier Fabry-Perot measurement.

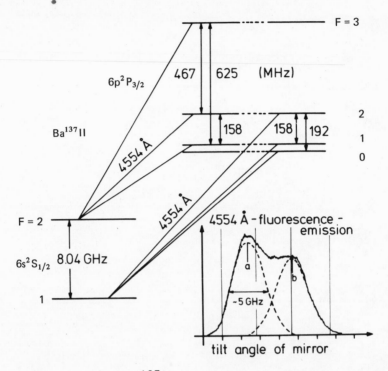

Fig. 8: Level scheme of Ba^{137}II indicating the selective laser excitation from F=2 groundstate (a) to the upper three and from F=1 groundstate (b) to the lower three hfs components of the excited 6p^{2}P$_{3/2}$ state. The hfs groundstate selection is achieved by the atomic fluorescence emission response.

The selective excitation from the two groundstate hfs-components raised the question as to whether high resolution fluorescence spectroscopy is possible on fast beams. It is easy to derive an approximative formula for the expected linewidth (Fig. 10); one can estimate from the laser and atomic beam properties that resolutions $\lambda/\Delta\lambda \gtrsim 10^5$ can be obtained, in particular when the focusing lens in Fig. 6 is eliminated which makes the time uncertainty spread (last term) negligible. In fact by using the argon laser in single mode a linewidth of 1.8 GHz was expected. In the experiment it was possible to clearly resolve the groundstate hfs splitting of Ba^{137}II with a linewidth of 1.64 GHz which corresponds to a resolution of $\lambda/\Delta\lambda = 2.5\cdot10^6$. This width is not comparable to those obtained by saturation spectroscopy. However, one should realize that the present technique can be readily applied to ionic beams and that it offers directly a Doppler tuning range of $\pm \sim 10$ Å.

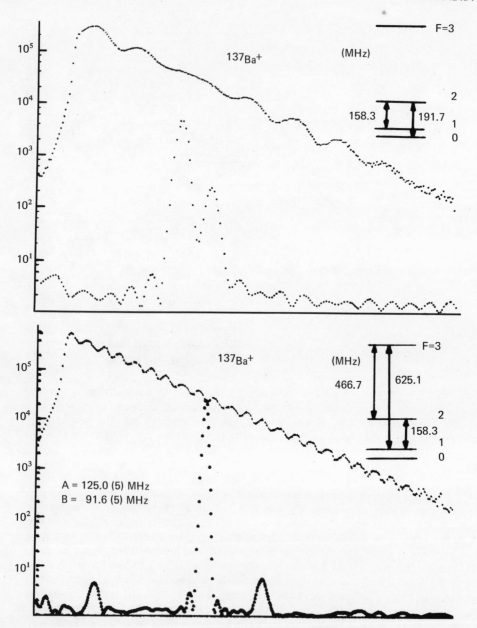

Fig. 9: Observed quantum beats in the λ 4554 Å of Ba[137]II and their
Fourier transforms. The upper part was obtained with setting (b) and
the lower part with setting (a) (Fig. 8) of the Doppler tuning angles.
The inserted level schemes indicate the measured beat frequencies.
Please note that time and frequency scales are different in the top
bottom part.

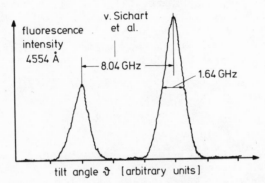

$$\Delta v = v_0 \cdot \beta \left[\cos \vartheta \, \frac{\Delta v}{v} + \sin \vartheta \, (\Delta \varphi + \Delta \vartheta) \right] + \frac{v}{2 \pi \, \Delta x}$$

velocity straggling $\qquad$ beam div. $\qquad$ Laser div.
$\sim 5 \cdot 10^{-4}$ $\qquad$ $\sim 10^{-3}$ $\qquad$ $\sim 5 \cdot 10^{-4}$

$\Delta x = 2 \, mm$; $\vartheta = 31°$: $\boxed{\Delta v \approx 1.8 \; GHz}$

fluorescence intensity 4554 Å

v. Sichart et al.

— 8.04 GHz —

1.64 GHz

tilt angle ϑ [arbitrary units]

Fig. 10: Formula for the linewidth of the atomic beam fluorescence emission response. v_0 is the emission frequency, $\beta = v/c$, v is the beam velocity, ϑ is the angle of intersection of laser and atomic beam, and Δx is the length of the intersection region. With the numbers given a linewidth $\Delta v \approx 1.8$ GHz is expected for a 335 keV Ba^+ beam. The experimental result is shown in the lower part.

CONCLUSION

        After presenting these four types of experiments the present situation of fast-beam spectroscopy can be summarized as follows:

(i)   The zero-field quantum-beat technique after foil or laser excitation is approaching a relative accuracy of 10^{-3} and can be used for the direct measurement of zero-field splittings from $\sim$ 30 MHz to $\sim$ 10 GHz. It therefore represents at least an alternative to other high-resolution techniques and offers unique features for the study of ionic systems.

(ii)   Selective laser excitation in combination with the unprecedented experimental conditions of fast beam spectroscopy is expected to set new standards in precision lifetime measurements ($\Delta \tau / \tau < 1\%$) of excited states.

(iii)   High-resolution laser fluorescence spectroscopy on fast beams can be further improved and will then offer a possibility for measuring level splittings which cannot be obtained or resolved by quantum beats. It is expected, therefore, that this method will become valuable for the study of particular cases. However, one has to realize that standard spectroscopic techniques compete with its present resolution.

ACKNOWLEDGEMENT

This progress report is mainly based on the work by W. Wittmann, A. Gaupp, L. Henke , M. Kraus, J. Plöhn, M. Gaillard and J. Macek. Permission for using their material prior to publication is gratefully acknowledged.

This work has been supported by the "Sonderforschungsbereich 161 der Deutschen Forschungsgemeinschaft".

REFERENCES

1   S. Bashkin, Atomic Physics II, edited by G.K. Woodgate and P.G. Sandars (Plenum , New York, 1871) p. 43.
2   S. Bashkin et al., Phys. Rev. Lett. 15 284 (1965)
3   L. Kay, Phys. Letters 5 36 (1963)
    S. Bashkin, Nucl. Instr. Meth. 28 88 (1964)
    Beam-Foil Spectroscopy, edited by S. Bashkin (Gordon and Breach, New York, 1968)
    Beam-Foil Spectroscopy, Nucl. Instr. Meth. 90 (1970)
    Beam-Foil Spectroscopy, Nucl. Instr. Meth. 110 (1973).
4   H.J. Andrä, Physics Scripta 9 257 (1974), and references quoted therein.
5   P.A. Franken, Phys. Rev. 121 508 (1961)
6   E.B. Aleksandrov, Opt. Spectry. (USSR) 14 232 (1963) and 16 209 (1964)
7   O. Nedelec et al., Compt. Rend. 257 3110 (1963)
    J.N. Dodd et al., Proc. Phys. Soc. 84 176 (1964)
    A. Corney and G. Series, Proc. Phys. Soc. 88 207 (1964)
    T. Hadeishi and W.A. Nierenberg, Phys. Rev. Lett. 14 891 (1965)
    J.N. Dodd et al., Proc. Phys. Soc. 92 497 (1967)
8   J. Macek, Phys. Rev. Lett. 23 1 (1969)
9   J. Macek and D.H. Jaecks, Phys. Rev. A4 2288 (1971)
10 H.G. Berry et al., J. de Phys. 33 947 (1972)
    D.G. Ellis, J. Opt. Soc. Am. 63 1232 (1973)
    U. Fano and J. Macek, Rev. Mod. Phys. 45 553 (1973)
    J. Bosse and H. Gabriel, Z. Phys. 266 283 (1974)
    See also appendix of ref. 4.
11 W. Wittmann et al., Z. Phys. 257 279 (1972)
12 H.J. Andrä, Phys. Rev. Lett. 25 325 (1970)
13 H.J. Andrä, Nucl. Instr. Meth. 90 343 (1970)
14 D.J. Lynch et al., Phys. Rev. Lett. 26 1211 (1971)

15 D.J. Burns and W.H. Hancock, Phys. Rev. Lett. 27 370 (1971)
16 P. Dobberstein et al., Z. Phys. 257 272 (1972)
17 K. Tillmann et al., Phys. Rev. Lett. 30 155 (1973)
18 D.J. Burns and W.H. Hancock, J. Opt. Soc. Am. 63 241 (1973)
19 H.G. Berry and J.L. Subtil, Phys. Rev. Lett. 27 1103 (1971)
20 H.J. Andrä et al., Nucl. Instr. Meth. 110 301 (1973)
21 H.G. Berry et al., Phys. Rev. A7 1609 (1973)
22 O. Poulsen and J.L. Subtil, Phys. Rev. A8 1181 (1973)
23 O. Poulsen and J.L. Subtil, J. Phys. B7 31 (1974)
24 P. Ceyzeriat et al., to be published
25 J. Lifsitz and R.H. Sands, Bull. Am. Phys. Soc. 10 1214 (1965)
26 A.N. Jette et al., Phys. Rev. A9 2337 (1974)
27 H.J. Andrä et al., Nucl. Instr. Meth. 110 453 (1973)
28 H.J. Andrä et al., Phys. Rev. Lett. 31 501 (1973)
29 F. v. Sichart et al., Z. Phys. 236 97 (1970)
30 S. Haroche et al., Phys. Rev. Lett. 30 948 (1973)

CASCADE AND STEPWISE LASER SPECTROSCOPY OF ALKALI ATOMS

W. Happer

Columbia University

New York, New York U.S.A.

A little more than one hundred years ago atomic spectroscopy began at Heidelberg with the work of Bunsen and Kirchnoff on the beautiful colors of flames seeded with salts of sodium and potassium. Intrigued by certain peculiar lines in the spectra of Dürkheim mineral water, Bunsen evaporated forty tons of water to isolate the first small samples of rubidium and cesium salts. Remarkably, the properties of those same alkali atoms continue to fascinate us today, and I am going to talk about the surprising results of some recent experimental investigations of the fine structure and hyperfine structure of alkali atoms. Most of this work has been done by two talented young men, Rajendra Gupta of Columbia University and Sune Svanberg of Göteborg University. They have been assisted by a number of able students whom I will mention later.

We may think of an alkali atom as a single valence electron revolving about a positively charged core with an inert gas configuration. The optical spectra, which are illustrated in Fig. 1, have been known for many years. The high angular momentum states, which correspond to non-penetrating electron orbits, have energies which are very nearly the same as those of an electron in a hydrogen atom, but the penetrating, low-angular-momentum states are more tightly bound than the corresponding states of hydrogen because they are less well shielded by the core electrons from the attractive force of the nucleus.

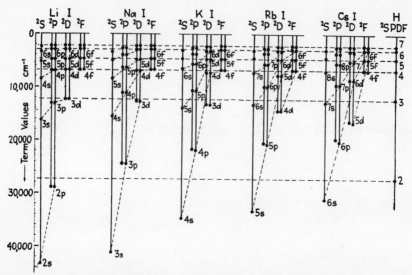

Fig. 1. The low-lying energy levels of alkali atoms
(from H. E. White, <u>Introduction</u> <u>to</u> <u>Atomic</u> <u>Spectra</u>
McGraw Hill, 1934).

Each non-S state is split into two fine-structure
components due to the interaction of the magnetic moment of the moving valence electron with the electric
fields generated by the nucleus and the core electrons.
A simple calculation shows that the fine structure interval should be proportional to the 3/2 power of the
binding energy of the valence electron. As one can see
from Fig. 2, the P-state and D-state fine structure intervals[1] follow the 3/2 power law very nicely but the
F state fine structure intervals do not follow this law
and in fact do not even have the correct sign. In the
lighter alkali atoms, rubidium potassium and sodium,
the D-state fine structure intervals also fail to follow
the 3/2-power law and many of the fine structure intervals are inverted. These fine structure anomalies have
been known for a very long time and they also occur for
alkali-like positive ions.[2] Although attempts to
explain the anomalies with configuration interaction
calculations seem to have been more or less successful[3],
they have been limited to very elaborate and complex
calculations for one or two states. It would be very
helpful to have a less cumbersome theory which could
account for the striking systematic variation of the
fine structure from state to state within the same atom.

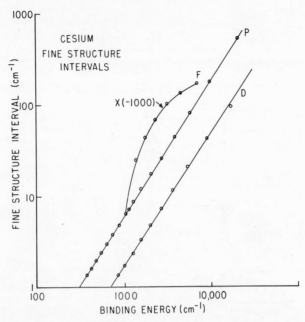

Fig. 2. Fine structure intervals of cesium (from ref.1).

Atomic fine structure is due to the interaction of the magnetic moments of the valence electrons with the electric and magnetic fields within the atom. We are also familiar with another interaction, atomic hyper-structure, which involves the magnetic moments of the electrons. Thus, it is natural to ask what sort of hyperfine interactions exist in the states with anom-alous fine structure? Had you asked this question three years ago you would have received no answer since essen-tially nothing was known. The situation in 1971 is illustrated in Fig. 3. Although the figure represents the state of knowledge for cesium, the situation was qualitatively the same for the other alkali atoms. The hyperfine structure of the ground state, the lowest S state has been known for many years with exquisite pre-cision as a result of atomic beam measurements.[4] The excited P states had been measured mainly here in Heidelberg by H. Bucka, G. zu Putlitz, A. Schenk and H. A. Schüssler[5]. No excited S state hyperfine structu-res were known and the results of only two rather rough measurements of D state hyperfine structures had been reported.[6,7] One does not have to seek far for the reasons that Fig. 3 looks as it does. With few excep-tions the hyperfine structures of the alkali atom are too small to be observed by conventional techniques of

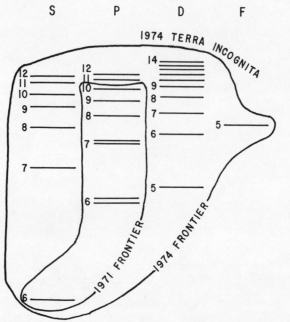

Fig. 3. The expansion in knowledge of the hyperfine
structure of cesium due to the development of cascade
and stepwise-laser spectroscopy.

optical spectroscopy and the only really suitable method
for measuring excited state hyperfine structure is the
optical double resonance technique invented by Kastler
and Brossel[8] and perfected by Kopfermann's group here at
Heidelberg.[5] However, since in conventional optical
double resonance techniques one excites ground state
atoms with resonance light, and one observes the fluorescence of the optically excited state it is clear that
only the P states of the alkali atoms can be studied.
The selection rules for electric dipole transitions
allow the ground-state alkali atoms to be optically excited into the P states but not into S states, D states
or any other excited states.

When S. Chang, R. Gupta and I became interested in
extending our knowledge of excited-state hyperfine
structure several years ago, we knew that Pebay Peyroula[7]
and his associates had already tried to get around the
electric dipole selection rules by exciting alkali atoms
with a beam of electrons. This allowed them to produce
aligned D state atoms, but the results of their work
seemed to indicate that electron excitation of the alkali atoms would be quite difficult. We ourselves

tried for some time to develop ways to use electron excitation of alkali atoms but with only marginal success.
We also considered stepwise excitation, since Smith and
Eck[9] had demonstrated that it was possible to use stepwise excitation to observe fine structure level crossings in the D states of sodium and lithium. However,
because of the limitations in intensity of conventional
lamps, stepwise excitation seemed to be a useful technique for only a few carefully chosen excited states.
The rapid advances in dye laser technology were soon to
change the situation and I shall discuss the technique
of stepwise excitation with cw dye lasers presently.

On surveying the impressive achievements of our
European colleagues in measuring the P-state hyperfine
structures, we noticed that many of the high lying P
states had substantial branching ratios to neighboring
S and D states, as is illustrated in Fig. 4 for rubidium. The S and D states undergo further radiative
decays and the fluorescence has wavelengths which are
suitable for detection with sensitive photomultiplier
tubes. Furthermore, when an atom undergoes spontaneous

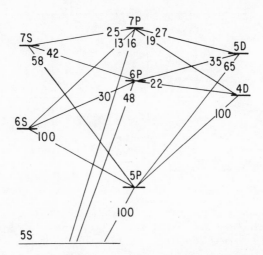

Fig. 4. Branching ratios (%) for spontaneous decay of the
excited states of rubidium. There is substantial branching of the higher P states to neighboring S and D states.

decay, a surprisingly large amount of polarization is
retained by the atom as is illustrated in Fig. 5.   It
may seem paradoxical that a random process like spontaneous decay can have so little effect on the polarization of an atomic state.   The reason that so much
polarization is retained is that the electric dipole
selection rules allow changes in the azimuthal quantum
number by only one unit or none ($\Delta m = \pm 1, 0$), and, consequently, many stages of cascading are required for
atoms which are originally located in a single Zeeman
sublevel to diffuse throughout a Zeeman multiplet.

Thus, by making a very minor change in the experimental arrangement of conventional optical double
resonance experiments, that is by looking at photons
emitted after one or more cascades from the initially
excited P state, it is possible to observe many low-
lying S and D states of the alkali atoms.   A typical
experimental arrangement[10] is shown in Fig. 6.

An important practical advantage of cascade experiments is that the problem of instrumentally scattered

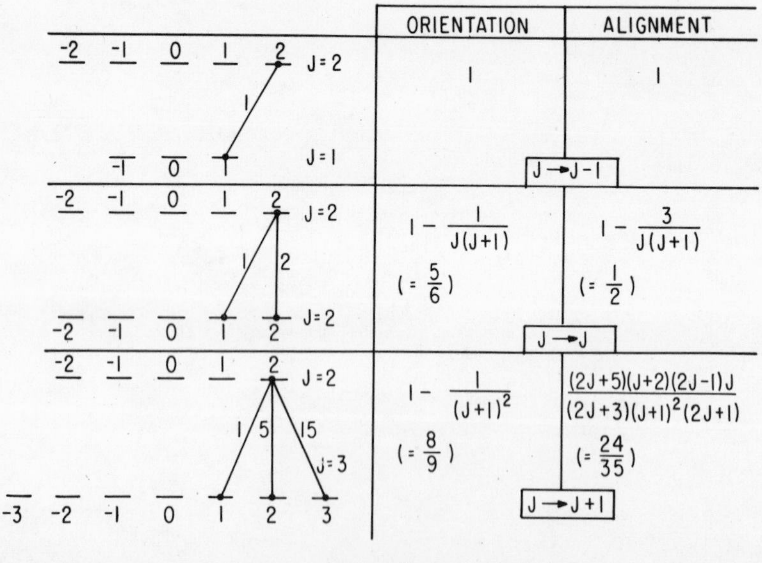

Fig. 5. Polarization, as a fraction of the maximum possible polarization, of a state produced by spontaneous
decay of a fully polarized excited state. Substantial
amounts of both orientation and alignment are transferred by spontaneous decay. Numerical values for the
fractional polarization are indicated for decay of a
state with angular momentum J = 2.

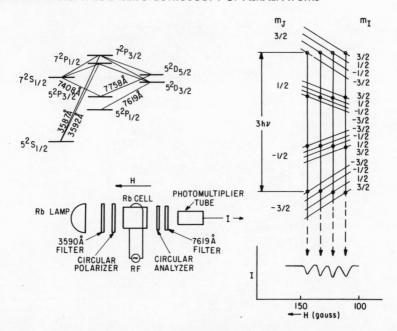

Fig. 6. Schematic diagram of a cascade radiofrequency spectroscopy experiment. The atoms are excited into the 7P state by circularly polarized light from a conventional lamp. Fluorescent light from the $7S_{1/2}$, $5D_{3/2}$ or $5D_{5/2}$ state is detected and sets of $(2I + 1)$ magnetic resonance peaks are observed at high magnetic fields. (from ref. 10)

lamp light can often be virtually eliminated by a suitable choice of optical filters. This is because the wavelength of the exciting light is necessarily much shorter than the wavelength of the detected light.

There are certain disadvantages to cascade experiments, too. For instance, it is very difficult to observe high-field level crossing signals, although zero-field level crossing signals (the Hanle effect) are easily observed. An example of this is illustrated in Fig. 7, which shows the high-field level crossing of the $6D_{3/2}$ state of cesium as observed by C. Tai.[11] Instead of the high-field level crossing signal being a factor of ten or so smaller than the zero-field signal, as would be the case if no cascades were involved, the high-field signal is more than two orders of magnitude smaller than the zero-field signal, and it can be detected only with long integration times. Another pecularity of the cascade level crossing signal of Fig. 7 is the fact that the zero-field level crossing signal is a

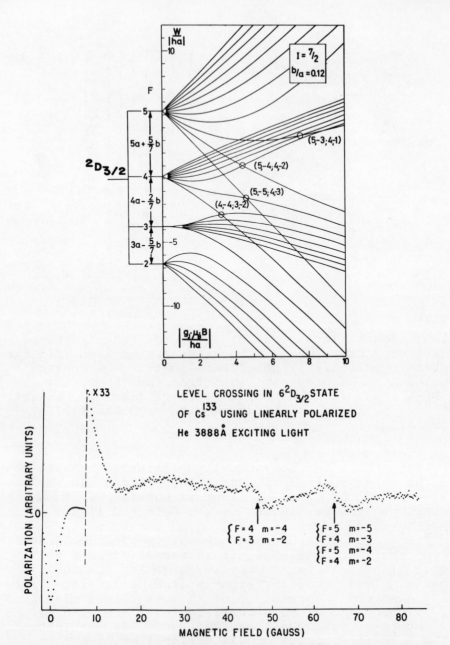

Fig. 7. Zero-field and high-field level crossing signals in a cascade experiment. The high-field level
crossing signals are much smaller than they would be in
the absence of a cascading step, and they are antisymmetric instead of symmetric like the zero-field
signal. (from ref. 11)

symmetric curve while the high-field level crossings
are anti-symmetric. If no cascading were present both
high-field and low-field resonances would have the same
symmetry (cf. Fig. 16).

Physically, the degradation of high-field level
crossing signals in cascade experiments comes about because the signals are associated with transverse atomic
polarization. Of course, transverse polarization will
usually rotate about an external magnetic field, and
if the rotation rate exceeds the natural decay rate the
transverse polarization will average to zero, as Hanle[12]
showed many years ago. When energy levels cross, certain
components of the transverse polarization do not rotate
and therefore do not average to zero. However if the
transverse polarization is generated by exciting the
atoms with transversely polarized light, and if one or
more stages of cascading occur before the fluorescence
is observed then it is necessary that level crossings
occur at the same magnetic field (to within the natural
line widths) for all atomic states through which the
atom cascades before its fluorescence is detected.
Although this automatically happens at zero-field, it
is a very rare occurance at high-fields.

It is interesting to note that anti-levelcrossing
signals are quite compatible with cascade experiments.
This is because anti-crossing signals are associated
with changes in longitudinal polarization. Longitudinal
polarization, in contrast to transverse polarization,
is transferred with fairly high efficiency from state
to state by cascading (cf. Fig. 5), even at high magnetic fields.

One important type of anti-crossing always occurs
near zero magnetic field, where hyperfine sublevels of
the same azimuthal quantum number m repel each other
due to the hyperfine interactions. For instance, as
shown on Fig. 8, we may think of the Breit–Rabi diagram
of a $^2S_{1/2}$ state as a collection of atomic energy levels
which anti-cross in the neighborhood of zero magnetic
field. If one observes the fluorescent light emitted
by excited S state atoms which have been excited by
cascades from optically excited P state atoms, it will
be found that the circular polarization of the fluorescence increases as the magnetic field increases.
Some typical experimental curves obtained by Gupta and
Chang[13] are shown in Fig. 9. The "width" of the curves
in Fig. 9 is very nearly the same as the width of the

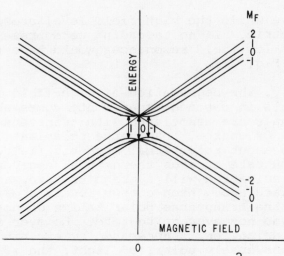

Fig. 8. The Breit-Rabi diagram for a $^2S_{1/2}$ state with nuclear spin I = 3/2. The diagram may also be thought of as a zero-field level crossing which has been split into crossings and anti-crossings by the magnetic hyperfine interaction. More complicated zero-field anti-crossings can be seen in the energy level diagram of Fig. 7.

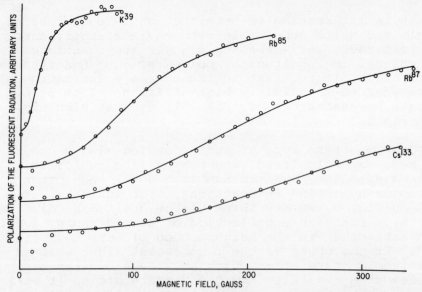

Fig. 9. Decoupling effects on the circular polarization of the fluorescence from the second excited S state of alkali atoms. The signals can be thought of as anti-crossing signals associated with Breit-Rabi diagrams like those of Fig. 8. (from ref. 13)

anti-crossing region of the appropriate Breit-Rabi diagram (cf. Fig. 8) for the state.

We prefer to think of these zero-field anti-crossing signals as a result of the decoupling of the electronic spin J from the nuclear spin I (cf. Fig. 10)
since this allows one to form a satisfying physical
picture of the process in terms of the vector model of
the coupling of angular momenta in an atom.

Although they are relatively imprecise, decoupling
experiments do have a number of important advantages
over more precise cascade magnetic resonance experiments.
A not inconsiderable advantage for short lived states
is that large radiofrequency fields are not required.
The signals are substantially larger for a cascade decoupling experiment and there is no search problem when
one is investigating the structure of a completely un-

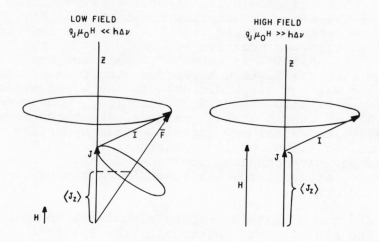

Fig. 10. The vector model of decoupling. The excited
atoms are produced with a polarized electron and an unpolarized nucleus. The circular polarization of the
fluorescence is due to the electron polarization. In
the absence of a magnetic field much of the electron
polarization can be transferred to the nucleus before
the atom decays. A large magnetic field can decouple
the electron from the nucleus and thereby increase the
electron polarization and the polarization of the fluorescent light. (from ref. 13)

known state. Once approximate values for the hyperfine
structure have been obtained from cascade decoupling
experiments, one can search over a relatively narrow
range for radiofrequency resonances, which determine
the hyperfine structure to much better precision.

A second important advantage of decoupling experiments is that they allow one to determine the signs of
the hyperfine coupling constants. When we completed
the first precision measurement of the magnetic dipole
coupling constant for a D state of rubidium (cf. Fig.6),
we wrote a small paper to describe our result and we
sent a copy to Dr. Allen Lurio for his comments. Lurio
pointed out to us that we had not really measured the
sign of the coupling constant; our radiofrequency resonances were just as consistent for a negative as for a
positive coupling constant. Since any simple theory
would predict a positive coupling constant, we at first
considered this to be an amusing academic point, but we
nevertheless began to wonder if a way might not be found
to prove that the sign was positive as expected. It
turned out that the decoupling signals were sensitive to
the sign of the magnetic coupling constant (cf. Fig. 11),
and to our surprise we found that the coupling constant
of the $D_{5/2}$ state was negative although that of the
$D_{3/2}$ state was positive as expected. This was the first
indication that the D-state hyperfine structure of the
alkali atoms was not as predictable as one might expect
for an atom with only one valence electron. Subsequent
measurements of many D state hyperfine structures have
so far failed to reveal any normal D-state hyperfine
structures. Negative coupling constants for the $D_{5/2}$
states and positive coupling constants for $D_{3/2}$ states
are the rule, at least for the lower lying states of
rubidium and cesium.

Why are the D state hyperfine structures so anomalous in the alkali atom? After all, the D states
correspond to the less penetrating electron orbits and
it would seem reasonable to expect the properties of
such states to be more normal than the properties of the
P states, which correspond to more penetrating orbits.
A similar anomaly was discovered some time ago by Brog,
Wieder and Eck[17] in the P states of lithium, where the
A values for the $P_{1/2}$ states are positive as expected
while those of the $P_{3/2}$ states are negative. These
anomalies in lithium were attributed to polarization
of the core of the lithium atom by the valence electron.[18]
Is core polarization also responsible for the anomalous

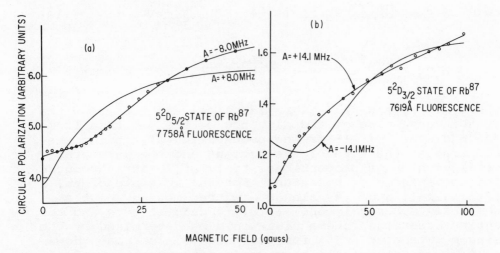

Fig. 11. Decoupling signals which prove that the sign of the $D_{3/2}$ A value is positive while the sign of the $D_{5/2}$ A value is negative in the 5D state of rubidium. (from ref. 10)

D-state hyperfine structure? Mr. K. Liao of Columbia University has recently provided the answer to this question; core polarization is responsible for the anomalous D state hyperfine structure but in contrast to the situation for the P states of lithium, the core polarization is markedly anisotropic. From considerations of symmetry one can show that the most general magnetic hyperfine interaction for a ^{2}D state is

$$H = a_\ell \ \vec{L} \cdot \vec{I} + a_c \ \vec{S} \cdot \vec{I} + a_d \left\{ \vec{S} \cdot \vec{I} - 3 \frac{\vec{S} \cdot \vec{LL} \cdot \vec{I} + \vec{I} \cdot \vec{LL} \cdot \vec{S}}{2L(L+1)} \right\} \quad (1)$$

where a_ℓ, a_c and a_d are independent parameters. For a one electron atom the parameters are

$$a_\ell = \frac{2\mu_B \mu_I}{I} \ \langle r^{-3} \rangle_\ell \quad (2)$$

$$a_c = \frac{\mu_B \mu_I}{I} \ g_s \ \frac{8\pi}{3} \ |\psi(0)|^2 = \frac{2\mu_B \mu_I}{I} \langle r^{-3} \rangle_c \qquad (3)$$

$$a_d = \frac{4\mu_B \mu_I}{I} \langle r^{-3} \rangle_d \ \frac{L(L+1)}{(2L-1)(2L+3)} \qquad (4)$$

where $\langle r^{-3} \rangle_\ell$ and $\langle r^{-3} \rangle_d$ are identical and represent the
mean inverse cube of the distance of the electron from
the nucleus, $|\psi(0)|^2$ is the probability of finding the
electron at the site of the nucleus and the parameter
$\langle r^{-3} \rangle_c$, which is proportional to $|\psi(0)|^2$, is introduced
for convenience of notation. For the ^2D state of an
ideal alkali atom (a valence electron moving about a
perfectly spherical core) $|\psi(0)|^2$ would be zero and the
entire hyperfine interaction would be determined by the
single parameter $\langle r^{-3} \rangle$ for the valence electron. However, in a real alkali atom we do not expect equations
(2), (3) and (4) to be correct since various configuration interactions will allow the core electrons to
participate in the hyperfine interaction. In fact it
is already clear from our experiments that equation (1)
cannot be correct if $\langle r^{-3} \rangle_c$ is zero and $\langle r^{-3} \rangle_\ell$ and
$\langle r^{-3} \rangle_d$ are identical, for this would imply that the A
values for both the $D_{3/2}$ and the $D_{5/2}$ states would be
of the same sign while we find experimentally that their
signs are opposite. Consequently we will no longer
think of $\langle r^{-3} \rangle_\ell$ and $\langle r^{-3} \rangle_d$ as representing, even approximately, the inverse cube of the distance of the electron
from the nucleus, but we shall regard $\langle r^{-3} \rangle_\ell$, $\langle r^{-3} \rangle_c$
and $\langle r^{-3} \rangle_d$ as three convenient parameters [19] which
determine the hyperfine Hamiltonian (1).

Unfortunately, we are able to measure only two
independent quantities at low magnetic fields, the
A-values for the $^2D_{3/2}$ and $^2D_{5/2}$ states. This is not
enough to determine the three independent parameters of
the magnetic hyperfine interaction. Sufficient additional information to completely determine the hyperfine
interaction can be obtained by making measurements at
high magnetic fields when L and S no longer couple to
form a total resultant angular momentum J. The way
these measurements were obtained for the 4^2D state of
rubidium is illustrated in Fig. 12. Since the 4D state
lies so low in energy its fluorescence lies too far in
the infrared to be detected directly and it is necessary
to monitor the D state by observing the 7947 Å and 7800 Å
resonance lines which are emitted after the D state decays
to the 5P state. This problem is characteristic of all

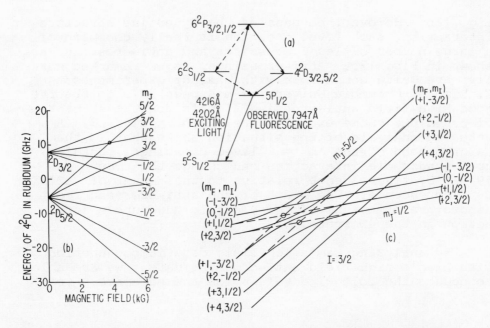

Fig. 12.   The states involved in a high-field cascade anti-level-crossing experiment.   Atoms are excited by polarized 4216 Å and 4202 Å resonance radiation and the intensity of the 7947 Å fluorescence is observed. Changes of the fluorescent intensity occur for fields in the neighborhood of certain fine structure level crossings.   However, anti-level-crossings, induced by the hyperfine interaction, are responsible for the observed signals.   Some typical anti-level-crossing signals are shown in Fig. 13.   (from ref. 20)

of the lowest lying S and D states of the alkali atoms. L. Lam of Columbia University has developed a double cascade method to measure these states.[10]   The fine structure of the 4D state is inverted and the fine structure interval is .44cm^{-1}.   Consequently a field of a few thousand gauss is sufficient to decouple L and S. However, radiofrequency spectroscopy for a short-lived excited state at a few thousand gauss is difficult since it involves high-power microwave fields.   Even if you can afford to buy the microwave sources, it is not easy to prevent the cells which contain the atoms from discharging in the intense electromagnetic field.   Since cascade excitation and detection are unavoidable at the present time for the 4D state, it is not possible to exploit the nice level crossings which are illustrated on

Fig. 12b. However, because of the hyperfine structure
interaction, each level crossing is really composed of
a large number of level crossings and anti-crossings as
shown in Fig. 12c. Anti-crossings, as we remarked earlier are quite compatible with cascade experiments.
K. Liao of Columbia University has been able to observe
good high-field anti-crossing signals for the 4D state
of rubidium. Some examples are shown in Fig. 13. A
detailed analysis of the situation shows that in the
neighborhood of the $\Delta m_J = 2$ level crossings, the electric quadrupole interaction can change the amount of
7947 Å D_1 fluorescence relative to the amount of 7800 Å
D_2 fluorescence. The width of the anti-crossing resonances are a measure of the electric quadrupole interaction in the 4D state.

By analyzing Lam's low-field data in conjunction
with Liao's high-field anti-crossing data[20] we are able
to deduce the following values for the parameters $\langle r^{-3} \rangle$

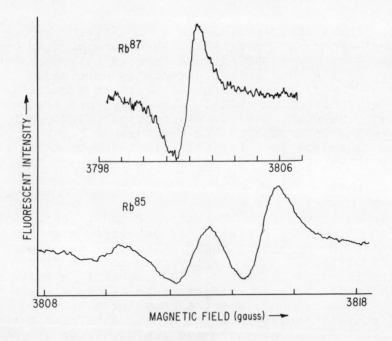

Fig. 13. Anti-level-crossing signals for the 4D state
of rubidium. The signal is proportional to the rate of
change of the 7947 Å fluorescent light with magnetic
field. The states involved in the experiment are shown
on Fig. 12. (from ref. 20)

(in units of $10^{24} cm^{-3}$)

$$\langle r^{-3} \rangle_\ell = +0.20(2)$$

$$\langle r^{-3} \rangle_c = -4.3(1)$$

$$\langle r^{-3} \rangle_d = -0.36(8)$$

Since $\langle r^{-3} \rangle_c$ is large and negative we see that there is a large negative core polarization for the 4D state of rubidium. This is reminiscent of the situation for the P states of lithium.[17,18,]   However, a new and surprising fact emerges from this data. The sign of the spin dipole interaction is reversed ($\langle r^{-3} \rangle_d$ is negative). This is the first instance of a negative spin dipole interaction that I know of, and it is interesting to note that very recent calculations by Lindgren[21] also lead to a negative value of $\langle r^{-3} \rangle_d$ for some of the higher D states of rubidium.

C. Tai[11] has demonstrated another interesting way in which one can determine the signs of low-field hyperfine coupling constants from cascade experiments. Under certain conditions it may be possible to produce nuclear polarization as well as electronic polarization in the initial step of excitation. An example is shown in Fig. 14. Because of the nuclear polarization, the (2I+1) high-field magnetic resonance signals will have different amplitudes. By comparing the (2I+1) high-field resonances of one state with the (2I+1) high-field resonances of another state which is involved in the same cascade process, one can determine if the magnetic coupling constants of the two states have the same signs (in which case the amplitudes of the (2I+1) resonances of each state exhibit the same pattern with increasing magnetic field) or if the coupling constants of the two states have opposite signs (in which case the patterns are mirror images of each other). From the data of Fig. 15 one can see that the coupling constant of the $5D_{3/2}$ state of cesium has the same sign as that of the $8P_{1/2}$ state, while the coupling constant for the $5D_{5/2}$ state has the opposite sign.

Although cascade excitation with conventional lamps is very useful for the low-lying states of the alkali atoms, it cannot be easily extended to higher excited states since the oscillator strengths of the higher resonance lines decrease very rapidly with increasing prin-

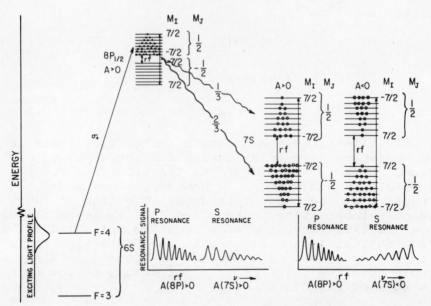

Fig. 14. Determination of the sign of magnetic dipole
hyperfine coupling constants A from intensity ratios.
In Cs^{133} it is possible to excite the $8P_{1/2}$ state with
the 3888Å line from a helium lamp. Because the helium
light can only excite atoms out of the F=4 sublevel of
the ground state, it can produce nuclear as well as
electronic polarization in the $8P_{1/2}$ state. The nuclear
polarization causes the high-field magnetic resonance
amplitudes to differ. States with similar intensity
ratios have the same signs for their A values. If the
intensity ratios are mirror images of each other the
signs are opposite. (from ref. 11)

cipal quantum number. In the summer of 1972 events con-
spired to allow us to reach much higher excited states.
Dr. Sune Svanberg arrived to spend a year with us at
Columbia University and he brought with him tremendous
enthusiasm for the potentialities of cw dye lasers in
atomic spectroscopy. Such lasers were just becoming
useful, commercially available experimental tools.
Since we had very nearly exhausted the easy cascade experiments for conventional lamps we decided to have a
go at stepwise excitation of alkali atoms with a conventional lamp and a dye laser. Some typical results of
Svanberg and Tsekeris are shown in Fig. 16. The alkali
atoms are excited from the ground state to the first

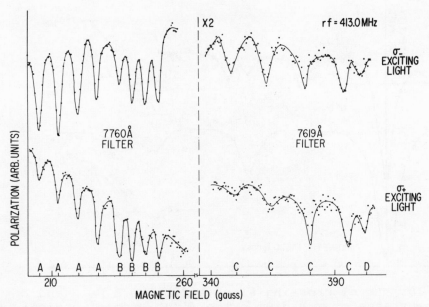

Fig. 15. Cascade magnetic resonance signals for the 7P and 5D states of rubidium; A, B, C and D refer to resonances in the $7P_{3/2}$, $5D_{5/2}$, $5D_{3/2}$ and $7P_{1/2}$ states respectively. From the symmetry of the intensity ratios it can be seen that the A values for the $7P_{3/2}$ and $5D_{3/2}$ state have the same signs while the A value for the $5D_{5/2}$ state has the opposite sign; cf. Fig. 14. The asymmetry was produced by exciting Rb^{87} with a Rb^{85} lamp. (from ref. 11)

excited state with the strong first resonance lines of a conventional lamp. The dye laser is used to excite the P-state atoms into a higher S or D state. Since no cascading steps intervene between the optical excitation of the D state and the detection of the fluorescent light, the $6D_{3/2}$ and $7D_{3/2}$-state level crossing signals can be detected without difficulty, in contrast to the situation illustrated in Fig. 7, where an intervening step of cascading seriously degrades the level crossing signal. Laser excitation can also be used in conjunction with cascading to extend its usefulness. For example, magnetic resonance signals from the F-state of cesium are shown in Fig. 17. Several steps of cascading in addition to stepwise laser excitation were used in this work.

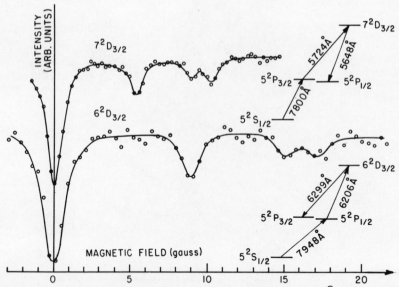

Fig. 16. Level crossing signals for the $6^2D_{3/2}$ and $7^2D_{3/2}$ states of rubidium. The first step of optical excitation is provided by photons from a conventional resonance lamp. The second step is induced with a cw dye laser. The energy level diagram for the $D_{3/2}$ states is similar to the one shown in Fig. 7. (from ref. 22)

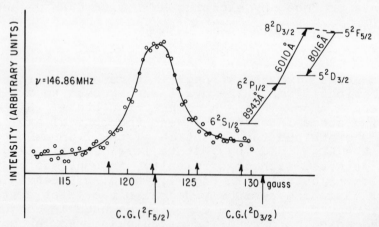

Fig. 17. Magnetic resonance signal in the 5F state of Cs^{133}. Stepwise excitation with a conventional resonance lamp and a cw dye laser, followed by one step of cascading, was used to produce the F states. (from ref. 22)

The actual experimental apparatus used by Svanberg
and Tsekeris[23] for stepwise laser excitation is sketched
in Fig. 18. The absorption cell temperature is adjusted
to maximize the number of P-state atoms and radiation
trapping of the resonance lines seems to increase the
effectiveness of the lamp. Because the number of P-state
atoms is still relatively low and the laser intensity is
very high it is difficult to suppress the instrumentally
scattered laser light enough to allow detection of the
fluorescence at the laser wave length, and it is often
advantageous to use fluorescent light at different wavelengths from the laser light, as is illustrated in Fig.
16 and Fig. 17, since optical filters may then be used
to greatly suppress the scattered laser light. However,
Svanberg has recently been able to observe fluorescence
at the laser wavelength from atoms optically excited in
an atomic beam.

When the laser is operating on a single mode, the
often-used approximation of white light excitation is
not valid for analyzing the experimental results. However, no special analysis is necessary when the level
crossing or radiofrequency resonances are well resolved.[25]

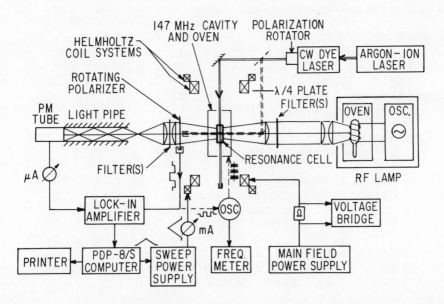

Fig. 18. Experimental arrangement for stepwise laser
excitation. (from ref. 23)

In any case the theory of narrow line excitation is already well known[24] and seems to be applicable to single
mode laser excitation. In multimode operation the laser
excitation is nearly the same as excitation by a powerful noisy conventional lamp. In all experiments reported so far the laser excitation rates have been well below the saturation values, and light shifts have been
negligible.

     Basically the laser experiments I have described
today have used the laser as a bright conventional lamp.
All of the elegant and powerful techniques of level-
crossing spectroscopy, radiofrequency spectroscopy and
related techniques have been retained. For convenience,
I shall refer to experiments of this type as perturbed
atomic fluorescence experiments, since their most characteristic feature is the detection of changes in the
intensity of atomic fluorescence which are brought about
by a perturbation of the environment (the external fields,
the atomic density, the foreign gas pressure, etc.) of
the excited atoms. Such experiments have an ancient
history which extends back at least as far as the work
of Hanle[12]. Although other laser techniques like saturated absorption spectroscopy[25] and two-photon excitation[26] have important potential for a limited class of
problems, it seems unlikely that they can seriously compete with perturbed atomic fluorescence experiments for
the investigation of the fine structure and hyperfine
structure of excited atomic states. For instance, of
the 54 previously unknown hyperfine structure parameters
which have been measured in the past two years for excited alkali atoms, all but one, the hyperfine splitting
of the 5S state of sodium[26] were measured by perturbed
atomic fluorescence techniques.

     The principal advantages of perturbed atomic fluo-
rescence methods over other methods are:

   1.   Small frequency intervals can be measured
        directly by inducing radiofrequency trans-
        itions between the corresponding atomic
        sublevels, by causing the sublevels to
        cross in an a precisely-known external field
        or by some similar method. It is not nec-
        essary to measure the difference between
        two optical frequencies.

   2.   The signals' amplitudes depend linearly on
        the light intensity. Saturation of the
        optical absorption line is neither necessary
        nor desirable.

3. There is negligible Doppler broadening
    of the resonances.

I would like to conclude this talk by summarizing
some of the results of recent studies of excited alkali
atoms. An overview of the contemporary experimental
situation can be gained from Fig. 19 through Fig. 23
where presently known magnetic dipole coupling constants
are plotted as a function of the binding energy of the
excited electron. Most of the measurements of a given
symmetry in the same atom can be connected by straight
lines with slopes of approximately 1.50. This shows that
the magnetic coupling constants of all states, except the
highly anomalous D states of the alkali atoms, are very
nearly proportional to the 3/2-power of the electron
binding energy. More precise values of the hyperfine
structure parameters are listed in Tables I through V.
At this writing, the number of known coupling constants
has been increased by approximately a factor of three
over what was known in 1971.

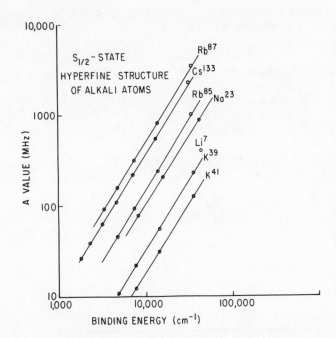

Fig. 19. A summary of the S-state magnetic dipole
coupling constants known in 1974. Note the remarkably
regular dependence of the A-values on the electron binding energy.

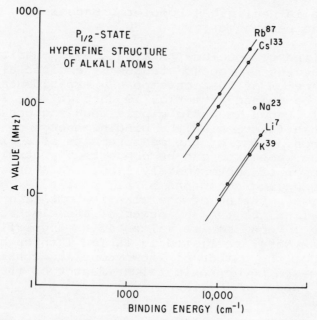

Fig. 20.   A summary of the $P_{1/2}$-state magnetic dipole
coupling constants known in 1974.

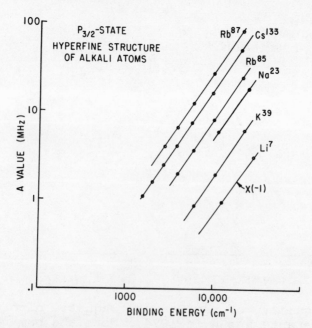

Fig. 21.   A summary of the $P_{3/2}$-state magnetic dipole
coupling constants known in 1974.

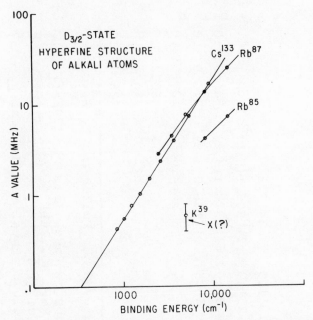

Fig. 22.   A summary of the $D_{3/2}$-state magnetic dipole
coupling constants known in 1974.

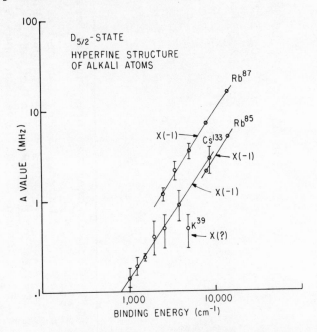

Fig. 23.   A summary of the $D_{5/2}$-state magnetic dipole
coupling constants known in 1974.

Table I. $^2S_{1/2}$-State Magnetic Dipole Coupling Constants A (in MHz) for Alkali Atoms.

State	Li7	Na23	K39	K41	Rb85	Rb87	Cs133
2s	401.74[a]						
3s		885.8[a]					
4s		202(3)[c]	230.85[a]	127.00[a]			
5s		78(5)[e]	55.50(60)[b]	30.75(75)[b]	1011.9[a]	3417.3[a]	
6s			21.81(18)[b]	12.03(40)[b]	239.3(1.2)[b]	809.1(5.0)[b]	2298.1[a]
7s			10.85(15)[d]		94.00(64)[b]	318.1(3.2)[b]	546.3(3.0)[b]
8s					45.5(2.0)[b]	158.0(3.0)[b]	218.9(1.6)[b]
9s						91.2(8)[f]	109.5(2.0)[b]
10s							63.2(3)[d]
11s							39.4(2)[d]
12s							26.30(15)[f]

a. Mostly atomic beam measurements referenced by G. H. Fuller and V. W. Cohen, Nuclear Moments.

b. R. Gupta, W. Happer, L. K. Lam and S. Svanberg, Phys. Rev. A8, 2792 (1973).

c. K. H. Liao, R. Gupta and W. Happer, Phys. Rev. A8, 2811 (1973).

d. P. Tsekeris et al, Physics Letters, 48A, 101 (1974).

e. N. Bloembergen, M. D. Levenson and M. M. Salour, Phys. Rev. Letters 32, 867 (1974).

f. P. Tsekeris and R. Gupta, private communication.

Table II. $^2P_{1/2}$-State Magnetic Dipole Coupling Constants A (in MHz) for Alkali Atoms.

State	Li[7]	Na[23]	K[39]	Rb[85]	Rb[87]	Cs[133]
2p	46.17(35)[a]					
3p	13.5(2)[a]	94.3(2)[a]				
4p			28.85(30)[a]			
5p			8.99(15)[a]	120.7(1)[d]	409.1(4.0)[d]	
6p				39.11(3)[e]	132.56(3)[a]	292.1(3)[a]
7p				17.65(2)[c]	59.92(9)[a]	94.5(9)[a]
8p						42.97(10)[b]

a. Mostly optical double resonance measurements referenced by A. Rosén and I. Lindgren, Physica Scrupta 6, 109 (1972).

b. C. Tai, R. Gupta and W. Happer, Phys. Rev. 8, 1661 (1973).

c. D. Feiertag and G. zu Putlitz, Z. Phys. 208, 447 (1968).

d. B. Senitzki and I. I. Rabi, Phys. Rev. 103, 315 (1956).

e. D. Feiertag and G. zu Putlitz, Z. Physik 261, 1 (1973).

Table III. $^2P_{3/2}$-State Magnetic Dipole Coupling Constants (A) (in MHz) for Alkali Atoms.

State	Li⁷	Na²³	K³⁹	K⁴¹	Rb⁸⁵	Rb⁸⁷	Cs¹³³
2p	-3.07(13)[a]						
3p	-0.96(13)[a]	18.65(10)[a]					
4p		6.006(30)[a]	6.09(4)[a]	3.40(2)[d]			
5p			1.972(10)[a]	1.08(2)[d]	25.029(16)[c]	84.852(30)[a]	
6p			0.866(8)[a]		8.25(10)[c]	27.70(2)[a]	50.31(5)[a]
7p					3.71(1)[e]	12.57(1)[a]	16.609(5)[a]
8p					1.99(2)	6.747(14)[a]	7.626(5)[a]
9p						4.05(10)[b]	4.129(7)[a]
10p							2.485(10)[a]
11p							1.600(15)[b]
12p							1.11(4)[b]

a. Mostly optical double resonance and level crossing measurements referenced by A. Rosén and I. Lindgren, Physica Scrupta 6, 109 (1972).

b. G. Belin and S. Svanberg, Physics Letters 47A, 5 (1974).

c. Mostly optical double resonance and level crossing measurements referenced by G. zu Putlitz, Ergeb. Exakten Naturwiss 37, 105 (1965).

d. J. Ney, Z. Phys. 223, 126 (1969).

e. G. zu Putlitz and K. V. Venkataramu, Z. Phys. 209, 470 (1968).

Table IV. $^2D_{3/2}$-State Magnetic Dipole Coupling Constants A (in MHz) for Alkali Atoms.

State	K^{39}	Rb^{85}	Rb^{87}	Cs^{133}
3d				
4d		7.3(5)[e]	25.1(9)[e]	
5d	±0.6(2)[c]	4.18(10)[a]	14.43(12)[a]	
6d			7.84(5)[b]	16.38(5)[d]
7d			4.53(3)[b]	7.4(2)[c]
8d			2.85(3)[c]	3.98(12)[b]
9d				2.37(3)[b]
10d				1.52(3)[c]
11d				1.055(15)[c]
12d				0.758(12)[c]
13d				0.556(8)[c]
14d				0.425(15)[c]

Table V. $^2D_{5/2}$-State Magnetic Dipole Coupling Constants A (in MHz) for Alkali Atoms.

State	K^{39}	Rb^{85}	Rb^{87}	Cs^{133}
4d		-5.2(3)[e]	-16.9(6)[e]	
5d	±0.5(2)[c]	-2.14(12)[d]	-7.44(6)[a]	±22.2(5)
6d			±3.6(7)[b]	-3.0(1.0)[d]
7d			-2.2(5)[b]	
8d			-1.2(2)[c]	-0.9(4)[c]
9d				-0.5(4)[c]
10d				-0.4(2)[c]
11d				±0.24(6)[c]
12d				±0.19(5)[c]
13d				±0.14(4)[c]

a. R. Gupta et al, Phys. Rev. Letters 29, 695 (1972).

b. S. Svanberg, P. Tsekeris and W. Happer, Phys. Rev. Letters 30, 817 (1973).

c. S. Svanberg, Preprint, Chalmers University of Technology, 1974. S. Svanberg and G. Belin, J. Phys. B 7 L 82 (1974)

d. C. Tai, Thesis, Columbia University 1974 (unpublished).

e. K. H. Liao et al, Phys. Rev. Letters 32, 1340 (1974)

Of course other excited state parameters can also
be measured by cascade and stepwise laser spectroscopy.
For instance, Svanberg and Hogervorst have measured a
number of Stark shift parameters for excited alkali
atoms,[27] electric quadrupole hyperfine structure constants can be deduced from level crossing and magnetic
resonance data,[22] and Gupta and Tai have measured atomic
lifetimes by the cascade Hanle effect.[11]

We can expect further dramatic advances in our ca-
pabilities to investigate more highly excited atomic
states as laser technology improves. For instance, even
a modest extension of cw dye laser wavelengths toward
the blue regions of the spectrum will make many more excited S and D states accessible to stepwise excitation
since the high lying states of an atom are so closely
spaced. Also it seems certain that cw lasers will soon
be available to excite the second resonance lines of the
heavier alkali atoms. Since the second excited P states
cascade into the lowest S and D states of an alkali atom,
large populations could be built up in the lowest S and
D states and these could serve as initial states for
laser excitation with a second cw tunable laser. Extensions of tunable cw lasers into the infrared would also
be very helpful for stepwise excitation out of higher
excited states. The future seems very bright for cascade
and stepwise laser spectroscopy.

REFERENCES

1. C. E. Moore, Atomic Energy Levels, National Stan-
    dard Reference Data System, National Bureau of
    Standards 35, 1971 Washington.

2. P. Risberg, Ark. Fys. 9, 483 (1955).
    B. Isberg, Ark. Fys. 35, 551 (1968).
    Y. G. Toresson, Ark. Fys. 17, 179 (1960).

3. D. R. Beck and H. Odabasi, Ann. Phys. (New York)
    67, 274 (1971); M. Phillips, Phys. Rev. 44, 644
    (1933).

4. I. I. Rabi, J. R. Zacharias, S. Millman and
    P. Kusch, Phys. Rev. 53, 318 (1938); Phys. Rev.
    53, 495 (1938); Phys. Rev. 55, 526 (1939);
    G. H. Fuller and V. W. Cohen, Nucl. Data, Sect. A,
    5, 433 (1969).

5. G. zu Putlitz, Ergeb. Exakten Naturwiss. 37, 105
    (1965).

6. H. Kopfermann, Nuclear Moments, p. 101 (Academic
    Press, New York, 1958).

7. Y. Archambault, J. P. Descoubes, M. Priou, A. Omont
    and J. C. Pébay-Peyroula, J. Phys. Radium, 21
    677 (1960).

8. J. Brossel and A. Kastler, Compt. Rend. 229, 1213
    (1949).

9. R. L. Smith and T. G. Eck, Phys. Rev. A2, 2179
    (1970).

10. R. Gupta, S. Chang, C. Tai and W. Happer, Phys.
    Rev. Letters 29, 695 (1972); R. Gupta, W. Happer,
    L. K. Lam and S. Svanberg, Phys. Rev. 8, 2792 (1973).

11. C. Tai, Thesis, Columbia University, 1974 (unpub-
    lished).

12. W. Hanle, Z. Physik 30, 93 (1924).

13. R. Gupta, S. Chang and W. Happer, Phys. Rev. 6,
    529 (1972).

14. A. Ellet and N. P. Heydenburg, Phys. Rev. 46, 583
     (1934); N. P. Heydenburg, Phys. Rev. 46, 802 (1934).

15. R. Prepost, V. W. Hughes and K. Ziock, Phys. Rev.
     Letters 6, 19 (1961).

16. M. Pavlovic and F. Laloë, J. Phys. (Paris) 31,
     173 (1970).

17. K. C. Brog, H. Wieder and T. Eck, Phys. Rev. 153,
     91 (1967); G. C. Ritter, Can. J. Phys. 43, 770
     (1965); R. C. Isler, S. Marcus and R. Novick,
     Phys. Rev. 187, 76 (1969).

18. D. A. Goodings, Phys. Rev. 123, 1706 (1961).

19. J. S. M. Harvey, Proc. Roy. Soc. 285, 581 (1965).

20. K. H. Liao, L. K. Lam, R. Gupta and W. Happer,
     Phys. Rev. Letters 32, 1340 (1974).

21. I. Lindgren (private communication).

22. S. Svanberg, P. Tsekeris and W. Happer, Phys. Rev.
     Letters 30, 817 (1973). S. Svanberg, Proc. of the
     Laser Spectroscopy Conference at Vail, Colorado,
     1973 Plenum, New York (1974)

23. P. Tsekeris and S. Svanberg (to be published).

24. A. Gallagher and A. Lurio, Phys. Rev. 136A, 87
     (1964).

25. T. W. Hänsch et al., Phys. Rev. Letters 30, 1336
     (1973).

26. M. D. Levenson and N. Bloembergen, Phys. Rev.
     Letters 32, 645 (1974).

27. W. Hogervorst and S. Svanberg, Phys. Lett. 48 A,
     89 (1974) and Z. Physik (to be published).

NEW RESULTS OBTAINED ON ELECTRONIC EXCITED STATES OF THE H_2 MOLE-
CULE BY METHODS INVOLVING COHERENCE OR DOPPLER-FREE EFFECTS

J.C. Pebay-Peyroula

Laboratoire de Spectrométrie Physique - Université

Scientifique et Médicale de Grenoble - 38041 GRENOBLE

1. INTRODUCTION

A few years ago the knowledge of excited levels of simple molecules resulted only from studies of optical spectra, the ultimate
resolution obtained being the one which could be provided by the
best interferential systems, this being situated at about 1000 MHz.
One should note here that the identification of levels, through the
intermediary of optical spectra can be extremely complex and tedi-
ous ; and that there are still even in the simplest molecules excited states which have not yet been completely identified. The interest of high resolution experiments is considerable : Parameters
which are not accessible by methods of optical spectroscopy may
thus be determined and will thereby provide theoreticians with
fundamental information.

For many years it seemed very tempting to generalize with regard to molecules, all radiofrequency-spectroscopy experiments performed on the excited states of atoms. The main obstacle is the
multiplicity of the molecular optical lines which make difficult
not only the excitation of a particular level but also the detection of the expected phenomena. By using conventional methods of
excitation, the progress of detection-technology opened the path
for magnetic depolarization and magnetic resonance experiments on
a few simple molecules (1,2,3). The use of lasers has permitted a
very rapid expansion of this type of experiment : we may cite as
examples experiments done on I_2 (4), on BaO (5). In addition the
use of lasers has led to the development of saturated spectroscopy
which was also the origin of a large amount of information on the
molecular excited states. A number of review articles (7,8) give

683

an overall picture of these topics.

In the case of H_2, the optical transitions towards the fundamental state are in the far U.V. (λ < 1000 Å) and in the present
state of the technique no experiment of Doppler-free spectroscopy
using optical excitation or absorption is possible. The experiments
described in this paper all require an <u>electronic excitation</u>. The
majority need a special excitation leading to an anisotropy of the
distribution of magnetic moments. The symmetry of this anisotropy
leads one to call it <u>alignment</u>.

Some readers may be surprised to hear about H_2 in a conference
on atomic physics. We will see that from all points of view the
work which we quote represents an extrapolation of methods of thin-
king and of techniques used for many years by atomic physicists. The
exactness of certain results corresponds to that which we are accus-
tomed to in atomic physics and allows us to expect theoretical
knowledge of H_2 with ever-increasing accuracy.

2. ELECTRONIC EXCITATION

The creation of aligned excited states can be made using an
optical excitation with polarized light. It has been shown that an
excitation by a slow electron beam could also produce this alignment detected in particular by emission of polarized light. In the
case of atoms this effect was discovered nearly fifty years ago by
Skinner (9) and a simple explanation of these phenomena and their
relation with the conservation of angular momenta is given by Lamb
(10). In the field of atomic physics this technique was used for
the realization of magnetic resonance (11,12), magnetic depolarization (13) and level-crossing (14) experiments and has allowed us
to obtain a large amount of information about excited levels (15).
The first application to molecular levels was done by Jette and
Cahill (16). The experiments described in this article use systematically electronic excitation. This has been applied also to
other molecular systems (O_2^+, N_2^+) (17). The fundamental study of
the phenomenon of anisotropic excitation such as it has been done
by Mac Farland (18) requires the use of an electron beam whose
energy is perfectly defined, the intensity is then very weak and
very few atoms are excited. In the experiments described above in
which we seek to create an alignment affecting the maximum number
of atoms possible, in order to obtain the best signal possible,
accuracy in the definition of energy is useless. The use of an
electron beam, accelerated by a voltage of about 30 volts with an
intensity of a few mA in a simple triode has proved the best solution. We may note that the electrons can also be produced in a radiofrequency discharge between parallel plates : with a good choice
of frequency and H_2 pressure the motion of the electrons can have

a preferential direction. P. Baltayan and O. Nédelec have evaluated in simple cases (united atom approximation, Born approximation) the alignment produced by the electronic excitation of H$_2$ (19).

Contrary to what one may expect from an optical excitation, difficulties are inherent in this method : (i) weak emission of the excited molecular level, (ii) emission from all excited levels ; the separation of the line emitted by different rotational levels necessitates a monochromator of high resolution, this separation is easier for molecules with low moment of inertia such as H$_2$ (iii) risk of cascades ; the excited level being examined can be populated by deexcitation of an upper level ; in this case the experimental results concerning the level studied are strongly perturbed and (iv) limitation of the static magnetic field strength which can perturb the electron beam. A precise analysis of these effects has been carried out in the atomic case (20). The experimental techniques are described in a review paper (15a).

3. HYDROGEN MOLECULE

Figure 1 represents in simplified form the electronic and vibrationnal states of the hydrogen molecule. The scale of the figure does not allow us to represent the rotational separations. One can see that the distance between the levels of vibrations are of the same size as the electronic separations, consequently the H$_2$ spectrum does not present the classical aspect of molecular spectra with heads of bands clearly visible. The identification of the optical spectrum was a difficult problem spread over about forty years. Until only a few years ago the experimental knowledge of H$_2$ was essentially limited (i) to the energetic position of the levels (we shall discuss the problem of the respective positions of singlet and triplet states in section 5), (ii) to some values of Landé's factor obtained by Zeeman effect which are sometimes imprecise and of an elementary interpretation because of the lack of detailed study of coupling (iii) to some precise values relative to the metastable level 2s 2p $^3\Pi_u$ obtained by magnetic resonance by a beam technique (21).

Corresponding to the progress in experimentation which has allowed the identification and characterization of energy levels, quantum calculations for the hydrogen molecule have, since 1960, reached a high degree of accuracy. A first stage in theoretical computations consists in the calculation of energy for the various states of the molecule. The wavefunctions used by Kolos and Wolniewicz (22) and also by Rothenberg and Davidson (23) permit one to obtain a precision of one cm^{-1} (10^{-5} in relative value) within the fundamental state and a precision of some 10 cm^{-1} (10^{-4} in relative value) for excited states. These calculations were made using an expansion in

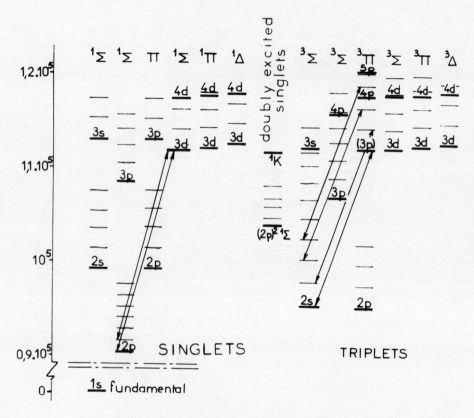

<u>Fig. 1</u> : Energy levels of the H₂ molecule.

elliptic coordinates and the excited states of H_2 have been described by a many-configuration wavefunction into natural orbitals.

A second step, theoretical computations of fine or hyperfine
structures has proved complex : the greatest difficulty arises from
the two electron fine structure operators for which no method of
calculation was until recently available for the molecular case.
Lichten's experimental results on the structure of the level
2s 2p $^3\Pi_u$ (21) have given rise to theoretical calculations by rather approximative methods in the case of the Born-Oppenheimer
approximation and supposing that the mixing with the other states
is negligible (Fontana (24), Chiu (25)), the agreement between
these theories and the values of Lichten is only qualitative for
the values of the fine structure and only very imperfect relative
to the lifetimes. Lombardi (26) recently transposed methods developed in the atomic case (27) to the H_2 molecule using 50 configuration wavefunctions of Rothenberg and Davidson : the agreement
with the experiment data of Lichten (21) and Freund and Miller (42)
(48) for the $2^3\Pi_u$ and $3^3\Pi_u$ level extends between 1 % and 5 %.

4. RESULTS ON THE H_2 MOLECULES OBTAINED BY METHODS INVOLVING ANISO-
   TROPY OF POPULATION OR COHERENT EXCITATION OF ZEEMAN SUBLEVELS.

As seen in section 2 excitation by slow electrons can produce
anisotropy in the repartition of angular moments. This may be described either simply by a difference in the populations of the
Zeeman sublevels or, in certain conditions of excitation -electro-
nic beam perpendicular to the magnetic field-, by a process of partially coherent excitation of Zeeman sublevels. This allows one to
carry out magnetic depolarization (or Hanle effect), magnetic resonance, level-crossing and magnetic repolarization experiments. For
a detailed study of these experiments the reader is referred to a
few review articles (28) and we shall limit ourselves to recalling
the essential ideas :

In the magnetic depolarization experiment (Hanle effect), one
applies a static magnetic field at a right angle to the alignment
direction (Fig. 2). In the classical explanation, the field produces a Larmor precession of the whole excited state, which, when
averaged over the lifetime, gives a rotation of the plane of polarization and a partial depolarization of the emitted light. In the
simple case where the fine and/or hyperfine coupling is not partially decoupled by the magnetic field (i.e., is either fully coupled,
or fully decoupled), the width of the Hanle curve (percentage of
polarization of the emitted light as a function of the magnetic
field) gives the g τ product (g is the Landé g factor and τ the lifetime of the excited level).

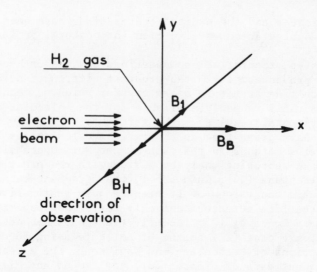

<u>Fig. 2</u> : Experimental set-up. B_H represents the direction of the
static magnetic field applied in the magnetic depolarization or level-crossing experiment, B_B and B_1 are the
static and rf fields applied in the magnetic resonance
experiment.

In the magnetic resonance experiments, one applies a static
magnetic field parallel to the alignment direction and a rf field
perpendicular to the static field (Fig. 2). The polarization of the
emitted light is determined by the difference of population of the
Zeeman sublevels. When the frequency of the rf field is equal to
the energy difference between the $\Delta m = 1$ Zeeman sublevels, transitions are induced between those sublevels. The resulting modification of population of the sublevels gives a modification of the
polarization of the emitted light. The position of magnetic resonance gives information about the Zeeman diagram (g values, fine
or hyperfine structure), the extrapolation of the width to zero
r.f. field strength leads to the lifetime of the level.

In the level-crossing experiments a coherent excitation of a
set of Zeeman sublevels is used. Interference phenomena, manifes-
ting themselves by a change of intensity of the light detected, can
be observed when two levels such as $\Delta M = 1$ or $\Delta M = 2$ (15b) cross.
From the value of magnetic field corresponding to the crossings, information on the Zeeman diagram is also obtained. Hanle effect can
be described as a zero field level-crossing.

In the 4(e) section, we describe a particular method giving information on hyperfine structure smaller than the natural linewidth.

N(v) → N₀(v₀)	λ	g values	τ₀ (n sec)	Zero-field Separations (MHz) (b)	(d)	(c)
1(0) → 0(0)	5994	J=1 1.249 ± 0,01 (a) (1.251) (b)	31 ± 3 (a)	α 160 ± 5 β 2100 ± 600	160 ± 6	163.4 ± 2.8 2366. ± 5.4
1(1) → 0(1)	6098			α 160 ± 12 β	163 ± 6	164.7 ± 1.7 2253.4 ± 3.3
1(2) → 0(1)	6201	J=2 1.249 ± 0,01 (a) (1.251) (b)	29.4 ± 3.2 (c)	α 195 ± 12 β	185 ± 6	185.7 ± 1.2 2066.8 ± 2.2
1(3) → 0(3)	6303			α 230 ± 25 β	212 ± 6	210.5 ± 1.9 1873.2 ± 3
1(0) → 2(0)	6023	J=1 0.77 ± 0.01 (a) (0,751) (b)	31.5 ± 3 (a)			
1(1) → 2(1)	6127	J=3 0.77 ± 0.01 (a) (0,778) (b)				
1(2) → 2(2)	6230	J=2 0.48 ± 0.02 (a) (0.473) (b)				

Table 1 : Experimental data on the (1s 3p) $^3\Pi_u$ level of para H₂.
(a) M.A. Maréchal (31), (b) Jost (32), (c) Freund and Miller (44), (d) Baltayan (38)
(b) Theoretical value in the case of b coupling.

$N(v) \rightarrow N(v_0)$	λ	g			τ (n sec)	
$1(0) \rightarrow 0(0)$	4627	0.885	± 0.01	(a)	15.6 ± 0.8	(a)
		0.89077	± 0.0001	(b)	21 ± 4	(b)
$1(0) \rightarrow 0(1)$	4928	0.87	± 0.03	(a)		
$2(0) \rightarrow 1(1)$	4932	0.575	± 0.04	(a)		
$3(0) \rightarrow 2(1)$	4934	0.44	± 0.06	(a)		
$4(0) \rightarrow 3(1)$	4933	(no separated lines)				
$5(0) \rightarrow 4(1)$	4928	0.3	± 0.08	(a)		

Table 2 : Experimental data on the $(1s\ 3d)^1\Sigma$ level on para H_2
(a) M.A. Maréchal (33), (b) Freund and Miller (41)

Level	Type of collision	Cross-section $T = 300°K$
$(1s\ 3p)^3\Pi_u$ $N=1$ $v=0$	depolarizing	$170 \pm 20\ \text{Å}^2$
$(1s\ 3d)^1\Sigma$ $N=1$ $v=0$		$190 \pm 30\ \text{Å}^2$
3^1K $N=2$ $v=2$ $N=2$ $v=3$	quenching	

Table 3.

a) Effects Involving a Particular Level of Given J Value ($\Delta J = 0$ transitions).

The first experiments on H$_2$ were performed by Van der Linde and Dalby (29) (magnetic depolarization) and by M.A. Maréchal et al. (30) (magnetic resonance). The work of exploitation was done on the (1s 3p) $^3\Pi_u$, (1s 4p) $^4\Pi_u$, (1s 3d) $^1\Sigma$ levels and on the doubly excited (2pσ 2pσ) 3^1K level for different rotational and vibrational states. The excitation cell is of the same type as that used in atomic physics experiments (15). The line studied is isolated by a monochromator or an interference filter. The signal is generally weak and a data storage sometimes reaching forty-eight hours for a resonance curve is necessary. Some characteristic results obtained are mentionned in tables 1, 2 ; all results, given in detail, are to be found in ref. (41)(31)(32)(33), examples of experimental curves are given in Fig. 3 and Fig. 4.

The following comments must be made :

. The width of the magnetic depolarisation curve leads to the product gτ, τ being the coherence time of the level. As a result of enlargement due to collisions, exploitation of the curves requires an extrapolation at zero pressure ; furthermore the magnetic depolarization effect is very sensitive to cascade effects. Consequently, except in a particular case where either g or τ is known, exploitation of the magnetic depolarization for spectroscopy is of little interest. Its role can be valuable, as a first stage to evaluate the alignment produced by excitation.

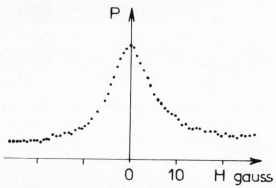

Fig. 3 : Experimental curve of magnetic depolarization observed on the λ = 4627 Å transition (level 1s 3d $^1\Sigma$), N=1, v=0.

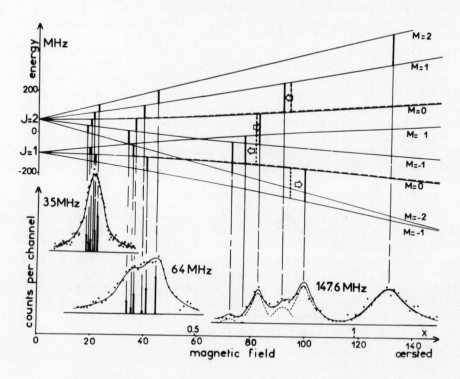

Fig. 4 : Energy-level diagram for J=1 and 2 as a function of magnetic field for the level 3p $^3\Pi_u$. The positions and the
intensities of the magnetic resonances are shown for 35,
64 and 147.6 MHz. The dashed lines represent the $M_J=0$
energy levels and the transitions arising from these levels, calculated by neglecting the J=0 level, while the
solid lines include this perturbation. The arrows show
the direction of the displacement in these transitions
due to the perturbation, the J=0 energy being assumed
positive in this diagram. In the lower part of this figure the points are experimental, the reconstituted curves are drawn in solid lines.
(from reference 32).

. The position of the magnetic resonance lines leads, in weak field conditions, to Landé's factor. In the presence of fine or hyperfine structures and in areas corresponding to intermediary fields the different radiofrequency transitions $\Delta m = 1$ are observed in different fields. The resonance lines are then deformed and their position no longer has any meaning. Working from their shape it is then possible to reconstitute the Zeeman diagram and to obtain the zero field energy difference. Such a study has been made in the case of level $^3\Pi_u$ N=1 of para H_2. By calling the E_J energies respectively $-\frac{5}{8}\,\alpha$, $+\frac{3}{8}\,\alpha$, β for the levels J=1, J=2, J=0 (α is the zero field separation between the levels J=1, J=2) α is about ten times weaker than β and consequently the influence of the level J=0 can be considered as a perturbation. Analysis of the shape of the resonances lines (32), Fig. 4, has led to the results given in Table 1.

. The determination of lifetimes from the width of resonance curves requires a double extrapolation to zero r.f. field strength and to zero pressure. This necessitates a very long time of experimentation.

The experimental studies of enlargement of the curves of magnetic resonance or magnetic depolarization in relation to the pressure permit one to determine the cross-sections of collisions between an excited state and the ground state. A study has been carried out (30)(33) in the case of the levels (1s 3p) $^3\Pi_u$, (1s 3d)$^1\Sigma$ and (2pσ 2pσ) 3^1K. The behaviour of the width and the amplitude of the magnetic depolarization curve leads one to attribute the characteristics given in the table 3.

The theoretical interpretation in the case of the level 3p $^3\Pi_u$ has been undertaken (33) by using the methods and the approximations defined, in particular, in the atomic case (34)(35)(36). The results thus obtained seem to agree with the experimental results but a precise comparison, necessary for a complete justification of the hypothesis, is difficult : it requires, in particular, knowledge of the temperature of the gas in the triode. The latter has been evaluated to be 800°K, but no realy significant measurement has been undertaken.

b) Level-Crossings

The level-crossing technique, a priori appealing because of its simplicity, has proved difficult in experiments : the intensity of the expected signals is weaker than that of magnetic depolarization effects and the influence of the magnetic field on the electronic excitation beam produces distortions of the observed signals. The level crossings $\Delta M = 2$ and $\Delta M = 1$ detected on the 3p $^3\Pi_u$ and 4p $^3\Pi_u$ levels have allowed us to determine the values

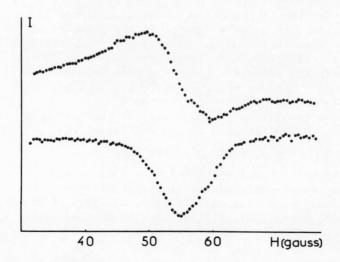

<u>Fig. 5</u> : Level crossing $\Delta M = 2$ observed on the $\lambda = 5994$ Å transi-
tion (level 3p $^3\Pi_u$, N=1, v=0)

of the separation α (38)(39)(40) - cf Fig. 5 and table 1.

c) <u>Experiments Involving Transitions $\Delta J = 1$</u>

Freund and Miller carried out a very important piece of work
on magnetic resonances involving $\Delta J = 1$ transition observed on the
3p $^3\Pi_u$ state of para H_2 and of ortho H_2 and of D_2 and on the 4p $^3\Pi_u$
state of para H_2 (42)(49). We shall limit ourselves to reporting
here a few results obtained on hydrogen ; the corresponding values
of D_2 are to be found in the refs (45)(47).

Fig. 6 presents for para H_2 the aspect of the Zeeman diagram,
the arrows indicating the different resonances observed. We note
in the area of intermediate fields ($\curvearrowright$2 KG) $\Delta M_J = 1$, $\Delta M_S = 2$ transitions which are forbidden in the strong magnetic field for which
M_S is a good quantum number. In the strong magnetic field the transition corresponds to $\Delta M_J = 1$, $\Delta M_S = 0$, $\Delta M_N = 1$. Fig. 7 represents
the experimental set-up : the cell is placed in a resonant cavity
excited by an electromagnetic wave of 9.2 GHz frequency. The position of the resonances observed for different values of the field
(Fig. 8) permits the deduction of values of the different molecular parameters. These parameters are defined when one writes the
Hamiltonian of the molecule in the presence of a field, in the
Born-Oppenheimer approximation conditions :

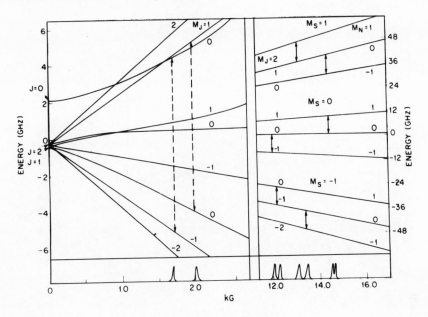

Fig. 6 : Energy levels of the 3p $^3\Pi_u$, N=1 at low magnetic field
(left) and high magnetic field (right). The observed
transitions are indicated by arrows.
From T.A. Miller, R.S. Freund, ref. (8)

$$
\begin{aligned}
H = \quad & B\,N^2 && \text{rotational energy of the molecules}\\[4pt]
+ \quad & A'\,\vec{L}.\vec{S} && \left.\text{spin-orbit coupling}\right\}\\[4pt]
+ \quad & \left(\frac{1}{2\sqrt{6}}\right) D'\,(3\,S_z^2 - \vec{S}^2) && \text{spin-spin coupling}\quad\Big\}\ \text{f.s.}\\[4pt]
+ \quad & a\,\vec{I}_T\,\vec{L} && \text{orbital interaction}\\[4pt]
+ \quad & a_F\,\vec{I}\,\vec{S} && \text{Fermi contact interaction}\Big\}\ \text{h.f.s.}\\[4pt]
+ \quad & C\,T^1(\vec{I}_T)T^1\big[T^1(\vec{S}),T^2(\vec{N})\big] && \text{Dipolar interaction}\\[4pt]
+ \quad & g_S\,\mu_0\,\vec{S}\,\vec{B} + g_L'\,\mu_0\,\vec{L}.\vec{B} && \text{Interaction of the field with the}\\
- \quad & \mu_0\,g_I\,\vec{I}_T\,\vec{B} && \text{magnetic moments}\\[4pt]
+ \quad & \mathbf{\chi}'\,B^2(3\,N_z^2 - \vec{N}^2) && \text{Quadratic effect}
\end{aligned}
$$

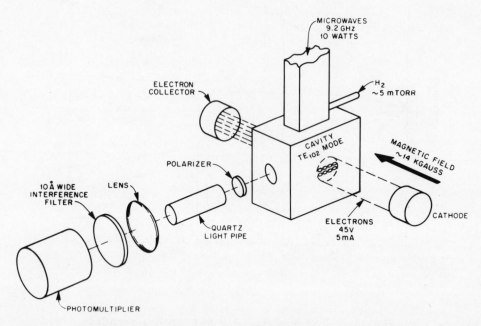

<u>Fig. 7</u> : Experimental set-up for the observation of ΔJ = 1 tran-
           sitions.
           From T.A. Miller and R.S. Freund, ref. (8).

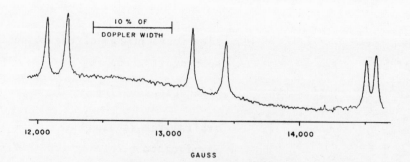

<u>Fig. 8</u> : ΔJ = 1 observed transitions on $^3\Pi_u$ level of para H_2 (N=1
           v=0) at high magnetic field.
           From T.A. Miller and R.S. Freund, ref. (8).

In the case of the N=1 level, $D' = B_0 \mp \sqrt{6}\, B_2$ - for the symmetric, + for the antisymmetric, Λ-doublet component. For the $^3\Pi_u$ N=1 level we can write :

$$E_{J=0} - E_{J=2} = \beta + \frac{3}{8}\alpha = -\frac{1}{2}A' + \frac{9}{20}(-B_0 - \sqrt{6}\, B_2)$$

$$E_{J=2} - E_{J=1} = \alpha = A' + \frac{3}{10}(-B_0 - \sqrt{6}\, B_2)$$

A program of computations has permitted the determination of these different constants. The values of α and β for the $^3\Pi_u$ levels are indicated in the table 1.

We should note that, from the Hamiltonian, the transitions in strong field are not dependant on the electronic spin and the a_F contact interaction constant ; on the other hand these parameters can be determined from resonances observed in intermediate fields. Landé's g_S factor of the electron bound to the molecule has been determined precisely : Fig. 9 indicates the values obtained.

d) Comments

Experimental results which we have given in the preceding paragraphs provide new data for the molecule H_2. They permit, in particular, the ab initio testing of theories which are often sophisticated and they can propose information about wavefunctions. It is not possible to sum up in a few lines the work of interpretation done over the last few years. The main ideas can be schematized in the following way :

The knowledge of the Landé's factors in weak field completing former data of Dieke in strong field allow one to envisage a detailed discussion in the case of coupling and of mixing of states. The case of the 3d $^1\Sigma$ state has been studied in detail by Van der Linde and Dalby (55) by Freund and Miller (41) and by Jost (37), in the hypothesis of an intermediary coupling b-d. Comparison with the experishows that the 3d $^1\Sigma$ state must be mixed with states of the same symmetry such as 3s $^1\Sigma$ and 3^1K.

The experimental results of Freund and Miller have shown that different molecular parameters vary according to the vibrational state of the level considered. This shows that Born-Oppenheimer's hypothesis is no longer justified here. This breakdown of Born-Oppenheimer's hypothesis is due to the mixing with other states such as 3p $^3\Sigma_u^+$ and 4p $^3\Sigma_u^+$ for which numerous vibrational states are close to the 3p $^3\Pi_u$ states. A theoretical analysis has permitted the determination of "deperturbed" constants, characteristic of the $^3\Pi_u$ state (cf. table 4). In general, comparison with Lombardi's

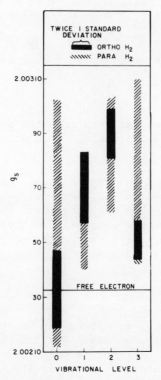

<u>Fig. 9</u> : Experimental values of g_S for ortho and para H_2 compared
to the free-electron value.
From ref. (48).

ab initio results -cf section 3- is very satisfactory, the difference is about 5 % except for the spin-orbit coupling constant
for which the difference is about 25 %.

. The quadratic term appearing in the Hamiltonian only occurs
in the position of the observed resonances by the anisotropy of
magnetic susceptibility. It is also necessary to consider the anisotropy of the polarizability of the molecule because of the electric fields ; these fields result on the one hand from electric
devices of the cell, and on the other hand from relativistic effects resulting from the motion of the molecule in the magnetic
field ; the addition of a known electric field permits the separation of the two effects. Information about the anisotropy of
susceptibilities in the excited states studied may result from
these experiments.

	(a)	(b)
A	$- 842 \pm 2$ MHz	$- 1053$ MHz
B_0	$- 377 \pm 3$	$- 398$
B_2	$- 1089 \pm 1$	$- 1116$
a	6.9 ± 1	6.0
a_F	465.3 ± 0.4	461.8
c $-$ 3d	72 ± 3	75.3

Table 4 : Deperturbed constants for the $(1s\ 3p)\,^3\Pi_u$ level of H_2 (a)
          from reference (47) and (8) and corresponding values cal-
          culated by Lombardi (26).
          (Nota : c $-$ 3d represents the dipolar constant in the hy-
          perfine interaction in the case of the $(1s\ 3p)\,^3\Pi_u$ level.

. The difference between Landé's g_S factors in relation to
Landé's free electron factor results from a certain number of effects whose order of magnitude is about some 10^{-5} : relativistic
and diamagnetic corrections, and corrections linked to the perturbations producing the breakdown of Born-Oppenheimer approximation.
However the interpretation of the results presented in Fig. 9 does
not at present appear obvious.

e) Repolarization Experiments

The experiments described above have permitted the determination
of hyperfine structures in cases in which their effect produces a
separation larger than the natural linewidth. Other methods such as
quantum beats, antilevel-crossings also find a limit within the
natural width. The effects of repolarization give an experimental
signal whose width is a function of the hyperfine structure. These
effects have been known for a long time and have recently been used
in alkaline atoms to measure h.f.s wider than the natural width (50).
Two experiments using optical pumping have permitted the measurement
of structures narrower than the natural width (51)(52). In the case
of H_2 the method of repolarization described below was used on the
level $(1s\ 3d)\,^1\Sigma$ (33)(53).

Let us describe the principle of the method, by using for simplicity's sake the case of an excited atomic state with the electronic moment J, with the nuclear spin I and with the total moment F :

The anisotropic excitation by the electron beam, based on electrostatic interactions does not act on the spin which keeps an isotropic distribution, and the anisotropic character is transferred to $\vec{J}$. After an average time of the order of the lifetime τ of the excited state, the emitted light corresponding to an electric dipole transition presents a polarization state P proportional to the $\vec{J}$ distribution at $t = \tau$. We wish to see how this distribution is influenced by the hyperfine interaction (A) and by a static magnetic field B_z, parallel to the direction of the initial anisotropy. For the sake of simplicity we take the case where $\vec{J}$ is oriented ($<J_z> \neq 0$). In zero magnetic field two cases exist :

-i) The h.f. coupling is zero (A=0) ; the $\vec{J}$ distribution does not change during the lifetime : $<J_z>_{t=\tau} = <J_z>_{t=0}$; the polarisation ratio P(A=0) is then proportional to the initial $\vec{J}$ distribution : $<J_z>_{t=0}$. The same result would be obtained if I=0.

-j) h.f. coupling exists (A $\neq$ 0) ; $\vec{I}$ and $\vec{J}$ precess around their resultant $\vec{F}$. At the moment of emission the $\vec{J}$ orientation, $<J_z>_{t=\tau}$, is given by projecting the initial orientation successively on $\vec{F}$, then on oz. This mean value is represented on Fig. 10 for an arbitrary $\vec{I}$ direction. This quantity is therefore smaller than the initial orientation $<J_z>_{t=0}$. This corresponds to the fact that part

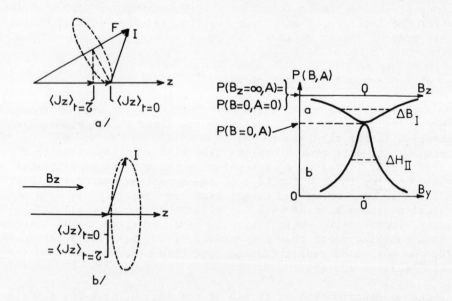

Fig. 10 : Vectorial model of the magnetic repolarization experiment.

Fig. 11 : Magnetic repolarization (a) and magnetic depolarization (b).

of the $\vec{J}$ orientation has been transferred to the spin through the
h.f. coupling. This result implies that the polarization ratio of
the emitted light, proportional to $\langle J_z \rangle_{t=\tau}$, is weaker when a h.f.s
exists.

If a magnetic field is applied along oz, the h.f. coupling is
perturbed and the polarization ratio P in the two preceding cases
is affected as follows :

. A=0, the magnetic field does not act on $\vec{J}_z$ and the polarization ratio P (B$_z$, A=0) is then independent of B$_z$ and proportional
to $\langle J_z \rangle_{t=0}$; we therefore have P(B$_z$, A=0) = P(B$_z$=0, A=0).

. A $\neq$ 0, B$_z$ tends to decouple $\vec{I}$ and $\vec{J}$. At low magnetic field
(g μ_B B$_z$ << A) the resulting momentum F slowly precesses around
the small magnetic field, and the mean z component of $\vec{J}$ is therefore smaller than the initial value $\langle J_z \rangle_{t=0}$. When B$_z$ increases $\vec{I}$
and $\vec{J}$ precess faster around B$_z$ than around $\vec{F}$, which is no longer
a good quantum number, and P increases with B$_z$. At high field
(g μ_B B$_z$ >> A), here symbolized by B$_z$ = ∞) the initial orientation,
parallel to B$_z$ is tightly bound to the oz axis (complete decoupling
of $\vec{I}$ and $\vec{J}$) and at all times, the mean z component of $\vec{J}$ is equal
to the initial value (Fig. 10 b).

In consequence, the polarization P(B$_z$ = ∞, A) related to
$\langle J_z \rangle_{t=\tau}$ is equal to the polarization P(B$_z$= 0, A=0) related to
$\langle J_z \rangle_{t=0}$.

Consequently the degree of polarization observed will be increased in accordance with the magnetic field ; the width of the
curve P = f(B$_z$) is a function of $\frac{A}{\Gamma}$, Γ being the inverse of the
lifetime.

The Fig. 11 represents in "a" the state of polarization of
the light emitted in the direction oy, the field B being directed
along oz in the same direction as the initial alignment created
by the excitation ; curve "b" represents an effect of magnetic
depolarization, the field being in the direction oy perpendicular
to the initial alignment.

The shape of the curve of repolarization was studied in detail
by Lombardi and Mme Maréchal (53) by using the formalism of the density matrix in order to describe the state of polarization of the
light emitted as a function of the magnetic field and of the direction of observation. This formalism leads to formulae expressed in
the form of a development in powers of A/Γ. The effect of depolarization corresponds to a term of the second degree in A/Γ.

Several transitions of the $(1s\ 3d)^1\Sigma \rightarrow (1s\ 2p)^1\Sigma$ band have been studied by this technique. For the 3d $^1\Sigma$ N=1 v=0 I=1 no re-polarization effect was detected which gives an upper limit of possible h.f. structure to be 0.15 MHz. In the case of the 3d $^1\Sigma$ N=1 v=1 I=1 level taking the g value of 0.606 given by Dieke (54) and the Γ value obtained by a magnetic depolarization experiment a value of A = 1 ± 0.17 MHz is obtain, which is about six times smaller than the natural width.

The origin of the h.f.s. of the N=1, v=1 $(1s\ 3d)^1\Sigma$ level is not known at present. If the level were a pure $^1\Sigma$ state (corresponding to a pure "b" Hund's coupling) no h.f.s. would occur, and the g-factor would be of the order of magnitude of a nuclear g-factor (10^{-3}). The weak h.f.s. can therefore have two different origins :

. a $^1\Sigma$ - $(^1\Pi, ^1\Delta)$ mixture, corresponding to an intermediate "b-d" Hund's coupling (the nuclear spin orbit in the $\Lambda \neq 0$ state is then responsible for the h.f.s. in the $^1\Sigma$ state).

. a singlet-triplet mixture (the contact interaction in the triplet state is then responsible for the h.f.s. in the singlet state) ; a similar phenomenon has been observed in atomic He (14).

The results obtained recently of the singlet-triplet separation (cf. section V) and future progress in the knowledge of coupling and mixing will probably permit one to be more precise about this origin.

5. LEVEL ANTICROSSING TECHNIQUE

The experiments previously mentioned use alignment created in the excited states by electronic excitation. Level anticrossing experiments have been carried out on hydrogen. These produce an optical signal whose width is linked to the natural width, without Doppler effect and this does not necessitate an anisotropic excitation but only a difference of population between the levels investigated.

a) Anticrossing Experiments

We should briefly recall the principle of anticrossing experiments :

Two radiating levels are populated differently by various excitation mechanisms, and the mechanism of detection can discriminate, at least partially, between light emitted by the two levels. These levels would cross for a value of an external (usually magnetic) field, unless a coupling between these two levels by some repulsive interaction V occurs. Roughly speaking, at the anticrossing point, the population of one level oscillates back and forth between the two levels, producing an even distribution. The resultant, resonance like, signal is centered on the anticrossing point. Its amplitude is proportional both to the population difference of the two levels and to the difference of the detection efficiency for the same levels. It has a width given by the field needed to separate the two levels by an amount of the order of V (or the mean value of their inverse lifetimes if V is smaller than this quantity). The interaction V can be an electric field or the hyperfine interaction within a fine-hyperfine multiplet (63)(57)(58). A general study of anticrossings is made by Wieder and Eck (63); we should note that in a general manner effects of coherence and effects of population occur ; in the experiments described below the effects of coherence do not exist.

b) Singlet-Triplet Separation

Since no intercombination lines between the singlet and triplet states have been detected so far, there was no direct method of fixing the singlet and triplet states of H_2 within an absolute common scale of energy. Different values have thus far been proposed for the energy of the state 2s $^3\Sigma_g^+$ v=0, N=0, the energy of the basic state (1s 1s)$^1\Sigma_g$ being taken as the origin ; these values are represented on the lines a, b, c, d, e of the table 5; a, b, d are experimental values obtained from the ionization potential.

a - Beutler and Junger (59)	95 085 cm^{-1}
b - Dieke (54)	95 226.0 cm^{-1}
c - Kolos (60)	95 079.7 cm^{-1}
d - Beutler and Junger (corrected)	95 073.2 cm^{-1}
e - Colbourn (61)	94 945 cm^{-1}
f - Freund and Miller (58) Jost (57)	95 076.4 cm^{-1}

Table 5 : Values of the energy of the 2s $^3\Sigma_g^+$ state of H_2.

A technique of anticrossing has permitted to obtain a direct value of this separation. In these experiments the coupling is due either to spin-orbit interaction (which, as is well known, couples singlet and triplet levels causing transition from LS to jj coupling) or to hyperfine Fermi contact interaction. These studies were carried out simultaneously by Jost and Lombardi (57) and by Freund and Miller (58), following another experiment on Helium (62).

The anticrossings which have been studied are for the pairs of levels quoted in column "a" (table 6). The separations obtained from the values of the position of the anticrossings are given in the columns b and c. The experimental set-up is very simple : A cell filled with H_2 (p ω 1 torr) is excited by a R.F. discharge and placed in a magnetic field. Jost and Lombardi used, at the "Service National des Champs Intenses" at Grenoble, a Bitter coil permitting the obtention of 150 kG. With such a value of the magnetic field one can expect to observe anticrossings for about fifty pairs of levels. The figure 12 gives as an example the experimental curves obtained for the 3d $^1\Sigma$ - 3d $^3\Sigma$ pairs by Jost.

(a)	(b)	(c)
3d $^1\Sigma$ - 3d $^3\Sigma$ v=0 N=2	$- 1,24 \pm 0,1$ cm^{-1}	
3d $^1\Sigma$ - 3d $^3\Sigma$ v=0 N=3	$+ 3,5 \pm 0,2$ cm^{-1}	
3d $^1\Delta$ - 3d $^3\Delta$ v=1 N=3	$+ 1,16 \pm 0,1$ cm^{-1}	$1,14 \pm 0,2$ cm^{-1}
4d $^3\Pi$ - 4d $^1\Pi$ v=1 N=4		$1,42 \pm 0,2$ cm^{-1}

Table 6 : (a) pairs of singlet-triplet levels studied
          (b)(c) triplet minus singlet separation measured (b)
          by Jost and Lombardi (57) (c) by Freund and Miller (58).

The values obtained and quoted on the table 6 are in accord, permitting us to conclude that the triplet levels must be lowered by $157,7 \pm 0,2$ cm^{-1} in relation to Dieke's data. The absolute energy of the level $a^3\Sigma_g^+$ shown on the last line "f" of table 6 is deduced from this.

One should note that:

. From the position of anticrossings it has been possible to obtain a number of certain values of g factors (57)(58).

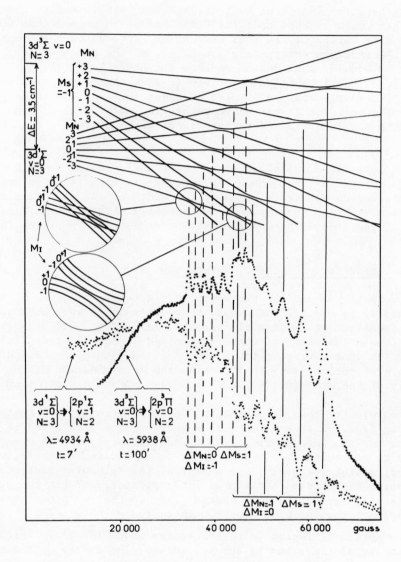

Fig. 12 : Anticrossing signals both for the 4934 Å R₂ singlet and
for 5938 Å R₂ triplet ortho lines, and the corresponding
Zeeman energy level diagram. The hyperfine sublevels are
sketched very roughly only in the circles because fine
and hyperfine structures are not known. The signal amplitude is a few percent.
(From ref. (57)).

. A greatly simplified evaluation of the singlet—triplet separation has recently been made (64) (one may be surprised that it
was not done earlier ...). This evaluation is based on the comparison of the exchange energy in the case of, on the one hand, the levels 3d ^{1}D and 3d ^{3}D of Helium and, on the other hand, of the levels
3d and 4d of H_2. Thus it was possible to show that Dieke's value for
the triplet energy must be lowered by about 160 cm^{-1} and to determine
clearly the pairs of levels whose separation must be about some cm^{-1}
permitting the detection of anticrossings with attainable magnetic
fields.

. A systematic exploitation of these anticrossing experiments
ought to provide a large amount of information about fine or hyperfine interaction constants by using the width of the anticrossings.
In particular Freund and Miller have recently obtained approximative
values of the fine structure constants of the 1s 3d $^3\Pi_g$ level (65).

c) Dissociation of the H_2 Molecule

Information about the processes of dissociation of the H_2 molecule leading to excited states of the hydrogen atom were obtained
by the technique of level anticrossings. The latter were observed
on the Zeeman sublevels of the H atom mixed by the electric field,
seen by the atom moving in the magnetic field. Study of the width
of the level anticrossing thus permits the determination of the
velocity of the atom (66)(67).

The velocity thus determined proved much greater than that
resulting from the thermal motion. A process of dissociation is
proposed for the interpretation : the collision between an electron
and the H_2 molecule takes the molecule to an excited dissociative
or pre-dissociative level ; the energy of the molecular excited
state being higher than the sum of internal energy of the two atoms,
the excess appears as kinetic energy given to the two atoms. When
the energy of the incident electron increases we note :

. a threshold region (electron energy about 18 eV) in which
the velocity of the atoms is about 8 km/sec. The energy of the pre-
dissociative level involved is consequently estimated to be
E = 17.23 ± 0.04 eV.

. at the threshold the variation of the width and amplitude of
the anticrossings leads to attribute the formation of the H atom to
two processes, one involving a pre-dissociative level giving the
atom a velocity of about 9 km/sec, the other involving the continuum
of a non-bound doubly excited level and giving a velocity of about
35 km/sec.

ACKNOWLEDGMENTS

The major part of the work presented here was carried out by R.S. Freund and T.A. Miller at the Bell Laboratories (Murray Hill) and by P. Baltayan, R. Jost, M. Lombardi, Mme M.A. Maréchal and Mlle O. Nédélec at the Laboratoire de Spectrométrie Physique (Grenoble). The author is grateful to his colleagues in Grenoble for the many discussions which he has had with them over the past few years and in particular for the advice given by R. Jost during the writing of this paper. R.S. Freund and T.A. Miller have always sent preprints of their work to the researchers at Grenoble and to the present author and they had a fruitful exchange on the occasion of the Aussois conference (8). The author would like to thank them.

REFERENCES

1. St.J. Silven, Th. Bergeman, W. Klemperer, J. Chem. Phys. 52, 4385 (1970)

2. K.R. German, R.N. Zare, Phys. Rev. Lett. 23, 1207 (1970)

3. R.L. de Zafra, A. Marshall, H. Metcalf, Phys. Rev. A3, 1557 (1971)

4. M. Broyer, J.C. Lehmann, Phys. Lett. 40 A, 43 (1972)

5. R.W. Field, R.S. Bradford, D.O. Harris, H.P. Broida, J. Chem. Phys. 56, 4712 (1972)

6. W. Demtröder, High Resolution Spectroscopy with Lasers, Physics Report, 5, 224 (1973)

7. T.W. Hänsch, Spectroscopy with tunable lasers, p. 579

    J.L. Hall, Saturated Absorption Spectroscopy with Application to the 3,39 μm Methane Transition, p. 615

    W. Demtröder, Recent Advances in the Spectroscopy of Small Molecules, p. 647

    Atomic Physics 3, Proceedings of the Third International Conference on Atomic Physic, 1972, Boulder - Plenum-Press, 1973

8. Proceedings of the International Conference "Doppler Free Methods of Spectroscopy on Excited Levels of Simple Molecular Systems". Aussois (France), May 1973 - Published by the C.N.R.S. (15, Quai Anatole France - 75700 PARIS) 1974.

9. H.W.B. Skinner, E.T.S. Appleyard, Proc. Roy. Soc. A 177, 224 (1927)

10. W.E. Lamb, Phys. Rev. 105, 559 (1957)

11. J.C. Pebay-Peyroula, J. Brossel, A. Kastler, Compt. Rend. Acad. Sci., Paris 244, 57 (1957), 245, 840 (1957)

12. H.G. Dehmelt, Phys. Rev. 103, 1125 (1956)

13. Mme A. Faure, Melle O. Nédelec, J.C. Pebay-Peyroula, Compt. Rend. Acad. Sci. 256, 5088 (1963)

14. J.P. Descoubes, B. Decomps, J. Brossel, Compt. Rend. Acad. Sci. 258, 4005 (1964)

15. J.C. Pebay-Peyroula, Spectroscopy of Atomic Excited States by Electronic Impact Excitation, p. 348

    J.P. Descoubes, Fine Structure and Hyperfine Structure of ^{4}He and ^{3}He, p. 340
    Proceedings of the International Symposium on the Physics of the one and two electrons atoms. Munich, September 1968
    North-Holland, Amsterdam.

16. A.N. Jette, P. Cahill, Phys. Rev. 160, 35 (1967)

17. J. Dufayard, M. Lombardi, O. Nédelec, Compt. Rend. Acad. Sci. Paris, 276, 471 (1973)

18. R.H. Mc Farland, Phys. Rev. 133, A 986 (1964)

19. P. Baltayan, O. Nédelec, J. Phys. B 4, 1332 (1972)

20. O. Nédelec, Thesis Grenoble (1966)

21. W. Lichten, Phys. Rev. 120, 848 (1962)
               Phys. Rev. 126, 1020 (1962)

22. W. Kolos, L. Wolniewicz, J. Chem. Phys. 41, 3663 (1964)
    43, 2429 (1965), 45, 509 (1966), 48, 3672 (1968)

23. S. Rothenberg, E.R. Davidson, J. Chem. Phys. 45, 2560 (1966)

24. L.Y. Chow Chiu, J. Chem. Phys. 40, 2276 (1964)

25. P. Fontana, Phys. Rev. 125, 220 (1962)

26. M. Lombardi, J. Chem. Phys. 58, 797 (1973)

27. R.L. Matcha, C.W. Kern, D.M. Schrader, J. Chem. Phys. 51, 2152 (1969)
    R.L. Matcha, C.W. Kern, Phys. Rev. Lett. 25, 981 (1970)
    J. Phys. B4, 1102 (1971)

28. G. Zu Putlitz, Double Resonance and Level-Crossing Spectroscopy, Atomic Physic 1. Proceedings of the First International Conference on Atomic Physics, June 1968 - New-York.

    A. Kastler, Science 158, 214 (1967)

    B. Budick, Advances in Atomic and Molecular Physics, 3, (1967)

    W. Happer, Optical Pumping, Rev. of Mod. Phys. 44, 169 (1972)

29. F.W. Dalby, J. Van der Linde, Colloque AMPERE XV, North Holland, Amsterdam 1969

30.   Mme M.A. Maréchal, A. Jourdan, Phys. Letters 30 A, 31 (1969)

31.   Mme M.A. Maréchal, R. Jost, M. Lombardi, Phys. Rev. A5,
      732 (1972)

32.   R. Jost, Mme M.A. Maréchal, M. Lombardi, Phys. Rev. A5,
      740 (1972)

33.   Mme M.A. Maréchal, Thèse Grenoble, 1973.

34.   P.W. Anderson, Phys. Rev. 76, 647 (1949), Phys. Rev. 86,
      809 (1952)

35.   C.J. Tsao, B. Curnutte, J. Quant. Spectrosc. Rad. Transf.
      2, 41 (1962)

36.   A. Omont, J. Phys. 26, 26 (1965)

37.   R. Jost, Chem. Phys. Letters, 17, 393 (1972)

38.   P. Baltayan, O. Nédelec, Physics Letters, 37 A, 31 (1971)

39.   P. Baltayan, Physics Letters, 42 A, 435 (1973)

40.   P. Baltayan, Thèse Grenoble, 1973

41.   R.S. Freund, T.A. Miller, J. Chem. Phys. 56, 2211 (1972)

42.   T.A. Miller, R.S. Freund, J. Chem. Phys. 56, 3165 (1972)

43.   T.A. Miller, R.S. Freund, J. Chem. Phys. 58, 2345 (1973)

44.   R.S. Freund, T.A. Miller, J. Chem. Phys. 58, 3565 (1973)

45.   R.S. Freund, T.A. Miller, J. Chem. Phys. 59, 4073 (1973)

46.   T.A. Miller, J. Chem. Phys. 59, 4078 (1973)

47.   T.A. Miller, R.S. Freund, J. Chem. Phys. 59, 4093 (1973)

48.   R.S. Freund, T.A. Miller, J. Chem. Phys. 59, 5770 (1973)

49.   R.S. Freund, T.A. Miller, J. Chem. Phys. 60, 3195 (1974)

50.   R. Gupta, S. Chang, W. Happer, Phys. Rev. 173, 76 (1968)

51.   J.C. Lehmann, J. Phys. 25, 809 (1964)

52.   F. Stoeckel, M. Lombardi, J. Phys. 34, 951 (1973)

53. Mme M.A. Maréchal-Mélières, M. Lombardi, to be published in J. Chem. Phys.

54. G.H. Dieke, J. Mol. Spectr. 2, 494 (1968)

55. J. Van der Linde, F.W. Dalby, Can. J. of Phys. 50, 287 (1972)

56. R. Jost, Thèse de 3ème Cycle, Grenoble 1972

57. R. Jost, M. Lombardi, Determination of the Singlet-Triplet Separation of H$_2$ by a "level anticrossing technique" To be published in Phys. Rev. Lett.

58. T.A. Miller, R.S. Freund, Singlet-Triplet Anticrossings in H$_2$ - To be published

59. H. Beutler, H.O. Junger, Z. Phys. 101, 285 (1936)

60. W. Kolos, Chem. Phys. Lett. 1, 19 (1967)

61. E.A. Colbourn, J. Phys. B 6, 2618 (1973)

62. T.A. Miller, R.S. Freund, Foch Tsai, T.J. Cook, B.R. Zegarski, Phys. Rev. A 9, 2474 (1974)

63. H. Wieder, T.G. Eck, Phys. Rev. 153, 103 (1967)

64. R. Jost, Private communication

65. T.A. Miller, R.S. Freund, Anticrossings and Microwave transitions Between Electronic States of H$_2$ - To be published.

66. L. Julien, M. Glass-Maujean, J.P. Descoubes, J. Phys. B 6, L 196 (1973)

67. M. Glass-Maujean, Thèse Paris, 1974

PHOTOIONIZATION AND AUTOIONIZATION OF EXCITED RARE GAS ATOMS*

R. F. Stebbings, F. B. Dunning, and R. D. Rundel

Dept. of Space Physics & Astronomy, Rice University

Houston, Texas 77001

Recent developments in the technology of tunable lasers have provided great opportunity and challenge for experimenters. At the 3rd Atomic Physics Conference a comprehensive review of laser developments and applications was given by Hänsch[1]. This paper is concerned with one such application, namely the study of ionization resulting from the interaction of laser radiation with excited rare gas atoms. Such processes have considerable application to the understanding of astrophysical and laboratory plasmas and in the physics of planetary atmospheres. However the emphasis in this paper will be exclusively upon the fundamental processes themselves.

1. APPARATUS

The apparatus, which is described in detail elsewhere[2], is shown schematically in Figure 1. A beam of ground state atoms formed by a multichannel glass array is merged with a magnetically confined beam of ∿ 70 eV electrons which produce excitation and ionization. Charged particles and atoms excited to high Rydberg levels are removed from the beam by a combination of electric and magnetic fields, and the resulting beam then comprises atoms in the ground state and in long lived metastable states. The unexcited atoms make no contributions to any of the measured signals and their presence may thus be ignored. Some ultraviolet radiation emerges from the excitation region along the beam axis but has been demonstrated to have negligible influence on the experiments performed.

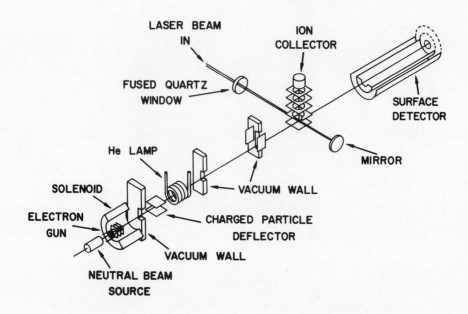

Figure 1   Schematic diagram of apparatus.

The metastable atom beam is intersected at right angles by a laser beam and then impacts upon a metal surface from which secondary electrons are ejected. The absolute flux of the metastable atoms is determinable if γ_m, the secondary emission coefficient for the metastable-surface combination, is known[3].

If the electron gun is operated in a pulsed mode, the velocity distribution of the metastable atoms may be obtained[4] from their arrival-time spectrum at the detector surface. This information, together with knowledge of the absolute metastable flux, enables the number density of metastable atoms at the interaction region to be determined.

The laser[5] used in these studies is shown schematically in Fig. 2. A cell through which dye flows is transversely pumped by the pulsed output beam of a nitrogen laser. A 60° prism performs the dual function of beam expander and polarizer. The cavity is bounded at one end by a diffraction grating, which is used for wavelength selection, and at the other end by a quartz flat through which the output beam emerges. This beam, which is linearly polarized and has a half-angle of divergence of 2 mrad, is coupled by a 100mm focal length lens into a non-linear crystal

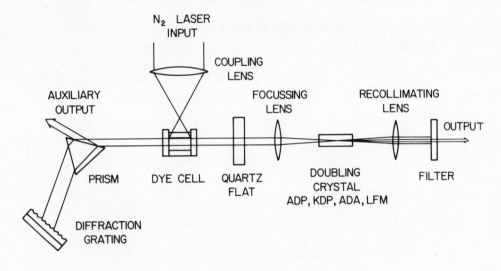

Figure 2   Schematic diagram of laser.

used for second harmonic generation. An Eppley thermopile is used
to measure the mean laser power and hence the mean photon flux.
The power output as a function of wavelength is shown in Fig. 3.
In the fundamental ($\lambda > 3500$ Å) the linewidth is of the order of
$1.5 - 2.0$ Å while for the second harmonic radiation ($\lambda < 3200$ Å)
it is typically of order 0.7 Å.

The laser beam enters the vacuum system through a fused
quartz window and is then reflected by a mirror so as to make two
passes through the metastable beam. Ions formed in the interaction region are extracted by an electric field and detected with
a Johnston particle multiplier. To discriminate photoions from
ions resulting from collisions of metastable atoms with the background gas the multiplier output is fed to two scalers. One of
these is gated to count ions produced during the laser pulse,
while the other is gated for an equal time interval when no laser
pulse is present. The difference in the two scaler count rates is
then due solely to photoions. Discrimination is also made against
ions resulting from photon impact with surfaces and against spurious effects associated with the firing of the laser.

The cross section for photoionization Q is then given by

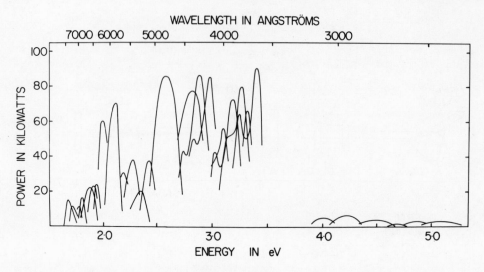

Figure 3   Laser output power.   Each curve represents
           a particular dye.

$$Q = \frac{S}{FN_m \ell k} \tag{1}$$

where S is the mean photoion count rate, which is typically 0.1 to
2 sec^{-1}. F is the mean photon flux and ℓ is the path length of
the photons through the metastable beam. The metastable atom
density N_m is of order $10^3 cm^{-3}$. The efficiency k with which the
ions are detected is not measured here but is estimated to be 0.95
$\pm$ 0.05[6].

2.   RESULTS FOR HELIUM

a)   Photoionization of He($2^{1,3}$S) Atoms.

When a ground state helium beam is excited by 70 eV electrons
both 2^1S and 2^3S metastable atoms are produced. However if this
mixed beam is irradiated by light from a helium discharge lamp the
2^1S atoms will be quenched via transitions of the type 2^1S $\rightarrow$ n^{1}P $\rightarrow$
1^1S. The 2^3S atoms are not quenched as the 2^3S state is the
lowest state of the triplet system. Data appropriate to each
metastable species may thus be derived from observations with the
helium lamp alternately on and off.

The measured absolute cross sections[7] for near threshold
photoionization of 2^1S and 2^3S atoms are shown in Fig. 4 together

with the results of several calculations. The error bars represent one standard deviation of the mean of the observed count
rates. The possible systematic error is estimated to be ± 14%.
For 2^1S atoms experiment and theory are in excellent agreement.

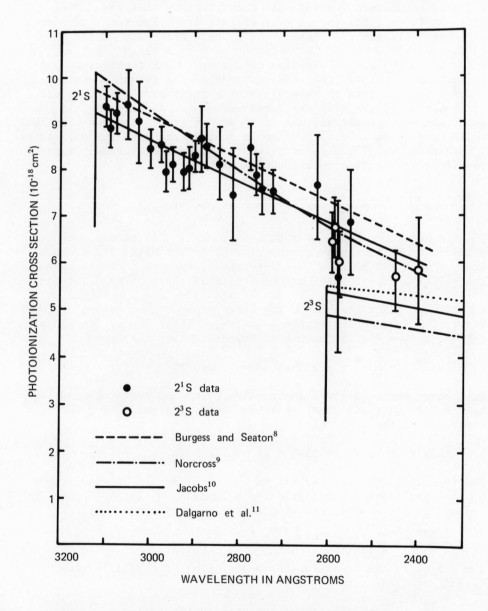

Figure 4   Photoionization cross sections as a function of
wavelength for He(2^1S) and He(2^3S).

For 2^3S the calculated values fall slightly below the measurements but lie within the overall range of experimental uncertainty. The reason for the small discrepancy is not entirely clear, although recent work of Piper, Velazco and Setser[12] on collisional quenching of metastable atoms stimulated a reexamination of the procedure utilized to measure γ. At this time it seems that the value obtained for $\gamma (2^1S)$ was correct whereas the value for $\gamma(2^3S)$ may have been slightly too high. Thus a small reduction of the value of $\gamma(2^3S)$ would not be inconsistent with this recent work and would bring the photoionization measurements and theory into better agreement. In that the accuracy of the recent photoionization calculations is thought to be high it may be argued that in equation (1) the quantity least well known is N_m. In this event the most profitable use of these 2^3S ionization measurements would perhaps be to utilize them in conjunction with the calculated Q to derive $\gamma(2^3S)$, a quantity whose reliable measurement appears to have consistently eluded experimenters.

b) Photoionization of $He(n^{1,3}P)$ atoms.

In addition to the measurements on $He(2^{1,3}S)$ atoms, the absolute cross sections for photoionization of $He(n^{1,3}P)$ atoms, for n = 3, 4 and 5, have been determined, each at a single wavelength. The experimental method is similar to that used for the $(2^{1,3}S)$ atoms. However, instead of measuring the ion signal as a function of laser frequency, the laser is now tuned to one of the absorption frequencies of the metastable helium atoms, for example $2^3S \rightarrow 3^3P$ (see Fig. 5) such that transitions of the type

$$He(2^3S) + h\nu \ (\ \lambda = 3888.6 \ \overset{\circ}{A}) \rightarrow \ \ He(3^3P) \tag{2}$$

will occur during a given laser pulse. Some of the atoms excited to the 3^3P state will then be subsequently ionized during the remainder of the same pulse

$$He(3^3P) + h\nu \ (\ \lambda = 3888.6 \ \overset{\circ}{A}) \rightarrow \ \ He^+ + e \tag{3}$$

The resulting ions are counted in the usual manner and the number of ions per laser pulse (typically between 0.005 and 0.1) is related to the $He(3^3P)$ photoionization cross section Q by

$$S = \ \ Q\ell k \ \int\limits_{\text{pulse}} F(t)N_p(t)dt \tag{4}$$

where $F(t)$ is the instantaneous photon flux, and $N_p(t)$ is the instantaneous density of the 3^3P atoms.

$N_p(t)$ is not directly measureable but its value may be inferred if the laser beam is so intense that saturation of the $2^3S \rightarrow 3^3P$

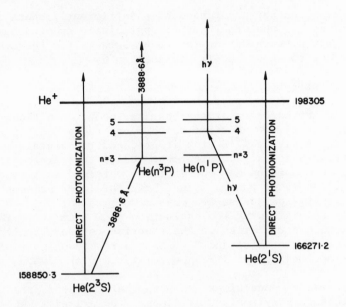

Figure 5   Abbreviated term diagram for helium.

transition is virtually instantaneous. That this condition was
fully realized was demonstrated by observing that the ion production
rate S was proportional to the photon flux F over a wide range of
photon fluxes. Had saturation not been achieved the number
density of 3^3P atoms excited would have been proportional to F and
the ion production rate proportional to F^2.

For linearly polarized light only $\Delta m_\ell = 0$ transitions occur
so that at saturation the number densities of 2^3S and 3^3P atoms
will be equal. If the natural lifetime of the P state is long
compared with the duration (~ 5 ns) of the laser pulse, as is the
case for the triplet levels studied, $N_P(t)$ will be essentially
constant with a value equal to half the original metastable atom
density N_m, because only a very small fraction of the P state
atoms are ionized. Thus

$$N_P(t) = \frac{N_m}{2} \tag{5}$$

and

$$S = Q\,\ell\,k\,\frac{N_m}{2}\,\int_{pulse} F(t)dt \tag{6}$$

where the integral is simply the total photon flux per pulse.

The n^1P atoms on the other hand have sufficiently short
natural lifetimes τ_n that they suffer appreciable spontaneous
decay to the ground state during the laser pulse. For these atoms
$N_P(t)$ is thus not constant but is represented by

$$N_P(t) = \frac{N_m}{2} \exp(-t/2\tau_n) \tag{7}$$

The factor 2 in the exponent arises because, under conditions of
saturation, for every two ^1P atoms which decay radiatively to the
ground state one additional ^1P atom is produced to maintain
equality of the 2^1S and n^1P populations. Using this expression
the singlet atom densities may therefore be obtained as a function
of time during the laser pulse. Since both the laser pulse shape
and the total energy per pulse are measured it is possible to
determine $F(t)$ thus allowing the evaluation of Q from (4). The
major systematic uncertainties in the experiments are associated
with the measurement of $F(t)$ and N_m. Consideration of possible
sources of error leads to an RMS uncertainty in Q of $\sim 20\%$.

The experimental conditions are such that the atoms are
excited and subsequently ionized by linearly polarized light
before the polarization transferred to them on excitation is
altered by collisions, external magnetic fields or, in the case of
triplets, spin-orbit interactions. As a consequence the measured
cross sections are not to be identified with the cross sections
appropriate to unpolarized light and unpolarized atoms, given by

$$Q = 4\pi^2\alpha^2 a_o^2 \left(\frac{df_S}{dE} + \frac{df_D}{dE}\right) \tag{8}$$

where α is the fine structure constant, a_o the Bohr radius and
$\frac{df_S}{dE}$ and $\frac{df_D}{dE}$ are the S and D partial wave oscillator strengths for
photoionization from the unpolarized $n^{1,3}P$ states. Instead, it
has been shown by Jacobs[13], using the Wigner-Eckart theorem, that
the measured cross section must be compared with

$$Q = 4\pi^2\alpha^2 a_o^2 \left(3 \frac{df_S}{dE} + 1.2 \frac{df_D}{dE}\right) \tag{9}$$

Values of Q determined from equation (9) using the partial
wave oscillator strengths obtained by Jacobs[13] with a close
coupling procedure, and by Hartquist and Lane[14] using a quantum
defect calculation, are given in Table 1, together with the
experimental data[15]. Agreement between experiment and theory is
satisfactory in all cases. It should be noted that no similar
polarization effects are present in the $(2^{1,3}S)$ photoionization
measurements because the metastable atoms are unpolarized.

TRIPLET SYSTEM

n	excitation wavelength Å	lifetime μ sec	oscillator strength for He 2^3S-n^3P transition	cross section - cm^2			
				present experiment	quantum defect calculation	close coupling calculation length approx[2]	velocity approx[2]
3	3888·6	0·105	$6·7 \times 10^{-2}$	$8·7 \pm 2·0 \ 10^{-18}$	$7·9 \times 10^{-18}$	$7·5 \times 10^{-18}$	
4	3187·7	0·198	$2·3 \times 10^{-2}$	$2·1 \pm 0·4 \ 10^{-18}$	$2·1 \times 10^{-18}$	$1·8 \times 10^{-18}$	$2·1 \times 10^{-18}$
5	2945·1	0·341	$1·14 \times 10^{-2}$	$8·6 \pm 1·8 \ 10^{-19}$	$8·8 \times 10^{-19}$		

SINGLET SYSTEM

n	excitation wavelength Å	lifetime n sec	oscillator strength for He 2^1S-n^1P transition	cross section - cm^2			
				present experiment	quantum defect calculation	close coupling calculation length approx[2]	velocity approx[2]
3	5015·6	1·8	$1·5 \times 10^{-1}$	$1·0 \pm 0·2 \ 10^{-17}$	$1·1 \times 10^{-17}$		
4	3964·7	4·1	$5·07 \times 10^{-2}$	$2·2 \pm 0·4 \ 10^{-18}$	$2·4 \times 10^{-18}$	$2·4 \times 10^{-18}$	$2·5 \times 10^{-18}$
5	3613·6	7·8	$2·21 \times 10^{-2}$	$8·8 \pm 1·8 \ 10^{-19}$	$9·7 \times 10^{-19}$		

Table 1　　Summary of results for helium.

3. RESULTS FOR ARGON, KRYPTON AND XENON

In the heavier rare gases, the photoionization cross-sections are strongly affected in the region near threshold by the presence of autoionizing states. Figure 6 shows an energy-level diagram for xenon. The ground state of the atom is $5p^6(^1S_0)$ while that of the ion is a doublet, $5p^5 \, ^2P_{3/2, 1/2}$; thus configurations of the type $5p^5nl$ give rise to two series, whose limits are respectively the $^2P_{3/2}$ and $^2P_{1/2}$ states of Xe$^+$. States belonging to the series whose limit is $^2P_{1/2}$ will, for large values of n, lie above the $^2P_{3/2}$ ionization limit. These states are thus degenerate with continuum states of Xe$^+(^2P_{3/2})$ + e, and can autoionize. Examples are the states $5p^5 \, (^2P_{1/2}) \, np'$ with $n \geq 7$, as shown in the figure.

The presence of such autoionizing states is manifested by structure in the photoionization cross-section near threshold. In the present experiments, photoionization from the metastable levels of several rare gas atoms was studied. These metastable levels belong to the configuration $np^5(n + 1)s$, and are often designated $^3P_2$ and $^3P_0$. Such a designation is somewhat misleading, however, since the coupling is not Russell-Saunders. The terms of

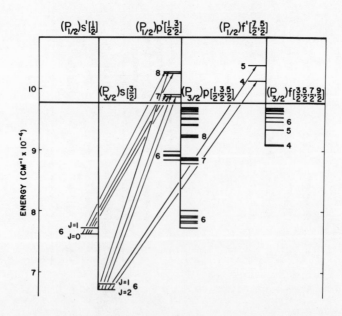

Figure 6   Term diagram for xenon.

this configuration are better described using the $j\ell$ coupling
scheme, in which the orbital angular momentum ℓ of the excited
electron is strongly coupled to the total angular momentum j of
the core, producing a resultant angular momentum k. Then k is
weakly coupled to the spin s of the excited electron to give the
total angular momentum J. Terms are thus designated by $p^5(^2P_{3/2}$
or $_{1/2})n\ell$ $[k]_J$. In this notation, the metastable levels are
$np^5(^2P_{3/2})(n + 1)s$ $[3/2]_2$ and $np^5(^2P_{1/2})(n + 1)s'$ $[1/2]_0$. The
other two terms of this configuration, s $[3/2]_1$ and $s'[1/2]_1$,
decay to the ground state by an electric dipole transition.

The autoionizing states of interest in this work are those
which are connected to either of the metastable levels by an
electric dipole transition. The requirement that parity must
change eliminates those configurations for which the excited
electron has an orbital angular momentum which is an even multiple
of $\hbar$. Consideration of requirements on ΔJ further limits these
possibilities, so that the only dipole-allowed transitions to
autoionizing states are

$$p^5(^2P_{3/2})s \ [3/2]_2 \rightarrow p^5(^2P_{1/2}) \ np' \ [3/2]_{1,2}$$

$$\rightarrow r^5(^2P_{1/2}) \ np' \ [1/2]_1$$

$$\rightarrow p^5(^2P_{1/2}) \ nf' \ [7/2]_3$$

$$\rightarrow p^5(^2P_{1/2}) \ nf' \ [5/2]_{2,3}$$

$$p^5(^2P_{1/2})s' \ [1/2]_0 \rightarrow p^5(^2P_{1/2}) \ np' \ [3/2]_1$$

$$\rightarrow p^5(^2P_{1/2}) \ np' \ [1/2]_1$$

Two unusual features of these transitions should be noted:
(1) the presence of "core-switching" transitions, in which the
initial state has a $^2P_{3/2}$ core while the final state has a $^2P_{1/2}$
core, and (2) transitions in which the s electron in the initial
state is excited to an f level. Similar transitions have previously been observed spectroscopically[16].

Figures 7, 8, and 9 show experimental photoionization data
for argon, krypton and xenon. For these atoms the number densities
N_m of the different metastable species are not well known and
thus absolute cross sections have not been determined. Relative
cross sections are obtained as the photoion signal per unit metastable atoms flux. It should be noted that if the metastable
states are populated according to their statistical weights, the
density of the ns $[\frac{3}{2}]_2$ atoms will be five times that of the
ns'$[\frac{1}{2}]_0$ atoms. Using argon as an example, ionization first appears
at a wavelength of 3071.6 Å, corresponding to direct transitions

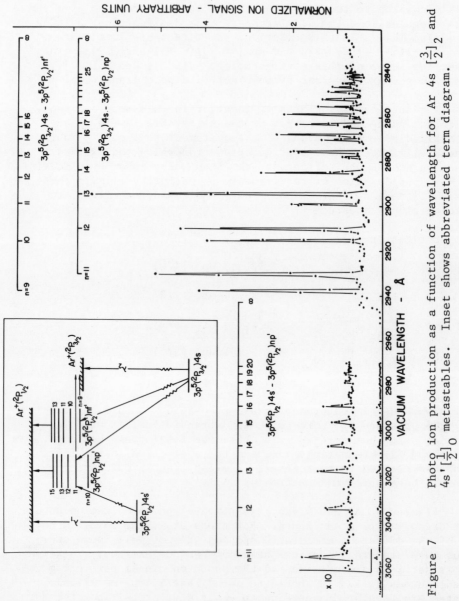

Figure 7   Photo ion production as a function of wavelength for Ar 4s $[\frac{3}{2}]_2$ and 4s'$[\frac{1}{2}]_0$ metastables. Inset shows abbreviated term diagram.

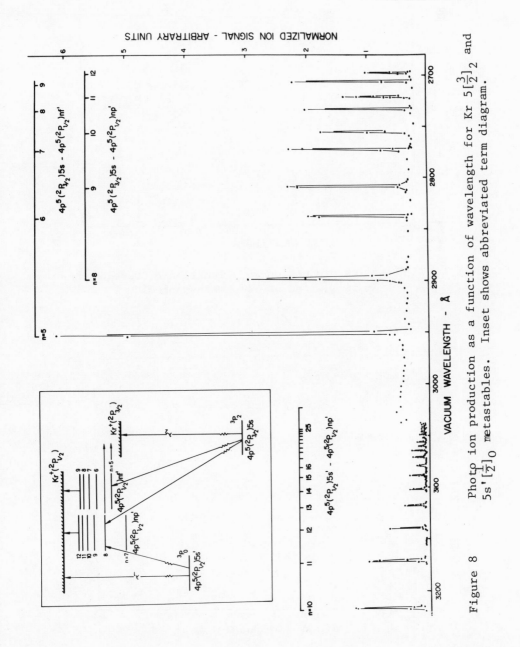

Figure 8   Photo ion production as a function of wavelength for Kr $5[\frac{3}{2}]_2$ and $5s'[\frac{1}{2}]_0$ metastables. Inset shows abbreviated term diagram.

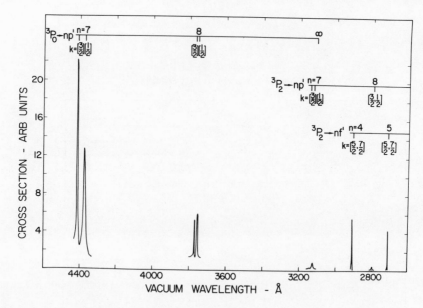

Figure 9   Photo ion production as a function of wavelength
                for Xe $6s[\frac{3}{2}]_2$ and $6s'[\frac{1}{2}]_0$ metastables.

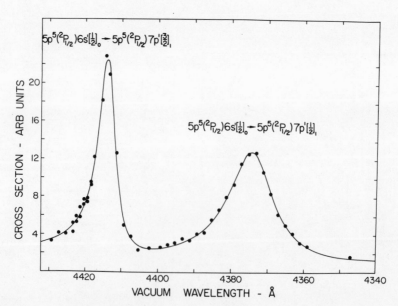

Figure 10   Typical autoionization structure in xenon metastable
                photoionization.

from 4s' $[1/2]_0$ into the $^2P_{3/2}$ continuum. At a wavelength of 3056 Å
a narrow peak in the photoionization appears, due to a transition
to 11p' $[3/2]_1$ or 11p' $[1/2]_1$ (the two cannot be resolved in the
present experiment). Further peaks appear, resulting from excitation to higher p' states, up to the series limit at 2942 Å. At
2944 Å, a step increase in the cross section appears, marking the
threshold for direct photoionization of the other metastable
state, the 4s $[3/2]_2$. Between 2944 Å and 2825 Å a series of
peaks appear in the photoionization cross section due to excitation
of 4s $[3/2]_2$ metastables to the various autoionizing p' and f'
levels described above.

        In both argon and krypton, all the autoionization peaks are
quite narrow. Due to the laser linewidth of ∿ 0.7 Å, the peak
shapes observed are probably characteristic of the laser rather
than the observed transitions. Thus for these atoms the primary
information that can be extracted from the photoionization data[17]
concerns the energy levels, and hence the quantum defects of
autoionizing levels, together with a lower bound on their lifetimes against autoionization.

        In xenon, however, the situation is substantially different.
The observed autoionizing states in xenon are more strongly
coupled to the continuum, giving them a shorter autoionization
lifetime, hence the peaks are broader. In addition, the fine
structure separations in xenon are larger, so that individual
states can be resolved. As a typical example, the transitions
6s' $[1/2]_0 \rightarrow$ 7p' $[3/2]_1$ and 6s' $[1/2]_0 \rightarrow$ 7p' $[1/2]_1$, are shown
with an expanded scale in Figure 10.

        A theoretical treatment of the peak shapes due to autoioniza-
tion has been carried out by Fano[18] and by Fano and Cooper[19].
For a single autoionizing state far removed in energy from other
states the photoionization cross section is given by

$$Q(\varepsilon) = Q_A \left(\frac{(q + \varepsilon)^2}{1 + \varepsilon^2} \right) + Q_B \tag{11}$$

where

$$\varepsilon = (E - E_r)/\tfrac{1}{2}\,\Gamma \tag{12}$$

is the departure of the photon energy from that needed to excite
the idealized (i.e., zero width) autoionizing state, located at
E_r, measured in units of the half-width of the transition, $\tfrac{1}{2}\,\Gamma$.
Q_A and Q_B are, respectively, cross sections for transitions to
continuum states which do or do not interact with the autoionizing
state, and q is a line-shape parameter. Values of Γ, Q_A, Q_B,
and q can be assumed to be independent of energy in the immediate

Transition	Transition Energy (cm^{-1})	Autoionization Lifetime ($\times 10^{-13}$ sec)	Continuum Oscillator Strength* ($\times 10^{-2}$ eV)	Discrete Oscillator Strength* ($\times 10^{-5}$)
$5p^5(^2P_{1/2})6s'[\frac{1}{2}]_0 \rightarrow 5p^5(^2P_{1/2})7p'[\frac{3}{2}]_1$	22,658	1·8	1·5	44
$\rightarrow 5p^5(^2P_{1/2})7p'[\frac{1}{2}]_1$	22,863	0·69	1·2	70
$\rightarrow 5p^5(^2P_{1/2})8p'[\frac{3}{2}]_1$	26,529	4	1·6	3·9
$\rightarrow 5p^5(^2P_{1/2})8p'[\frac{1}{2}]_1$	26,627	2	1·2	11
$5p^5(^2P_{3/2})6s[\frac{3}{2}]_2 \rightarrow 5p^5(^2P_{1/2})7p'[\frac{3}{2}]_{1,2}$	31,814	—	0·54	0·21
$\rightarrow 5p^5(^2P_{1/2})7p'[\frac{1}{2}]_1$	31,995	0·6	0·47	6·4
$\rightarrow 5p^5(^2P_{1/2})8p'[\frac{1}{2}]_1$	35,757	—	0·33	1·4
$\rightarrow 5p^5(^2P_{1/2})4f'[\frac{7}{2}]_3[\frac{5}{2}]_{2,3}$	34,355	7	0·26	5·5
$\rightarrow 5p^5(^2P_{1/2})5f'[\frac{7}{2}]_3[\frac{5}{2}]_{2,3}$	36,864	10	0·24	2·8

* Derived assuming $\bar{\delta}$(Xe) = 0·005, and statistical population of the two metastable levels. Estimated uncertainty is a factor of 3.

Table 2   Summary of results for xenon.

vicinity of an autoionization peak. Thus by fitting experimental
data to the form of equation (11), experimental values of these
parameters may be derived.

The value of the photoionization cross section for large $|\varepsilon|$
will be $Q_A + Q_B$. In the vicinity of $\varepsilon = 0$, there is a minimum at
$\varepsilon = -q$, where $Q = Q_B$, and a maximum at $\varepsilon = q^{-1}$, where $Q = (q^2 + 1)Q_A$
$+ Q_B$. For large values of q, the minimum is broad and shallow,
and the maximum is strongly pronounced; in fact, as $q \to \infty$, the
lineshape approaches a Lorentzian.

When experimental lineshapes have been fitted and values of
E_r, Γ, Q_A, Q_B, and q obtained, the following information can be
derived:
> (1) Energy levels.
> Energy levels, and hence quantum defects, are obtained
from E_r. It should be noted that the maximum cross section
occurs not at E_r, but at

$$E_{max} = E_r + (\Gamma/2q) \tag{13}$$

> (2) Lifetimes
> The lifetime τ against autoionization of the upper
state is given by

$$\tau = \hbar/\Gamma \tag{14}$$

The value of Γ represents the square of the energy
matrix element connecting the upper state to the continuum.

> (3) Continuum oscillator strengths.
> The oscillator strength per unit energy interval,
df/dE, of the transition from the lower state to the continuum
is given by

$$\frac{df}{dE} = (\frac{2\,m_e}{h^2\,\alpha})\ (Q_A + Q_B) \tag{15}$$

> (4) Discrete state oscillator strengths.
> The oscillator strength f of the transition between the
lower and upper discrete states is given by

$$f = (\frac{m_e\,c}{2he^2})\,\Gamma\,(q^2-1)\ Q_A \tag{16}$$

The nine autoionizing transitions which have been studied in
xenon have been fitted to the lineshape described above using a
non-linear least-squares method[20]. Results are shown in Table 2.

It should be noted that, although the present autoionization experiments have observed only the even parity p and f levels, the technique is by no means restricted to these levels. The s, d, and g autoionizing levels may be studied using two lasers (or one laser operating simultaneously at two frequencies). One laser is used to excite a transition from one of the metastable levels to an intermediate p level. The second laser then excites transitions from the p level to the three odd parity levels s, d, and g. Work of this type is currently under way.

REFERENCES

*   Research supported by NASA contract NGR 44-006-156 and by NSF Grant GP 39024.

1. T. W. Hänsch, Proc. 3rd Int. Conf. on Atomic Physics, 579 (1972).
2. J. P. Riola, J. S. Howard, R. D. Rundel, and R. F. Stebbings, J. Phys. B, 7, 376 (1974).
3. R. D. Rundel, F. B. Dunning, J. S. Howard, J. P. Riola, and R. F. Stebbings, Rev. Sci. Inst., 44, 60 (1973).
4. R. D. Rundel, F. B. Dunning, and R. F. Stebbings, Rev. Sci. Inst., 45, 116 (1974).
5. F. B. Dunning, F. K. Tittel, and R. F. Stebbings, Optics Comm., 7, 181 (1973); F. B. Dunning, E. D. Stokes, and R. F. Stebbings, Optics Comm., 6, 63 (1972).
6. R. D. Rundel, K. L. Aitken, M.F.A. Harrison, J. Phys. B, 2, 954 (1969).
7. R. F. Stebbings, F. B. Dunning, F. K. Tittel, and R. D. Rundel, Phys. Rev. Lett., 30, 815 (1973).
8. A. Burgess and M. J. Seaton, Mon. Not. R. Astr. Soc., 120, 121 (1960).
9. D. W. Norcross, J. Phys. B, 4, 652 (1971).
10. V. Jacobs, Phys. Rev. A, 4, 939 (1971).
11. A. Dalgarno, H. Doyle, and M. Oppenheimer, Phys. Rev. Lett., 29, 1051 (1972).
12. L. G. Piper, J. E. Velazco, and D. W. Setser, J. Chem. Phys., 59, 3323 (1973).
13. V. Jacobs, Phys. Rev. Lett., 32, 1399 (1974).
14. T. Hartquist and N. F. Lane (private communication).
15. F. B. Dunning and R. F. Stebbings, Phys. Rev. Lett., 32, 1286 (1974).
16. C. J. Humphreys and E. Paul, Jr., J. Opt. Soc. Am., 60, 200 (1970); 60, 1302 (1970).
17. F. B. Dunning and R. F. Stebbings, Phys. Rev., 9, 2378 (1974).
18. U. Fano, Phys. Rev., 124, 1866 (1961).
19. U. Fano and J. W. Cooper, Phys. Rev., 137, A1364 (1965).
20. D. W. Marquardt, J. Soc. Indust. Appl. Math, 11, 431 (1963).

STRUCTURE INFORMATION FROM ATOMIC-BEAM MAGNETIC-RESONANCE STUDIES

OF MANY-ELECTRON ATOMS*

William J. Childs

Argonne National Laboratory

Argonne, Illinois

INTRODUCTION

In this talk there will be time only for a very brief review
of the role that atomic-beam magnetic-resonance (ABMR) has played
in the recent development of our understanding of atomic structure.
I have chosen to give a few examples of the different types of contributions such research can make, rather than to present extensive
tabulation of results.

We cannot here touch on the enormous contribution of ABMR work
to nuclear physics since the pioneering work of Rabi, Zacharias,
Millman, and Kusch[1] back in 1938-1940. In the realm of atomic
physics, studies using such techniques make possible direct and
unambiguous measurement of the electronic angular momentum J,
g factor g_J, and hyperfine interaction constants A, B, C, ... of an
atomic state. In addition, many other details of atomic structure,
such as electron-coupling schemes, admixtures, $\langle r^{-3} \rangle$ values, tests
for configuration interaction including core polarization, the
presence of relativistic effects and more can be deduced from the
quantities measured directly.

THEORETICAL BACKGROUND

Before we consider particular atomic-beam experiments and their
results, it is desirable to review the theoretical background briefly.

---

* Work performed under the auspices of the U.S. Atomic Energy
  Commission.

Nearly all of the transitions observed in ABMR are magnetic-dipole, and occur between different sublevels of a single atomic state $|\alpha \, SLJ\rangle$. The transition energy is therefore simply the difference between appropriate eigenvalues of the matrix containing the hyperfine and Zeeman Hamiltonians. Even when the theories are put in their simplest form, such a treatment normally allows quantitative description of a large number of precision observations in terms of a very small number of hyperfine constants and a g factor. In taking account of the dipole hyperfine structure, if one distinguishes between the orbital and spin-dipole contributions of orbiting ($\ell > 0$) electrons to the field at the nucleus, one can understand crudely the J-dependence of the dipole A-factors observed for the different members of a multiplet. Bauche and Judd[2] in their study of Pu have shown that allowing for the effect of core polarization (polarization of inner, closed s-shells by the outer partially filled electron shell) can markedly improve our understanding of the J-dependence. In 1965, however, Harvey,[3] in examining the hfs of O and F, showed that the two parameter approach ($\langle r^{-3} \rangle$ for electrons in the partially filled shell and a second parameter for core polarization) was not consistent with his experimental results. He found that for consistency, one had to allow different values of $\langle r^{-3} \rangle$ for the orbital and spin-dipole contributions to the A factors, thus requiring a total of three parameters.

In order to make the inclusion of relativistic effects in many-electron hyperfine calculations more convenient, Sandars and Beck[4] in 1965 introduced an effective operator that, when interacting between nonrelativistic LS states produced the same result as the true hyperfine operator between real states. They started with the general form of the hyperfine Hamiltonian

$$\mathcal{H}_{hfs} = \sum_{k \geq 1} T_n^{(k)} \cdot T_e^{(k)} \quad , \tag{1}$$

in which $T_n^{(k)}$ and $T_e^{(k)}$ refer respectively to operators in the nuclear and electronic spaces, and chose for the electronic part the effective operator

$$T_e^{(k)}(eff) = \sum_{k_s, k_\ell} P^{(k_s, k_\ell)k} \, U^{(k_s, k_\ell)k} \tag{2}$$

in which the P's are constants. The $U^{(k_s, k_\ell)k}$ are double tensor operators of rank k_s in spin space, k_ℓ in orbital space, and k in combined or J space.[5] They showed quite generally that there are at most three such operators U for each order k of the hyperfine interaction, and they showed precisely which three are required. Of the various ways of expressing their result for the magnetic-dipole interaction, the most commonly used for an electron configuration of the type $n\ell^N$ is

$$\mathcal{H}_{hfs}(M\text{-}1) = \underset{\sim}{I} \cdot \underset{i}{\Sigma} \, [a^{01}\underset{\sim}{\ell}_i - \sqrt{10}\, a^{12}\Big\{\underset{\sim}{s}\, \underset{\sim}{C}^{(2)}\Big\}^{(1)}_i$$

$$+ \, a^{10}\, \underset{\sim}{s}_i] \, , \tag{3}$$

in which the superscripts on the a's (which are basically the nuclear dipole moment μ times the P's of Eq. (2)) give the values of k_s and k_ℓ for the associated operator. It should be noted that there are three operators, as required by Sandars and Beck, and three a's. The presence of the contact term a^{10} is important, and it does not necessarily vanish for $\ell \neq 0$ electrons.

The corresponding expression for the electric-quadrupole hyperfine interaction may be written, for ℓ^N configurations $(\ell > 0)$, as

$$\mathcal{H}_{hfs}(E\text{-}2) = \underset{\sim}{T}^{(2)} \cdot \underset{i}{\Sigma} \, [P^{(02)2}\underset{\sim}{U}^{(02)2}_i + P^{(13)2}\underset{\sim}{U}^{(13)2}_i$$

$$+ \, P^{(11)2}\underset{\sim}{U}^{(11)2}_i] \, , \tag{4}$$

in which the first term, of rank two in the (electronic) orbital space, is equivalent to the second-order spherical harmonic of the nonrelativistic quadrupole Hamiltonian. The other two terms are entirely relativistic in origin, are normally much smaller than the first term, and are spin-dependent. In writing the quadrupole Hamiltonian, the P's are sometimes replaced by parameters $b^{(k_s,k_\ell)}$ which are basically the same but with the nuclear quadrupole moment Q included. They are thus analogous to the $a^{(k_s,k_\ell)}$ of the dipole interaction.

Sandars and Beck also showed, for a pure ℓ^N configuration, exactly what values are required for the a's, b's, or P's for the effective operator theory to give the correct relativistic result. They showed that each radial quantity $a^{(k_s,k_\ell)}$ or $b^{(k_s,k_\ell)}$, for a given value of the appropriate nuclear moment μ or Q, is given as a definite linear combination of three relativistic radial integrals, each of which has $\langle r^{-3} \rangle$ as its nonrelativistic limit. For the dipole quantities, for example, we may indicate the equivalence (not one-to-one) as

$$\begin{pmatrix} a^{01} \\ a^{12} \\ a^{10} \end{pmatrix} \longleftrightarrow \mu \begin{pmatrix} F_{++} \\ F_{--} \\ F_{+-} \end{pmatrix} \, , \tag{5}$$

in which the dipole integrals $F_{jj'}$ are

$$F_{jj'} \propto \int_0^\infty \frac{P_j Q_{j'} + Q_j P_{j'}}{r^2} \, dr \rightarrow \langle r^{-3} \rangle_{jj'} \qquad (6)$$

and in which P_j and Q_j are the large and small components of the relativistic radial wave function. As indicated in Eq. (6), the nonrelativistic limit of $F_{jj'}$ is $\langle r^{-3} \rangle$ for all j, j'. Because of the way we chose to write the dipole hyperfine Hamiltonian (Eq. (3)), it is convenient to define an effective $\langle r^{-3} \rangle$ value for each operator in the expression by the relation

$$a^{k_s, k_\ell} = \frac{2\mu_B \mu_N}{h} \frac{\mu_I}{I} \langle r^{-3} \rangle^{k_s, k_\ell} \qquad (7)$$

in which all symbols have the conventional meaning.

In applying the theory to a real atom, it is clear that we must know in detail the composition of the atomic states studied in order to be able to evaluate the expectation values of the hyperfine Hamiltonians. If the eigenvector of a state is accurately known, and if the nuclear dipole moment μ is known, the A factor for the state may be expressed as a linear combination of the three quantities

$$\langle r^{-3} \rangle^{01}, \ \langle r^{-3} \rangle^{12}, \ \langle r^{-3} \rangle^{10}, \qquad (8)$$

and if configuration interaction is ignored, these three quantities can be calculated explicitly. Differences between the "observed" and calculated $\langle r^{-3} \rangle$ values can then be regarded as measures of other structural effects. The situation for the quadrupole structure is similar except for the lack of directly measured values of Q. For the magnetic-octupole and electric-hexadecapole hyperfine interactions seen by atomic-beam workers, the $\langle r^{-3} \rangle$ values are replaced by $\langle r^{-5} \rangle$ values.

We also see that since the theory is essentially a three-parameter description, it is highly desirable to make measurements in as many states of a configuration as possible so that the parameters will be over-determined.

APPLICATIONS OF THEORY TO OBSERVED HYPERFINE STRUCTURE

An interesting atom to which to apply the theory is the rare-earth Sm, for which Woodgate[5], Robertson et al.[6], and others[7] have made extensive ABMR measurements of the hyperfine structure in six atomic states of two stable isotopes.

The eigenvectors of the $4f^6 6s^2 \; ^7F_{0,1,2,3,4,5,6}$ ground multiplet were determined by Judd and Lindgren[8] and by Conway and Wybourne[9] by fitting the parameterized fine-structure eigenvalues to the observed level energies. Although the purity they found is above 94% for the states of interest, it is essential to keep careful account of all the small admixtures when analyzing the hyperfine structure theoretically.

The g_J values measured by ABMR[10] are so precise that relativistic and diamagnetic effects had to be taken into account[7-9] before comparison of the predicted values with experiment. It is found[7] that the differences between theoretical and experimental g values occur only in the sixth figure for each of the six $J \neq 0$ states. This degree of agreement is very unusual, however; differences of up to several percent often occur even for atoms for which multi-configuration eigenvectors are available.

From the Sandars-Beck effective-operator theory, using the best available eigenvectors and the known dipole moment, one may express the A factors for the six ^7F states ($J \neq 0$) as linear combinations of the three $\langle r^{-3} \rangle$ values for the 4f-electron shell. For example, the result for the $^7F_3$ state of ^{147}Sm is

$$A(^7F_3) = -11.1346 \; \langle r^{-3} \rangle^{01} + 2.7979 \; \langle r^{-3} \rangle^{12} -11.0206 \; \langle r^{-3} \rangle^{10}$$

$$= -50.2396 \text{ MHz} , \tag{9}$$

where the experimental result, corrected for second-order hyperfine structure, is given on the right. Since we have one of these equations for each of the six states of ^7F, the three $\langle r^{-3} \rangle$ values are overdetermined, and we can obtain the results of Table I by a least-squares fit[7]. In the second column, we notice first that a nonzero value is obtained for $\langle r^{-3} \rangle^{10}$ (often an indication of core polarization), and second, that there is a slight difference between the values required for the orbital and spin-dipole $\langle r^{-3} \rangle$. Rosén[11] has used an optimized relativistic Hartree-Fock-Slater technique to calculate these integrals ab initio, and his results are seen to be in very good agreement with experiment. The difference between the calculated and experimental values of the ratio $\langle r^{-3} \rangle^{01}/\langle r^{-3} \rangle^{12}$ was found to be relatively insensitive to the method of calculation, and is interpreted as evidence of configuration interaction. The excellent agreement between theory and experiment for $\langle r^{-3} \rangle^{10}$ strongly argues that the contact-like term arises almost entirely from relativistic effects rather than from core polarization or other form of configuration interaction. This is in sharp contrast to the situation in the 3d-shell, as we shall see. Table I also shows the striking success modern self-consistent relativistic calculations can have in explaining the quadrupole hyperfine structure. It should be noted that the two entirely relativistic contributions to the quadrupole hyperfine structure, though small, are understandable at least crudely.

Quantity	Experimental Value (a.u.)	Calculated Value (a.u.)	Difference (%)
$\langle r^{-3}\rangle^{01}$	6.339(5)	6.225	1.8
$\langle r^{-3}\rangle^{12}$	6.420(5)	6.709	-4.5
$\langle r^{-3}\rangle^{01}/\langle r^{-3}\rangle^{12}$	0.987	0.928	6.4
$\langle r^{-3}\rangle^{10}$	-0.215(5)	-0.230	-7.0
$\langle r^{-3}\rangle^{13}/\langle r^{-3}\rangle^{02}$	0.024	0.030	
$\langle r^{-3}\rangle^{11}/\langle r^{-3}\rangle^{02}$	-0.010	-0.014	

Table I. Comparison of $\langle r^{-3}\rangle$ values deduced from ABMR hfs studies of ^{147}Sm and those calculated by Rosén with a relativistic, optimized Hartree-Fock-Slater technique.

Not all atoms are so well behaved (or understood) as Sm, however. The nearby rare-earth Ce has 47 levels below 7000 cm^{-1}, of which 33 have been seen[12] by ABMR. In Tb, nearly 20 levels lie below 5000 cm^{-1}, and the ABMR technique has been used[13] to study the hfs of the lowest 17. In Tb, the configurations $4f^{9}6s^{2}$ and $4f^{8}5d6s^{2}$ both lie low, and in fact the atomic ground state was only identified by Klinkenberg and van Kleef[14] in 1970. Interpretation of the hfs results for the even-parity states has been difficult because of the extreme complexity of obtaining good eigenvectors from the many thousands of levels expected from $4f^{8}5d6s^{2}$ and $4f^{8}5d^{2}6s$. Saying that the problem is only a practical one rather than a fundamental one isn't entirely satisfying, and doesn't lead to improved understanding. In spite of such difficulties, Arnoult and Gerstenkorn's[15] severely truncated eigenvectors have had surprising success in accounting for the dipole hyperfine constants, and theoretical work has continued.[16]

One of the most interesting features of the Tb analysis is that the value of the ^{159}Tb nuclear quadrupole moment could be extracted[13] from the hyperfine structure of the 4f and 5d electron shells independently, and the results differed by more than 30%. The intrinsic quadrupole moment has been measured[17] by Coulomb excitation studies, however, and if the usual projection factor for rotational states is accepted, the resulting "experimental" shielding factors[18] are in good agreement with those calculated by Sternheimer[19].

CONFIGURATION INTERACTION AND HYPERFINE STRUCTURE

Wybourne[20] and others have shown that if it is not too strong, the effect of configuration interaction on hyperfine structure is simply to distort the values of $\langle r^{-3} \rangle$ deduced from the data. The importance of configuration interaction in the hyperfine structure of the 3d-shell atoms has long been apparent. A large amount of experimental ABMR data has been obtained[21-27] in a number of laboratories, and the results have been summarized by Winkler[28], Childs[29], and Armstrong[30]. Two effects in particular stand out. The first is that the value of the ratio

$$\alpha = \frac{\langle r^{-3} \rangle^{01}}{\langle r^{-3} \rangle^{12}}$$

is observed to be much further from 1 than could be explained[18] by relativistic effects alone, and undergoes substantial changes as one goes through the 3d-shell. Bauche-Arnoult[31] has studied the problem with some success by perturbation theory; it is clearly associated with configuration interaction.

The second effect of interest is the surprisingly large value found for the core-polarization constant $\chi \propto a^{10}/(\mu/I)$. The experimental results are summarized in Fig. 1, in which the core-polarization constant is plotted. The uncertainties in the points are difficult to assign because they arise principally from uncertainties in the intermediate-coupling eigenvectors used and are not experimental in origin. The magnitudes of χ are all much too large[18] to be due to the contact-like relativistic term of the Sandars-Beck effective operator, but are due to core polarization. Freeman and Watson[32], Bagus, Liu, and Schaefer[33], and others have had some success in accounting for the general trend through the 3d-shell with spin-unrestricted Hartree-Fock calculations, but the magnitudes of the calculated values are only 35-75% of the observed values for most 3d-shell atoms. A many-body calculation by Kelly[34] for Fe shows rather good quantitative agreement with experiment.

Some studies of core-polarization effects in the 3p-shell have been made. Although experimental ABMR results for some 4d- and 5d-shell atoms are becoming available, principally from Bonn, the theoretical situation here is more difficult. It may well be that core polarization and other configuration-interaction effects cannot be adequately taken into account without making the calculations relativistically.

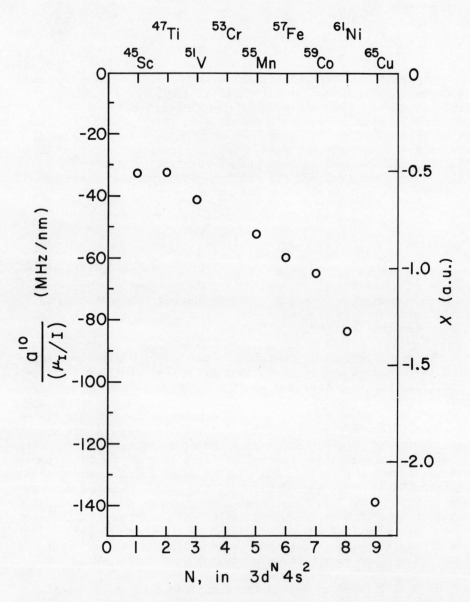

Figure 1. Observed variation of the core-polarization constant
χ in the ground term of 3d-shell atoms.

ABMR measurements by Ting[35] of the hyperfine constants in the $5d6s^2$ $^2D_{3/2,5/2}$ ground term of ^{139}La show that the effects of configuration interaction are very strong. In the absence of configuration interaction, the value calculated from the theory for the ratio

$$A(^2D_{5/2})/A(^2D_{3/2})$$

is 0.429, while the experimental value is 1.291. Although it is simple to analyze this result on the assumption that the discrepancy is due to core polarization, the result of so doing leads to very unphysical values of the radial parameters. It is certainly not due to relativistic effects. Stein[36], Wilson[37], and more recently Ben Ahmed et al.[38] have made 3-configuration least-squares fits to the levels of $5d6s^2$, $5d^26s$, and $5d^3$, and find explicitly strong configuration-interaction admixtures ($\sim 15\%$) in the ground multiplet. With such eigenvectors it has been possible to fit the hyperfine structure observed[39] by ABMR for 13 low levels of ^{139}La reasonably well, although the $\langle r^{-3} \rangle$ values deduced are only in fair agreement with Hartree-Fock calculations[37]. An interesting effect in La is the apparent variation of $\langle r^{-3} \rangle_{5d}$ from state to state, seen both in the experimental hfs[39] and in Hartree-Fock calculations by Wilson[37]. The agreement is only qualitative and the effect should be investigated further. Off-diagonal hyperfine structure (that between two different atomic states) has been used[39] with ABMR in La to see how radial parameters vary from state to state, and further work of this kind would be very interesting.

RELATIVISTIC EFFECTS IN HYPERFINE STRUCTURE

Making relativistic calculations of atomic hyperfine structure can be very difficult, especially if configuration-interaction effects are to be included, and experimental measurements of purely relativistic effects are very valuable as a guide for the further development of the theory and of calculational procedures. Perhaps the best place to look for relativistic effects in hyperfine structure is where the nonrelativistic limit predicts no structure. The inherently high precision of the ABMR technique is well suited for this purpose.

Although it is easy to find states of ℓ^N configurations for which the nonrelativistic theory predicts a zero A factor, a nonzero result can always be due to configuration interaction rather than to relativistic effects. For example, a nonzero A value is observed for a number of atomic ground states that arise from half-filled shells, like d^5 $^6S_{5/2}$. Nonzero A factors have also been seen in the $^3P_1$ state of the configuration p^2, for which the orbital and spin-dipole contributions are predicted to cancel exactly nonrelativistically.

Evans, Sandars, and Woodgate[40] recognized that a significant part of the very small quadrupole hyperfine constant B for the ground S state of half-filled shells could be relativistic in origin, and they made accurate ABMR measurements for ^{55}Mn, $3d^54s^2$ $^6S_{5/2}$, and ^{151,153}Eu, $4f^76s^2$ $^8S_{7/2}$. Their calculations showed that the purely relativistic contribution due to the term $P^{(11)2}_{}U^{(11)2}$ of eq. (4) was much larger than the ordinary contribution of intermediate coupling. Explanation of the exact value observed for the B factor (the quadrupole moments are known from other experiments) has been the subject of several theoretical investigations[40,41], and depends on accurate evaluation of the relevant quadrupole relativistic radial integrals. The same problem arises in the $4d^55s$ $^7S_3$ ground state of Mo, and has been investigated[42] in several ABMR experiments.

Relativistic effects can sometimes be seen even when the non-relativistic theory does not predict a zero result. Thus in the p^2 configuration of Sn and Pb, the two states with J = 2 can be regarded in first approximation as linear combinations of the two LS basis states $^3P_2$ and $^1D_2$

$$|^1D_2'\rangle = \sqrt{1 - \alpha^2} \; |^1D_2\,^{LS}\rangle + \alpha \; |^3P_2\,^{LS}\rangle$$

$$|^3P_2'\rangle = -\alpha \; |^1D_2\,^{LS}\rangle + \sqrt{1 - \alpha^2} \; |^3P_2\,^{LS}\rangle, \qquad (10)$$

where α is the mixing coefficient. If one sets up the matrix of the electrostatic and spin-orbit interactions for these states in Sn, for example, one finds[43] that to fit the observed energies one needs

$$\alpha \text{ (energy)} = 0.320,$$

as shown in Table II. The ABMR measurements[43,44] of the g factors of the J = 2 states show equal and opposite perturbations from the LS-limit values, and this perturbation leads directly to the independent result[43]

$$\alpha \text{ (Zeeman)} = 0.323,$$

in nearly perfect agreement. A nonrelativistic analysis of the observed dipole hfs however leads to the somewhat different result

$$\alpha \text{ (hfs, nonrelativistic)} = 0.472.$$

When the hyperfine structure is analyzed using the correct relativistic radial integrals, the value found[43] for α is

$$\alpha \text{ (hfs, relativistic)} = 0.366,$$

which is very much closer to the value found from the other two observables. The effects are much more pronounced[45] in the heavier Pb atom. The situation has been analyzed in detail for both atoms by Lindgren and Rosén[18].

Observable Used	Value of α Obtained
Energy separation	0.320
Zeeman effect	0.323
Dipole hfs	
(nonrelativistic analysis)	0.472
(relativistic analysis)	0.366

Table II. Dependence of the admixture coefficient α for the $J = 2$ states of Sn on the observable used.

HYPERFINE STRUCTURE OF ACTINIDE ATOMS

Nearly all of the problems faced by theoreticians attempting to calculate atomic hyperfine structure become much more severe for the heaviest elements, and for this reason experimental results are badly needed. Not only does the multiplicity of levels become very large with a resulting complexity of all eigenvectors, but relativistic effects become much more important. Still worse is the fact that it may be difficult or impossible to separate the calculation of mixing effects from relativistic effects.

Considerable ABMR work has been done by Marrus, Armstrong and others at Berkeley[46] on Pa, Np, Pu, and Am, and such work has been important for recognition[2] of core polarization in the actinides. More recently, Goodman, Diamond, Stanton, and Fred[47] at Argonne have measured the g factor of the atomic ground state of ^{254}Fm, and in this way were able to identify it as $5f^{12}7s^2 \; ^3H_6$. The same authors have now measured[48] the hyperfine structure of ^{253}Es and shown that the measured dipole hfs constant A is consistent with the independently measured dipole moment μ and Lewis's[49] relativistic Dirac-Slater radial hyperfine integrals. The eigenvector used for analysis of the hfs was obtained by fitting the experimental g_J value.

SUMMARY AND CONCLUSIONS

I have tried in this brief review to show by examples some of
the various ways ABMR experiments can yield information about details
of the structure of many-electron atoms. I have only hinted at the
important interplay and mutual stimulation of the experimental results and the theoretical development. All contributions to nuclear
physics have been ommitted entirely.

Apart from the quantities measured directly, the experiments
have been of great value for the insight they give for calculation
of $\langle r^{-3} \rangle$ and $|\psi_s(0)|^2$ values by ab initio methods, and in recent
years with the inclusion of relativistic effects in such calcu-
lations. Hyperfine structure is very sensitive to impurities in
a state, including those due to core polarization and other forms
of configuration interaction, and is therefore well-suited for such
investigations. The development of the detailed theory of the
Zeeman effect, including corrections[8,9] for diamagnetic and relativistic effects, was closely correlated with high-precision ABMR
measurements. The value of atomic-beam measurements of quadrupole
hyperfine structure has been important in understanding shielding
effects.

In the years just ahead many areas of investigation are open
for atomic-beam research. Comprehensive studies of 4d- and 5d-shell
atoms are needed in order to sort out the effects of configuration
interaction from those of relativity. Only with such information
is it possible to know which simplifications in calculation can
safely be made. Much work remains to be done in the region of the
actinides, where all aspects of atomic structure become so complex.
For lighter atoms, the use of laser excitation to populate metastable states opens up a wide field of research. If hyperfine
studies were made of enough states in a single element, it might
well be possible to achieve a much more comprehensive understanding
of the structure.

Finally, the further development of self-consistent field and
many-body techniques for calculations can be expected to suggest and
to profit from specific ABMR measurements throughout the periodic
table.

REFERENCES

1. I. I. Rabi, J. R. Zacharias, S. Millman, and P. Kusch, Phys. Rev. 53, 318 (1938); P. Kusch, S. Millman, and I. I. Rabi, Phys. Rev. 57, 765 (1940).

2. J. Bauche and B. R. Judd, Proc. Phys. Soc. (London) 83, 145 (1964).

3. J. S. M. Harvey, Proc. Roy. Soc. (London) A285, 581 (1965).

4. P. G. H. Sandars and J. Beck, Proc. Roy. Soc. (London) A289, 97 (1965).

5. G. K. Woodgate, Proc. Roy. Soc. (London) A293, 117 (1966).

6. R. G. H. Robertson, J. C. Waddington, and R. G. Summers-Gill, Canad. J. Phys. 46, 2499 (1968).

7. W. J. Childs and L. S. Goodman, Phys. Rev. A6, 2011 (1972). It may be noted that in making the least-squares fit to the dipole A factors for Table I, the additional constraint $\langle r^{-3} \rangle^{01} - \langle r^{-3} \rangle^{10} = 6.554(23) \, a_0^{-3}$ (obtained experimentally at strong field in ref. 5) was used.

8. B. R. Judd and I. Lindgren, Phys. Rev. 122, 1802 (1961).

9. J. G. Conway and B. G. Wybourne, Phys. Rev. 130, 2325 (1963).

10. F. M. J. Pichanick and G. K. Woodgate, Proc. Roy. Soc. (London) A263, 89 (1961); see also ref. 7.

11. A. Rosén, J. Phys. B (Atom. Molec. Phys.) 2, 1257 (1969).

12. W. J. Childs and L. S. Goodman, Phys. Rev. A1, 1290 (1970).

13. W. J. Childs, Phys. Rev. A2, 316 (1970); unpublished ABMR work on Tb has also been done by K. H. Chan and P. J. Unsworth.

14. P. F. A. Klinkenberg and Th. A. M. van Kleef, Physica 50, 625 (1970).

15. C. Arnoult and S. Gerstenkorn, J. Opt. Soc. Am. 56, 177 (1966).

16. C. Bauche-Arnoult and J. Bauche, to be published.

17. M. C. Olesen and B. Elbek, Nucl. Phys. 15, 134 (1960); B. Elbek, Thesis, Univ. of Copenhagen, 1963; see also Nuclear Data Tables A7, 527 (1970).

18. I. Lindgren and A. Rosén, Case Stud. Atom. Phys., 4, 93 (1974).

19. R. M. Sternheimer, Phys. Rev. 146, 140 (1966); private communication, (1970).

20. B. G. Wybourne, Spectroscopic Properties of Rare Earths (Interscience, New York, 1965), pp. 148-151.

21. W. J. Childs, Phys. Rev. A4, 1767 (1971); 156, 71 (1967).

22. W. J. Childs and L. S. Goodman, Phys. Rev. 156, 64 (1967); 148, 74 (1966); 170, 50 (1968); 170, 136 (1968).

23. W. J. Childs and B. Greenebaum, Phys. Rev. A6, 105 (1972).

24. H. Gebauer, private communication, (1973).

25. K. H. Channappa and J. M. Pendlebury, Proc. Phys. Soc. (London) 86, 1145 (1965).

26. G. K. Woodgate and J. S. Martin, Proc. Phys. Soc. (London) 70A, 485 (1957).

27. W. Fischer, Z. Physik 161, 89 (1961).

28. R. Winkler, Phys. Letters 23, 301 (1966).

29. W. J. Childs, Phys. Rev. 160, 9 (1967).

30. L. Armstrong, Jr., Theory of the Hyperfine Structure of Free Atoms (Wiley-Interscience, New York, 1971).

31. C. Bauche-Arnoult, Proc. Roy. Soc. (London) A322, 361 (1971).

32. A. J. Freeman and R. E. Watson, Phys. Rev. 131, 2566 (1963); 123, 2027 (1961).

33. P. S. Bagus, B. Liu, and H. F. Schaefer, III, Phys. Rev. A2, 555 (1970).

34. H. P. Kelly, Phys. Rev. A2, 1261 (1970).

35. Y. Ting, Phys. Rev. 108, 295 (1957).

36. J. Stein, J. Opt. Soc. Am. 57, 333 (1967).

37. M. Wilson, Phys. Rev. A3, 46 (1971).

38. Z. Ben-Ahmed, C. Bauche-Arnoult, and J. F. Wyart, private communication, (1974).

39. W. J. Childs and L. S. Goodman, Phys. Rev. A3, 25 (1971).

40. L. Evans, P. G. H. Sandars, and G. K. Woodgate, Proc. Roy. Soc. (London) A289, 108 (1965); A289, 114 (1965).

41. M. A. Coulthard, Proc. Phys. Soc. (London) 90, 615 (1967); J. P. Desclaux, private communication, (1971).

42. J. M. Pendlebury, and D. B. Ring, J. Phys. B5, 386 (1972); S. Büttgenbach, M. Herschel, G. Meisel, E. Schrödl, W. Witte, and W. J. Childs, Z. Physik 266, 271 (1974).

43. W. J. Childs, Phys. Rev. 4, 439 (1971).

44. W. J. Childs and L. S. Goodman, Phys. Rev. 134, A66 (1964).

45. A. Lurio and D. A. Landman, J. Opt. Soc. Am. 60, 759 (1970).

46. See, for example, J. Faust, R. Marrus, and W. A. Nierenberg,
     Phys. Letters 16, 71 (1965); L. Armstrong, Jr. and R. Marrus,
     Phys. Rev. 144, 994 (1966).

47. L. S. Goodman, H. Diamond, H. Stanton, and M. S. Fred, Phys.
     Rev. A4, 473 (1971).

48. L. S. Goodman, H. Diamond, and H. Stanton, private com-
     munication, (1974).

49. W. B. Lewis, private communication (1972).

MANY-BODY EFFECTS IN ATOMIC HYPERFINE INTERACTION

Ingvar Lindgren

Department of Physics
Chalmers University of Technology
Göteborg, Sweden

I. INTRODUCTION

Our understanding of the atomic hyperfine interaction has improved considerably during the last ten years, due to important progress experimentally as well as theoretically. The advent of optical resonance and laser methods has made it possible to study a large number of shortlived excited states, and the further development of the atomic-beam method has made many metastable states accessible for experimental study. These experiments have clearly demonstrated that the simple single-particle model is unsufficient to explain the atomic hyperfine interaction, which means that many-body effects have to be taken into account. Our methods of handling many-body problems theoretically have also improved drastically lately. Techniques such as the linked-diagram expansion and the effective-operator formalism have here been found to be extremely useful. The rapid development of electronic computers have furthermore made it possible to apply these fairly complicated methods not only to the lightest atoms. The two previous talks have been mainly concerned with the experimental development in this field, and, therefore, I shall in my talk concentrate on the theoretical side of the problem and try to explain how the experimental results can be interpreted in the light of recent many-body calculations.

II. ATOMIC MODELS

In order to treat a many-body problem like the atomic one, it is necessary to start with some model, which is relatively easy to handle and which can serve as a starting point for further calculations. For atomic (as well as many other) systems a natural choice

is the <u>independent-particle model</u> (IPM). In this model all particles
are assumed to move independently of each other in the "external"
field (e.g. the Coulomb field from the nucleus) and the <u>average</u>
field from the other particles of the system. In this model all
kinds of <u>correlations</u> between the particle motions are neglected.
In the "<u>unrestricted</u>" Hartree-Fock (UHF) model the wavefunction for
the system consists of an antisymmetric combination (Slater determi-
nant) of single-particle functions, which are varied without any
restrictions until the energy minimum is reached. Therefore, this
model represents in some sense the best IPM model. All effects
beyond this model (apart from relativistic effects) are defined as
correlation effects.

Also the IPM model is in its general form too complicated to
be handled at present for most atomic systems. Therefore, one
normally goes one step further and assumes not only that the particles move independently of each other but also that the average field
the particles move in is spherically symmetric. This is the <u>central-
field model</u> (CFM), which is known to be a good approximation for
atomic systems. In this model the orbitals can be separated into
radial and spin-angular parts, the latter being independent of the
shape of the central field. This gives rise to the well-known shell
structure. Minimizing the total energy with this restriction leads
to the "<u>restricted</u>" Hartree-Fock (HF) model, which therefore can be
regarded as the best CFM. In going from this model to the more
general IPM, the shell structure is partially broken up, and effects
of this kind are referred to as <u>polarization</u> effects.

In the language of perturbation theory the polarization effects
can be described by means of <u>single excitations</u>, while the correlation effects require <u>multiple excitations</u>. This follows directly
from Brillouin's theorem [1], which states that the matrix elements
of the Hamiltonian between the ground and singly-excited states are
identically zero in the UHF model.

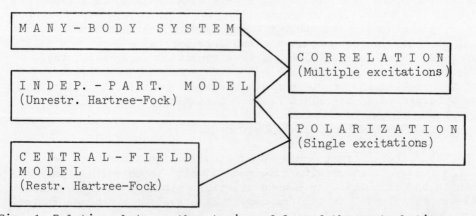

Fig. 1. Relations between the atomic models and the perturbations.

The relations between the models and the corresponding physical effects discussed so far are illustrated in Fig. 1.

III. HYPERFINE INTERACTIONS

The magnetic dipole and electric quadrupole interactions can, using conventional notations, be represented by the operators

$$h^{dip} \propto \Sigma \{ \underset{\sim}{l} \; r^{-3} - \sqrt{10} \; (\underset{\sim}{s} \; \underset{\sim}{c}^{(2)})^{(1)} \; r^{-3} + \frac{8\pi}{3} \; \underset{\sim}{s} \; \delta(\underset{\sim}{r}) \} \cdot \underset{\sim}{\mu}_I \qquad (1a)$$

$$h^{quadr} \propto \Sigma \; \underset{\sim}{c}^{(2)} \; r^{-3} \cdot \underset{\sim}{Q}^{(2)}. \qquad (1b)$$

The summation is performed over all electrons of the system, but in the CFM the closed shells give no net contribution. The last term of the dipole operator, the <u>contact</u> term, contributes only for s electrons, while the remaining <u>orbital</u> and <u>spin-dipole</u> terms, as well as the <u>quadrupole term</u>, contribute only for non-s electrons.

It has been known for a long time that the simple CFM often predicts hyperfine interactions in poor agreement with experiments. This is most conspicuous in cases where the result is independent of the central field. One such example is the hyperfine interaction of atomic systems with a half-filled shell, e.g., Mn^{++} with a $3d^5$ configuration. According to Hund's rules, such a system has $L = 0$ in the ground state, which makes the charge distribution spherically symmetric. In the extreme CFM the hyperfine interaction would then be identically zero, while the experimental interaction is found to be quite appreciable. Abragam and Pryce [2] suggested that this might be due to the polarization of the inner shells, because of the exchange interaction with the highly polarized open shell. The radial functions for the closed shells would then depend on the m_s quantum number, and the hyperfine interactions from the two orientations would not cancel each other completely. We call this effect <u>spin polarization</u>. Abragam and Pryce tried to estimate this effect for some transition metal ions by simple configurational mixing (3s-ns excitations). The effect they found was one order of magnitude too small, which we now know was due to the fact that they only considered excitations to <u>bound</u> states. A more successful approach was later used by Heine, Wood and Pratt [3], who performed spin-polarized HF (SPHF) calculations, i.e., a form of UHF, where the radial functions for different m_s are allowed to be different. Such a calculation is equivalent to mixing into the ground state all possible states, where a single s electron is excited into any other s state, bound as well as unbound. As an example of SPHF calculations we consider the ground state of the alkali atoms (Table I). The unpolarized result is obtained from a relativistic

Table I. Comparison between experimental and theoretical hyperfine
          constants for the ground state of the alkali atoms (in MHz).

	Non-rel HF	Rel HF	Rel UHF	Exptl
^{7}Li	284.7	284.9	379.0	401.8
^{23}Na	622.5	631.1	774.4	885.8
^{39}K	144.7	150.9	187.8	230.9
^{87}Rb	1881.6	2265.5	2783.4	3417.3
^{133}Cs	982.8	1518.5	1854.3	2298.2

HF (Fock-Dirac) calculation, while the polarized result is taken
from a non-relativistic SPHF calculation and corrected for relativistic effects in an approximative way [4]. This comparison shows
that the polarization can have an appreciable effect upon the atomic
hyperfine interaction but also that effects beyond UHF, i.e., correlation effects, can be quite significant.

For systems with an open shell of non-s electrons also other
types of UHF calculations have been performed, for instance by
Freeman and Watson [5]. In these calculations the m_l restriction
is relaxed, and we may refer to them as orbital-polarized HF (OPHF).
This technique has been used particularly to estimate the polarization effect on the quadrupole interaction (Sternheimer shielding).

The UHF is evidently an improvement upon the restricted HF
method, but it does not form any suitable starting point for more
complete many-body calculations. Therefore, we shall not consider
this technique further here but instead turn back to the simpler
CFM.

In the CFM the summations in (1) can, as mentioned, be restricted to the open shells. In a more general model, such as UHF, however, the summations must be extended to all electrons, since the
"closed" shells are no longer completely isotropic. One disadvantage
with such a procedure is that the contributions from inner shells
are obtained as small differences between very large numbers. An
alternative procedure is to replace the "true" hyperfine operator by
an "effective" or "equivalent" operator which by definition operates
only within the open shells. The spin polarization of the closed
shells discussed above can in this way be represented by a "contact"
term, also when there are no s electrons in the open shells. As we
shall see below, also orbital polarizations and most of the correlation effect can be included by allowing the radial factors of the
orbital, spin-dipole and quadrupole terms to be different. We then
get effective hyperfine operators of the following form

$$h_{eff}^{dip} \propto \Sigma \{ \underset{\sim}{l} <r^{-3}>_l - \sqrt{10} \ (\underset{\sim}{s} \ \underset{\sim}{C}^{(2)})^{(1)} \ <r^{-3}>_{sd} + \underset{\sim}{s} <r^{-3}>_s \} \cdot \underset{\sim}{\mu}_I \qquad (2a)$$

$$h_{eff}^{quadr} \propto \Sigma \ \underset{\sim}{C}^{(2)} \ <r^{-3}>_q \cdot \underset{\sim}{Q}^{(2)} \qquad\qquad\qquad\qquad (2b)$$

where the summation is limited to open shells. It has been shown by
Sandars and Beck [6] that also relativistic effects can in the dipole
case be expressed by means of the same operator, while in the quadrupole case two more terms are required. It was first shown by Harvey
[7] that the experimental hyperfine data could be very well fitted
with this type of operator. Harvey measured several hyperfine constants, diagonal as well as non-diagonal, within the same configuration, and his parameters for oxygen and fluorine are shown in Table
II together with the corresponding relativistic HF values. This
comparison shows that the orbital and spin-dipole parameters are
considerably different, which cannot be explained as a relativistic
effect. This is even more conspicuous for the contact term, which
should be close to zero in the relativistic CFM. It should be
emphasized that the number of experimental constants is in these
cases larger than the number of parameters and that the fit is much
better than could be expected from purely statistical grounds. This
demonstrates that the parameters used have physical significance.
Effective operators of this type are nowadays frequently used in
analyzing experimental hyperfine data [4]. As we shall see below,
the effective operator is also a very useful tool in the theoretical
analysis, and the comparison between experimental and theoretical
results can therefore be made in a very convenient way in terms of
the radial parameters.

IV. MANY-BODY THEORY

As mentioned, the UHF model contains polarization effects but
no correlation between the electrons. In order to estimate the
latter it is therefore necessary to go beyond that model. Essentially
two approaches are here available, namely

Table II. Comparison between experimental and theoretical hyperfine
parameters for oxygen and flourine (in atomic units).

	Oxygen		Flourine	
	Exptl	RHF	Exptl	RHF
$<r^{-3}>_l$	4.56(4)	4.96	7.34(2)	7.57
$<r^{-3}>_{sd}$	5.18(1)	4.99	8.14(3)	7.63
$<r^{-3}>_s$	0.48(4)	-0.01	0.60(4)	-0.02

a) variational

b) perturbational.

In the variational procedure the wavefunction is constructed out of a limited basis set, and the total energy of the system is minimized within this set. In the perturbational procedure a complete basis set is available, but the calculation must be terminated at a certain order. The choice of the basis set is the main difficulty of the former procedure, and the inclusion of all important higher-order effects is the main problem of the latter. The calculations performed by Nesbet and others [8] are of variational type, while those of Kelly, Das et al. [9] are of perturbational type. In order to make connections with the effective operators discussed above it is for the present purpose more convenient to use the perturbational approach, and we shall therefore not consider the variational procedure further here.

Most perturbative calculations are nowadays based on the Linked-Diagram Expansion (LDE), developed by Brueckner, Goldstone and others [10] and applied to atomic systems by Kelly, Das and coworkers [9]. This expansion can be combined in a very nice way with the effective-operator formalism, as has, for instance, been shown by Brandow [11] and Sandars [12].

In a perturbative calculation one starts with a zero-order or "model" Hamiltonian, H_o, with known eigenvalues and eigenfunctions,

$$H_o \psi_o^a = E_o^a \psi_o^a \; .$$

The difference between the "true" and the model Hamiltonians is the perturbation

$$V = H - H_o \; .$$

The Hilbert space of the system is by means of the eigenfunctions of H_o partitioned into two parts, a model space (D) and a complementary space (Q), in such a way that degenerate eigenfunctions belong to the same subspace. The projection operators for the two subspaces are

$$P = \sum_D \; | \; \psi_o^a \rangle \; \langle \psi_o^a \; | \; ; \quad Q = 1 - P = \sum_Q \; | \; \psi_o^i \rangle \; \langle \psi_o^i |.$$

The states which "originate" from the model space, i.e., states which would go into this space, if the perturbation is turned off slowly, can be described by the effective operator. If the model

space has d dimensions, there are in general d such states,
Ψ^α ($\alpha = 1, 2...d$). The projections of these functions onto the
model space

$$\Psi_0^\alpha = P\Psi^\alpha \quad (\alpha = 1, 2,...d) \tag{3}$$

are called the model functions. (Note that these are in general
distinct from the limiting functions mentioned above.) We can also
define a "waveoperator", Ω, which transforms the d model functions
back into the corresponding true wavefunctions

$$\Psi^\alpha = \Omega\Psi_0^\alpha \quad (\alpha = 1, 2,...d). \tag{4}$$

So far, Ω is only a mathematical symbol, but we shall later see how
it can be evaluated. First, we shall use it to express effective
operators in a general way.

From the Schrödinger equation

$$H\Psi^\alpha = E^\alpha\Psi^\alpha \tag{5}$$

we can now get by operating with P from the left and using (3, 4)
the formal equation

$$PH\Omega\Psi_0^\alpha = E^\alpha\Psi_0^\alpha \quad (\alpha = 1, 2,...d).$$

The operator on the left-hand side,

$$H_{eff} = PH\Omega = PH_0P + PV\Omega, \tag{6}$$

is called the effective Hamiltonian, and it satisfies the eigenvalue equation

$$H_{eff}\Psi_0^\alpha = E^\alpha\Psi_0^\alpha \quad (\alpha = 1, 2,...d). \tag{7}$$

This means that d of the eigenvalues of H_{eff} coincide with those of
H, and the corresponding eigenfunctions of H_0 are the model functions
defined above (3). Therefore, we can obtain the exact energies of the
states corresponding to the model states by solving the effective-
operator equation (7) instead of the full Schrödinger equation (5).
It should be observed that no approximations have been made so far.
The only sacrifice in going from (5) to (7) is that only a limited
number of energies are obtained. Furthermore, it should be emphasized

that the effective Hamiltonian (6) is the same for the entire model
space. Therefore, if this operator is known with sufficient accuracy,
it yields directly the energy structure of all the corresponding
states. This is an advantage of this technique over, for instance,
the variational procedure, where a separate calculation has to be
performed for each state.

In atomic problems the model Hamiltonian is always some kind
of central-field operator,

$$H_o = \sum_i \left[-\frac{1}{2} \nabla_i^2 - \frac{Z}{r_i} + U(r_i) \right] = \sum_i h_o(i), \qquad (8)$$

and the zero-order functions are Slater determinants of single-
electron orbitals, satisfying the equation

$$h_o \phi_a = \varepsilon_a \phi_a . \qquad (9)$$

The zero-order energy is then

$$E_o = \sum \varepsilon_a \qquad (10)$$

summed over occupied orbitals. Obviously, all states belonging to
the same configuration (specified by the nl quantum numbers) are
degenerate in this order. Normally, the model space consists of a
single configuration, which means that the zero-order function is
a linear combination of degenerate determinants, but it is also
possible to choose a model space consisting of several configurations (several open shells) [13]. Perturbations from states outside
the model space are incorporated into the effective Hamiltonian in
an approximative way. By extending the model space, certain admixtures are treated exactly at the expense of solving a somewhat
larger secular equation (7).

We shall now see how the wave operator (4) and the effective
Hamiltonian (6) can be evaluated, and for this purpose we shall use
the Rayleigh-Schrödinger perturbation formalism. The wave operator
can be shown to satisfy the equation [13]

$$\Omega H_o - H_o \Omega = V\Omega - \Omega V\Omega \qquad (11)$$

and with

$$\Omega = \Omega^{(0)} + \Omega^{(1)} + \Omega^{(2)} + \dots$$

one obtains the recursion formula

$$\Omega^{(0)} = P ,$$

$$\left[\Omega^{(n)}, H_o \right] = QV\Omega^{(n-1)} - \sum_{m=1}^{n-1} \Omega^{(n-m)} V\Omega^{(m-1)} \tag{12}$$

for the components of different orders. As will be explained further below, the <u>linked-diagram theorem</u> implies that the second term on the right-hand side of (12) essentially cancels against so called <u>unlinked</u> parts of the first term, so that the theorem can be expressed

$$\left[\Omega^{(n)}, H_o \right] = \{ V\Omega^{(n-1)} \}_L \tag{13}$$

where index "L" stands for the <u>linked</u> part. In other words, the wave operator of a certain order is obtained by operating with the perturbation on the wave operator of next lower order, keeping only the linked part. The commutator on the left-hand side yields the appropriate energy denominator. The model space can in this treatment contain several zero-order energies (configurations). In the degenerate case (single configuration), the left-hand side can be replaced by $(E_0 - H_0)\Omega^{(n)}$, where E_0 is the zero-order energy of the model space, and the LDE can then be written in the more usual form [9]

$$\Omega^{(n)} = \left\{ \frac{Q}{E_o - H_o} V\Omega^{(n-1)} \right\}_L = \left\{ \frac{Q}{E_o - H_o} V \right\}_L^n P . \tag{14a}$$

or

$$\Psi^\alpha = \Omega\Psi_o^\alpha = \left\{ 1 + \frac{Q}{E_o - H_o} V + \frac{Q}{E_o - H_o} V \frac{Q}{E_o - H_o} V + \dots \right\}_L \Psi_o^\alpha . \tag{14b}$$

In the applications considered here we shall start from a single configuration and, therefore, we shall confine the following discussion to the expansion (14).

By inserting (14) into (6) we get the following expansion for the effective Hamiltonian

$$H_{eff} = P \left\{ H_o + V + V \frac{Q}{E_o - H_o} V + V \frac{Q}{E_o - H_o} V \frac{Q}{E_o - H_o} V + \dots \right\} P . \tag{15}$$

The first two terms represent the true Hamiltonian and the remaining terms the perturbations from states outside the model space. The latter terms compensate for the fact that the effective Hamiltonian operates only within a limited subspace (the model space).

In a hyperfine problem one is not interested in the total
energy splitting of the states but only in the part that is due to
the small hyperfine interaction. This can easily be handled with
the technique described above, simply by replacing V by V + h, where
h is the hyperfine interaction, and by considering terms linear in
h. We then get from (15) an "effective" hyperfine interaction

$$
h_{eff} = \left\{ h + h\frac{Q}{E_o-H_o}V + V\frac{Q}{E_o-H_o}h + h\frac{Q}{E_o-H_o}V\frac{Q}{E_o-H_o}V + \right.
$$

$$
\left. + V\frac{Q}{E_o-H_o}h\frac{Q}{E_o-H_o}V + V\frac{Q}{E_o-H_o}V\frac{Q}{E_o-H_o}h + \ldots \right\}_L \tag{16}
$$

where we have included all first-, second- and third-order terms
(characterized by the total number of interactions). The energy
splitting caused by the hyperfine interaction is then given by

$$
E_h^{\alpha} = \langle \Psi_o^{\alpha} | h_{eff} | \Psi_o^{\alpha} \rangle . \tag{17}
$$

Consequently, a calculation of this type contains two steps, firstly,
the determination of the correct zero-order states Ψ_o^{α}, and, secondly,
evaluation of the effective operator, h_{eff}. The zero-order states
are formally eigenfunctions of H_{eff} (7), but if V is the electrostatic interaction, H_{eff} cannot mix states with different SLM_SM_L.
In many cases it is therefore a trivial problem to find these zero-
order states, and the only real problem is the determination of the
effective operator.

In evaluating the effective operator (16) essentially two different methods can be used. Firstly, one can generate a "complete"
set of eigenfunctions of H_o (including unbound states) $\{\Psi_o^i\}$, and use
the expansion

$$
\frac{Q}{E_o-H_o} = \sum_{exc\ i} \frac{|\Psi_o^i\rangle\langle\Psi_o^i|}{E_o-E_o^i} . \tag{18}
$$

(When this is inserted into (10) we recognize some terms of the
ordinary Rayleigh-Schrödinger perturbation expansion in matrix form.)
This is the technique used by Kelly, Das et al. [9]. Secondly, one
can solve inhomogeneous, differential equations of the type

$$
\begin{cases} (E_o-H_o)F = QV\Psi_o & (19a) \\ (E_o-H_o)f = Qh\Psi_o & (19b) \end{cases}
$$

where the parameter α is left out for simplicity.

By means of the solutions of these first-order equations

$$
\begin{cases}
F = \dfrac{Q}{E_o - H_o} \, V\Psi_o & \text{(20a)} \\[3mm]
f = \dfrac{Q}{E_o - H_o} \, h\Psi_o & \text{(20b)}
\end{cases}
$$

the hyperfine splitting (17) to third order can be expressed as

$$
E_h^{(1)} = \langle \Psi_o | h | \Psi_o \rangle \qquad\qquad\qquad \text{(21a)}
$$

$$
E_h^{(2)} = \langle \Psi_o | h | F \rangle + \langle \Psi_o | V | f \rangle \qquad\qquad \text{(21b)}
$$

$$
E_h^{(3)} = \langle f | V | F \rangle + \langle F | h | F \rangle + \langle F | V | f \rangle \qquad \text{(21c)}
$$

In practice, of course, the equations (19) are separated into radial and angular parts, and only the radial equations have to be solved numerically.

The equation (19b) is a single-particle equation, since h is a one-body operator. This equation has for long time been used by Sternheimer [14] in calculating the quadrupole shieldings. Since the two terms of (21b) are equal, the entire second-order effect can be obtained by means of the single-particle function f. Since f contains only single excitations, the second-order effect is entirely a polarization effect. In order to get the third-order contribution, it is with this technique necessary to solve also the electrostatic equation (19a). Since V is a two-body operator, it gives rise to single and double excitations in first order. The latter, which cause correlation effects, lead to a pair equation, and solving this equation is the main difficulty in this approach. In recent years efficient methods for handling this problem have been developed [15].

V. DIAGRAMMATIC REPRESENTATION

The various contributions to the effective operator (16) can be conveniently represented by means of diagrams. Since this operator acts only within the open shell, the free lines must all be open-shell lines. Diagrams with n incoming and n outgoing lines are said to represent an effective n-body operator, and the total operator can evidently be split up into zero-body, one-body, two-body,.... m-body parts, where m is the number of electrons in the open shell. The zero-body operator does not contribute to the hyperfine splitting, and the two-body and higher-order interactions can be expected to be

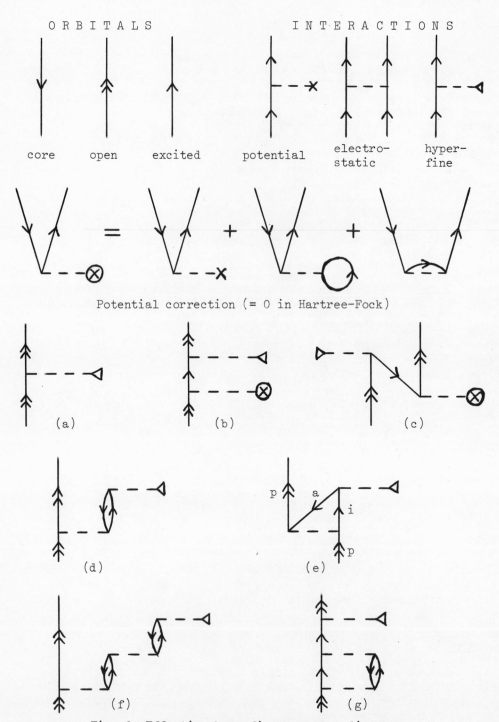

Fig. 2. Effective hyperfine-operator diagrams.

small. Therefore, it is in the hyperfine case often sufficient
to consider only effective one-body operators. Fig. 2 shows
some of the lowest-order diagrams of these operators. (a) is the
first-order contribution to the energy (21a), i.e. the hyperfine
interaction with the zero-order wavefunction or model function,
ψ_0^α. (b) and (c) are "potential-correction" diagrams, which are
required, if the occupied orbitals do not fulfill the HF condition.
(d) and (e) are the second-order contributions, due to first-order
perturbations of the model function (polarization). (f) is an
example of a higher-order polarization effect, and (g) is a correlation diagram of lowest order. Polarization diagrams are characterized by the fact that there is only one excitation at a time,
while correlation diagrams contain simultaneous excitations of
several electrons.

As an illustration of the evaluation of a diagram we consider
the polarization diagram (e). This has the value

$$\frac{<a|h|i><pi|g|ap>}{\varepsilon_a-\varepsilon_i}$$

where h and g represent the hyperfine and electrostatic interactions, respectively. This should be summed over all core orbitals,
a, and all excited orbitals, i. This sum is conveniently separated
into different parts, one for each shell of the core and for each
angular momentum of i. The radial part of such a sum is easily
expressed by means of the perturbed, radial functions (20). The
angular part of the diagrams can be evaluated in a very elegant
way using the technique of angular-momentum graphs, introduced by
Jucys et al. [16]. In the case considered above one gets

$$= R(-1)^k \begin{Bmatrix} 1_p & \kappa & 1_p \\ 1_a & k & 1_i \end{Bmatrix}$$

(22)

where R is the radial integral and the following symbol is a 6-j
symbol. k and κ are the (spatial) ranks of the interactions. Similarly, it can be shown that all diagrams of an effective one-body
operator can be expressed as a factor times the corresponding first-
order diagram. But this means that all these diagrams can be
represented by an operator having the same tensorial structure as
the true operator and operating only on the unperturbed functions.
All higher-order effects are simply included in a multiplicative
constant. This is the idea behind the effective operator, and it
explains why it in the hyperfine case has the same form (2) as the
ordinary hyperfine operator (1). The radial as well as the angular

factor in (22) depends on the tensor rank of the operator, and,
therefore, the parameters of the different terms of the effective
operator become   different.

    A serious problem in all perturbative calculations is the
choice of the model Hamiltonian H_0. In principle, of course, the
results should be independent of this choice, but since the calculations have to be terminated at a certain order, this is not so
in practice. Therefore, it is important to choose H_0 so that the
convergence is sufficiently rapid. If the basis functions are of
the HF type, many diagrams would vanish. However, there are several
problems connected with such a choice. In order to be able to
separate the calculations into radial and angular parts, H_0 must
be of central-field type, i.e. restricted HF. But this is not well-
defined in the open-shell case. The configurational average, which
is often used in HF calculations [4, 17], leads to a non-Hermitian
model Hamiltonian and is therefore unsuitable as a basis for many-
body calculations. If there is a single electron outside closed
shells (as in the examples considered below), it is natural to generate the orbitals in the HF potential of the ion core. In such a case

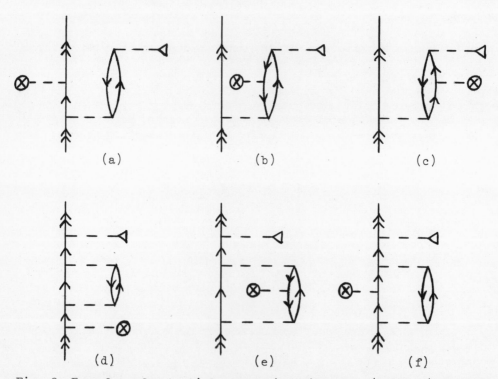

Fig. 3. Examples of potential-correction diagrams. (a,b,d,e) vanish
        if occupied HF orbitals are used. (c) can be included by
        modifying the single-particle functions with the potential
        correction.

all potential-correction diagrams (Fig. 3) vanish. If the excited
states are generated by means of inhomogeneous, differential equations (19), it is difficult to work with exact exchange, since this
requires that the equations are solved iteratively. Instead, different local potentials are here often used. The occupied orbitals
can still be of HF type, which means that no potential correction
of these orbitals is needed (Fig. 3a,b,d,e). Diagrams of the type
in Fig. 3c can be included by perturbing the single-particle
functions a few times by the potential correction. In this procedure, which is used in our calculations, the only potential correction diagrams that are omitted are those of the type in Fig. 3f,
where a pair function is perturbed. These diagrams appear for the
first time in the fourth-order perturbation.

VI. EXAMPLES OF MANY-BODY CALCULATIONS

A number of accurate many-body calculations of the atomic
hyperfine interaction have been performed in recent years. As
illustrative examples I have chosen different states of alkali
atoms, where much information is available, experimentally as well
as theoretically. These systems are particularly simple, since
there is only one electron outside closed shells, in which case the
many-body theory is well developed. Furthermore, there is presently
a great interest in the alkali atoms, due to new experimental information.

Many-body calculations have also been performed on atoms with
several electrons outside closed shells. Such calculations, however,
are complicated by technical difficulties (such as the choice of a
suitable central potential and the evaluation of effective many-
body operators), which I shall not have time to go into here. From
a physical point of view, however, the alkali atoms are general
enough to illustrate various interesting many-body effects.

A. Ground States of Alkali Atoms

As mentioned, the experimental hyperfine interactions in the
ground states of the alkali atoms deviate appreciable from the CFM
predictions, and the IPM (UHF) can only partially explain this deviation (see Table I). For Li and Na more complete calculations have
been performed by Schulman and Lee [18] and by Lee, Dutta, and Das
[9]. Both calculations are of perturbative type and are carried to
third order. Schulman and Lee use the inhomogeneous, differential
equations given above (19), while Lee et al. expand the perturbations in a complete set of basis functions (18). The contact interaction represents a mathematical difficulty in the former technique,

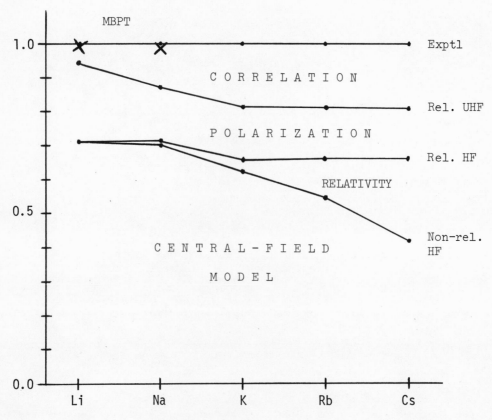

Fig. 4. Different contributions to the hyperfine interaction for the
        ground state of the alkali atoms. The crosses represent full
        many-body calculations.

but this can be overcome by means of special substitution [18]. The
results obtained for Li and Na, after correcting for relativistic
effects [4], are 399.5 and   870 MHz, respectively, which is in
very good agreement with the experimental results. In Fig. 4 the
various contributions to the hyperfine interaction are illustrated.
It can be observed that the polarization and the correlation
effects are of the same order of magnitude and that their relative
importance is essentially independent of Z, while, of course, the
relativistic effect increases very rapidly for high Z.

B. P States of Alkali Atoms

     As a second example we consider the lowest p states of Li and
Na. For the 2p state of Li the diagonal dipole interaction constants,

$a_{1/2}$ and $a_{3/2}$, have been measured, as well as the off-diagonal constant, $a_{1/2\ 3/2}$. This makes it possible to extract experimental values of all three parameters of the effective dipole operator (2a). This is therefore a good test case for many-body calculations. Fig. 5 shows a comparison between the experimental results and two such calculations, one performed by Nesbet [8] and one at our laboratory [19]. Nesbet's calculation is of variational type, while ours is a perturbative one. In our calculations we have used the effective-operator form of the linked-diagram expansion and the technique of inhomogeneous, differential equations described above. Two different local potentials are used, "Optimized Hartree-Fock-Slater" (OHFS) and "Hartree-Slater" (HS), respectively [4]. After applying the potential corrections (Fig. 3), the two potentials yield essentially identical results. The agreement between our results and those of Nesbet is found to be extremely good, and, since the methods used are entirely different, this is a good check of the reliability of the calculations. Furthermore, the calculated orbital and spin-dipole parameters fall entirely within the experimental limits.

For the 3p state of Na the off-diagonal dipole constant is not known, and, therefore, it is not possible to obtain separate values of the orbital and spin-dipole parameters. The contact parameter can be determined, however, as well as an average of the remaining two. As can be seen from Fig. 6, the agreement between the experimental average and the calculated one is quite good. It is reasonable to expect then that the quadrupole interaction obtained in the same calculation should be quite accurate.

C. D States of Alkali Atoms

The experimental hyperfine investigations of the alkali atoms have for a long time been limited to the ground state and excited p states. As we have just heard, however, also excited s, d, f... states have recently become accessible for experiments, due to new techniques invented by Happer, Svanberg et al. [20]. Strange results have been found in the measurements on the d states, indicating very large perturbations. It is therefore a challenging task to try to calculate the hyperfine interaction in such a state, in order to see if the present technique is capable of explaining the results.

Unfortunately, there is hardly any experimental information available for the d states of the lighter alkali atoms, Li, Na, K. The large number of electrons and possible relativistic effects make a full many-body calculation for the heavier alkali atoms quite hard and uncertain. Nevertheless, such calculations are presently being undertaken at our laboratory, and here I shall give some preliminary results.

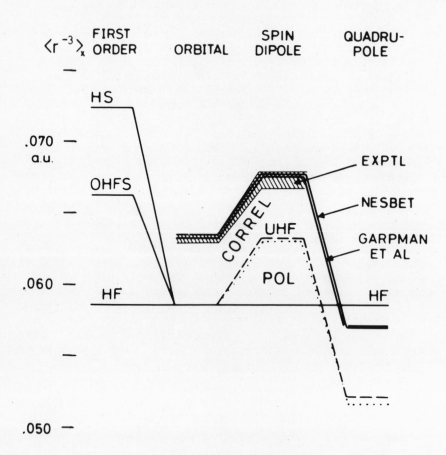

Fig. 5.

Comparison between experimental and theoretical hyperfine parameters for the 2p state of Li. The dotted line represents the
third-order polarization result and the dashed line the unrestricted HF result (equivalent to polarization to all orders). The
results of two full many-body calculations are shown, performed
by Nesbet [8] and by Garpman et al. [19]. The shadowed region
represents the experimental values with their uncertainties.

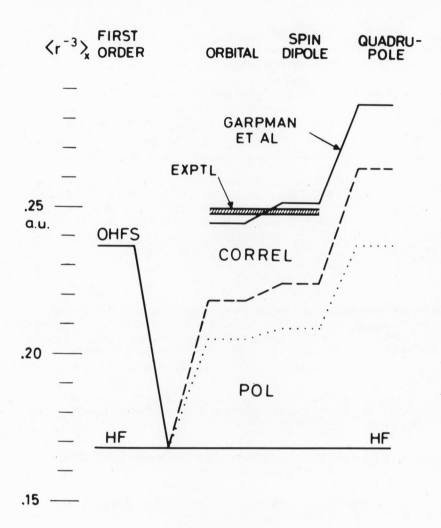

Fig. 6.

Hyperfine parameters for the 3p state of Na. The dotted and
dashed lines represent the second- and third-order polarization
results, respectively, and the full line the results of a many-
body calculation [19]. The shadowed region represents the average of the orbital and the spin-dipole parameters.

One difficult problem in the calculation on high angular-
momentum states is the choice of the central potential. In principle,
the dependence on the potential is eliminated by the "potential-
correction" procedure, described above. However, these states are
so sensitive to the potential that it might be difficult to handle
this "correction". This is also experienced by Sternheimer [14],
who has recently calculated quadrupole shielding factors for such
states. Sternheimer uses two local potentials (without potential
corrections), which in some cases yield vastly different results,
although both potentials are fitted to experimental parameters.
This shows that one has to be extremely careful in this type of
calculation.

In our calculations we have normally used the HS potential as
a starting point. The occupied orbitals are modified a few times
by the potential correction, using the single-particle equation
(19b). This is equivalent to using HF orbitals generated in the
potential of the ion core, which eliminates diagrams of the type in
Fig. 3a,b,d,e. The excitations are generated by means of the one-
and two-particle equations (19) using the local HS potential, and
the single-particle functions are modified a few times by the potential correction to eliminate diagrams of the type in Fig. 3c. As in
the calculations of the p states, we have checked this procedure by
using different potentials. We have then found that our second-order
results for the d states are independent of the potential within
1 % (for the p states the agreement is considerably better). Therefore, we believe that our second-order results are quite accurate
and well-defined.

Table III gives our first- and second-order results for some d
states of Rb. The first-order values are obtained with the zero-
order functions, without (HS) and with (HF) potential modifications,
respectively. Note that the two values can differ by as much as a

Table III. Second-order hfs parameters for d states of Rb (a.u.).

Rb	FIRST ORDER		SECOND ORDER			
	HS	HF	Orbital	Spin-dipole	Contact	Quadru-pole
4d	0.1310	0.0413	0.0105	−0.0245	−0.204	0.1158
5d	0.0858	0.0259	0.0089	−0.0109	−0.122	0.0524
6d	0.0415	0.0148	0.0057	−0.0051	−0.069	0.0269
7d	0.0210	0.0090	0.0036	−0.0027	−0.041	0.0155
8d	0.0119	0.0058	0.0024	−0.0016	−0.026	0.0097

factor of three, which illustrates how sensitive these states are
to the potential! The second-order values of the parameters in the
effective operators (2) (after potential modifications) are given
in the following columns. The potential modifications change the
orbital and spin-dipole parameters by 25-50% and the contact and
quadrupole parameters considerably less (~10%).

In Fig. 7 I have compared our second-order results with the
experimental values. For the 4d state Happer et al. [20] have been
able to determine the orbital and spin-dipole parameters separately,
while for the other states only an average of the two (with the
relative weights 7:1, respectively) is known. The most drastical
effect is, of course, the negative value of the spin-dipole term,
which also appears in our second-order calculation. Using a value
of the quadrupole moment of [87]Rb of 0.13b, one can get experimental
values for some of the quadrupole parameters,as shown in the figure.
There is also in this case a qualitative agreement with our second-
order results, in spite of the large perturbations.

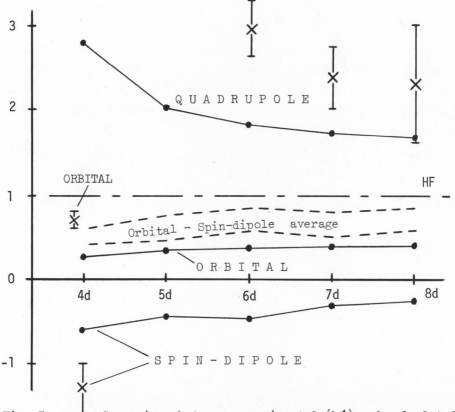

Fig. 7.   Comparison between experimental (✕) and calculated
second-order (●) parameters for d states of Rb. The dashed lines
represent the experimental limits for the orbital-spin-dipole average.

As mentioned, the polarization effect on the quadrupole interaction has also been calculated by Sternheimer [14] for some d states
of Rb. Sternheimer's results are given in terms of his correction
constant, R, and they are therefore not directly comparable to ours.
This constant is defined

$$R = - \frac{\delta <r^{-3}>}{<r^{-3}>_{St}} \qquad (23)$$

where $\delta <r^{-3}>$ is the second-order change in the quadrupole $<r^{-3}>$
value and $<r^{-3}>_{St}$ is the first-order value with Sternheimer's
function. Sternheimer does not calculate the potential corrections
discussed above, and he also omits the k = 0 part of the electrostatic interaction. These terms cancel in the semi-empirical procedure for evaluating the quadrupole moment (see further below), but
in order to estimate the quadrupole interaction ab initio they
should be included. It turns out that the k = 0 terms are relatively
small for the d states, and, therefore one can estimate his quadrupole parameter using the relation

$$<r^{-3}>_{q} \approx <r^{-3}>_{HF} - R<r^{-3}>_{St} \qquad (24)$$

(It should be noted that this relation is not valid for p states,
due to the large k = 0 terms). With the values given by Sternheimer
and the HF values in Table III one obtains the quadrupole parameters 0.133, 0.051 and 0.025, respectively, for the 4, 5 and 6d
states. The agreement with our values (Table III) is good, but this
is partly fortuitous, since individual terms are found to be quite
different.

For the 4d state of Rb we have also performed a third-order
calculation, similar to that for the p states of Li and Na. The re-

Table IV. Comparison between experimental and theoretical parameters
for the 4d state of Rb (in atomic units).

Rb 4d	Orbital	Spin-dipole	Contact	Quadrupole
Hartree-Fock	0.041	0.041		0.041
Incl. pol. 2nd order	0.010	-0.024	-0.20	0.116
- " - 3rd -"-	-0.002	-0.055	-	0.127
Incl. correl. 3rd order	(0.04)	(-0.02)	-	(0.14)
Experimental	0.030(3)	-0.053(12)	0.63(2)	0.12(10)

sults, which are very preliminary, are summarized in Table IV. In this case a very large number of pair functions have to be generated, and the calculations are therefore very time consuming. So far, only the correlation among the n=4 shells have been studied, but also inner shells can be expected to give significant contributions. Anyhow, the correlation effects found are large enough to explain the discrepancy between the polarization and the experimental results.

VII. EVALUATION OF QUADRUPOLE MOMENTS

The purpose of many hyperfine investigations is to determine nuclear quadrupole moments. In order to evaluate this quantity from the experimental b factor, one has to know the quadrupole parameter $<r^{-3}>_q$ of the effective operator (2b). If no accurate many-body calculation is available, this parameter has to be estimated by some other means. In the light of the calculations described above, we shall consider different ways of doing this.

One way of estimating the $<r^{-3}>_q$ parameter is obviously to use a theoretical $<r^{-3}>$ value, corrected for the second-order effects as described above. This is equivalent to including the polarization effects of lowest order (diagrams Fig. 2b,c,d,e), but neglecting higher-order polarization effects (Fig. 2f) and all correlation effects (Fig. 2g). It is demonstrated above that the correlation affects the hyperfine interaction appreciably, and, therefore, this procedure can not be expected to be particularly accurate.

The conventional procedure in evaluating the quadrupole parameter is to use the experimental dipole or spin-orbit interaction and the quadrupole shielding corrections calculated by Sternheimer [14]. But these corrections contain only the polarization effect on the quadrupole interaction. If, for instance, the experimental dipole interaction is used in evaluating the quadrupole parameter, the correct polarization correction to apply is, of course, the difference between the polarization effects on the quadrupole and the dipole interactions. As mentioned, Sternheimer leaves out the k = 0 part of the electrostatic interaction in his calculations, the reason being that it cancels in the semi-empirical procedure. However, also the k ≠ 0 terms can have an appreciable effect on the dipole (as well as on the quadrupole) interaction, as we have seen particularly for the d states above. Therefore, it is important to take into account the polarization effect on the dipole interaction, if this is used to extract an $<r^{-3}>$ value. The perturbation in the magnetic case is calculated in exactly the same way as in the quadrupole case, and this correction is therefore very easy to make.

Similarly, if the spin-orbit parameter is used, the corresponding polarization effect should be included in the correction factor.

This is more difficult to do, however, since the spin-orbit interaction is quite different from the hyperfine interaction.

If the second-order correction is applied in the correct way, the main approximation in the semi-empirical procedure is that the difference between the correlation effects on the dipole and the quadrupole interactions is neglected. As we can see from the calculations shown above, this seems to be a fairly accurate procedure for the p states of the alkali atoms, since the correlation has approximately the same effect on the dipole and the quadrupole interactions. For excited d states, on the other hand, the situation is more complicated. The correlation effect on the various parameters is here quite different, and the semi-empirical procedure is less accurate.

Let us take the 6d state of ^{87}Rb as an example [20]. From the experimental a factors, $a_{3/2} = 7.84(5)$ and $a_{5/2} = -3.6(7)$ MHz, one obtains an average $<r^{-3}>$ value (after eliminating the contact part) of 0.0105(18) a.u. The corresponding second-order value is (from Table III) $\frac{7}{8}$ 0.0057 - $\frac{1}{8}$ 0.0051 = 0.0043, while the quadrupole value is 0.0269. The correction to apply is then 0.0269 - 0.0043 = 0.0226, which together with the dipole average, 0.0105, yields a quadrupole parameter $<r^{-3}>_q = 0.0331$. With the experimental b factor, 0.53(6) MHz, this gives a quadrupole moment of 0.17b, which should be compared with the "true" value of 0.13b. The discrepancy is reasonable and due to correlation effects.

If the correction factor given by Sternheimer, R = -0.2018, is used in the example above, the result would be $<r^{-3}>_q = 0.0126$ and Q = 0.45b, in serious disagreement with the results from the p series. The reason for this is that Sternheimer's correction factor is related to his first-order value $<r^{-3}>_{St}$ in (23), which is very large (0.04946) compared with the dipole average (0.0105).

VII. CONCLUSIONS

It is well-known that the polarization has a considerable effect on the atomic hyperfine interaction. More surprising is, perhaps, that the correlation effect usually is of the same order of magnitude as the polarization effect. This means that the unrestricted HF model, which contains polarization effects but no correlation, is not particularly suited for atomic hyperfine calculations. To calculate the correlation effect is quite complicated, but it seems that the present many-body technique is capable of doing so in most cases, except for heavy elements, where relativistic effects are important. Most calculations have been performed on atoms with an open s or p shell, and there the agreement with experiments is impressive. Calculations on atoms with an open d shell are harder, particularly for excited states, because these states are very sensitive to perturbations, and one cannot expect the same accuracy as for s and p electrons.

One important problem in the hyperfine analysis is the evaluation of the nuclear quadrupole moment. From the complete calculations one can draw some conclusions about the best procedure to use in cases where less information is available. The semi-empirical procedure, where a value of $<r^{-3}>$ is obtained from the dipole interaction, seems to be fairly accurate for p electrons, if the correct polarization correction is applied. For d electrons the same procedure seems to work, but more information is needed, before any more definite conclusions can be drawn.

References

1. L. Brillouin, Actualités Sci. Ind. No 159 (1934)

2. A. Abragam and M.H.L. Pryce, Proc. Roy. Soc. A205, 135 (1951)

3. V. Heine, Phys. Rev. 107, 1002 (1957)
   J.H. Wood and G.W. Pratt, Phys. Rev. 107, 995 (1957)

4. I. Lindgren and A. Rosén, Case Studies in Atomic Physics, 4, 93 (1974)

5. A.J. Freeman and R. Watson, Phys. Rev. 132, 706 (1963)

6. P.G.H. Sandars and J. Beck, Proc. Roy. Soc. A289, 97 (1965)

7. J.S.M. Harvey, Proc. Roy. Soc. A285, 581 (1965)

8. R.K. Nesbet, Adv. Chem. Phys. 14, 1 (1969); Phys. Rev. A2, 661 (1970)

9. H.P. Kelly, Adv. Chem. Phys. 14, 129 (1969)
   J.D. Lyons, R.T. Pu, and T.P. Das, Phys. Rev. 178, 103 (1969)
   T. Lee, N.C. Dutta, and T.P. Das, Phys. Rev. A1, 995 (1970)

10. K.A. Brueckner, Phys. Rev. 100, 36 (1955)
    J. Goldstone, Proc. Roy. Soc. A239, 267 (1957)

11. B.H. Brandow, Rev. Mod. Phys. 29, 771 (1967)

12. P.G.H. Sandars, Adv. Chem. Phys. 14, 365 (1969)

13. I. Lindgren, J. Phys. B. (to appear)

14. R.M. Sternheimer, Phys. Rev. A3, 837 (1971); A9, 1783 (1974)

15. V. McKoy and N.W. Winter, J. Chem. Phys. 48, 5514 (1968)
    N.W. Winter, A. Laferrière and V. McKoy, Phys. Rev. A2, 49 (1970)
    J.I. Musher and J.M. Schulman, Phys. Rev. 173, 93 (1968)

16. A.P. Yutsis, I.B. Levinson, and V.V. Vanagas, Mathematical Apparatus of the Theory of Angular Momentum, (Israel Program for Sci. Transl., Jerusalem, 1962)

17. J.C. Slater, Phys. Rev. 165, 655 (1968)
    C.H. Froese-Fisher, Chem. Phys. 45, 1417 (1966); Computer
    Physics Comm. 1, 151 (1970)

18. J.M. Schulman and W.S. Lee, Phys. Rev. A5, 13 (1972)

19. S. Garpman, I. Lindgren, J. Lindgren and J. Morrison, submitted
    to Phys. Rev. and unpublished works.

20. R. Gupta, S. Chang and W. Happer, Phys. Rev. A6, 529 (1972)
    R. Gupta, W. Happer, L.K. Lam and S. Svanberg, Phys. Rev. A8,
    2792 (1973)
    S. Svanberg, P. Tsekeris and W. Happer, Phys Rev. Lett 30,
    817 (1973)
    K.H. Liao, L.K. Lam, R. Gupta and W. Happer, Phys Rev. Lett.
    32, 1340 (1974). See also the paper by W. Happer, p. 651 in
    this volume, for further references.

AUTHOR INDEX

SUBJECT INDEX

Adiabatic approximation for
electron scattering 438f
Adiabatic transitions in
atomic collisions 252
Alignment
of excited atoms in a
dc-gas discharge 34ff
- "hidden" alignment 41f
- effect of reabsorp-
tion 39
in gaslaser discharge 31
Angular momentum, basic
theory 14f
Anti-level crossings
in cascade spectroscopy
659, 702ff
in crossed electric and
magnetic fields 24
Antiproton
mass, magnetic moment 178
proton-antiproton inter-
action 187
Approximations in thermal
collision processes
adiabatic 536
impact 530, 538
with fixed nuclear
spin 532
secular, for electronic
transistions 533
Approximations for electron
transfer collisions 400ff
Born, first and second 401
Brinkman-Kramers 403
continuum distorted
wave 423

linear combination of
atomic states 412
Atomic beam magnetic
resonance 731ff
actinide atoms 741
3d-shell atoms 737 f
hyperfine structure
measurements 731ff
rare earth atoms 735ff
Atomic beam techniques 619ff
magnetic filtering 624
Atomic spectra, theory 2,13
Atomic states picture in
electron transfer 398ff
close-coupling 410f
Auger electrons from
atomic collisions 263
Autodetachment of
H⁻ excited states 493ff
Autoionization of excited
rare gas atoms 713ff
lifetime 729
line shapes 727f
Balmer line Hα,Dα 95
Beat signals in spontaneous
laser radiation 30ff
Bolometric detection of
ions, non-destructive 579
Born approximation
in electron-atom
collisions 473
for low energy electron
scattering 436ff
in electron transfer
collisions 400ff